# Introduction to Applied Algebraic Systems

# Introduction to Applied Algebraic Systems

Norman R. Reilly

OXFORD
UNIVERSITY PRESS
2010

Oxford University Press, Inc., publishes works that further
Oxford University's objective of excellence
in research, scholarship, and education.

Oxford  New York
Auckland   Cape Town   Dar es Salaam   Hong Kong   Karachi
Kuala Lumpur   Madrid   Melbourne   Mexico City   Nairobi
New Delhi   Shanghai   Taipei   Toronto

With offices in
Argentina   Austria   Brazil   Chile   Czech Republic   France   Greece
Guatemala   Hungary   Italy   Japan   Poland   Portugal   Singapore
South Korea   Switzerland   Thailand   Turkey   Ukraine   Vietnam



Published by Oxford University Press, Inc.
198 Madison Avenue, New York, New York 10016

www.oup.com



Library of Congress Cataloging-in-Publication Data

Reilly, Norman R.
Introduction to applied algebraic systems / Norman R. Reilly.
    p. cm.
ISBN: 978-0-19-536787-4
1. Algebra. I. Title.
QA155.R3656 2008
512–dc22   2008017220

9 8 7 6 5 4 3 2 1

Printed in the United States of America
on acid-free paper

# Preface

In the last forty to fifty years, algebraic systems have played a critical role in the development of the computer and communication technology that surround us in our daily lives. This has made algebra much more relevant to the modern student. Algebra is no longer just a pretty subject with an interesting history, but a discipline that plays a critical role in modern commerce, communication and entertainment. Some traditional approaches to undergraduate algebra can easily leave a student with the impression that the development of algebra ended with the 19th century. It is the goal of this text to show that the subject is alive, vibrant, exciting, and more relevant to modern technology than it has ever been. This justifies a shift in the emphasis and selection of topics in any modern introduction to algebra. In the storage of data, whether in computers, on compact disks or dvd's, it would be prohibitively expensive to try to develop error free systems. It is more cost effective to provide for the correction of a small number of errors or corruptions of the data. Likewise, in the communication of data, such as from satellites where signals may be weak and it might not be cost effective to constantly resend messages, it is again preferable to devise systems to correct small errors and small losses of data. These needs have led to the development of error correcting codes, codes which attempt to accept slightly faulty or corrupted data and correct it. At the forefront of these techniques are algebraic codes. Algebraic codes take advantage of the properties of finite fields. In another direction, the need for security in the transmission and storage of information has led to the the development of ever more advanced encryption schemes, schemes that must become ever more sophisticated as computing power grows exponentially year by year and techniques available to cryptanalysts also become more sophisticated. Some of the most widely used and most advanced of these encryption techniques, such as RSA encryption and elliptic curve encryption, are based on number theory, group theory and algebraic geometry.

In this book, we provide an introduction to algebraic systems that keeps these modern applications of algebra in mind while providing a solid introduction to the basic algebraic structures of groups, rings, fields and algebraic geometry. In a nutshell, the goal of this text is to provide a strong introduction to algebraic systems motivated by modern applications, and to build a foundation in algebra appropriate for the exploration of modern developments. Sections are included on topics such as bar codes, RSA encryption, error correcting codes, Enigma encryption and elliptic curves and their applications. Through the sections devoted to applications, we provide a window into those applied areas and provide evidence of the critical importance of the basic concepts of abstract algebra to those applied areas. In the line of applications, we do not overlook some more established areas of application of algebraic systems and include sections on applications of groups to counting techniques. Mathematical software is playing an increasing role in both research and in undergraduate programs. In that direction, computational algebra is a rapidly growing field, and one of its most important tools is Groebner bases for ideals in polynomial rings. In large measure this is due to the fact that they can be efficiently computed with the help of the Buchberger algorithm. Apart from being computationally important, these notions are interesting in and of themselves and facilitate the development of the introductory theory of algebraic geometry.

The book is divided into seven chapters. Chapter 1 provides a thorough introduction to number theory, which is the foundation of the subject and provides a model for much that follows. There are also sections illustrating applications of number theory to bar codes and encryption. Chapter 2 introduces rings, polynomials, fields, Galois fields, minimal polynomials and concludes with a section on Bose-Chauduri-Hocquenghem and Reed-Solomon error correcting codes. The motivation for departing from the usual practice of introducing groups before rings and fields is the strong parallel that exists between the basic elements of number theory and the techniques used in the construction of field extensions; the Euclidean Algorithm, the factorization into irreducibles, the importance of the congruence modulo a prime compared with the congruence on the ring of polynomials over a field modulo an irreducible polynomial. All this makes it possible to approach field theory as a natural extension of basic number theory. Chapter 3 introduces groups with an emphasis on permutations and permutation groups. The definitions of even and odd permutations are introduced via permutation matrices, thereby avoiding the sometimes awkward verification that the concepts are well defined. The important ideas of orbits and stabilizers are explored followed by an introduction to the counting applications developed by Cauchy, Frobenius and Pólya. The notions of normal subgroups and homomorphisms are delayed until Chapter 4 for important pedagogical reasons. First there is a great deal of very interesting material that has no need of these concepts and, secondly, there is insufficient time in a one semester course that covers rings and fields as well as groups to demonstrate the importance of these ideas. The result can be that they are left dangling with no apparent utility. Chapter 4 then continues with the characterization of finite abelian groups, homomorphisms and normal subgroups, conjugacy (including the "theorem that won World War II") and the striking Sylow theorems. The classes of nilpotent and solvable groups are also introduced. The chapter concludes with a description of the Enigma machine as an illustration of the potential sophistication that can be built around permutation ciphers.

Polynomials and their zeros are at the very heart of algebra and are usefully investigated in the context of polynomial rings and affine spaces. Chapter 5 is devoted to the associated topics of ideals, affine varieties, Groebner bases, parameters and intersection multiplicities. Chapter 6 introduces elliptic curves. Elliptic curves are important in smart card encryption and prime factorization. Here, although space does not permit a complete treatment of some results such as Bézout's theorem, sufficient detail is provided for a student to develop a good understanding of the basics. The details of the remarkable Nine-Point Theorem and the arithmetic on an elliptic curve are included. The final chapter, Chapter 7, contains a pot-pourri of topics that round off the discussion of previous chapters. The topics include elliptic curve cryptosystems, prime factorization using elliptic curves, the link between Fermat's Last Theorem, and elliptic curves, and the chapter concludes with a discussion of Pell's equation.

What to teach in a course based on this text? An important feature of the text is its flexibility. Chapter 1 is suitable on its own as a short (12–13 lectures) introduction to Number Theory for a general audience at the first or second year college level. Chapters 1, 2 and 3 make an excellent self-contained half-year course (in the range of 35–40 lectures), suitable for second or third year college or university students, introducing the basic structures of modular arithmetic, groups, rings and fields, with plenty of substantial material. Chapters 1 through 7 are suitable for a whole year course (70–80 lectures) in algebra. Chapters 4 through 7 are suitable for third and fourth year undergraduate students and can also be used as an introductory graduate text for students who have taken only a half year undergraduate course in algebra or who have taken a more traditional undergraduate course in algebra.

How should one teach from this text? There is a considerable amount of material in each of the sections with a considerable amount of detail. There are also places where a detailed verification is left to the reader, perhaps because the argument is just a repetition of an earlier argument. So to teach every detail and fill in every missing argument may not be the best way to present the subject. I would recommend that an instructor exercise some discretion as to what is covered in complete detail and what is covered lightly. One of my own rules of thumb is to ask myself three questions about each proof, especially longer proofs:

(i)  Is there any clear benefit to the student from going through these calculations?

(ii)  Will it help in understanding the result itself?

(iii)  Are there ideas in the proof that are especially clever, useful or interesting in and of themselves?

If the answers to all three of these questions is no, then what is the point of wading through the gory details?

For example, is there value in going through the details of the Fundamental Theorem of Arithmetic? Well, that might depend on the level of the students. In an elementary class, it is an interesting application of induction; in a more advanced class, it might all seem too trivial. Is it worthwhile to cover the equivalence between equivalence relations and partitions in detail, or would it be better to discuss it on an intuitive level and leave the details to be read by those students who are most interested in that? Then in Chapter 2 one of the great advantages of treating rings and fields before groups is that the parallel between modular arithmetic and the construction of factor rings of polynomial rings modulo a polynomial (ideal) is so strong that the vast majority of students have no difficulty accepting the concepts without rigorously checking all the details. In other words, an instructor can easily focus students' efforts and interests on the most important and most interesting topics.

There are many worked examples in the text to illustrate the theory and each section is followed by numerous exercises (almost 850 in total) ranging from routine to challenging. The longer and more difficult exercises are marked by either one or two asterisks. A Solutions Manual is available to course instructors from the author.

An appropriate background for this book would be an introductory course in linear algebra, determinants and matrices.

# Acknowledgments

I would like to acknowledge my sincere gratitude to Drs Edmond Lee and Michael Monagan who read parts of the manuscript very carefully and who, in addition to making many corrections, made many useful remarks and suggestions that influenced both content and presentation. I would also like to thank the anonymous reviewers whose rigorous reading of the manuscript and thoughtful advice resulted in major improvements in the text.

# Contents

# Introduction to Applied Algebraic Systems

# 1

## Modular Arithmetic

In this chapter we study the most basic algebraic system of all—the integers. Our familiarity with the integers from everyday affairs might lead us to the impression that they are uninteresting. Nothing could be further from the truth, and in this chapter we will barely scratch the surface of a subject that has intrigued mathematicians for more than 2000 years.

### 1.1 Sets, functions, numbers

We begin with some basic ideas and useful notation concerning sets, functions, and numbers.

A *set* is a collection of objects that we call the *elements* or *members* of the set.

We are relying, here, on a rather naive definition of a set. The concept of a set is extremely subtle and goes to the very foundations of mathematics. We will be satisfied here if we can develop a sound intuitive understanding.

Sets can be described in (at least) three ways:

(1) *An exhaustive list:* For example,

$A = \{1,3,6,8\}$
$B = \{$Beijing, Moscow, New York, Ottawa$\}$
$C = \{$Jean, John, Mary$\}$.

(2) *A partial list:*

$$A = \{1, 2, 3, \ldots\}$$
$$B = \{0, 1, -1, 2, -2, \ldots\}$$
$$C = \{1, 1/2, 1/3, 1/4, \ldots\}.$$

(3) *A rule or condition:*

$A = \{$integers $n \mid n$ is divisible by $2\}$
$B = \{$integers $n \mid$ there exists an integer $k$ with $n = 3k^2 + 2k + 1\}$
$C = \{$persons $P \mid P$ is studying calculus and has a birthday in January$\}$.

To indicate that $a$ is a member of the set $A$ we write $a \in A$. To indicate that $a$ is not a member of the set $A$ we write $a \notin A$. For example, if

$$A = \{x \mid x \text{ is the name of a Canadian province}\}$$

then *Manitoba* $\in A$ but *California* $\notin A$.

In various situations it can be helpful to reduce the length of a formula or condition by using the following notation:

"$\exists$" means "there exists".
"$\forall$" means "for all".

For example, we might write

$$\{A \mid A \text{ is a set of integers and } \exists \, a \in A \text{ that is a multiple of } 5\}$$

or

$$\{A \mid A \text{ is a set of integers and } \forall a \in A \; \exists \text{ an integer } n \text{ with } a = 2^n + 1\}.$$

The *empty* set, which we denote by $\emptyset$, is the set with no elements. If $A$ is a set and $A \neq \emptyset$ then we say that $A$ is *nonempty*. It is quite possible to be discussing the empty set without being aware of it (initially, at least). For instance, in the 17th century, P. de Fermat wrote that the set

$$A = \{n \geq 3 \mid \exists \, x, y, z \in \mathbb{N} \setminus \{0\} \text{ with } x^n + y^n = z^n\}$$

is empty. However, he provided no proof, and it was only in the last decade of the 20th century that Fermat's claim was finally confirmed by Andrew Wiles. So for more than 300 years it remained an open question whether $A$ is empty.

As another example, we might be interested in the set of integers

$$A = \{a \mid \exists \text{ a positive integer } b \text{ with } a^3 + 2a + 2 = b^2\}$$

only to discover at some point that there is no such integer $a$ so that $A = \emptyset$.

Let $A$ and $B$ be sets. Then $A$ is a *subset* of $B$ if and only if every element of $A$ is an element of $B$. When this is the case, we write $A \subseteq B$. The two extreme cases are

$$\emptyset \subseteq B \text{ and } B \subseteq B.$$

If $A$ is a subset of $B$, but not equal to $B$, then we say that $A$ is a *proper* subset of $B$ and write $A \subset B$. Two sets $A$ and $B$ are *equal* if and only if they have the same members. Thus, to show that two sets, $A$ and $B$, are equal, we must show that every member of $A$ is a member of $B$ and vice versa—that is, that $A \subseteq B$ and $B \subseteq A$.

**Example 1.1.1**  Consider the list of names

   Alfred, Jean, John, Li, Mark, Peter, Stephen.

Let

   $A =$ set of names in the list containing the letter $e$
   $B =$ set of names in the list containing two vowels.

Then $A = B$ since every name containing the letter $e$ has two vowels and every name with two vowels contains the letter $e$.

   If $A$ and $B$ are sets, then

$A \cup B = \{x \mid x \in A \text{ or } x \in B\}$
$A \cap B = \{x \mid x \in A \text{ and } x \in B\}$
$A \setminus B = \{x \mid x \in A \text{ and } x \notin B\}.$

For any finite set $A$,

$$\mid A \mid = \text{ the number of elements in } A$$

and is sometimes referred to as the *cardinality* of $A$. There are various important sets of numbers:

   $\mathbb{N} = \{1, 2, 3, \ldots\}$ the natural numbers or positive integers
   $\mathbb{N}_0 = \{0, 1, 2, 3, \ldots\}$ the nonnegative integers
   $\mathbb{N}_n = \{1, 2, 3, \ldots, n\}$
   $\mathbb{Z} =$ the set of all integers
   $\mathbb{Q} =$ the set of rational numbers
   $\mathbb{R} =$ the set of real numbers

For $a, b \in \mathbb{Z}$ with $a < b$ we write

$$[a, b] = \{a, a + 1, a + 2, \ldots, b\}.$$

All these number systems which we now take for granted so easily, represent major steps forward in the historical development of mathematics. Even the lowly number zero had a difficult introduction. Some historians trace its origins back to Sumeria, a region covering parts of modern-day Iran and Iraq, about 2500 years ago. However, even by the time of the ancient Greeks, it had not been fully accepted and was viewed with some suspicion. A great source for anyone interested in the origins of our place-valued number system, zero, negative integers, arithmetic, and so forth, is the book review by J. Dauben [Dau].

Complex numbers will also be part of our disussions. Recall that complex numbers are just those numbers of the form $a + ib$ (or $a + bi$) where $a$ and $b$ are real numbers and $i^2 = -1$. We denote the set of all complex numbers by $\mathbb{C}$. Thus

$$\mathbb{C} = \{a + ib \mid a, b \text{ are real numbers}\}.$$

Complex numbers combine under addition and multiplication just like real numbers, except that every time we encounter $i^2$ we replace it by $-1$. Note that $\mathbb{C}$ contains $\mathbb{R}$, since any real number $a$ can be written as $a + i0$.

Just as we represent the real numbers as points on a line, it is often convenient to represent complex numbers as points on a plane: $a + ib \rightarrow (a, b)$.

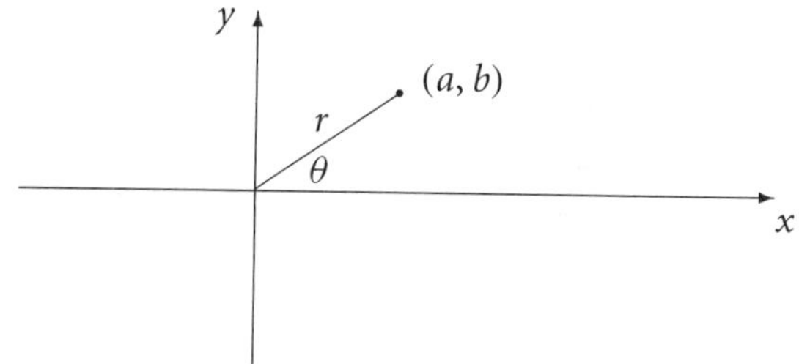


Then the *polar coordinates* for $(a, b)$ are

$$r = (a^2 + b^2)^{1/2}, \quad \theta = \tan^{-1}(b/a).$$

The original coordinates, in terms of the polar coordinates, are given by

$$a = r \cos\theta, \quad b = r \sin\theta.$$

In terms of the original complex number $a + ib$, we define

$$|a + ib| = (a^2 + b^2)^{1/2}$$

to be the *absolute value* of $a + ib$ and the *argument* of $a + ib$ is

$$\arg(a + ib) = \theta = \tan^{-1}(b/a).$$

Of course, in terms of polar coordinates, $|a+ib| = r$. We can express $z = a+ib$ in terms of the polar coordinates of $(a, b)$ as follows:

$$z = a + ib$$
$$= r(a/r + i\,b/r)$$
$$= |z|\,(\cos\theta + i\,\sin\theta).$$

This way of writing a complex number is particularly convenient for multiplication. For instance, if $z = |z|(\cos\theta + i\,\sin\theta)$ and $z' = |z'|(\cos\theta' + i\,\sin\theta')$ then

$$zz' = |z| \cdot |z'| \cdot (\cos(\theta + \theta') + i\,\sin(\theta + \theta')).$$

In particular, it follows that for every $z = |z|\,(\cos\theta + i\,\sin\theta) \in \mathbb{C}$ and every $n \in \mathbb{N}$,

$$z^n = |z^n|\,(\cos n\theta + i\,\sin n\theta).$$

We include in our basic set of axioms for numbers, the following.

**Axiom 1.1.2 (Well-ordering axiom for $\mathbb{N}$)** *Every nonempty subset of $\mathbb{N}$ has a least element.*

**Example 1.1.3** Let

$$A = \{n \in \mathbb{N} \mid 2^{2^n} - 1 \text{ is divisible by } 13\}.$$

By the well-ordering axiom, we know that either $A$ is the empty set or that $A$ has a least element.

Let $A$ and $B$ be sets. Then

$$A \times B = \{(a, b) \mid a \in A, b \in B\}$$

is the *Cartesian* or *direct product* of $A$ and $B$. More generally, if $A_1, \ldots, A_n$ are sets, then

$$A_1 \times A_2 \times \cdots \times A_n = \{(a_1, a_2, \ldots, a_n) \mid a_i \in A_i\}$$

is the Cartesian product of the sets $A_i$. Familiar examples of this would be $\mathbb{R}^2$ (the coordinatization of the plane) and $\mathbb{R}^n$ (the coordinatization of real

$n$-space). For finite sets, we can give a complete listing of the members. For example, if $A = \{1, 2\}$ and $B = \{3, 5, 7\}$, then

$$A \times B = \{(1, 3), (1, 5), (1, 7), (2, 3), (2, 5), (2, 7)\}.$$

Much of mathematics is devoted to the study of functions.

**Definition 1.1.4** A *function* or a *mapping* $f$ from a nonempty set $A$ to a set $B$ is a rule that associates with each element of $A$ a *unique* element of $B$. We write $f : A \to B$. The set $A$ is the *domain* of $f$, $B$ is the *range* of $f$, and $\{f(x) \mid x \in A\}$ is the *image* of $f$.

**Example 1.1.5**

(1) $A =$ set of all people who have reached age 10
  $B = \mathbb{R}$
  $f : A \to B$ where
  $f(a) =$ height in centimeters of the person $a$ at age 10

(2) $A =$ set of all people living in California
  $B =$ set of all first names
  $f : A \to B$
  $f(a) =$ first name of person $a$

(3) $A = B =$ the set of all nonnegative real numbers
  $f : A \to B$
  $f(x) = x^2 + \sqrt{x} + 1$

(4) $A = \mathbb{R}^m, B = \mathbb{R}^n$, where the elements of $A$ and $B$ are viewed as column vectors. Let $M$ be an $n \times m$ matrix.
  $T : A \to B$
  $T(v) = Mv.$

If $A$ and $B$ are two statements, then we will say that $A$ *implies* $B$ and write $A \Rightarrow B$ if the statement $B$ holds whenever the statement $A$ holds. For example, if $A$ is the statement "$x$ is the sum of two odd numbers" and $B$ is the statement that "$x$ is an even number", then $A \Rightarrow B$. Extending this notation, we also write $A \Leftrightarrow B$ (which we read as "$A$ if and only if $B$") whenever both $A \Rightarrow B$ and $B \Rightarrow A$.

Let $f : A \to B$ be any mapping. Then

$f$ is *injective*  if for all $x, y \in A, x \neq y$ implies that $f(x) \neq f(y)$
$f$ is *surjective* if for all $y \in B$, there exists an $x \in A$ with $f(x) = y$
$f$ is *bijective*  if it is both injective and surjective.

In these circumstances we say that $f$ is an *injection, surjection,* or *bijection,* respectively.

Note that $f$ is injective if and only if it satisfies the condition

$$\text{for all } x, y \in A, \quad f(x) = f(y) \implies x = y.$$

We will often rely on this equivalent statement when we wish to establish that a function is injective.

Less formally, the term *one-to-one* (sometimes written as 1-1) may be used instead of injective and *onto* may be used in place of surjective. If $f : A \to B$ is a bijection, it can be helpful to think of $f$ as giving a correspondence between the elements of $A$ and those of $B$. For this reason we occasionally refer to a bijection as being a *one-to-one correspondence* or *1-1 correspondence*.

We shall see in the exercises that the properties of being injective or surjective are indeed independent. However, in one very important situation it turns out that they coincide. This is a very useful fact.

**Lemma 1.1.6** *Let $A$ and $B$ be finite nonempty sets with $|A| = |B|$. Let $f : A \to B$. Then $f$ is injective if and only if $f$ is surjective.*

**Proof.** Let

$$C = \{f(x) \mid x \in A\}$$

be the image of $f$. Then it is easily seen that

$$|C| = |A| \iff f \text{ is injective}$$

while

$$|C| = |B| \iff f \text{ is surjective.}$$

Hence,

$$\begin{aligned}
f \text{ is injective} &\iff |C| = |A| \\
&\iff |C| = |B| \\
&\iff f \text{ is surjective.} \quad \square
\end{aligned}$$

In the sections to follow, we will often want to determine whether some function is injective or surjective. So let us consider an example to see exactly how this is done. Let $f : \mathbb{R}^3 \to \mathbb{R}^3$ be the function defined by

$$f(x_1, x_2, x_3) = (x_1 - x_2, x_2 - x_3, x_3 - x_1).$$

Is $f$ injective? To determine this we must consider any elements $u, v \in \mathbb{R}^3$ and check whether

$$f(u) = f(v) \implies u = v.$$

Toward this end, let $u = (x_1, x_2, x_3)$, $v = (y_1, y_2, y_3) \in \mathbb{R}^3$. Then

$$f(u) = f(v) \iff (x_1 - x_2, x_2 - x_3, x_3 - x_1) = (y_1 - y_2, y_2 - y_3, y_3 - y_1)$$
$$\iff x_1 - x_2 = y_1 - y_2, \; x_2 - x_3 = y_2 - y_3, \; x_3 - x_1 = y_3 - y_1.$$
$$(1.1)$$

However, these equations do not imply that $u = v$. For instance, let $a$ be any nonzero real number and let $y_i = x_i + a$. Clearly the equations in (1.1) are satisfied, yet $(x_1, x_2, x_3) \neq (y_1, y_2, y_3)$. In particular, if we take $a = 1$ and $u = (0, 0, 0)$, then $v = (1, 1, 1)$ and clearly $f(0, 0, 0) = f(1, 1, 1)$. Thus, $f$ is not injective.

Is $f$ surjective? We could cheat here and invoke a little linear algebra. Since $f$ is a linear transformation and we have seen that its kernel is not the zero subspace, it follows that the dimension of $\mathrm{Im} f$ is less than 3. Therefore, $\mathrm{Im} f \neq \mathbb{R}^3$ and $f$ is not surjective. However, let us still consider the question from first principles. We have that

$$f \text{ is surjective} \iff \forall \, (a, b, c) \in \mathbb{R}^3, \; \exists \, (x_1, x_2, x_3) \in \mathbb{R}^3$$
$$\text{with } f(x_1, x_2, x_3) = (a, b, c).$$

Now

$$f(x_1, x_2, x_3) = (a, b, c) \iff (x_1 - x_2, x_2 - x_3, x_3 - x_1) = (a, b, c)$$
$$\iff x_1 - x_2 = a, \; x_2 - x_3 = b, \; x_3 - x_1 = c.$$

However, adding these three equations together we find that $0 = a + b + c$. Consequently, any element $(a, b, c) \in \mathbb{R}^3$ that does not satisfy this condition will not be in the image of $f$. For example, $(1, 1, 1) \notin \mathrm{Im} f$. Therefore, $f$ is not surjective.

If $g : A \to B$ and $f : B \to C$ are two functions, then the *composite* $f \circ g$ is the function from $A$ to $C$ defined by

$$f \circ g(x) = f(g(x)) \qquad (x \in A).$$

For example, if $f, g : \mathbb{R} \to \mathbb{R}$ are functions defined by

$$f(x) = 1 + 2x + 3x^2, \quad g(x) = (1 + x^2)^{1/2}$$

then $f \circ g : \mathbb{R} \to \mathbb{R}$ is defined by

$$f \circ g(x) = 1 + 2g(x) + 3g(x)^2 = 1 + 2(1 + x^2)^{1/2} + 3(1 + x^2)$$
$$= 4 + 2(1 + x^2)^{1/2} + 3x^2.$$

The composition of functions satisfies a very important rule.

**Lemma 1.1.7** *Let $f : C \to D$, $g : B \to C$ and $h : A \to B$ be functions. Then*

$$f \circ (g \circ h) = (f \circ g) \circ h.$$

**Proof.** The proof is very simple but it illustrates how we must proceed if we wish to show that two functions are equal. We must show that the "rule" defining each function leads to the same result for all elements in the domain. The domain in this case is $A$. So let $x \in A$. Then

$$f \circ (g \circ h)(x) = f((g \circ h)(x)) = f(g(h(x)))$$

while

$$(f \circ g) \circ h(x) = (f \circ g)(h(x)) = f(g(h(x))).$$

Since we obtain the same result in both cases, the two functions $f \circ (g \circ h)$ and $(f \circ g) \circ h$ must be equal.  $\square$

The rule or law in Lemma 1.1.7 is called the *associative law*. The importance of Lemma 1.1.7 lies in the fact that we no longer need to concern ourselves with the positioning of brackets when composing several functions and so we may write simply

$$f \circ (g \circ h) = f \circ g \circ h$$

unambiguously.

To simplify the notation further, we abbreviate $f \circ g$ to $fg$ and write

$$f \circ f = f^2, \quad f \circ f \circ f = f^3$$

and so forth.

A very simple function, but nonetheless an important function, on any nonempty set $A$ is the *identity function $I_A$* defined by

$$I_A(a) = a \quad \text{for all} \quad a \in A.$$

We will also write $I_A = 1_A$.

Sometimes functions will "cancel each other out" in the following sense: Two functions $f : A \to B$ and $g : B \to A$ are said to be *inverses* ($f$ the *inverse* of $g$ and $g$ the *inverse* of $f$) if

$$f \circ g = I_B \quad \text{and} \quad g \circ f = I_A.$$

It is a simple exercise to show that if $f$ and $g$ are inverses, then they are both bijections. Conversely, suppose that $f : A \to B$ is a bijection. Then for every element $b$ in $B$, there exists a unique element $a$ in $A$ with $f(a) = b$. Therefore we can define a mapping $g : B \to A$ by the following: for all $b \in B$, $g(b)$ is the unique element $a$ in $A$ such that $f(a) = b$. It is easy to see that $f \circ g = I_B$ and that $g \circ f = I_A$, in other words, $f$ and $g$ are inverse mappings. Consequently, we have the useful observation that a mapping $f : A \to B$ is a bijection if and only if $f$ has an inverse.

## Exercises 1.1

1. Use the notation $\{\cdots \mid \cdots\}$ to describe each of the following sets.

    (i)  $\{1, 2, 3, \ldots, 100\}$.
    (ii)  The set of all even integers.
    (iii)  $\{3, 6, 9, \ldots\}$.

2. Let $A = \{a, b, c, d\}$. List all possible subsets of $A$.

3. Let $A$ be a nonempty set with $\mid A \mid = n$. Show that $A$ has $2^n$ subsets.

4. For each of the following pairs of sets, decide whether $A \subset B$, $B \subset A$, $A = B$, or none of these hold.

    (i)  $A =$ set of even positive integers, $B = \{x \in \mathbb{N} \mid x^2 \text{ is even}\}$.
    (ii)  $A = \{1, 3, 5, 7, 9\}$, $B = \{1, 2, 3, 4, 5, 6, 7, 8, 9, 10\}$.
    (iii)  $A = \{x \in \mathbb{N} \mid \exists\, y \in \mathbb{N} \text{ with } x = y^2\}$,
        $B = \{x \in \mathbb{N} \mid \exists\, y \in \mathbb{N} \text{ with } x = y^3\}$.
    (iv)  $A = \{x \in \mathbb{N} \mid 1 < x < 20 \text{ and } x - 1 \text{ is divisible by 3}\}$,
        $B = \{x \in \mathbb{N} \mid 1 < x < 20 \text{ and } x^2 - 1 \text{ is divisible by 3}\}$.

*5. Let $A$, $B$, and $C$ be subsets of a set $X$. Show that

$$A \cap (B \cup C) = (A \cap B) \cup (A \cap C)$$

and that

$$A \cup (B \cap C) = (A \cup B) \cap (A \cup C).$$

(These equalities are known as the *distributive laws*.)

In each of the Exercises 6 through 14, determine whether the function $f : A \to B$ is injective, surjective, or bijective.

6. $A = \mathbb{R}, B = \mathbb{R}, f(x) = 2x - 3.$

7. $A = \mathbb{R}, B = \mathbb{R}, f(x) = x^2.$

8. $A = \mathbb{R}, B = \mathbb{R}, f(x) = e^x.$

9. $A = \mathbb{R}, B = \mathbb{R}, f(x) = x^3 - 2x + 1.$

10. $A = \mathbb{R}^2, B = \mathbb{R}^3, f(x_1, x_2) = (x_1 + x_2, x_1 - x_2, 2x_1 - 3x_2).$

11. $A = \mathbb{R}^3, B = \mathbb{R}^2, f(x_1, x_2, x_3) = (x_1 + x_3, x_2 - x_3).$

12. $A = \mathbb{R}^2, B = \mathbb{R}^2, f(x_1, x_2) = (2x_1 + x_2, x_2).$

13. $A = \mathbb{R}^3, B = \mathbb{R}^3, f(x_1, x_2, x_3) = (x_1 - x_3, x_2 + x_3, x_1 - x_2 - 2x_3).$

14. $A = \mathbb{R}^3, B = \mathbb{R}^3, f(x_1, x_2, x_3) = (x_1 + x_2 + x_3, x_2 - x_3, x_2).$

In Exercises 15 and 16, determine whether the functions $f, g : \mathbb{R}^3 \to \mathbb{R}^3$ are inverses (of each other).

15. $f(x_1, x_2, x_3) = (x_1 + x_2, \ x_2 + x_3, \ x_1 + x_3);$
$g(x_1, x_2, x_3) = \frac{1}{2}(x_1 - x_2 + x_3, \ x_1 + x_2 - x_3, \ -x_1 + x_2 + x_3).$

16. $f(x_1, x_2, x_3) = (x_1 + x_2 + x_3, \ x_2, \ x_3);$
$g(x_1, x_2, x_3) = (x_1 - x_2 - x_3, \ -x_1 + x_2 - x_3, \ -x_1 - x_2 + x_3).$

17. Let $A = B = \{1, 2\}$ and $C = \{1, 2, 3\}$. List all the elements of $A \times B$ and $A \times B \times C$. Is $A \times B$ a subset of $A \times B \times C$? Is $\{1\} \times B \times \{2, 3\}$ a subset of $A \times B \times C$?

18. Let $A$ and $B$ be nonempty sets and $f : A \to B$. Show that $I_B f = f = f I_A.$

*19. Let $A = \{1, 2, 3, 4\}$ and let $T_A$ denote the set of mappings from $A$ to $A$. An element $f \in T_A$ is an *idempotent* if $f^2 = f$. Let $f \in T_A$ be such that $f(1) = 2, f(2) = 3, f(3) = 4, f(4) = 3.$

   (i)  Find the smallest integer $n$ such that $f^n$ is an idempotent.
   (ii)  Find $g \in T_A$ such that $fgf = f$ and $gfg = g.$
   (iii)  Show that $fg$ and $gf$ are idempotents.
   (iv)  Find $h, k \in T_A$ such that $h \neq I_A$ and $hk = kh = I_A.$

*20. Let $A$ be a nonempty set and $g : A \to B, f : B \to C$ be mappings. Establish the following:

   (i)  $fg$ injective $\implies g$ is injective.
   (ii)  $fg$ surjective $\implies f$ is surjective.

Give examples to show the following:

   (iii)  $fg$ injective need not imply that $f$ is injective.
   (iv)  $fg$ surjective need not imply that $g$ is surjective.

## 1.2 Induction

A tremendously important tool in the study of structures that involve natural numbers is the method of argument called *induction*. Imagine that we wish to show that a certain statement is true for all integers $n$ that are greater than or equal to some fixed integer $a$. For example, the claim might be that for all positive integers $n \geq 1$,

$$1^2 + 2^2 + \ldots + n^2 = \frac{1}{6}n(n+1)(2n+1).$$

One way to proceed would be to assume that the claim is false and then to show that this leads to a contradiction. If the claim is false, then, by the Well Ordering Axiom, there must be a smallest value of $n$ for which the claim does not hold, that is,

$$1^2 + 2^2 + \ldots + n^2 \neq \frac{1}{6}n(n+1)(2n+1).$$

However, when $n = 1$, we have that the left side of the equation reduces to $1^2 = 1$ and the right side reduces to $\frac{1}{6} \cdot 1 \cdot 2 \cdot 3 = 1$. Thus, the claim is true for 1 and we may assume that the claim holds for the sum of $1, 2, \ldots$, up to $n - 1$ squares. Since the claim holds for the sum of $n - 1$ squares, we must have:

$$1^2 + 2^2 + \ldots + (n-1)^2 = \frac{1}{6}(n-1)(n)(2n-1)$$

so that

$$1^2 + 2^2 + \ldots + (n-1)^2 + n^2 = \frac{1}{6}(n-1)(n)(2n-1) + n^2$$

$$= \frac{1}{6}n(n+1)(2n+1).$$

But this contradicts our assumption that the claim failed for $n$. Hence the claim must hold for all nonnegative integers.

This method of proof is formalized in what is known as the principle of induction. Notice that in the above argument, we were able to assume that the claim was true for $1, 2, \ldots, n - 1$, but only actually used the validity of the claim for $n - 1$. This is a common occurrence in these types of arguments and is reflected in two equivalent forms of the principle.

**First Form of the Principle of Induction 1.2.1** *Let A be a subset of* $\mathbb{N}_0$ *with the following properties:*

(1) *A contains the integer a.*

(2) *If A contains n where $n \geq a$, then A contains $n + 1$.*

*Then A contains all integers greater than or equal to a.*

The element $a$ is referred to as the *base* of the induction argument, (2) is referred to as the *induction step* and the assumption in (2) that $n \in A$ is called the *induction hypothesis*. The induction principle is really just another form of the well-ordering axiom.

**Example 1.2.2** Show by induction that

$$1 \cdot 2 + 2 \cdot 3 + 3 \cdot 4 + \cdots + n(n + 1) = \frac{1}{3}n(n + 1)(n + 2) \qquad (1.2)$$

for all positive integers $n$.

Let $A$ denote the set of positive integers $n$ for which the equation (1.2) holds. We wish to show that $A = \mathbb{N}$.

When $n = 1$, we have

$$\text{Left side of } (1.2) = 1 \cdot 2 = 2$$

$$\text{Right side of } (1.2) = \frac{1}{3} \, 1 \cdot 2 \cdot 3 = 2.$$

Thus, equation (1.2) holds when $n = 1$. In other words, $1 \in A$ and we can take $a = 1$ as the base for our induction.

We now adopt the induction hypothesis that equation (1.2) holds for some integer $n$, where $n \geq 1$, and we wish to complete the induction step by showing that the equation (1.2) holds when we replace $n$ with $n + 1$. The left side then becomes

$$1 \cdot 2 + 2 \cdot 3 + \cdots + n(n + 1) + (n + 1)(n + 2)$$

$$= \frac{1}{3}n(n + 1)(n + 2) + (n + 1)(n + 2) \quad \text{by the induction hypothesis}$$

$$= \frac{1}{3}(n + 1)(n + 2)(n + 3)$$

$$= \frac{1}{3}(n + 1)((n + 1) + 1)((n + 1) + 2).$$

Thus, equation (1.2) also holds when $n$ is replaced with $n+1$. By the first form of the principle of induction, this implies that

$$A = \{n \in \mathbb{N} \mid n \geq 1\}.$$

In other words, $A = \mathbb{N}$, as required.

**Second Form of the Principle of Induction 1.2.3** *Let $A$ be a subset of $\mathbb{N}_0$ with the following properties:*

(1)  *A contains the integer $a$.*

(2)  *If $n \geq a$ and $A$ contains every integer $k$ with $a \leq k \leq n$, then $A$ contains $n+1$.*

*Then $A$ contains all integers greater than or equal to $a$.*

Again, the number $a$ is the base for the induction argument, the assumption that "$n \geq a$ and $A$ contains every integer with $a \leq k \leq n$" is the induction hypothesis and the verification of the conclusion in (2) is the induction step.

We will illustrate the second form of the principle of induction by using it to prove an important result concerning integers in Section 1.4.

The principle of induction is an important tool in studying properties of integers. However, induction is not the only method that we can apply to establish statements concerning integers. This point is well illustrated by the following treatment of progressions.

An *arithmetic progression* is a sequence $a_1, a_2, a_3, \ldots$, often written as $\{a_n\}$, of integers such that the difference between any two successive numbers in the sequence is a constant For example, the sequence $2, 5, 8, 11 \ldots$ is an arithmetic progression with constant difference 3. A general arithmetic progression takes the form

$$a, (a + d), (a + 2d), \ldots, (a + (n - 1)d) \ldots.$$

The $n$th term of the progression is $a + (n - 1)d$ and the sum of the first $n$ terms is

$$S = a + (a + d) + (a + 2d) + \cdots + (a + (n - 1)d).$$

The number $d$ is referred to as the *common difference.* We can find a compact expression for $S$ if we rewrite it as follows:

$$S = a + (a + d) + (a + 2d) + \cdots + (a + (n - 1)d)$$
$$S = (a + (n - 1)d) + (a + (n - 2)d) + (a + (n - 3)d) + \cdots + a.$$

Adding, we obtain

$$2S = (2a + (n-1)d) + (2a + (n-1)d) + \cdots + (2a + (n-1)d) \quad (n \text{ terms})$$
$$= n(2a + (n-1)d)$$

so that

$$S = \frac{n}{2}(2a + (n-1)d). \tag{1.3}$$

This leads to many useful formulae. If we take $a = d = 1$, then we obtain

$$1 + 2 + \cdots + n = \frac{n(n+1)}{2}$$

whereas if we take $a = 1$ and $d = 2$ we obtain

$$1 + 3 + \cdots + (2n-1) = n^2.$$

A *geometric progression* is a sequence $\{a_n\}$, of integers such that the ratio between any two successive numbers in the sequence is a constant For example, the sequence $2, 6, 18, 54 \ldots$ is an geometric progression with constant ratio 3. A general geometric progression takes the form

$$a, ar, ar^2, \ldots, ar^{n-1} \ldots.$$

The $n$th term of the progression is $ar^{n-1}$ and the sum of the first $n$ terms is

$$S = a + ar + ar^2 + \cdots + ar^{n-1}.$$

The number $r$ is referred to as the *common ratio*. Multiplying by $r$ we obtain

$$rS = ar + ar^2 + \cdots + ar^n$$

so that

$$S - rS = a - ar^n$$

and

$$S = a\frac{(1 - r^n)}{1 - r}$$

provided $r \neq 1$. This also has some interesting special cases. If we take $a = 1$, $r = 2$, then we obtain

$$1 + 2 + 2^2 + \cdots + 2^{n-1} = \frac{1 - 2^n}{1 - 2} = 2^n - 1$$

whereas if $a = 1$ and $r = 3$ we find

$$1 + 3 + 3^2 + \cdots + 3^{n-1} = \frac{1}{2}(3^n - 1).$$

An interesting sequence of numbers with many strange connections is the *Fibonacci sequence* $\{a_n\}$ of natural numbers and it is defined inductively as follows:

$$a_1 = 1, \quad a_2 = 1, \quad a_{n+2} = a_n + a_{n+1}.$$

The first few terms of the sequence are

$$1, \ 1, \ 2, \ 3, \ 5, \ 8, \ 13, \ \ldots .$$

Each term of the sequence, after the first two terms, is the sum of the two preceding terms. Leonardo Fibonacci was born in Pisa, Italy, in about 1170. As a result of traveling extensively around the Mediterranean, he became familiar with the Hindu-Arabic system of numerals with which we are so familiar, but which was not yet in general use in Europe. He authored four mathematical texts. The Fibonacci numbers first appear as the solution to a rabbit population problem in *Liber Abaci*, by Fibonacci, which first appeared in 1202. The Fibonacci sequence shows up in other strange ways, such as the number of petals in certain flowers, and has connections to the "Golden Ratio".

**Exercises 1.2**　Let $n$ be a positive integer. Establish the following by induction.

1. $1 + 5 + 9 + \cdots + (4n + 1) = (2n + 1)(n + 1).$

2. $2 + 4 + 6 + \cdots + 2n = n(n + 1).$

3. $3 + 3 \cdot 4 + 3 \cdot 4^2 + \cdots + 3 \cdot 4^{n-1} = 4^n - 1.$

4. $1^2 + 3^2 + 5^2 + \cdots + (2n - 1)^2 = \frac{n}{3}(2n - 1)(2n + 1).$

5. $1^3 + 2^3 + 3^3 + \cdots + n^3 = \left[\frac{n(n+1)}{2}\right]^2.$

6. $1^4 + 2^4 + 3^4 + \cdots + n^4 = \frac{1}{30}n(n + 1)(2n + 1)(3n^2 + 3n - 1).$

7. Show that the $n$th Fibonacci number $a_n$ satisfies $a_n < 2^n$.

8. Show that for all $n \geq 2$, the Fibonacci numbers satisfy:

   (i)　$a_n a_{n+1} - a_{n-1} a_n = a_n^2.$
   (ii)　$a_{n-1} a_{n+1} - a_n^2 = (-1)^n.$

9. Show that for all $n \geq 1$, the Fibonacci numbers satisfy:

$$a_1^2 + a_2^2 + \cdots + a_n^2 = a_n a_{n+1}.$$

*10. Show that if $A$ is a set with $n$ elements, then $A$ has $2^n$ subsets.

## 1.3 Divisibility

For any integers $x$ and $y$, we say that $x$ is a *divisor* of $y$ or that $x$ *divides* $y$ if there exists an integer $q$ with $y = qx$. We write $x \mid y$ to indicate that $x$ divides $y$. Note that for all integers $x$ we have $0 = 0 \cdot x$. Thus, every integer divides zero.

**Lemma 1.3.1** *Let $a, b, c \in \mathbb{Z}$. If $a \mid b$ and $a \mid c$, then $a \mid b \pm c$.*

**Proof.** Exercise.  $\square$

Note that the converse of Lemma 1.3.1 does *not* hold. For example, 5 divides $10 = 2 + 8$. But 5 does not divide 2 or 8.

We all know from elementary arithmetic that if we divide one integer $a$ by another integer $b \neq 0$, then we obtain a quotient and a remainder. For example, if $a = 181$ and $b = 17$ then

$$a = 10b + 11.$$

This idea is formalized in the next result.

**Lemma 1.3.2 (The Division Algorithm)** *Let $a$ and $b$ be integers with $b > 0$. Then there exist unique integers $q$ and $r$, with*

$$a = bq + r \quad \text{and} \quad 0 \leq r < b.$$

**Proof.** We break the proof into two parts. First, the existence of the integers $q$ and $r$, and then their uniqueness.

**Existence.** Let $R = \{x \in \mathbb{N}_0 \mid \exists\, y \in \mathbb{Z} \text{ with } a = by + x\}$. Since we can write $a = b \cdot 0 + a$ if $a \geq 0$ and $a = b \cdot a + (1 - b)a$ if $a < 0$, where $(1 - b)a \in \mathbb{N}_0$, it follows that $R \neq \emptyset$. By the well-ordering principle, we then know that $R$ has a least member $r$, say. By the definition of $R$, there must be an integer $q$ with $a = bq + r$.

If $r \geq b$, then

$$a = bq + r = bq + b + (r - b) = b(q + 1) + (r - b)$$

where $0 \leq r - b < r$. This contradicts the minimality of $r$. So we must have $0 \leq r < b$.

**Uniqueness.** Now suppose that $q'$ and $r'$ are also integers such that

$$a = bq' + r' \quad \text{where} \quad 0 \leq r' < b.$$

Then,

$$0 = a - a = bq + r - (bq' + r')$$
$$= b(q - q') + (r - r')$$

so that

$$r' - r = b(q - q').$$

But $0 \leq r, r' < b$ so that $-b < r - r' < b$ and the only way that we can have $b$ dividing $r' - r$ is if $r' - r = 0$. Thus, $r = r'$, which then implies that $b(q - q') = 0$. Since $b \neq 0$, we conclude that $q - q' = 0$ and therefore $q = q'$. This establishes the uniqueness of $q$ and $r$.  $\square$

By the *greatest common divisor* of two integers $a$ and $b$, at least one of which is nonzero, we mean the largest integer $d$ such that $d \mid a$ and $d \mid b$. Since 1 divides every integer, it is clear that the greatest common divisor exists and is an integer lying between 1 and the smaller of $\mid a \mid$ and $\mid b \mid$. It is customary to abbreviate the words greatest common divisor to g.c.d. and to denote the g.c.d. of the integers $a$ and $b$ by $\gcd(a, b)$ or simply $(a, b)$.

**Example 1.3.3**  Find the g.c.d. of 36 and 94.

**Answer:**  The positive divisors of 36 are 1, 2, 3, 4, 6, 9, 12, 18, 36, and the positive divisors of 45 are 1, 3, 5, 9, 15, 45. Since the largest number dividing both 36 and 45 is 9, the greatest common divisor of 36 and 45 is 9.

When working with small numbers, it is an easy matter to find the greatest common divisor. All that we need to do is to find all the divisors of both the integers and select the largest divisor common to both numbers. This becomes less practical when dealing with really large numbers. The next result gives us an efficient algorithm that works well with large integers. While the array of equations may appear daunting at first sight, it should be less so once it is appreciated that it is only formulating a sequence of long divisions.

**Theorem 1.3.4 (The Euclidean Algorithm)**  *Let $a \in \mathbb{Z}$, $b \in \mathbb{N}$ and $q_i$ $(2 \leq i \leq n)$, $r_i$ $(2 \leq i \leq n)$ be defined by the equations*

$$
\begin{aligned}
a &= bq_2 + r_2 & 0 &\leq r_2 < b \\
b &= r_2 q_3 + r_3 & 0 &\leq r_3 < r_2 \ \text{if } r_2 > 0 \\
r_2 &= r_3 q_4 + r_4 & 0 &\leq r_4 < r_3 \ \text{if } r_3 > 0 \\
&\ \ \vdots \\
r_{n-2} &= r_{n-1} q_n + r_n & 0 &\leq r_n < r_{n-1} \ \text{if } r_{n-1} > 0 \\
r_{n-1} &= r_n q_{n+1}. & r_{n+1} &= 0
\end{aligned}
$$

*Then,*

(i) $r_n = (a, b)$.
(ii) *If $c \in \mathbb{Z}$, $c \mid a$ and $c \mid b$ then $c \mid (a, b)$.*
(iii) *There exist $x, y \in \mathbb{Z}$ with $(a, b) = xa + yb$.*

(The reason for starting the indices for $q_i$ and $r_i$ with $i = 2$ will become apparent in Theorem 1.3.6.)

**Proof.** By repeated application of the division algorithm, we know that there exist integers $q_2$ and $r_2$ such that the first equation holds, and then integers $q_3$ and $r_3$ such that the second equation holds, and so on. Let $r_1 = b$.

Since the integers $b = r_1, r_2, r_3, r_4, \ldots$ are nonnegative and decreasing, there must be some integer $n \geq 0$ with $r_{n+1} = 0$, $r_n \neq 0$.

Now from the last equation we deduce that $r_n \mid r_{n-1}$; from the second-to-last equation we deduce that $r_n \mid r_{n-2}$, from the third to last equation we deduce that $r_n \mid r_{n-3}$ and so on, so that eventually we obtain $r_n \mid a$ and $r_n \mid b$. Thus, $r_n$ is a common divisor of $a$ and $b$, and $r_n \geq 1$.

Now let $c \in \mathbb{Z}$ and $c \mid a$ and $c \mid b$. Reversing the argument of the preceding paragraph, it now follows from the first equation that $c \mid r_2$. From the second equation it then follows that $c \mid r_3$, from the third equation it follows that $c \mid r_4$, and so on, until we finally obtain that $c \mid r_n$. Therefore, $c \leq r_n$ and, consequently, we must have $r_n = \gcd(a, b)$, which establishes (i). Since we now know that $r_n = \gcd(a, b)$, on rereading the previous paragraph, we find that we have established (ii).

From the second-to-last equation, we have

$$r_n = r_{n-2} - r_{n-1} q_n.$$

Using the third-to-last equation, we can eliminate $r_{n-1}$ to obtain

$$r_n = r_{n-2} - q_n(r_{n-3} - q_{n-1} r_{n-2}).$$

Then we can eliminate $r_{n-2}$ and so on until we finally obtain

$$r_n = xa + yb \tag{1.4}$$

for some integers $x$ and $y$.  $\square$

**Example 1.3.5** Let $a = 710$, $b = 68$. Then

$$710 = 68 \times 10 + 30$$
$$68 = 30 \times 2 + 8$$

$$30 = 8 \times 3 + 6$$
$$8 = 6 \times 1 + 2$$
$$6 = 2 \times 3.$$

Therefore, $(710, 68) = 2$. Moreover, working our way back up this system of equations we have

$$
\begin{aligned}
2 &= 8 - 6 \\
&= (68 - 2 \times 30) - (30 - 3 \times 8) \\
&= 68 - 2(710 - 10 \times 68) - (710 - 10 \times 68) \\
&\quad + 3 \times (68 - 2 \times 30) \\
&= b - 2 \times a + 20 \times b - a + 10 \times b \\
&\quad + 3b - 6(710 - 10 \times 68) \\
&= b - 2a + 20b - a + 10b + 3b - 6a + 60b \\
&= -9a + 94b.
\end{aligned}
$$

For positive integers $a$ and $b$, we shall frequently refer to the fact that the Euclidean algorithm enables us to compute $\gcd(a, b)$. However, we shall see later that it is also important to have a good algorithm to calculate integers $x$, $y$ with $ax + by = (a, b)$. The method used in the previous example is clearly unmanageable in larger computations.

**Theorem 1.3.6 (The Extended Euclidean Algorithm.)** *Let $a, b \in \mathbb{N}$ and $q_i$, $r_i$ $(2 \leq i)$ be defined as in Theorem 1.3.4. Let*

$$
\begin{aligned}
x_0 &= 1, \quad y_0 = 0, \quad r_0 = a \\
x_1 &= 0, \quad y_1 = 1, \quad r_1 = b
\end{aligned}
$$

*and for $i \geq 2$ define*

$$x_i = x_{i-2} - q_i x_{i-1}, \quad y_i = y_{i-2} - q_i y_{i-1}.$$

*Then,*

$$(a, b) = x_n a + y_n b.$$

**Proof.** Exercise. (Hint: Show that $r_k = x_k a + y_k b$ for $0 \leq k \leq n$.)  $\square$

**Example 1.3.7** If we apply this extended Euclidean algorithm to the previous example with $a = 710$, $b = 68$, then we can tabulate the outcome step by step as follows:

| $i$ | $q_i$ | $r_i$ | $x_i$ | $y_i$ |
|---|---|---|---|---|
| 0 | — | 710 | 1 | 0 |
| 1 | — | 68 | 0 | 1 |
| 2 | 10 | 30 | 1 | −10 |
| 3 | 2 | 8 | −2 | 21 |
| 4 | 3 | 6 | 7 | −73 |
| 5 | 1 | 2 | −9 | 94 |
| 6 | 3 | 0 | 34 | −355 |

Since $r_6 = 0$, it follows as before that $(710, 68) = r_5 = 2$ and, from the extended Euclidean algorithm, that

$$2 = -9 \times 710 + 94 \times 68.$$

Two integers $a, b$ are *relatively prime* if $(a, b) = 1$ whereas integers $a_1, \ldots, a_k$ are *pairwise relatively prime* if they are *relatively prime* in pairs—that is, $(a_i, a_j) = 1$ for all $i, j$. For example, 8 and 15 are relatively prime whereas 8, 15, and 49 are pairwise relatively prime.

**Corollary 1.3.8** *Let $a, b \in \mathbb{Z}$. Then $a$ and $b$ are relatively prime if and only if there exist $x, y \in \mathbb{Z}$ with $ax + by = 1$.*

**Proof.** If $a$ and $b$ are relatively prime then $(a, b) = 1$ so that, by Theorem 1.3.4 (the Euclidean algorithm), there exist integers $x, y$ with

$$1 = (a, b) = ax + by.$$

Conversely, if there exist integers $x, y$ with $1 = ax + by$ and if $d = (a, b)$, then $d$ divides $a$ and $b$, and therefore $d \mid 1$. But, $d \geq 1$. Therefore we must have $d = 1$.   $\square$

**Example 1.3.9** $175 = 5^2 \times 7$ and $72 = 2^3 \times 3^2$. Thus, $(175, 72) = 1$. The Euclidean algorithm yields

$$175 = 72 \times 2 + 31$$

$$72 = 31 \times 2 + 10$$

$$31 = 10 \times 3 + 1$$

$$10 = 1 \times 10 + 0.$$

whereas back substitution or the extended Euclidean algorithm will yield

$$1 = 7 \times 175 + (-17) \times 72.$$

The next result gives us a really important characterization of the greatest common divisor of two integers and details the properties that we use most.

**Corollary 1.3.10** *Let a and b be integers, not both of which are zero. Let $d \in \mathbb{Z}$. Then $d = (a, b)$ if and only if it satisfies all the following conditions:*

    (i) $d \geq 1$.
    (ii) $d \mid a$ *and* $d \mid b$.
    (iii) *If* $c \in \mathbb{Z}$, $c \mid a$ *and* $c \mid b$, *then* $c \mid d$.

**Proof.** In light of Theorem 1.3.4 (the Euclidean algorithm) and the definition of the greatest common divisor, it is clear that the greatest common divisor satisfies the three conditions (i), (ii), and (iii).

Conversely, let $d$ satisfy conditions (i), (ii), and (iii). By (i) and (ii), $d$ is a positive common divisor of $a$ and $b$. If $c$ is any other positive common divisor of $a$ and $b$, then, by (iii), $c \mid d$ so that $c \leq d$. Therefore, $d$ is indeed the greatest common divisor. $\quad\square$

Dual to the concept of the greatest common divisor of two integers $a$ and $b$ we have the concept of *least common multiple* of two nonzero integers $a$ and $b$, which we define to be, as the words suggest, the least positive integer that is divisible by both $a$ and $b$. Clearly, $\mid a \mid \cdot \mid b \mid$ is divisible by both $a$ and $b$ so that there is indeed a least positive integer divisible by $a$ and $b$, and it must lie between 1 and $\mid a \mid\mid b \mid$. When convenient, we abbreviate *least common multiple* to *lcm* and write lcm$(a, b)$.

**Lemma 1.3.11** *Let $a, b$ be nonzero integers and $m \in \mathbb{N}$. Then $m$ is the least common multiple of $a$ and $b$ if and only if*

    (i) $m \geq 1$,
    (ii) $a, b \mid m$,
    (iii) *if* $n \in \mathbb{N}$ *and* $a, b \mid n$ *then* $m \mid n$.

**Proof.** Exercise. $\quad\square$

**Exercises 1.3**

    1. Show that $x \mid 0$ for all $x \in \mathbb{Z}$.

    2. Show that if $a \mid b$ and $b \mid c$ then $a \mid c$.

    3. Show that if $d \mid a$ and $d \mid b$ then $d \mid ax + by$ for all $x, y \in \mathbb{Z}$.

    4. Show that if $a > 0$, $b > 0$, and $a \mid b$ then $a \leq b$.

    5. Show that if $a, b \in \mathbb{Z}$ are such that $a \mid b$ and $b \mid a$, then $a = \pm b$.

    6. $3 \mid 2^{2n} - 1$.

7. $15 \mid 4^{2n} - 1$.

8. $2 \mid n^2 + 3n$.

9. $6 \mid n^3 + 3n^2 + 2n$.

10. Show that if $n$ is odd then $8 \mid n^2 - 1$.

11. Show that if $n$ is even then $6 \mid n^3 - n$.

12. Show that if $n$ is odd then $24 \mid n^3 - n$.

13. For each of the following pairs of integers $a$, $b$, use the extended Euclidean algorithm to find $d = \gcd(a, b)$ and also integers $x$, $y$ such that $d = ax + by$.

    (i)  175; 72.
    (ii)  341; 377.
    (iii)  1848; 525.

14. Let $a, b, c \in \mathbb{Z}$ and $d = (a, b)$. Show that there exist $x, y \in \mathbb{Z}$ with $ax + by = c$ if and only if $d \mid c$.

15. Find integers $x$, $y$ such that

$$36x + 75y = 21.$$

16. Show that if $a, b, c \in \mathbb{Z}$, $c \neq 0$, and $ca \mid cb$, then $a \mid b$.

17. Let $a, b \in \mathbb{Z}$, $c \in \mathbb{N}$. Show that $(ca, cb) = c(a, b)$.

18. Let $a, b \in \mathbb{Z}$ and $d = (a, b)$. Show that

$$\left( \frac{a}{d}, \frac{b}{d} \right) = 1.$$

19. Show that for all $a, n \in \mathbb{N}$,

$$(a, a + n) \mid n.$$

20. Let $a, b \in \mathbb{N}$. Show that if $(a, b) = 1$ then $(a + b, a - b)$ is either 1 or 2.

*21. Let $a, b$ be integers, not both of which are zero. Let $B$ be a $2 \times 2$ matrix with integer entries such that $\det(B) = \pm 1$ and

$$B \begin{bmatrix} a \\ b \end{bmatrix} = \begin{bmatrix} d \\ 0 \end{bmatrix}$$

where $d$ is a positive integer. Show that $(a, b) = d$.

22. In this exercise we develop some interesting properties of the numbers that appear in the extended Euclidean algorithm. Set $r_{n+1} = 0$. Now establish the following:

    (i) $(a, b) = x_n a + y_n b$.
    (ii) For

$$A_k = \begin{bmatrix} x_k & y_k \\ x_{k+1} & y_{k+1} \end{bmatrix},$$

$\det(A_k) = (-1)^k$.
    (iii) $(x_k, y_k) = 1$ for $1 \leq k \leq n + 1$.
    (iv)

$$A_k \begin{bmatrix} a \\ b \end{bmatrix} = \begin{bmatrix} r_k \\ r_{k+1} \end{bmatrix}.$$

    (v)

$$\frac{a}{b} = -\frac{y_{n+1}}{x_{n+1}}.$$

23. Just for the purposes of this exercise, for all nonzero integers $a, b, c$ define $(a, b, c)$ to be the greatest integer $d$ such that $d$ divides $a, b$ and $c$. Show that

    (i) $(a, b, c) = ((a, b), c)$.
    (ii) $\exists\, x, y, z \in \mathbb{Z}$ with

$$(a, b, c) = xa + yb + zc.$$

24. Show that any two consecutive Fibonacci numbers are relatively prime.

25. (i) Let $a, b \in \mathbb{N}$, $x, y \in \mathbb{Z}$ be such that $ax + by = 1$. Show that $(yn + a, -xn + b) = 1$ for all $n \in \mathbb{Z}$.
    (ii) Show that $(4n + 3, 7n + 5) = 1$ for all $n \in \mathbb{Z}$.

26. Show that, for all positive integers $a, b$,

$$ab = \gcd(a, b) \times \mathrm{lcm}(a, b).$$

## 1.4  Prime Numbers

A positive integer $p > 1$ is a *prime* or a *prime number* if the only positive integer divisors of $p$ are 1 and $p$ itself. The first few primes are 2, 3, 5, 7, 11, 13, 17, 19. Since the time of the ancient Greeks, the study of prime numbers and the search for new prime numbers has fascinated mathematicians. This interest has intensified in recent years as a result of the importance of large prime numbers in such modern applications as cryptography, which we will have a look at later. The largest known prime numbers currently are of the form $2^n - 1$, with $n$ on the order of 43 million. However, a new record is being set roughly every 2 years or so. The latest record can be found easily via an Internet search.

There is a simple, but important, characterization of prime numbers that we tend to take for granted.

**Lemma 1.4.1** *Let $p \in \mathbb{N}$, $p > 1$. Then $p$ is a prime number if and only if it satisfies the following condition:*

$$\text{(C)} \qquad \textit{for any } a, b \in \mathbb{Z}, \quad p \mid ab \implies p \mid a \textit{ or } p \mid b.$$

**Proof.** First suppose that $p$ satisfies (C). Let $x \in \mathbb{N}$ and $x \mid p$. Then there exists $y \in \mathbb{N}$ with $p = xy$. Clearly we must have $1 \leq x, y \leq p$. But, $p = xy$ implies that $p \mid xy$ and therefore, by (C), $p \mid x$ or $p \mid y$. If $p \mid x$ then necessarily $p = x$ and $y = 1$, whereas if $p \mid y$ then necessarily $p = y$ and $x = 1$. Therefore, either $x = 1$ or $x = p$. This means that $p$ must be prime.

Now let us assume that $p$ is a prime number and that $p \mid ab$ where $a, b \in \mathbb{Z}$. Let $d = (a, p)$. Then $d \mid p$ so that $d = 1$ or $p$. If $d = p$, then $p = d \mid a$, as required.

So suppose that $d = 1$. By Theorem 1.3.4, there exist $s, t \in \mathbb{Z}$ with

$$1 = sa + tp.$$

Then $b = sab + tpb$. Now $p \mid ab$ so that $p \mid sab$ and $p \mid tpb$. Therefore, $p \mid b$ as required.  $\square$

Our familiarity with the equivalence established in Lemma 1.4.1 can easily lead us to assume that these two properties are fundamentally (and obviously?) the same. However, this is far from true. Indeed, the two properties are not equivalent in certain number systems closely related to the integers.

The next result is one that is usually encountered (without proof) in high school. However, the proof is very interesting because it provides an excellent illustration of the second form of the principle of induction. Also note that the proof invokes Lemma 1.4.1, which is the reason that we could not use this result to help us prove Lemma 1.4.1.

**Theorem 1.4.2 (The Fundamental Theorem of Arithmetic)** *Let $n \in \mathbb{N}$, $n \neq 1$. Then there exist distinct primes $p_1, \ldots, p_k$ and integers $n_1, \ldots, n_k \in \mathbb{N}$ such that*

$$n = p_1^{n_1} \cdots p_k^{n_k}. \tag{1.5}$$

*Moreover, this expression is unique in the sense that if $q_1, \ldots, q_\ell$ are also distinct primes and $m_1, \ldots, m_\ell \in \mathbb{N}$ are such that*

$$n = q_1{}^{m_1} \cdots q_\ell{}^{m_\ell} \tag{1.6}$$

*then $k = \ell$, $\{p_1, \ldots, p_k\} = \{q_1, \ldots, q_k\}$, and if $p_i = q_j$, then $n_i = m_j$.*

**Proof.** Our first step is to prove the following claim by induction:

*Claim:* Every integer $n > 1$ can be expressed as a product of primes as in Equation (1.5).

If $n = 2$, then the result is clearly true, so that we may take 2 to be the base for our induction argument.

For our induction hypothesis, we make the assumption that the claim is true for all integers $k$ such that $2 \leq k \leq n$. To apply the second form of the principle of induction, we must now show that the claim holds for $n+1$—that is, that $n + 1$ can be written as a product of primes. If $n + 1$ is itself a prime, then there is nothing to prove because $n + 1$ is clearly expressible as a product of primes.

So suppose that $n + 1$ is not a prime. Then there must exist integers $a, b$ with $1 < a, b \leq n$ and $n+1 = ab$. By our induction hypothesis we know that $a$ and $b$ can be written in the form

$$a = a_1{}^{r_1} \cdots a_t{}^{r_t}, \quad b = b_1{}^{s_1} \cdots b_u^{s_u}$$

where $a_i, b_j$ are primes and $r_i, s_j \in \mathbb{N}$. It follows that

$$n + 1 = ab = a_1{}^{r_1} \cdots a_t{}^{r_t} \, b_1{}^{s_1} \cdots b_u^{s_u}.$$

In other words, $n + 1$ can be written as a product of primes. By the second form of the principle of induction, this establishes our claim.

It remains for us to tackle the question of uniqueness. So suppose that we have two expressions (1.5) and (1.6) for $n$ as a product of primes. Let us write them as

$$n = p_1 p_2 \cdots p_m = q_1 q_2 \cdots q_n \tag{1.7}$$

where $p_i, q_i$ are primes, but $p_i$ (respectively, $q_j$) are not necessarily distinct. Then

$$p_1 \mid n = q_1(q_2 \cdots q_n)$$

so that, by Lemma 1.4.1, either $p_1 \mid q_1$ or $p_1 \mid q_2 \cdots q_n$. Repeating this argument we find that there must be $q_j$ such that $p_1 \mid q_j$. But $q_j$ is itself a prime and so divisible by only 1 and $q_j$. Therefore, we must have $p_1 = q_j$. Dividing Equation (1.7) by $p_1 = q_j$, we obtain

$$p_2 \cdots p_m = q_1 \cdots q_{j-1} q_{j+1} \cdots q_n.$$

Repeating the previous argument, we find that each $p_i$ pairs off with a $q_j$, which it equals. Clearly, there can be no $q_j$'s left over and so we obtain the desired uniqueness.   $\square$

The fundamental theorem of arithmetic provides an easy verification of Euclid's discovery that there are infinitely many prime numbers. Suppose, on the contrary, that there were only finitely many prime numbers. Then we could list them as $p_1, p_2, \ldots, p_k$, say. Let $n = p_1 p_2, \cdots p_k$. Then $n + 1$ is, by the fundamental theorem of arithmetic, equal to a product of prime numbers. But clearly it is not divisible by any of $p_1, p_2, \ldots, p_k$. Hence, there must be another prime that is not contained in this list. Thus we have a contradiction and there must be infinitely many prime numbers.

The factorization of $n$ into a product of primes in Equation (1.5) is called the *prime factorization* of $n$. When the prime factorization of two integers is available, we may use it to calculate their greatest common divisor (as we have in fact been doing with small examples).

**Corollary 1.4.3** *Let $a, b \in \mathbb{N}$ and*

$$a = p_1^{r_1} \cdots p_m^{r_m}, \quad b = p_1^{s_1} \cdots p_m^{s_m}$$

*(where some $r_i, s_j$ may be zero) be the prime factorizations of $a$ and $b$. Then,*

$$(a, b) = p_1^{t_1} \cdots p_m^{t_m}$$

*where $t_i = \min(r_i, s_i)$, $1 \le i \le m$.*

**Proof.** Exercise.   $\square$

**Example 1.4.4** Let

$$a = 5^{10} 13^6 19^{11} 29^4 73^5, \quad b = 5^4 11^3 29^6 73^3 79^8.$$

Then,

$$a = 5^{10} 11^0 13^6 19^{11} 29^4 73^5 79^0, \quad b = 5^4 11^3 13^0 19^0 29^6 73^3 79^8$$

so that

$$(a, b) = 5^4 11^0 13^0 19^0 29^4 73^3 79^0 = 5^4 29^4 73^3.$$

In the next lemma we gather together some observations that are useful in calculations involving greatest common divisors. We could use the fundamental theorem of arithmetic to establish these results and we encourage you to do that. However, to develop more familiarity with other approaches, we provide alternative proofs.

**Lemma 1.4.5** *Let $a, b, c \in \mathbb{Z}$.*

   (i)  If $c > 0$, and $c \mid a, b$ then $(a, b) = c(a/c, b/c)$.
   (ii)  If $a \mid bc$, and $(a, b) = 1$ then $a \mid c$.
   (iii)  $(a, bc) = 1$ if and only if $(a, b) = 1$ *and* $(a, c) = 1$.
   (iv)  If $a \mid c$, $b \mid c$ and $(a, b) = 1$ then $ab \mid c$.

**Proof.** (i) Let $d = (a, b)$. Since $c \mid a, b$, it follows from Corollary 1.3.10 (iii) that $c \mid d$. Let $d = d_1 c$, $a = a_1 c$, and $b = b_1 c$. Then $d_1 c = d \mid a = a_1 c$, which implies that $d_1 \mid a_1$. Similarly, $d_1 \mid b_1$. Thus $d_1$ is a divisor of $a_1$ and $b_1$.

Now suppose that $x \in \mathbb{N}$ and $x \mid a_1, b_1$. Then $xc \mid a_1 c = a$ and $xc \mid b_1 c = b$ and so, by Corollary 1.3.10 (iii), $xc \mid d = d_1 c$. Therefore, $x \mid d_1$ and, by Corollary 1.3.10, $d_1 = (a_1, b_1)$ so that

$$(a, b) = d = cd_1 = c(a_1, b_1) = c(a/c, b/c).$$

(ii) The proof of this part illustrates a common technique that is used when working with greatest common divisors.

Since $(a, b) = 1$, it follows from Theorem 1.3.4 that there exist integers $x$ and $y$ with $ax + by = 1$. Then,

$$c = c \cdot 1 = axc + bcy.$$

Now $a \mid bc$, by hypothesis, and clearly $a \mid axc$. Consequently,

$$a \mid axc + bcy = c,$$

as required.

   (iii) and (iv). Exercises.   $\square$

**Exercises 1.4**

  1. Let $a, b \in \mathbb{N}$ have prime factorizations

$$a = p_1^{r_1} p_2^{r_2} \cdots p_m^{r_m}, \ b = p_1^{s_1} \cdots p_m^{s_m}$$

where some $r_i$, $s_j$ may be zero. Show that

$$\gcd(a, b) = p_1^{\min(r_1, s_1)} \cdots p_m^{\min(r_m, s_m)}$$

$$\operatorname{lcm}(a, b) = p_1^{\max(r_1, s_1)} \cdots p_m^{\max(r_m, s_m)}.$$

2. Show that for all $a, b \in \mathbb{N}$,

$$\gcd(a, b) \times \mathrm{lcm}(a, b) = ab.$$

3. Let $a, b, c \in \mathbb{N}$. Prove Lemma 1.4.5(iii), that is that

$$(a, bc) = 1 \iff (a, b) = 1 \quad \text{and} \quad (a, c) = 1.$$

4. Let $a, b, c \in \mathbb{N}$. Prove Lemma 1.4.5(iv), that is, that

$$a \mid c, b \mid c, (a, b) = 1 \implies ab \mid c.$$

5. Show that if $n \geq 2$ and $n$ is not a prime, then there exists a prime $p$ such that $p \mid n$ and $p \leq \sqrt{n}$.

6. Use Exercise 5 to determine whether 439 is a prime.

7. Show that for all $a, b, x \in \mathbb{N}$

$$x^{ab} - 1 = (x^a - 1)(x^{(b-1)a} + x^{(b-2)a} + \cdots + x^a + 1).$$

8. Use Exercise 7 to show that if $2^n - 1$ is a prime, then $n$ is one as well. (Prime numbers of the form $2^p - 1$ are known as *Mersenne primes*. At time of writing, the six largest known prime numbers are all Mersenne primes.)

9. Find a prime $p$ such that $2^p - 1$ is not a prime.

10. Show that for all $x, k \in \mathbb{N}$,

$$x^{2k+1} + 1 = (x + 1)(x^{2k} - x^{2k-1} + \cdots - x + 1).$$

11. Let $a, k, \ell \in \mathbb{N}$ be such that $a > 1$ and let $m = \ell(2k + 1)$. Show that $a^m + 1$ is not a prime. (Hint: Put $x = a^\ell$ and use Exercise 10.)

12. Show that if $m \in \mathbb{N}$ is such that $2^m + 1$ is a prime, then $m = 2^n$ for some $n \in \mathbb{N}$. (Hint: Use Exercise 11.)

13. Show that

$$2^{32} + 1 = (2^9 + 2^7 + 1)(2^{23} - 2^{21} + 2^{19} - 2^{17} + 2^{14} - 2^9 - 2^7 + 1)$$

and deduce that $2^{2^5} + 1$ is not a prime.

14. Show that $2^{2^n} + 1$ is a prime for $n = 0, 1, 2, 3,$ and 4. (Numbers of the form $2^{2^n} + 1$ are known as *Fermat numbers*.)

15. Show that for all $n \in \mathbb{N}$, none of the $n$ consecutive integers

$$(n + 1)! + 2, (n + 1)! + 3, \cdots, (n + 1)! + n + 1$$

is prime.

16. Show that for all $n \in \mathbb{N}$, the integer $n! + 1$ is not divisible by any prime number less than $n + 1$. Deduce that there must be infinitely many prime numbers.

17. Show that no rational number satisfies the equation $x^2 = 2$.

## 1.5 Relations and Partitions

By a *relation $R$* on a set $A$ we mean a subset of $A \times A$. If $(a, b) \in R$, then we also write $a\, R\, b$ and we say that $a$ is *related* to $b$. This definition of a relation is just a generalization to arbitrary sets of the familiar notion of relationships between people (e.g., having the same height, weight, birthday, being the brother/sister of) or places (e.g., having a smaller population, milder climate) or things (being denser, harder, darker, and so forth.) that we use every day.

**Example 1.5.1** Let $A = \mathbb{N}_6 = \{1, 2, 3, 4, 5, 6\}$ and

$$R = \{(1, 1), (1, 2), (1, 3), (1, 4), (1, 5), (1, 6)$$
$$(2, 2), (2, 4), (2, 6), (3, 3), (3, 6), (4, 4), (5, 5), (6, 6)\}.$$

Do you recognize this relation? It is just the relationship of "being a divisor of". Thus,

$$R = \{(a, b) \mid a, b \in \mathbb{N}_6 \ \text{and} \ a \mid b\}.$$

The following properties that a relation may (or may not) have are particularly important. Let $R$ be a relation on a set $A$. Then $R$ is

(1) *reflexive* if $a\, R\, a$ for all $a \in A$,

(2) *symmetric* if $a\, R\, b \implies b\, R\, a$,

(3) *transitive* if $a\, R\, b, \ b\, R\, c \implies a\, R\, c$.

**Example 1.5.2** Let the relations $R_1, R_2$ be defined on $\mathbb{Z}$ as follows:

$$a\, R_1\, b \iff a < b.$$
$$a\, R_2\, b \iff a \leq b.$$
$$a\, R_3\, b \iff a \ \text{and} \ b \ \text{have the same prime divisors.}$$

Then $R_1$ is transitive but not reflexive or symmetric, $R_2$ is reflexive and transitive but not symmetric, and $R_3$ is reflexive, symmetric, and transitive.

If a relation is reflexive, symmetric, and transitive, then it is called an *equivalence* relation. Equivalence relations will occur frequently in our discussions, with the following being one of the most fundamental.

Let $n \in \mathbb{N}$ and $a, b \in \mathbb{Z}$. Then we say that $a$ is *congruent* to $b$ *modulo* $n$ or that $a$ is congruent to $b$ mod $n$ if $n$ divides $a - b$. We adopt the notation

$$a \equiv b \bmod n \quad \text{or} \quad a \equiv_n b \quad \text{or} \quad a \equiv b$$

to indicate that $a$ is congruent to $b$ mod $n$.

If we take $n = 2$, then $a \equiv b$ mod 2 if and only if 2 divides $a - b$ in other words, if and only if $a$ and $b$ are both even or both odd.

**Theorem 1.5.3** *Let $n \in \mathbb{N}$. Then $\equiv_n$ is an equivalence relation on $\mathbb{Z}$.*

**Proof.** We must show that $\equiv_n$ is reflexive, symmetric, and transitive.

(i) $\equiv_n$ *is reflexive.* For all $a \in \mathbb{Z}$, $a - a = 0$ and $n \mid 0$. Therefore, $a \equiv_n a$.
(ii) $\equiv_n$ *is symmetric.* Let $a, b \in \mathbb{Z}$ and $a \equiv_n b$. This means that $n$ divides $a - b$. But then $n$ divides $b - a$ so that $b \equiv_n a$.
(iii) $\equiv_n$ *is transitive.* Let $a, b, c \in \mathbb{Z}$, $a \equiv_n b$ and $b \equiv_n c$. Then $n$ divides $a - b$ and $n$ divides $b - c$. This means that there exist $x, y \subset \mathbb{Z}$ with

$$nx = a - b$$
$$ny = b - c.$$

Hence,

$$n(x + y) = nx + ny = a - c.$$

In other words, $n$ divides $a - c$ and $a \equiv_n c$.   $\square$

Let $I$ be a nonempty set. For each $i \in I$, let $X_i$ be a nonempty set. Then $\mathcal{X} = \{X_i \mid i \in I\}$ is a *family* of sets *indexed* by $I$ or *with index $I$*.

A *partition* of a nonempty set $X$ is a family $\{X_i \mid i \in I\}$ of nonempty subsets of $X$ such that

(i) $X = \cup_{i \in I} X_i$.
(ii) $i \neq j \Rightarrow X_i \cap X_j = \emptyset$.

The sets $X_i$ are referred to as the *parts* of the partition.

**Example 1.5.4** Let $X = \{1, 2, 3, 4, 5, 6, 7, 8, 9, 10\}$, $X_1 = \{1, 3\}$, $X_2 = \{2, 4, 6, 8, 10\}$, $X_3 = \{5\}$, $X_4 = \{7, 9\}$. Then $\{X_i \mid i = 1, 2, 3, 4\}$ is a partition of $X$.

Equivalence relations and partitions are closely related concepts.

**Lemma 1.5.5** *Let $R$ be an equivalence relation on a nonempty set $X$. For each $x \in X$, let*

$$C_x = \{z \in X \mid x\,R\,z\}.$$

*Then,*

   (i) *for all $x \in X$, $x \in C_x$,*
   (ii) *for all $x, y \in X$,*

$$C_x \cap C_y \neq \emptyset \implies C_x = C_y,$$

  (iii) *$X = \bigcup_{x \in X} C_x$,*
  (iv) *$\mathcal{P}_R = \{C_x \mid x \in X\}$ is a partition of $X$.*

**Proof.** (i) Let $x \in X$. Then we have $x\,R\,x$, since $R$ is reflexive. Thus, $x \in C_x$ and $C_x \neq \emptyset$.

(ii) Let $x, y, z \in X$ and $z \in C_x \cap C_y$. Then

$$
\begin{aligned}
w \in C_y &\implies x\,R\,z,\ y\,R\,z,\ y\,R\,w \\
&\implies x\,R\,z,\ z\,R\,y,\ y\,R\,w \quad \text{since } R \text{ is symmetric} \\
&\implies x\,R\,y,\ y\,R\,w \quad\quad\ \text{since } R \text{ is transitive} \\
&\implies x\,R\,w \quad\quad\quad\quad\quad \text{since } R \text{ is transitive} \\
&\implies w \in C_x.
\end{aligned}
$$

Consequently, $C_y \subseteq C_x$. Repeating the previous argument with the roles of $x$ and $y$ interchanged, we can *similarly* show that $C_x \subseteq C_y$. Thus, $C_x = C_y$.

Parts (iii) and (iv) follow immediately from parts (i) and (ii).   $\square$

If $R$ is an equivalence relation on a nonempty set $X$ and $x \in X$, then

$$C_x = \{z \in X \mid x\,R\,z\}$$

is called the (*equivalence*) *class* of $x$ with respect to $R$. If $R$ is $\equiv_n$, then it is customary to adopt the notation

$$C_x = [x]_n \quad \text{or} \quad C_x = [x].$$

It is important to develop an understanding of these equivalence classes.

**Example 1.5.6** (1) The equivalence relation $\equiv_2$ has just two classes:

$$[0] = \{0, \pm 2, \pm 4, \ldots\} \quad \text{and} \quad [1] = \{\pm 1, \pm 3, \ldots\}.$$

(2) The equivalence relation $\equiv_3$ has three classes:

$$[0] = \{0, \pm 3, \pm 6, \ldots\},$$
$$[1] = \{\ldots, -5, -2, 1, 4, 7, \ldots\},$$
$$[2] = \{\ldots, -4, -1, 2, 5, \ldots\}.$$

(3) For $n \in \mathbb{N}$ and $a \in \mathbb{Z}$, the class of $a$ is

$$[a] = \{a + kn \mid k \in \mathbb{Z}\}.$$

Note that, for example,

$$[0]_2 = [2]_2 = [-2]_2 = [4]_2 = \ldots$$
$$[2]_3 = [-1]_3 = [5]_3 = [8]_3 = \ldots$$
$$[a]_n = [a + n]_n = [a - n]_n = \ldots.$$

**Lemma 1.5.7** *Let* $\mathcal{P} = \{X_i \mid i \in I\}$ *be a partition of* $X$. *Let a relation* $R = R_{\mathcal{P}}$ *be defined on* $X$ *by*

$$a \; R \; b \iff a \text{ and } b \text{ lie in the same part}$$
$$\text{of the partition } \mathcal{P}.$$

*Then* $R$ *is an equivalence relation on* $X$.

**Proof.** Clearly, $R$ is reflexive, symmetric, and transitive.    $\square$

**Theorem 1.5.8** *Let* $X$ *be a nonempty set,* $R$ *an equivalence relation on* $X$, *and* $\mathcal{P}$ *a partition of* $X$. *Then,*

$$R_{\mathcal{P}_R} = R \quad \text{and} \quad \mathcal{P}_{R_{\mathcal{P}}} = \mathcal{P}.$$

*Thus the mappings*

$$R \to \mathcal{P}_R \quad \text{and} \quad \mathcal{P} \to R_{\mathcal{P}}$$

*are inverse bijections from the set of equivalence relations to the set of partitions and from the set of partitions to the set of equivalence relations on* $X$.

**Proof.** Exercise.    $\square$

**Exercises 1.5**   In questions 1 through 8 a relation is defined on a set $X$. In each case, determine whether the relation is reflexive, symmetric, or transitive.

1. $X = \mathbb{Z}$, $a \, R \, b \iff a \mid b$.

2. $X = \mathbb{Z}$, $a \, R \, b \iff a + b = 10$.

3. $X = \mathbb{Z}$, $a \, R \, b \iff a - b > 0$.

4. $X = \mathbb{Z}$, $a \, R \, b \iff 3 \mid a + b$.

5. $X = \mathbb{Z} \times (\mathbb{Z} \setminus \{0\})$, $(a, b) \, R \, (c, d) \iff ad = bc$.

6. $X = \mathbb{R} \times \mathbb{R}$, $(a, b) \, R \, (c, d) \iff \sqrt{(a - c)^2 + (b - d)^2} \leq 1$.

7. $X = \mathbb{R} \times \mathbb{R}$, $(a, b) \, R \, (c, d) \iff ac + bd = 0$.

8. $X = \{1, 2, \ldots, 9\}$, $a \, R \, b \iff (a, b) > 1$.

9. Define the relation $R$ on $\mathbb{Z} \times \mathbb{Z}$ by

$$(a, b) \, R \, (c, d) \iff b - a = d - c.$$

   Show that $R$ is an equivalence relation and describe the classes of $R$ geometrically.

10. Define the relation $R$ on $\mathbb{R} \times \mathbb{R}$ by

$$(a, b) R \, (c, d) \iff a^2 + b^2 = c^2 + d^2.$$

   Show that $R$ is an equivalence relation and describe the classes of $R$ geometrically.

11. Define the relation $R$ on $X = \{1, 2, \ldots, 20\}$ by

$$a \, R \, b \iff a \text{ and } b \text{ have the same}$$
$$\text{prime divisors.}$$

   Show that $R$ is an equivalence relation. Describe the classes of the corresponding partition of $X$.

12. Describe a partition of the set of all prime numbers into four classes.


## 1.6 Modular Arithmetic

The relation $\equiv_n$ is the main focus of our attention in this chapter. Here we begin to develop some of its properties.

**Lemma 1.6.1**  *Let $n \in \mathbb{N}$, $a \in \mathbb{Z}$, and $a = qn + r$ where $0 \leq r < n$. Then $r$ is the unique element in $[a]_n \cap [0, n - 1]$. In particular, $a \equiv r \bmod n$.*

**Proof.**  We have

$$a - r = qn,$$

which means that $a \equiv_n r$ and $r \in [a]_n$. Hence, $r \in [a]_n \cap [0, n-1]$. To see that $r$ is unique in this regard, let $r' \in [a]_n \cap [0, n-1]$. Then, $a \equiv_n r'$ so that $n \mid a - r'$. Let $a - r' = q'n$ where $q' \in \mathbb{Z}$. Then we have

$$a = qn + r \quad \text{and} \quad a = q'n + r'$$

where $0 \le r, r' < n$. By the uniqueness property of the division algorithm, it follows that $q = q'$ and $r = r'$. Therefore, $r$ is unique.   $\square$

We will denote by $\mathbb{Z}_n$ the set of distinct classes of $\equiv_n$. Thus,

$$\mathbb{Z}_n = \{[0], [1], \ldots, [n-1]\}.$$

Note that $\mid \mathbb{Z}_n \mid = n$. We refer to $\mathbb{Z}_n$ as the (*set of*) *integers modulo n*. It is the next result that enables us to consider $\mathbb{Z}_n$ as an algebraic structure.

**Lemma 1.6.2** *Let $a, b, c, d \in \mathbb{Z}$ and $n \in \mathbb{N}$. Let $a \equiv_n b$ and $c \equiv_n d$. Then,*

   (i)  $a + c \equiv_n b + d$,
  (ii)  $a - c \equiv_n b - d$,
 (iii)  $ac \equiv_n bd$.

**Proof.** By the hypothesis, there exist $q, q' \in \mathbb{Z}$ such that

$$a - b = qn, \qquad c - d = q'n.$$

Then,

$$\begin{aligned}
(a + c) - (b + d) &= (a - b) + (c - d) \\
&= qn + q'n \\
&= (q + q')n.
\end{aligned}$$

Thus, $n \mid (a + c) - (b + d)$ and so $a + c \equiv_n b + d$. This establishes part (i), and the proof of part (ii) is similar.
    Also,

$$\begin{aligned}
ac - bd &= ac - bc + bc - bd \\
&= (a - b)c + b(c - d) \\
&= qnc + bq'n \\
&= (qc + bq')n.
\end{aligned}$$

Thus, $n \mid ac - bd$ and $ac \equiv_n bd$. This establishes part (iii).   $\square$

**Corollary 1.6.3** *Let $k, n \in \mathbb{N}$ and $a, b \in \mathbb{Z}$ be such that $a \equiv_n b$. Then*

(i) $\pm ka \equiv_n \pm kb$,
(ii) $a^k \equiv_n b^k$.

The observations in Lemma 1.6.2 can be described by saying that the equivalence relation $\equiv_n$ *respects* the operations of *addition* and *multiplication* in $\mathbb{Z}$. It is for this reason that we also refer to the relation $\equiv_n$ not just as the *equivalence* relation modulo $n$ but as the *congruence* relation modulo $n$, and that we refer to the classes of $\equiv_n$ as the *congruence* classes mod $n$.

The fact that $\equiv_n$ respects the addition and multiplication in $\mathbb{Z}$ makes it possible for us to define new operations of addition and multiplication in $\mathbb{Z}_n$ as follows: For all $[a]_n, [b]_n \in \mathbb{Z}_n$,

$$[a]_n + [b]_n = [a + b]_n$$

$$[a]_n[b]_n = [ab]_n.$$

Since the operations of addition and multiplication involve the combination of two classes, we refer to them as *binary operations*. It is easy to see that we can deal with multiples of elements and powers of elements in the natural way:

$$k[a]_n = [ka]_n \qquad (k \in \mathbb{Z})$$

$$([a]_n)^k = [a^k]_n \qquad (k \in \mathbb{N}).$$

When $[a]_n \neq [0]_n$, we define $([a]_n)^0 = [1]_n$. All the familiar laws of exponents will now hold.

**Example 1.6.4** Here are some examples of "arithmetic" in $\mathbb{Z}_n$:

$$[3]_7 + [6]_7 = [9]_7 = [2]_7$$

$$[10]_6 + [13]_6 = [23]_6 = [5]_6$$

$$[3]_7 \cdot [6]_7 = [18]_7 = [4]_7$$

$$[10]_6 \cdot [13]_6 = [130]_6 = [4]_6.$$

Notice that in this new "arithmetic" it does not matter whether we add $[3]_7$ and $[6]_7$ or $[10]_7$ and $[-15]_7$. Since $[3]_7 = [10]_7$ and $[6]_7 = [-15]_7$, the result will be the same:

$$[10]_7 + [-15]_7 = [-5]_7 = [2]_7.$$

Similarly, we may use $[-2]_6$ instead of $[10]_6$ and $[-11]_6$ in place of $[13]_6$:

$$[-2]_6 \cdot [-11]_6 = [22]_6 = [4]_6.$$

The answer is the same. In other words the rule for the addition (multiplication) of the classes $[a]_n$ and $[b]_n$ is independent of the choice of the "representatives" that we work with from $[a]_n$ and $[b]_n$. Thus, if $[a]_n = [a']_n$ and $[b]_n = [b']_n$, then

$$[a]_n + [b]_n = [a + b]_n = [a' + b']_n$$
$$= [a']_n + [b']_n$$

and likewise

$$[a]_n[b]_n = [a']_n[b']_n.$$

The new addition and multiplication depend only on the *classes* and not on any particular choice of representatives. We say that formally as follows:

**Lemma 1.6.5** *Addition and multiplication in $\mathbb{Z}_n$ are well defined.*

**Theorem 1.6.6** *For all $n \in \mathbb{N}$, the (binary) operations $+$ and $\cdot$ in $\mathbb{Z}_n$ obey the following laws:*

(1) $[a] + [b], [a] \cdot [b] \in \mathbb{Z}_n.$     Closure

(2) $[a] + ([b] + [c]) = ([a] + [b]) + [c],$
    $[a]([b][c]) = ([a][b])[c].$     Associativity

(3) $[a] + [b] = [b] + [a],$
    $[a] \cdot [b] = [b] \cdot [a].$     Commutativity

(4) $[a] + [0] = [a], [a] \cdot [1] = [a].$     Zero and identity

(5) $[a]([b] + [c]) = [a][b] + [a][c].$     Distributivity

(6) For all $[a] \in \mathbb{Z}_n, \exists$ a unique $[b] \in \mathbb{Z}_n$
    with $[a] + [b] = [0]$—namely, $[-a].$     Existence of inverse

**Proof.** Exercise.  □

It is worth noting that Theorem 1.6.6(1) holds trivially because of the discussion preceding the theorem. However, closure is not always so obvious or even true when introducing operations on a set. For example, if we consider the operation of division on the set of nonzero integers, then it is well defined but the result of dividing one integer by another may be a proper fraction rather than another integer.

When the modulus $n$ is clear, we write simply $a$ or $[a]$ for $[a]_n$. Thus, for example, we write

$$\mathbb{Z}_n = \{0, 1, 2, \ldots, n-1\}.$$

or,

$$\mathbb{Z}_7 = \{0, 1, 2, 3, 4, 5, 6\}.$$

This greatly simplifies the notation, but we must try to ensure that there is no ambiguity. For this reason, and to add emphasis, we will still use the class notation $[a]$ or $[a]_n$ when appropriate.

We call $[0]$ the *zero* element of $\mathbb{Z}_n$ and $[1]$ the *identity* or *unity* element of $\mathbb{Z}_n$. Note that these elements behave just like 0 and 1 in $\mathbb{Z}$:

$$[0] + [a] = [a] + [0] = [a]$$
$$[0] \cdot [a] = [a] \cdot [0] = [0]$$
$$[1] \cdot [a] = [a] \cdot [1] = [a]$$

for all $[a] \in \mathbb{Z}_n$.

For small values of $n$ we can completely describe the addition and multiplication by means of "tables".

**Example 1.6.7** (1) $\mathbb{Z}_2 = \{0, 1\}$.

| + | 0 | 1 |
|---|---|---|
| 0 | 0 | 1 |
| 1 | 1 | 0 |

| $\cdot$ | 0 | 1 |
|---|---|---|
| 0 | 0 | 0 |
| 1 | 0 | 1 |

(2) $\mathbb{Z}_3 = \{0, 1, 2\}$.

| + | 0 | 1 | 2 |
|---|---|---|---|
| 0 | 0 | 1 | 2 |
| 1 | 1 | 2 | 0 |
| 2 | 2 | 0 | 1 |

| $\cdot$ | 0 | 1 | 2 |
|---|---|---|---|
| 0 | 0 | 0 | 0 |
| 1 | 0 | 1 | 2 |
| 2 | 0 | 2 | 1 |

We refer to the rules of addition and multiplication in $\mathbb{Z}_n$ as *modular arithmetic*. Modular arithmetic can be used to simplify many calculations in regular arithmetic.

An interesting and sometimes useful rule of arithmetic states that an integer (written in standard form to the base 10) is divisible by 3 or 9 if and only if the sum of its digits is divisible by 3 or 9, respectively. For example, the sum of the digits of $n = 2543694$ is 33, which is divisible by 3 but not 9.

Therefore, $n$ is divisible by 3 but not 9. The explanation for this can be found easily by using modular arithmetic. We first note that

$$\begin{aligned} 10 &= 9 + 1 &\equiv 1 \bmod 9 \text{ or } 3 \\ 100 &= 99 + 1 &\equiv 1 \bmod 9 \text{ or } 3 \\ 1000 &= 999 + 1 &\equiv 1 \bmod 9 \text{ or } 3 \end{aligned}$$

and so on. Therefore, if $x = x_n x_{n-1} \cdots x_2 x_1 x_0$ is an integer with digits $x_n, x_{n-1}, \ldots, x_1, x_0$ then

$$\begin{aligned} x &= x_n 10^n + x_{n-1} 10^{n-1} + \cdots + x_1 10 + x_0 \\ &\equiv x_n \cdot 1 + x_{n-1} \cdot 1 + \cdots + x_2 \cdot 1 + x_1 \cdot 1 + x_0 \bmod 9 \text{ or } 3 \\ &\equiv x_n + x_{n-1} + \cdots + x_1 + x_0 \bmod 9 \text{ or } 3. \end{aligned}$$

Thus,

$$x \equiv 0 \bmod 9 \text{ or } 3$$

if and only if

$$x_n + x_{n-1} + \cdots + x_1 + x_0 \equiv 0 \bmod 9 \text{ or } 3.$$

Consequently 9 (or 3) divides $x$ if and only if it divides the sum $x_n + \cdots + x_2 + x_1 + x_0$ of its digits.

Another feature of modular arithmetic is that certain calculations that may appear quite daunting at first glance turn out to be relatively simple. The secret is to spot integers that can be reduced to much simpler integers. For example, suppose we want to calculate the least positive remainder of $2^{350}$ on division by 34. We calculate a few powers of 2:

$$2^2 = 4, \ 2^3 = 8, \ 2^4 = 16, \ 2^5 = 32.$$

Now $2^5 = 32 \equiv -2 \bmod 34$. Therefore, every time we see $2^5$ we can replace it by $-2$:

$$\begin{aligned} 2^{350} &= (2^5)^{70} \equiv (-2)^{70} = 2^{70} \equiv (2^5)^{14} \equiv (-2)^{14} \\ &= 2^{14} = (2^5)^2 \cdot 2^4 \equiv (-2)^2 \cdot 2^4 \\ &= 2^6 = 2^5 \cdot 2 \equiv (-2) \cdot 2 \\ &\equiv -4 \bmod 34 \\ &\equiv 30 \bmod 34. \end{aligned}$$

From this, it follows that 30 is the remainder when $2^{350}$ is divided by 34.

The same idea works for other products. For example,

$$7642 \equiv_9 7 + 6 + 4 + 2 = 19 \equiv_9 1$$
$$8741 \equiv_9 8 + 7 + 4 + 1 = 20 \equiv_9 2$$

so that

$$7642 \times 8741 \equiv_9 1 \times 2 = 2.$$

Thus, for example, the least positive remainder of $7642 \times 8741$ upon division by 9 is 2.

Modular arithmetic has many interesting applications, as will become apparent later. However, even at this stage we can see how it can be used to study questions that don't, at first sight, appear to have anything to do with modular arithmetic. By a *Diophantine equation*, we mean any equation with integer coefficients in variables $x, y, z, \ldots$ (such as $x^2 + 2y^3 + 3z^5 = 0$). The search for integer or rational solutions to Diophantine equations is a fascinating and important topic that we can only mention in passing here. A few of the exercises will indicate how modular arithmetic can sometimes help.

### Exercises 1.6

1. Which of the following integers are related with respect to the given modulus?

$$-14, -12, -5, 0, 1, 21, 32, 462$$

   (i) mod 3.
   (ii) mod 5.
   (iii) mod 6.

2. Match the products (in $\mathbb{Z}_{12}$)

$$[4][5], \ [5][6], \ [5][7], \ [9][9], \ [13][15]$$

   to the following classes

$$[3], \ [-4], \ [-3], \ [114], \ [-1].$$

3. Let $a = 11,736,412$ and $b = 12,543$. Without multiplying out or doing long division, show the following:

   (i) $ab \equiv 6 \bmod 10$.
   (ii) $ab \equiv 1 \bmod 5$.
   (iii) $ab \equiv 6 \bmod 9$.
   (iv) The remainder on division of $a$ by 9 is 7.

4. Show that

   (i) $2^{213} \equiv 1 \bmod 7$.
   (ii) $7^{108} \equiv 1 \bmod 12$.
   (iii) $33^{33} \equiv 8 \bmod 31$.

5. Show that, for all $n \in \mathbb{N}$,

$$10^n \equiv (-1)^n \bmod 11.$$

6. Show that if $a = x_n x_{n-1} \cdots x_2 x_1$ in base 10 notation, then

$$a \equiv x_1 - x_2 + x_3 - \cdots + (-1)^{n-1} x_n \bmod 11.$$

7. Show that

$$43,171,234 \equiv 7 \bmod 11.$$

8. Let $m$ be a positive integer. Show that 3 divides $2^{2m+1} + 1$ and $2^{2m} - 1$.

9. Show that if $2^{2^m} \equiv -1 \bmod p$, then $2^{2^{m+1}} \equiv 1 \bmod p$.

10. Show that $(2^{2^m} + 1, 2^{2^n} + 1) = 1$ for all $m, n \in \mathbb{N}$ with $m \neq n$. (Hint: Use Exercise 9.)

11. Construct addition and multiplication tables for $\mathbb{Z}_5$ and $\mathbb{Z}_6$.

12. For which nonzero elements $x \in \mathbb{Z}_{12}$ does there exist a nonzero element $y$ with $xy = 0$? What about in $\mathbb{Z}_5$ and $\mathbb{Z}_7$?

13. Let $p$ be a prime and $k$ be an integer such that $0 < k < p$. Show that

$$\binom{p}{k} \equiv 0 \bmod p$$

   where $\binom{p}{k}$ denotes the $k$th binomial coefficient.

14. Let $f$ and $g$ be the mappings of $\mathbb{Z}_5$ into itself defined by

$$f(x) = x^3 + 2x^2 + 3 \qquad (x \in \mathbb{Z}_5)$$
$$g(x) = x^3 - 3x^2 + 5x - 2 \quad (x \in \mathbb{Z}_5).$$

   Show that $f = g$.

15. Let $R$ be the equivalence relation on $\mathbb{Z}$ defined by

$$a \, R \, b \iff a \text{ and } b \text{ are divisible by the same primes.}$$

   Show that $R$ respects multiplication but not addition.

16. An element $a \in \mathbb{Z}_n$ is a *quadratic residue* if the equation $x^2 = a$ has a solution in $\mathbb{Z}_n$. List the quadratic residues in

    (i) $\mathbb{Z}_3$
    (ii) $\mathbb{Z}_6$
    (iii) $\mathbb{Z}_7$.

17. (i) Show that the only solution to the equation $x^2 + y^2 = 0$ in $\mathbb{Z}_3$ is $x = y = 0$.
    (ii) Let $k \in \mathbb{Z}$ be such that $(3, k) = 1$. Show that there are no integer solutions to the equation $x^2 + y^2 = 3k$.

18. (i) Show that the only solution to the equation $x^2 + y^2 = 0$ in $\mathbb{Z}_7$ is $x = y = 0$.
    (ii) Let $k \in \mathbb{Z}$ be such that $(k, 7) = 1$. Show that there are no integer solutions to the equation $x^2 + y^2 = 7k$.

## 1.7 Equations in $\mathbb{Z}_n$

Let $a \in \mathbb{Z}$, $n \in \mathbb{N}$, and

$$f(x) = a_k x^k + a_{k-1} x^{k-1} + \cdots + a_1 x + a_0$$

be a polynomial with integer coefficients. Suppose that $f(a) \equiv 0 \bmod n$. Then, for all $b \equiv a \bmod n$ we will have $f(b) \equiv f(a) \equiv 0 \bmod n$. This leads us to consider polynomials over $\mathbb{Z}_n$. Indeed, if we are going to calculate modulo $n$, then we should really treat the coefficients as elements of $\mathbb{Z}_n$ so that we could write

$$f(x) = [a_k]x^k + [a_{k-1}]x^{k-1} + \cdots + [a_1]x + [a_0]$$

as a polynomial with coefficients in $\mathbb{Z}_n$. Then, for all $[a] \in \mathbb{Z}_n$,

$$
\begin{aligned}
f([a]) &= [a_k][a]^k + [a_{k-1}][a]^{k-1} + \cdots + [a_1][a] + [a_0] \\
&= [a_k a^k + a_{k-1} a^{k-1} + \cdots + a_1 a + a_0] \\
&= [f(a)].
\end{aligned}
$$

Thus,

$$f(a) \equiv 0 \bmod n \iff f([a]) = 0 \quad \text{in} \quad \mathbb{Z}_n.$$

So we speak of the class $[a]$ as being a *solution of the congruence* or *zero of the congruence*

$$f(x) \equiv 0 \bmod n$$

or a *root* or *zero* of the polynomial $f(x) \in \mathbb{Z}_n[x]$.

The *number* of solutions (modulo $n$) to this congruence then means the number of distinct classes (that is, distinct elements of $\mathbb{Z}_n$) that are solutions. Naturally we begin the study of equations in $\mathbb{Z}_n$ with the consideration of linear equations.

**Theorem 1.7.1** *Let $a \in \mathbb{Z}$, $n \in \mathbb{N}$. Then the following statements are equivalent:*

    (i) $(a, n) = 1$.
    (ii) $\exists\, x \in \mathbb{Z}$ *with* $ax \equiv 1 \bmod n$.
    (iii) $\exists\, x \in \mathbb{N}$ *with* $ax \equiv 1 \bmod n$.
    (iv) $\exists\, [x] \in \mathbb{Z}_n$ *with* $[a][x] = [1]$.
    (v) $\exists$ *a unique* $[x] \in \mathbb{Z}_n$ *with* $[a][x] = [1]$.

**Proof.**  (i) *implies* (ii). By Corollary 1.3.8, there exist $x, y \in \mathbb{Z}$ with $ax + ny = 1$. In other words, $ax - 1 = n(-y)$ so that $ax \equiv 1 \bmod n$.

    (ii) *implies* (iii). Let $x \in \mathbb{Z}$ be such that $ax \equiv 1 \bmod n$ and let $k \in \mathbb{N}$ be such that $x + kn > 0$. Then,

$$a(x + kn) = ax + akn$$
$$\equiv 1 + 0 \bmod n$$
$$\equiv 1 \bmod n.$$

    (iii) *implies* (iv). By the definition of multiplication in $\mathbb{Z}_n$ it is clear that (iii) implies the existence of an $[x] \in \mathbb{Z}_n$ with $[a][x] = [1]$.

*Equivalence of* (iv) *and* (v). Clearly (v) implies (iv). Now suppose that we have elements $[x], [y] \in \mathbb{Z}_n$ such that $[a][x] = [1] = [a][y]$. Then,

$$[x] = [x] \cdot [1] = [x][a][y] = [a][x][y]$$
$$= [1]\,[y] = [y].$$

Thus the uniqueness in part (v) comes as a bonus.

    (iv) *implies* (ii). This is obvious.

    (ii) *implies* (i). Let $x \in \mathbb{Z}$ be such that $ax \equiv 1 \bmod n$. Then there must exist $q \in \mathbb{Z}$ with $ax - 1 = qn$ or $ax + n(-q) = 1$. By Corollary 1.3.8, this means that $(a, n) = 1$, as required.

    It follows that each of the statements implies all the other statements and therefore they are all equivalent.  $\square$

The results of Theorem 1.7.1 can be generalized as follows:

**Corollary 1.7.2** *Let $a \in \mathbb{Z}$ and $n \in \mathbb{N}$. Then the following statements are equivalent:*

    (i) $(a, n) = 1$.
    (ii) $ax \equiv b \bmod n$ *has a solution for all $b \in \mathbb{Z}$.*
    (iii) $[a][x] = [b]$ *has a unique solution in $\mathbb{Z}_n$, for all $[b] \in \mathbb{Z}_n$.*

**Proof.**  (i) *implies* (ii). By Theorem 1.7.1, there exists $z \in \mathbb{Z}$ with $az \equiv 1 \bmod n$. Consequently,

$$a(zb) = (az)b \equiv_n 1 \cdot b = b.$$

(ii) *implies* (iii). It follows immediately from (ii) and the definition of $\mathbb{Z}_n$ and the multiplication in $\mathbb{Z}_n$ that every equation of the form $[a][x] = [b]$ will have a solution in $\mathbb{Z}_n$.

To show that the solution will be unique, first let $[z]$ be a solution to the equation $[a][z] = [1]$. Now let $[b], [x], [y] \in \mathbb{Z}_n$ be such that

$$[a][x] = [b] = [a][y].$$

Then,

$$[x] = [1][x] = [z][a][x] = [z][a][y]$$
$$= [1][y] = [y]$$

and the solution is indeed unique.

(iii) *implies* (i). Taking $b = 1$, the hypothesis implies that there exists $[x] \in \mathbb{Z}_n$ with $[a][x] = [1]$. The claim then follows from Theorem 1.7.1.  $\square$

An element $[a]$ in $\mathbb{Z}_n$ is said to be *invertible* or to be a *unit* if there exists $[x] \in \mathbb{Z}_n$ such that $[a][x] = [1]$—that is, if it satisfies the condition in part (iv) of Theorem 1.7.1. Of course, when we identify $\mathbb{Z}_n = \{0, 1, \ldots, n - 1\}$, then we would speak of the element $a$ as being invertible (it being understood that we mean in $\mathbb{Z}_n$ or modulo $n$). Although we use part (iv) of Theorem 1.7.1 to define invertible elements, we use part (i) to find them.

**Example 1.7.3**  The invertible elements in $\mathbb{Z}_7$ are 1, 2, 3, 4, 5, and 6, whereas the invertible elements in $\mathbb{Z}_{10}$ are 1, 3, 7, and 9.

**Lemma 1.7.4**  *Let $p$ be a prime number. Then the invertible elements in $\mathbb{Z}_p$ are* $1, 2, 3, \ldots, p - 1$.

**Proof.**  Let $a \in \mathbb{Z}_p$. Since $p$ is a prime number, $(a, p) = 1$ if and only if $a \in \{1, 2, 3, \ldots, p - 1\}$. In other words, the invertible elements in $\mathbb{Z}_p$ are simply $1, 2 \ldots, p - 1$.  $\square$

If $[a] \in \mathbb{Z}_n$ is invertible, then we write $[a]^{-1}$ for the unique element of $\mathbb{Z}_n$ with the property that $[a][a]^{-1} = [1]$ and we call $[a]^{-1}$ the *inverse* of $[a]$. When $[a]$ is invertible, this provides an elegant way of writing the solution to the equation $[a][x] = [b]$. For we have

$$[a]([a]^{-1}[b]) = ([a][a]^{-1})[b] = [1][b] = [b]$$

so that the unique solution to the equation $[a][x] = [b]$ is given by

$$[x] = [a]^{-1}[b].$$

Finding the inverse of an invertible element $[a] \in \mathbb{Z}_n$ is extremely easy. Since $[a]$ is invertible, we know from Theorem 1.7.1 that $(a, n) = 1$ and so, by the Euclidean algorithm, there exist $x, y \in \mathbb{Z}$ with $ax + ny = 1$. From this we see that $ax \equiv 1 \bmod n$ which means that $[a][x] = [1]$ in $\mathbb{Z}_n$. Thus, $[x] = [a]^{-1}$ and we can use the extended Euclidean algorithm to calculate inverses in $\mathbb{Z}_n$.

**Lemma 1.7.5** *The set of units in $\mathbb{Z}_n$ is closed under multiplication.*

**Proof.** Let $a$ and $b$ be invertible elements in $\mathbb{Z}_n$. By Theorem 1.7.1, this means that there exist $x, y \in \mathbb{Z}$ with

$$ax \equiv_n 1 \quad \text{and} \quad by \equiv_n 1.$$

Consequently,

$$(ab)(xy) = (ax)(by) \equiv_n 1 \cdot 1 = 1.$$

Thus, $ab$ is also invertible.    $\square$

We will denote the set of units in $\mathbb{Z}_n$ by $\mathbb{Z}_n^*$. Thus

$$Z_n^* = \{[a] \in \mathbb{Z}_n \mid [a] \text{ is invertible}\}.$$

It is interesting to take a moment to see what it means for a nonzero element $a$ in $\mathbb{Z}_n$ *not* to be invertible. By Theorem 1.7.1, this means that we must have $d = (a, n) > 1$. Then $\frac{n}{d} \in \mathbb{N}$ and $1 \leq \frac{n}{d} < n$ so that $[\frac{n}{d}] \neq 0$. However, we also have $\frac{a}{d} \in \mathbb{Z}$, from which it follows that

$$[a]\left[\frac{n}{d}\right] = \left[a\,\frac{n}{d}\right] = \left[\frac{a}{d}\,n\right] = [0].$$

Thus there exists a nonzero solution to the equation

$$[a][x] = [0].$$

This, of course, is something that can never happen in "usual" arithmetic. We have a special name for such elements. If $a \in \mathbb{Z}_n$, $a \neq 0$ and there exists $b \in \mathbb{Z}_n$ with $b \neq 0$ such that $ab = 0$, then we say that $a$ is a *zero divisor*.

**Corollary 1.7.6** *Let $n \in \mathbb{N}$, $n > 1$.*

    (i) *$\mathbb{Z}_n$ has zero divisors if and only if $n$ is not a prime.*

    (ii) *Let $a \in \mathbb{Z}_n$, $a \neq 0$. Then $a$ is a zero divisor if and only if $(a, n) > 1$.*

    (iii) *Let $a \in \mathbb{Z}_n$, $a \neq 0$. Then $a$ is a zero divisor if and only if it is not invertible.*

**Proof.** Exercise. $\square$

**Example 1.7.7** The noninvertible elements in $\mathbb{Z}_{12}$ are 0, 2, 3, 4, 6, 8, 9, and 10, and we have $2 \cdot 6 = 3 \cdot 4 = 6 \cdot 10 = 8 \cdot 9 = 0$.

We are all so familiar with the *cancellation law* in the arithmetic of $\mathbb{Z}$ that we take it for granted that if $ax = ay$ and $a \neq 0$, then $x = y$. This law does not always hold in $\mathbb{Z}_n$. For instance, we have, in $\mathbb{Z}_{12}$, that $[3] \neq [6]$ and $[4] \neq 0$. But, $[4][3] = [12] = [0] = [24] = [4][6]$. However, invertible elements in $\mathbb{Z}_n$ can always be cancelled.

**Lemma 1.7.8** *Let $a, x, y \in \mathbb{Z}$, $n \in \mathbb{N}$, and $(a, n) = 1$. Then*

$$ax \equiv ay \bmod n \implies x \equiv y \bmod n.$$

*Equivalently, if $[a], [x], [y] \in \mathbb{Z}_n$ and $[a]$ is invertible, then*

$$[a][x] = [a][y] \implies [x] = [y].$$

**Proof.** Since $(a, n) = 1$, we know from Theorem 1.7.1 that there is an element $b \in \mathbb{Z}$ with $ab \equiv 1 \bmod n$. Consequently,

$$ax \equiv_n ay \implies bax \equiv_n bay$$

$$\implies 1 \cdot x \equiv_n 1 \cdot y$$

$$\implies x \equiv_n y. \quad \square$$

Our focus has been on solutions to $ax \equiv_n b$ where $(a, n) = 1$. However, solutions can also exist when $(a, n) > 1$.

**Lemma 1.7.9** *Let $a, b \in \mathbb{Z}_n$.*

    (i) *$ax \equiv b \bmod n$ has a solution if and only if $(a, n)$ divides $b$.*

    (ii) *If $d = (a, n)$ divides $b$, then $ax \equiv b \bmod n$ has $d$ solutions in the interval $1 \leq x \leq n$.*

**Proof.** Exercise. $\square$

One interesting application of modular arithmetic is the construction of certain kinds of patterns in arrays of numbers. A *latin square* is an $n \times n$ matrix of $n$ symbols arranged in such a way that each symbol appears exactly once in each row and column. This idea is explored further in exercises 15 and 16.

**Exercises 1.7**

1. Find all solutions to the following equations:

    (i) $x^2 + 2x + 1 = 0$ in $\mathbb{Z}_8$.
    (ii) $x^3 + 2x^2 + 4 = 0$ in $\mathbb{Z}_5$.

2. List the invertible elements in $\mathbb{Z}_6$, $\mathbb{Z}_8$, $\mathbb{Z}_{11}$, and $\mathbb{Z}_{12}$.

3. Find the inverse of each invertible element in $\mathbb{Z}_{15}$.

4. Use the fact that

$$23 \times 107 - 41 \times 60 = 1$$

to find the inverse of 23 in $\mathbb{Z}_{41}$.

5. Find the inverse of $17 \in \mathbb{Z}_{95}$.

6. Find the solution to each of the following congruences $ax \equiv b$ by finding $[a]^{-1}$ and calculating $[a]^{-1}[b]$:

    (i) $3x \equiv 7 \bmod 10$.
    (ii) $4x \equiv 5 \bmod 9$.

7. Let $a, b \in \mathbb{Z}_n$. Show that

    (i) $a$ invertible $\implies a^{-1}$ invertible and $(a^{-1})^{-1} = a$
    (ii) $a, b$ invertible $\implies ab$ invertible and $(ab)^{-1} = b^{-1}a^{-1}$.

8. Let $p$ be a prime and $a \in \mathbb{Z}_p$.

    (i) Show that $a = a^{-1} \iff a^2 = 1$.
    (ii) Deduce that $a = a^{-1} \iff a = 1$ or $p - 1$.

9. (Wilson's Theorem) Show that for all primes $p$,

$$(p - 1)! \equiv -1 \bmod p.$$

(Hint: Use Exercise 8 to pair off elements and their inverses in the product $(p - 1)!$.)

10. Solve the following system of equations in $\mathbb{Z}_3$:

$$x - 2y = 1$$
$$x - y = 1.$$

11. (i) Solve the following system of equations in $\mathbb{Q}$:

$$x - 2y = 1$$
$$2x - y = 1.$$

   (ii) Does the system in (i) have a solution in $\mathbb{Z}_3$? $\mathbb{Z}_5$?

12. Solve the following system of equations in $\mathbb{Z}_5$:

$$3x + y + 2z = 3$$
$$x + y + 2z = 2$$
$$2x + y + 4z = 3.$$

13. Evaluate the following determinant in

   (i) $\mathbb{Q}$,
   (ii) $\mathbb{Z}_4$,
   (iii) $\mathbb{Z}_5$:

$$\begin{vmatrix} 3 & 1 & 2 \\ 2 & 5 & 4 \\ -2 & 1 & 3 \end{vmatrix}$$

*14. Solve $81x \equiv 105 \bmod 879$.

15. Let $A = (a_{ij})$ be the $n \times n$ matrix where

$$a_{ij} = (i - 1) + (j - 1) \qquad (i, j \in \mathbb{Z}_n).$$

   Show that $A$ is a latin square.

16. Let $p$ be a prime. Let $A_t (t \in \mathbb{Z}_p{}^*)$ denote the $p \times p$ matrix over $\mathbb{Z}_p$ with the $(i, j)$th entry equal to $ti + j$ $(i, j \in \mathbb{Z}_p)$.

   (i)   With $p = 5$, calculate $A_1, A_2, A_3$.
   (ii)  Show that each $A_t$ (with arbitrary $p, t$) is a latin square.
   (iii) Show that for all $a, b \in \mathbb{Z}_p$ and any distinct $t, u \in \mathbb{Z}_p{}^*$, there is at most one position $(i, j)$ such that

$$(A_t)_{ij} = a \quad \text{and} \quad (A_u)_{ij} = b.$$

   (Such latin squares are said to be *orthogonal*.)

## 1.8 Bar codes

*Bar codes* are based on the simple yet clever observation that by representing the digit one by a bar line and the digit zero by a bar space it is possible to represent any sequence of ones and zeros (and therefore any number) by a sequence of bar lines and bar spaces.

Many items passing through our hands, from library books to grocery store items, are labeled with bar codes. These bar codes are simply coded sequences of numbers that can be read by optical scanners. They can be used to keep track of inventory, to speed the tallying of grocery bills, or as a security check. It is a technology that we encounter almost everywhere in our lives. The first bar code was developed by Bernard Silver and Norman Woodland in 1948. Bar codes are now only one example of a class of techniques that go under the title of *symbologies*. An interesting history of the development (and near failure) of bar codes other symbologies (such as the bull's-eye "circular" bar system) can be found in [Nel].

The code that appears on supermarket items is called the *universal product code (UPC)* and it consists of a sequence of 12 digits, each of which is drawn from $\{0, 1, 2, \ldots, 9\}$. In other words, it is a *vector* in $\mathbb{Z}_{10}{}^{12}$. The first 11 digits contain information concerning the producer and the product. The 12th digit is a *check digit*. The 12th digit is chosen to ensure that the dot product of the code with the *check vector* (31 31 31 31 31 31) is congruent to 0 mod 10.

In the first few pages of any recent book you will find the *International Standard Book Number (ISBN)*. For books published prior to 2007, this is a sequence of 10 digits chosen from $\{0, 1, 2, \ldots, 9, X\}$ where the symbol $X$ represents the number 10. The first nine digits carry information concerning the book and the publisher. The 10th digit is a check digit to ensure that when the dot product is formed with the check vector (10 987654321), the answer is congruent to 0 mod 11. This system is now referred to as ISBN-10. In 2007, anticipating a shortage in the availability of ISBN-10 code numbers to cope with the increasing volume of books, a new system, ISBN-13, was introduced. This is a sequence of 13 digits chosen from $\{0, 1, 2, \ldots, 9\}$. The dot product of the 13 digits with the check vector (1 3 1 3 1 3 1 3 1 3 1 3 1) must be congruent to 0 mod 10. The first 12 digits carry information concerning the book and the publisher. The last digit is a check digit. The 13th digit must equal the difference between 10 and the dot product of the first 12 digits with the check vector (1 3 1 3 1 3 1 3 1 3 1 3) and reduced modulo 10. For example, suppose that we wish to calculate the check digit for the ISBN-13 number: 978-0-19-536387. First we calculate

$$P = 9 \times 1 + 7 \times 3 + 8 \times 1 + 0 \times 3 + 1 \times 1 + 9 \times 3 + 5 \times 1 + 3 \times 3$$
$$+ 6 \times 1 + 3 \times 3 + 8 \times 1 + 7 \times 3 = 124.$$

Now we reduce 124 modulo 10 to obtain 4 and finally subtract 4 from 10 to obtain 6. If the result at this stage was 10, then we would reduce it modulo 10 to 0. However, the result this time is 6, so no further reduction is required and we have $x = 4$. Note that ISBN-13 will not detect all transpositions. In this example, the adjacent digits 3 and 8 differ by 5, so that if they are transposed, then we obtain $3 \times 1 + 8 \times 3 = 27$ instead of $3 \times 3 + 8 \times 1 = 17$ as the contributions from these components to the dot product. Since 27 and 17 are congruent modulo 10, the transposition will not alter the result of the calculation and therefore will be recognized.

**Example 1.8.1** (1) Consider the UPC number $(0\,47400\,11710\,5)$. The dot product with the check vector is

$$(047400117105) \cdot (31\,31\,31\,31\,31\,31)$$
$$= 0 + 4 + 21 + 4 + 0 + 0 + 3 + 1 + 21 + 1 + 0 + 5$$
$$= 0 \bmod 10.$$

(2) Consider the ISBN-10 number $(0672215012)$. The dot product of this vector with the check vector $(10\ 9\ 8\ 7\ 6\ 5\ 4\ 3\ 2\ 1)$ is

$$(0672215012) \cdot (10\,987654321)$$
$$= 0 + 54 + 56 + 14 + 12 + 5 + 20 + 0 + 2 + 2$$
$$= 165$$
$$\equiv 0 \bmod 11.$$

The purpose of the final check digit is to detect errors in reading the code. Suppose that the code in Example 1.8.1 (1) was read as $(047400171105)$. Note that the second 7 has been interchanged with the preceding 1. The dot product then becomes

$$(047400171105) \cdot (313131313131)$$
$$= 0 + 4 + 21 + 4 + 0 + 0 + 3 + 7 + 3 + 1 + 0 + 5$$
$$= 48$$
$$\equiv 8 \bmod 10.$$

A scanning device would signal that an unacceptable code was recorded and that the code should be reread or checked.

The error-detecting component of these systems consists of a check vector and a modulus. Clearly, some choices will be better than other choices.

The UPC system will certainly detect any single-digit error. For instance, suppose that the vector $a = (a_1\ a_2 \ldots a_i \ldots a_{12})$ is read as $b = (a_1\ a_2 \ldots b_i \ldots a_{12})$—that is, the $i$th digit is read as $b_i$ instead of $a_i$. Let us assume that $i$ is odd. A similar discussion applies if $i$ is even. Since $a$ is a proper UPC code vector, we have

$$3a_1 + a_2 + 3a_3 + \cdots + 3a_i + \cdots a_{12} \equiv 0 \bmod 10.$$

Consequently, when we apply the check vector to $b$ we obtain

$$
\begin{aligned}
3a_1 + a_2 + 3a_3 + \cdots + 3b_i + \cdots &= 3a_1 + a_2 + 3a_3 \\
&\quad + \cdots + 3a_i + \cdots + a_{12} + 3b_i - 3a_i \\
&\equiv 0 + 3(b_i - a_i) \bmod 10 \\
&\equiv 3(b_i - a_i) \bmod 10.
\end{aligned}
$$

But $0 \leq a_i,\ b_i \leq 9$. So, the only way that the result $3(b_i - a_i)$ can be congruent to $0 \bmod 10$ is if $b_i = a_i$. Thus, a single error will always be detected. Of course, multiple errors might compensate for each other.

A very common error, at least for the human eye, is the transposition of two digits. For example, 0195056612 is read as 0159056612. The UPC system will detect most transpositions. Let

$$
\begin{aligned}
a &= (a_1 a_2 \ldots a_i a_{i+1} \ldots a_{12}) \\
b &= (a_1 a_2 \ldots a_{i+1} a_i \ldots a_{12})
\end{aligned}
$$

where $a$ is a legitimate UPC code. Again we will assume that $i$ is odd, although a similar argument will apply when $i$ is even. When $i$ is odd, we have

$$
\begin{aligned}
b \cdot (3131 \ldots 31) &= 3a_1 + a_2 + \cdots + 3a_{i+1} + a_i + \cdots + a_{12} \\
&= 3a_1 + a_2 + \cdots + 3a_i + a_{i+1} + \cdots + a_{12} - 2a_i + 2a_{i+1} \\
&\equiv 2(a_{i+1} - a_i) \bmod 10.
\end{aligned}
$$

Thus the error will pass undetected if $2(a_{i+1} - a_i) \equiv 0 \bmod 10$, which is equivalent to $a_{i+1} - a_i \equiv 0 \bmod 5$ or, since $0 \leq a_i,\ a_{i+1} \leq 9$, to $\mid a_{i+1} - a_i \mid = 5$.

The ISBN-10 system will detect all single-digit errors and also all interchanges, even of digits that are not adjacent. Let

$$
\begin{aligned}
a &= (a_1 a_2 \ldots a_i \ldots a_j \ldots a_{10}) \\
b &= (a_1 a_2 \ldots a_j \ldots a_i \ldots a_{10})
\end{aligned}
$$

where $a$ is a legitimate ISBN-10 code. Then

$$
\begin{aligned}
b \cdot (10\,9\,8 \cdots 1) &= 10a_1 + \cdots + ma_j + \cdots + na_i + \cdots + a_{10} \\
&= 10a_1 + \cdots + ma_i + \cdots + na_j + \cdots a_{10} \\
&\quad + (m - n)a_j - (m - n)a_i \\
&\equiv (m - n)(a_j - a_i) \bmod 11.
\end{aligned}
$$

But $1 \leq n < m \leq 10$ so that $1 \leq m - n \leq 9$ and $(m - n, 11) = 1$. Thus $m - n$ is invertible mod 11 so that $(m - n)(a_j - a_i) \equiv 0 \bmod 11$ if and only if $a_j - a_i \equiv 0 \bmod 11$. Since $0 \leq a_i, a_j \leq 9$, $a_j - a_i \equiv 0 \bmod 11$ if and only if $a_i = a_j$. Thus, all transpositions will be detected. The error detecting capabilities of ISBN-13 are similar to those of the UPC code.

The error-detecting capabilities of bar codes are summarized in the following result.

**Theorem 1.8.2** *Let a bar-coding scheme consist of those vectors in $\mathbb{Z}_n^m$ for which the dot product with the check vector $c = (c_1, \ldots, c_m)$ is congruent to zero modulo $n$.*

 (i) *All single-digit errors will be detected if and only if $(c_i, n) = 1$ for all $1 \leq i \leq m$.*
 (ii) *All transpositions will be detected if and only if $(c_i - c_j, n) = 1$ for all $1 \leq i < j \leq m$.*

Armed with this information regarding the basic properties of bar code systems, let us now see how to create a system of our own. Suppose that we would like to label up to 2 million items (library cards, coffee club members, etc.). Then we want to have at least 7 digits plus a check vector, for a total of 8 digits. So we want to work with 8-tuples of the digits $0, 1, 2, \ldots, 9$. Next, we want to choose a modulus. Clearly, the modulus should be greater than 9 (otherwise the system would be unable to catch some single-digit errors), and in the previous discussion we saw that choosing a prime improves the error-checking ability. This makes 11 a very natural choice. Next, we want to choose a check vector $c = (c_1 c_2 \ldots c_8)$ such that $(c_i, 11) = 1$ and $(c_i - c_j, 11) = 1$. The vector

$$c = 12345678$$

will do just fine. Our set of code vectors will be

$$C = \{a = (a_1 a_2 \ldots a_8) \in \mathbb{Z}^8 \mid 0 \leq a_i \leq 9, \ a \cdot c = 0\}.$$

Each item or person is now given a 7-digit number to which a check digit is added, making 8 digits in total. For example, suppose that we give an item the code number 7654321. This is then extended to a vector in $C$ by adding a check digit $x$ to give $7654321x$, where $x$ must satisfy

$$7654321x \cdot 12345678 \equiv 0 \bmod 11.$$

That is,

$$7 \cdot 1 + 6 \cdot 2 + 5 \cdot 3 + 4 \cdot 4 + 3 \cdot 5 + 2 \cdot 6 + 1 \cdot 7 + x \cdot 8 \equiv 0 \bmod 11,$$

which reduces to $8x + 7 \equiv 0 \; mod \; 11$. Solving this equation, we find that $x = 6$. Thus, our item is given the code number 76543216.

To convert this into a bar code, we first represent each digit by a sequence of ones and zeros and then represent each one by a solid bar and each zero by a space. Bar code scanners measure the width of bars and spaces, so there is no problem in having several bars or several spaces running together consecutively. However, the scanner must be able to distinguish whether it is reading the number from left to right or from right to left. So we divide the number into left and right halves (LH and RH, respectively):

$$LH = 7654, \; RH = 3216.$$

We now convert the digits into sequences of zeros and ones using different encodings depending on whether the digit is in the left or the right half. For instance, the following scheme is used in the UPC bar codes:

| Digit | Left Half | Right Half |
| --- | --- | --- |
| 0 | 0001101 | 1110010 |
| 1 | 0011001 | 1100110 |
| 2 | 0010011 | 1101100 |
| 3 | 0111101 | 1000010 |
| 4 | 0100011 | 1011100 |
| 5 | 0110001 | 1001110 |
| 6 | 0101111 | 1010000 |
| 7 | 0111011 | 1000100 |
| 8 | 0110111 | 1001000 |
| 9 | 0001011 | 1110100 . |

Notice that, not only are the digits represented differently in the left and right halves, but even reading any right-half sequence backward is still different from any left-half sequence. However, all left-half sequences begin with at least one zero and all right-half sequences end with at least one zero, so that makes it necessary to place a marker sequence at the beginning and end so that the scanner will know where the actual number itself begins. The sequence 101 is used at both ends for this purpose. Last, the sequence 01010 is inserted in the middle to separate the two halves (note that the left-half sequences all end in a 1, whereas the right-half sequences all begin with a 1). In this way, our simple code number 76543216 becomes converted into the following string of ones and zeros, where spaces have been inserted just to make it possible to recognize more easily the component parts:

101 0111011 0101111 0110001 0100011 01010 1000010 1101100 1100110

1010000 101.

This is now converted into a bar code, representing each 1 by a bar line and each 0 by a space.

Bar codes of the type that we have been discussing are known as *linear bar codes*. Two-dimensional or matrix bar codes are also extensively used. *The Bar Code Book* by Palmer [Pal] provides a broad survey of bar code symbologies.

**Exercises 1.8**

1. If 100 200 3$x$0 405 is a UPC number, what is the value of $x$?

2. If 2$x$ 31 41 51 62 is an ISBN-10 number, what is the value of $x$?

3. Show that the UPC system will not detect all transpositions of digits.

4. Let $a$ and $b$ be UPC numbers and $x \in \mathbb{Z}_{10}$. Consider $a$ and $b$ as vectors in $\mathbb{Z}_{10}^{12}$ (i.e., if $a = a_1 a_2 \cdots a_{12}$ then consider $a$ as the vector $(a_1, a_2, \ldots, a_{12}) \in \mathbb{Z}_{10}^{12}$). Now show that $a + b$ and $xa$ are also UPC "vectors."

5. If 978-0-19-536543-$x$ is an ISBN-13 number, what is the value of $x$?

## 1.9 The Chinese Remainder Theorem

In Theorem 1.7.1 and Corollary 1.7.2 we considered the solutions of a single congruence. Here we investigate the simultaneous solution of a system of congruences with different, but relatively prime, moduli. The Chinese remainder theorem has a long history, with an early instance appearing as a problem posed by the Chinese mathematician Sun Tsu Suan-Ching in the curious form:"There are certain things whose number is unknown. Repeatedly divided by 3, the remainder is 2; by 5, the remainder is 3; and by 7, the remainder is 2. What will the number be?" (See [W], Puzzle 70) There appears to be some uncertainty as to whether Sun Tsu lived in the first or fourth century, so it may be that the earliest recorded instance is due to Nicomachus (see [Di] and [W]).

**Theorem 1.9.1** *Let $n_i \in \mathbb{N}$  $(i = 1, 2, \ldots, k)$ and $(n_i, n_j) = 1$ if $i \neq j$. Let $a_1, \ldots, a_k \in \mathbb{Z}$. Then the system of congruences*

$$\left.\begin{array}{rlr} x \equiv a_1 & \bmod & n_1 \\ x \equiv a_2 & \bmod & n_2 \\ & \vdots & \\ x \equiv a_k & \bmod & n_k \end{array}\right\} \tag{1.8}$$

*has a solution that is unique modulo $n = n_1 n_2 \cdots n_k$.*

**Proof.** We begin by considering a much simpler system of congruences:

$$\left.\begin{array}{rlr} x \equiv 1 & \bmod & n_1 \\ x \equiv 0 & \bmod & n_2 \\ & \vdots & \\ x \equiv 0 & \bmod & n_k. \end{array}\right\} \tag{1.9}$$

Clearly, the solutions to the second to $k$th congruences are exactly the multiples of $n_2 n_3 \cdots n_k$. Thus, the system (1.9) has a solution if and only if there exists a multiple $t\, n_2 n_3 \cdots n_k$ of $n_2 \cdots n_k$ that satisfies the first congruence. But $(n_1, n_2 \cdots n_k) = 1$ and so there do exist $s, t \in \mathbb{Z}$ with

$$s n_1 + t\, n_2 n_3 \cdots n_k = 1.$$

Then, $w_1 = t\, n_2 n_3 \cdots n_k$ is a solution to the system (1.9). Replacing $x \equiv 1$ mod $n_1$ in the system (1.9) by $x \equiv 1 \bmod n_2, \ldots, x \equiv 1 \bmod n_k$ in turn we obtain $k$ such systems for which we can find solutions $w_1, w_2, \ldots, w_k$, say. The numbers $w_1, \ldots, w_k$ are referred to as the *weights*. Then we have

$$\left.\begin{array}{llr} w_1 \equiv 1 \bmod n_1 & w_2 \equiv 0 \bmod n_1 & \cdots w_k \equiv 0 \bmod n_1 \\ w_1 \equiv 0 \bmod n_2 & w_2 \equiv 1 \bmod n_2 & \cdots w_k \equiv 0 \bmod n_2 \\ \quad\vdots & \quad\vdots & \quad\vdots \\ w_1 \equiv 0 \bmod n_k & w_2 \equiv 0 \bmod n_k & \cdots w_k \equiv 1 \bmod n_k \end{array}\right\}.$$

Let

$$w = a_1 w_1 + a_2 w_2 + \cdots + a_k w_k.$$

Then, for all $i$,

$$\begin{aligned} w &= a_1 w_1 + a_2 w_2 + \cdots + a_k w_k \\ &\equiv a_1 \cdot 0 + a_2 \cdot 0 + \cdots + a_i \cdot 1 + \cdots + a_k \cdot 0 \bmod n_i \\ &\equiv a_i \bmod n_i. \end{aligned}$$

Thus, $w$ is a solution to the system of congruences (1.8). Now suppose that $v$ is also a solution. Then, for each $i$, we have

$$v \equiv a_i \equiv w \bmod n_i$$

so that $n_i \mid v - w$. Since the $n_i$ are relatively prime, it follows that $n_1 \cdots n_k \mid v - w$, by Lemma 1.4.5. Hence, $v \equiv w \bmod n_1 \cdots n_k$ and so the solution $w$ is unique modulo $n_1 n_2 \cdots n_k$.

**Example 1.9.2**  Find a solution to the system of congruences

$$x \equiv 2 \bmod 3, \quad x \equiv 1 \bmod 5, \quad x \equiv 4 \bmod 8.$$

The first step is to find the weights. Consider the system

$$x \equiv 1 \bmod 3, \quad x \equiv 0 \bmod 5, \quad x \equiv 0 \bmod 8.$$

Any solution must be a multiple of 5 and 8, and so we simply search through these multiples. In this case, the very first one, $w_1 = 40$, will do since $40 \equiv 1 \bmod 3$. For the second weight, we consider the system

$$x \equiv 0 \bmod 3, \quad x \equiv 1 \bmod 5, \quad x \equiv 0 \bmod 8.$$

Searching the multiples of $24 = 3 \times 8$ we have

$$24, 48, 72, 96$$

and $w_2 = 96$ will fill the bill. Last, we consider

$$x \equiv 0 \bmod 3, \quad x \equiv 0 \bmod 5, \quad x \equiv 1 \bmod 8$$

and find a solution $w_3 = 105$ from the multiples of $15 = 3 \times 5: 15, 30, 45, 60, 75, 90, 105$.

We can now generate a solution to the given system:

$$
\begin{aligned}
w &= 2 \times 40 + 1 \times 96 + 4 \times 105 \\
&= 80 + 96 + 420 \\
&= 596 \\
&\equiv 116 \bmod 120.
\end{aligned}
$$

Thus, $w \equiv 116 \bmod 120$ is the required solution.

**Example 1.9.3**  Other problems involving congruences can sometimes be reduced to problems that can be solved using the Chinese remainder theorem.

For example, consider the problem of solving the congruence $11x \equiv 76 \bmod 120$. Since $120 = 3 \times 5 \times 8$, we have that $11x \equiv 76 \bmod 120$ if and only if

$$11x \equiv 76 \bmod 3, \quad 11x \equiv 76 \bmod 5, \quad \text{and} \quad 11x \equiv 76 \bmod 8. \qquad (1.10)$$

Since $11 \equiv 2 \bmod 3$, $76 \equiv 1 \bmod 3$, and so forth, this is equivalent to

$$2x \equiv 1 \bmod 3, \quad x \equiv 1 \bmod 5, \quad \text{and} \quad 3x \equiv 4 \bmod 8.$$

If we multiply the first congruence by 2 we obtain $x \equiv 2 \bmod 3$. Multiplying this by 2 will recover the congruence $2x \equiv 1 \bmod 3$. Therefore, the congruences $2x \equiv 1 \bmod 3$ and $x \equiv 2 \bmod 3$ are equivalent. Likewise (multiplying by 3), the congruences $3x \equiv 4 \bmod 8$ and $x \equiv 4 \bmod 8$ are equivalent. Consequently, the system $(1.10)$ is equivalent to the system

$$x \equiv 2 \bmod 3, \quad x \equiv 1 \bmod 5, \quad x \equiv 4 \bmod 8.$$

This is precisely the system that we solved in the previous example, so we know that the solution is $x \equiv 116 \bmod 120$.

Another approach to problems of this type is to notice that $(11, 120) = 1$ so that $[11]$ is invertible in $\mathbb{Z}_{120}$. It so happens that $[11] \times [11] = [121] = [1]$ in $\mathbb{Z}_{120}$—in other words, $[11]$ is its own inverse in $\mathbb{Z}_{120}$. Consequently,

$$11x \equiv 76 \bmod 120$$

if and only if

$$x \equiv 121x = 11 \times 11x \equiv 11 \times 76 \bmod 120$$
$$\equiv 836 \bmod 120$$
$$\equiv 116 \bmod 120$$

as before.

The Chinese remainder theorem has an interesting application that makes it possible to do large calculations involving integers in parallel. Suppose that $n$ is a large integer and that $n = n_1 n_2 \cdots n_k$ where the $n_i$ are relatively prime. Suppose further that we wish to calculate the value of some function $f(x)$ of the integer $x$ and we know that the calculation only involves integer values and that the answer lies between 0 and $n - 1$, although intermediate calculations may involve much larger integers.

One possible approach is the following. Perform $k$ separate calculations (in parallel, if so desired) to evaluate

$$r_i \equiv f(x) \bmod n_i \qquad (i = 1, \ldots, k)$$

where $0 \le r_i \le n_i - 1$. Then use the Chinese remainder theorem to solve the system of congruences:

$$y \equiv r_i \bmod n_i \qquad (i = 1, \ldots, k).$$

This approach has two features in its favor. The first is that if $n_i$ is very much smaller than $n$, then calculations mod $n_i$ are much easier and faster than calculations involving values on the order of $n$ or greater. The second feature is that the calculations modulo $n_i$ $(i = 1, \ldots, k)$ can be done in parallel. In particular, this leads to modern applications in situations such as RSA encryption (see section 1.12), where calculations that need to be performed in $\mathbb{Z}_{pq}$ where $p, q$ are large prime numbers can, instead, be done separately in $\mathbb{Z}_p$ and $\mathbb{Z}_q$.

**Example 1.9.4**  Let us evaluate the determinant

$$D = \begin{vmatrix} 121 & 123 & 125 \\ 121 & 124 & 130 \\ 121 & 126 & 146 \end{vmatrix}$$

using the Chinese remainder theorem. Let us work modulo $792 = 8 \times 9 \times 11$. We have

$$D \equiv 6 \bmod 8$$
$$D \equiv 6 \bmod 9$$
$$D \equiv 0 \bmod 11.$$

Using the Chinese remainder theorem, we find that the unique solution to this system of congruences (modulo 792) is 726. Provided that the correct value of $D$ is in the range $726 - 791$ to $726 + 791$, then we have found the correct value for $D$ as 726. In fact, this is the correct value.

**Exercises 1.9**

1. Solve the system

$$x \equiv 1 \bmod 3, \quad x \equiv 2 \bmod 5, \quad x \equiv 3 \bmod 7.$$

2. Solve the system

$$x \equiv 3 \bmod 6, \quad x \equiv 2 \bmod 5, \quad x \equiv 1 \bmod 11.$$

3. Solve the congruence $7x \equiv 36 \bmod 200$ in two ways:
   (i)  By finding $[7]_{200}^{-1}$
   (ii)  By using the Chinese remainder theorem

4. Solve the congruence $5x \equiv 16 \bmod 198$ in two ways:

  (i)  By finding $[5]_{198}^{-1}$
  (ii)  By using the Chinese remainder theorem

*5 Let $a, b \in \mathbb{Z}, m, n \in \mathbb{N}, d = (m, n)$. Show that the system of congruences

$$x \equiv a \bmod m, \quad x \equiv b \bmod n$$

has a solution if and only if $d$ divides $a - b$.

## 1.10 Euler's function

For all $n \in \mathbb{N}$, $\varphi(n)$ is defined to be the number of integers $a$ with $1 \le a \le n$ and $(a, n) = 1$. The function $\varphi$ is called the *Euler $\varphi$-function* (Euler's totient function or simply Euler's function) and it is extremely important in the study of properties of integers. The value of $\varphi(n)$ is easily calculated for small values of $n$:

| $n$ | $\{a \mid 1 \le a \le n \text{ and } (a, n) = 1\}$ | $\varphi(n)$ |
|---|---|---|
| 1 | $\{1\}$ | 1 |
| 2 | $\{1\}$ | 1 |
| 3 | $\{1,2\}$ | 2 |
| 4 | $\{1,3\}$ | 2 |
| 5 | $\{1,2,3,4\}$ | 4 |
| 6 | $\{1,5\}$ | 2 |
| 7 | $\{1,2,3,4,5,6\}$ | 6 |
| 8 | $\{1,3,5,7\}$ | 4 |

Clearly, $\varphi(p) = p - 1$ for all primes $p$. The following will help us in the main theorem.

**Lemma 1.10.1** *Let $a, b \in \mathbb{Z}, n \in \mathbb{N}$. If $a \equiv b \bmod n$, then $(a, n) = (b, n)$. In particular, $(a, n) = 1$ if and only if $(b, n) = 1$.*

**Proof.** Let $a \equiv b \bmod n$. Then there exists an integer $q \in \mathbb{Z}$ with $a - b = qn$ so that $b = a - qn$. Therefore, if $d \mid a$ and $d \mid n$, then we must also have $d \mid b$. Similarly, if $d \mid b$ and $d \mid n$, then $d \mid a$. Consequently,

$$(a, n) \mid (b, n) \quad \text{and} \quad (b, n) \mid (a, n)$$

which implies that $(a, n) = (b, n)$.   $\square$

The converse of Lemma 1.10.1 does *not* hold.

**Example 1.10.2** We have

$$(2, 6) = 2 = (4, 6)$$

but $2 \not\equiv 4 \bmod 6$.

The next result is particularly useful when calculating values of the Euler $\varphi$-function.

**Theorem 1.10.3** *Let $m, n \in \mathbb{N}$, and $(m, n) = 1$. Then $\varphi(mn) = \varphi(m)\varphi(n)$.*

**Proof.** Let $r = \varphi(m)$, $s = \varphi(n)$, and $t = \varphi(mn)$. Let

$$A = \{\ell \in \mathbb{N} \mid 1 \le \ell < m, \ (\ell, m) = 1\}$$
$$B = \{\ell \in \mathbb{N} \mid 1 \le \ell < n, \ (\ell, n) = 1\}$$
$$C = \{\ell \in \mathbb{N} \mid 1 \le \ell < mn, \ (\ell, mn) = 1\}.$$

Then $|A| = \varphi(m)$ and $|B| = \varphi(n)$, so that $|A \times B| = \varphi(m)\varphi(n)$ whereas $|C| = \varphi(mn)$. Thus we will achieve our objective if we can prove that $|A \times B| = |C|$. We will do exactly that by showing that there exists a bijective mapping $\theta : A \times B \to C$.

For each $a \in A$, $b \in B$ there is, by the Chinese Remainder Theorem, a solution to the system

$$x \equiv a \bmod m, \quad x \equiv b \bmod n \tag{1.11}$$

that is unique modulo $mn$. Since $(a, m) = 1 = (b, n)$, it follows from Lemma 1.10.1 that $(x, m) = (x, n) = 1$ and, therefore, that $(x, mn) = 1$. By Lemma 1.6.1, $x \equiv c \bmod mn$ for some unique element $c$ with $1 \le c < mn$. By Lemma 1.10.1, $(c, m) = 1$ so that $c \in C$. Therefore, $x \equiv c \bmod mn$ for some unique element $c \in C$ so that $c$ is the unique solution to the system (1.11) in $C$. Thus, we have a mapping $\theta$ of $A \times B$ into $C$:

$$\theta : (a, b) \to c$$

where $c$ is the unique solution to (1.11) in $C$.

Now let $(a, b)$, $(a', b')$ be distinct elements of $A \times B$. Then, either $a \ne a'$ or $b \ne b'$. So without loss of generality, we may assume that $a \ne a'$. Since $1 \le a, a' < m$, it follows that $a \not\equiv a' \bmod m$.

Suppose that $\theta(a, b) = c$, $\theta(a', b') = c'$. Then

$$c \equiv a \bmod m, \quad c' \equiv a' \bmod m.$$

Since $a \not\equiv a' \bmod m$ we must have that $c \ne c'$. Thus, $\theta(a, b) \ne \theta(a', b')$ and $\theta$ is injective.

On the other hand, for all $c \in C$, we have $(c, mn) = 1$ so that $(c, m) = 1 = (c, n)$ and, by Lemma 1.6.1, there exist $a \in A$, $b \in B$ with $c \equiv a \bmod m$ and $c \equiv b \bmod n$. Hence, $\theta(a, b) = c$ and $\theta$ is surjective.

Since $\theta$ is both injective and surjective, it follows that $\theta$ is a bijection and $|A \times B| = |C|$. Therefore,

$$\varphi(mn) = |C|$$
$$= |A \times B|$$
$$= |A|\,|B|$$
$$= \varphi(m)\,\varphi(n). \quad \square$$

**Corollary 1.10.4** *If $p$ and $q$ are primes then*

$$\varphi(pq) = \varphi(p)\varphi(q) = (p-1)(q-1).$$

**Example 1.10.5**

$$\varphi(35) = \varphi(5)\varphi(7) = 4 \times 6 = 24$$
$$\varphi(55) = \varphi(5)\varphi(11) = 4 \times 10 = 40.$$

If we write an integer in its prime factorization

$$n = p_1{}^{r_1} \cdots p_k{}^{r_k}$$

then Theorem 1.10.3 tells us that

$$\varphi(n) = \varphi(p_1{}^{r_1})\varphi(p_2{}^{r_2}) \cdots \varphi(p_k{}^{r_k}).$$

The next result shows us how to calculate $\varphi(p^r)$ for any prime $p$.

**Theorem 1.10.6** *Let $p$ be a prime and $n \in \mathbb{N}$. Then*

$$\varphi(p^n) = p^{n-1}(p-1).$$

**Proof.** Exercise.   $\square$

**Exercises 1.10**

1. Find $\varphi(m)$ for $m = 8, 11, 12, 16, 21$.

2. Prove Theorem 1.10.6.

3. Calculate $\varphi(125)$, $\varphi(1800)$, $\varphi(10800)$.

*4. Find all solutions $x$ to $\varphi(x) = 6$.

5. Show that $\varphi(n)$ is even for $n \geq 3$.

6. Find the inverse of 11 modulo $\varphi(35)$.

7. Let $m, n \in \mathbb{N}$. Show that the number of integers $\leq mn$ and relatively prime to $m$ is $n\,\varphi(m)$.

8. Show that if $m$ and $n$ have the same prime divisors, then
$\varphi(mn) = n\,\varphi(m)$.

## 1.11 Theorems of Euler and Fermat

Suppose that we have an element $a$ in $\mathbb{Z}_n$ and that we calculate successive powers of $a$ (in $\mathbb{Z}_n$): $a, a^2, a^3, \ldots$. Since $\mathbb{Z}_n$ is finite, it is evident that these powers cannot all be distinct. In this section we discuss two famous theorems that shed some light on exactly what happens to these powers.

**Theorem 1.11.1 (Euler's Theorem)**  *Let $a \in \mathbb{Z}$, $n \in \mathbb{N}$, and $(a, n) = 1$. Then*

$$a^{\varphi(n)} \equiv 1 \bmod n.$$

**Proof.**  Let

$$A = \{[r_1], \ldots, [r_k]\}$$

be the set of invertible elements in $\mathbb{Z}_n$. By Theorem 1.7.1, $k = \varphi(n)$. By Lemma 1.7.8,

$$(a, n) = 1 \implies [a][r_i] \neq [a][r_j] \text{ for } i \neq j. \tag{1.12}$$

By Lemma 1.7.5,

$$(a, n) = 1 \implies [a][r_i] \text{ is invertible for all } i. \tag{1.13}$$

From (1.12) and (1.13) it follows that

$$B = \{[ar_1], \ldots, [ar_k]\}$$

is a set of $k$ invertible elements. But $\mathbb{Z}_n$ only has $k$ invertible elements all together. Therefore, we must have $B = A$ and $B$ is just a listing of the elements of $A$ in some order. Consequently, since multiplication is commutative,

$$[r_1][r_2] \cdots [r_k] = [ar_1] \cdots [ar_k]$$

so that

$$[r_1 \cdots r_k] = [a^k r_1 \cdots r_k]$$
$$= [a^k][r_1 \cdots r_k].$$

By Lemma 1.4.5 (iii), we have $(r_1 \cdots r_k, n) = 1$ so that, by Lemma 1.7.8,

$$[1] = [a^k]$$

or

$$a^{\varphi(n)} = a^k \equiv 1 \bmod n. \quad \square$$

In the proof of Euler's Theorem, we multiplied the set of invertible elements by an invertible element, and, by a counting argument, we could claim that we still had the set of all invertible elements, just rearranged. To see how this works, consider the invertible elements in $\mathbb{Z}_{10}$—namely, $\{1, 3, 7, 9\}$. If we multiply each of these elements by 9, then we obtain $\{9, 7, 3, 1\}$, which is exactly the same set, but in a different order. Similarly in $\mathbb{Z}_{15}$, if we multiply the invertible elements $\{1, 2, 4, 7, 8, 11, 13, 14\}$ by 2, say, then we obtain $\{2, 4, 8, 14, 1, 7, 11, 13\}$, which again is just the same set in a different order.

**Example 1.11.2** If $n = 12$ then $\varphi(12) = 4$ and with $a = 5$ we have

$$a^1 = 5, a^2 = 25 \quad \equiv 1 \bmod 12$$
$$a^4 = (a^2)^2 \equiv 1^2 = 1 \bmod 12.$$

If $n = 30$ then $\varphi(n) = 8$ and with $a = 7$ we have

$$a^1 = 7, a^2 = \quad 49 \quad \equiv 19 \bmod 30$$
$$a^4 = (a^2)^2 \equiv 19^2 = 361 \equiv 1 \bmod 30$$
$$a^8 = (a^4)^2 \equiv 1^2 \equiv 1 \bmod 30.$$

**Theorem 1.11.3 (Fermat's Theorem)** *Let $p$ be a prime and $a \in \mathbb{Z}$.*

(i) $(a, p) = 1 \implies a^{p-1} \equiv 1 \bmod p.$
(ii) $a^p \equiv a \bmod p.$

**Proof.** Since $p$ is a prime we know that $\varphi(p) = p - 1$. Part (i) is then a special case of Theorem 1.11.1.

(ii) There are two cases.

(a) $(a, p) = 1$. Then $a^{p-1} \equiv 1 \bmod p$, by part (i), so that

$$a^p = a \cdot a^{p-1} \equiv a \bmod p.$$

(b) $p \mid a$. Then $p \mid a^p$ and $p \mid a^p - a$. Therefore,

$$a^p \equiv a \bmod p. \quad \square$$

Why, one might wonder, should a straightforward corollary to Euler's Theorem be graced with the name of Fermat? Perhaps the fact that Fermat (1601–1665) preceded Euler (1707–1783) by roughly a century explains why Fermat's Theorem could seem so striking at the time. The following converse of Fermat's Theorem suggests an important technique for testing the primality of an integer.

**Lemma 1.11.4** *Let $p \in \mathbb{N}$ be such that $p > 1$ and*

$$a^{p-1} \equiv 1 \bmod p, \text{ for all } a \text{ with } 1 < a < p.$$

*Then $p$ is a prime.*

**Proof.** Let $a$ be such that $1 < a < p$. Then the hypothesis of the lemma tells us that there exists $k \in \mathbb{Z}$ with

$$a^{p-1} - 1 = kp.$$

Consequently, $1 = a^{p-1} - kp$, from which it follows that $(a, p) \mid 1$. But we always have $(a, p) \geq 1$. Therefore, $(a, p) = 1$ for all integers $a$ with $1 < a < p$. In other words, $p$ is a prime. $\quad \square$

Experience has shown that testing whether $a^{p-1} \equiv 1 \bmod p$ for just a few values of $a$, such as $a = 2, 3, 5, 7$, will often demonstrate that $p$ is not a prime, if that is indeed the case. In fact, a question posed by Fermat himself can be answered using this test. He asked: Is $2^{2^n} + 1$ always prime? He was able to check that this is indeed true for $n = 0, 1, 2, 3, 4$, for which the values are 3, 5, 17, 257 and 65537, respectively, all of which are prime numbers. He was unable to resolve the case $n = 5$ where $2^{2^5} + 1 = 4294967297$. However, it turns out that $2^{2^5} + 1$ fails the test when $a = 3$. A different way of seeing that this number is not a prime using modular arithmetic is provided at the end of this section.

There are certain composite numbers known as *Carmichael numbers* that undermine the usefulness of the above test for the primality of a positive integer. A Carmichael number $n$ is a positive integer such that $a^{n-1} \equiv 1 \bmod n$, for all $a$ that are relatively prime to $n$. The smallest Carmichael number is $561 = 3 \times 11 \times 17$. For more background on this topic, see ([CLRS], Section 31.8). We will return to the topic of primality testing again briefly in Section 7.3.

From Euler's Theorem we know that for any invertible element $a$ in $\mathbb{Z}_n$, we have $a^{\varphi(n)} \equiv 1 \bmod n$. The least positive integer $m$ such that $a^m \equiv 1 \bmod n$ is called the *order* of $a \pmod n$.

**Example 1.11.5** The invertible elements in $\mathbb{Z}_{10}$ are 1, 3, 7, 9, and $\varphi(10) = 4$. We have

$$1^1 = 1 \equiv 1$$

$$3^2 = 9, 3^3 = 27 \equiv 7, 3^4 \equiv 7 \times 3 \equiv 21 \equiv 1$$

$$7^2 = 49, 7^3 \equiv 9 \times 7 = 63 \equiv 3, 7^4 \equiv 3 \times 7 \equiv 21 \equiv 1$$

$$9^2 = 81 \equiv 1.$$

Thus, the orders of 1, 3, 7, and 9 are 1, 4, 4, and 2, respectively.

Note that if we take powers of a noninvertible element then we will never arrive at 1. For example:

$$5^2 = 25 \equiv 5, 5^3 = 5^2 \times 5 \equiv 5 \times 5 \equiv 5, \text{ and so on.}$$

$$4^2 = 16 \equiv 6, 4^3 \equiv 4^2 \times 4 \equiv 6 \times 4 \equiv 24 \equiv 4, \text{ and so on.}$$

**Lemma 1.11.6** *Let* $a \in \mathbb{Z}$, $n \in \mathbb{N}$, *and* $(a, n) = 1$. *Then the order of a divides* $\varphi(n)$.

**Proof.** By Euler's Theorem we know that $a^{\varphi(n)} \equiv 1 \bmod n$. Let $m$ denote the order of $a$. Necessarily, $m \leq \varphi(n)$. Let

$$\varphi(n) = qm + r \qquad \text{where } 0 \leq r < m.$$

Then,

$$1 \equiv a^{\varphi(n)} \equiv a^{qm+r} \equiv (a^m)^q a^r \equiv 1^q a^r \equiv a^r.$$

By the minimality of $m$, we must have $r = 0$. Therefore, $\varphi(n) = qm$ and $m \mid \varphi(n)$. $\square$

These observations on the order of an element can help us resolve Fermat's problem in regard to the number $2^{2^5} + 1$. Suppose that $p$ is a prime divisor of $2^{2^5} + 1$. Then $2^{2^5} \equiv -1 \bmod p$ so that $2^{2^6} \equiv 1 \bmod p$, from which it follows that the order of 2 mod $p$ is exactly $2^6 = 64$. But, by Lemma 1.11.6, the order of 2 must divide $\varphi(p) = p - 1$. Hence $p \equiv 1 \bmod 64$. A searchof primes of the form $64x + 1$ soon arrives at $p = 641$ which does, in fact, divide $2^{2^5} + 1$. Thus $2^{2^5} + 1$ is not a prime.

### Exercises 1.11

1. Use Fermat's Theorem to simplify $2^{127} \bmod 31$.

2. Use Fermat's Theorem to calculate the remainder from $12^{519}$ divided by 47.

3. Use Euler's Theorem to simplify $3^{50} \bmod 35$.

4. Simplify $5^{90} \bmod 27$.

5. Find the remainder from $6^{124}$ divided by 11.

6. Let $a, b \in \mathbb{N}$ and $p$ be a prime. Use Fermat's Theorem to show that

$$(a \pm b)^p \equiv a^p \pm b^p \bmod p.$$

7. Show that if $a, b \in \mathbb{N}$ and $p$ is a prime such that $a^p \equiv b^p \bmod p$, then $a^p \equiv b^p \bmod p^2$.

8. Find the order of the following elements:

   (i) $7 \in \mathbb{Z}_{12}$
   (ii) $5 \in \mathbb{Z}_{16}$
   (iii) $2 \in \mathbb{Z}_{29}$

## 1.12 Public Key Cryptosystems

For more than 2000 years people have been devising schemes, or ciphers as they are often called, for encoding and decoding messages. The primary objective of such schemes is to prevent a message being intercepted and read by some unauthorized person. Until recent times, the messages would have been sent among politicians, leaders, and generals or other representatives of the armed forces; the intent of encoding messages was to prevent the enemy from intercepting them.

In today's world, messages concerning war and peace are not the only messages that are deemed important enough to be protected from eavesdroppers. Confidentiality is an important aspect of life. It is important that the transfer of all kinds of data be done in a secure manner—for example, health records, insurance data, and financial data, to name a few. Data security and data encryption are now an important part of the high-tech industry.

In a basic scheme, the sender and receiver would agree on some method of encryption. For example, a method used in Greece in about 100 B.C. was to arrange the letters of the (Greek) alphabet in an array and then each letter would be replaced by its coordinates in the array. A modern-day equivalent might look like

|   | A | B | C | D | E |
|---|---|---|---|---|---|
| A | A | B | C | D | E |
| B | F | G | H | I | K |
| C | L | M | N | O | P |
| D | Q | R | S | T | U |
| E | V | W | X | Y | Z |

(Note that the Greek alphabet had 24 letters compared with 26 in English.) The message "call home E.T." was then encrypted as

$$AC\ AA\ CA\ CA\ BC\ CD\ CB\ AE\ AE\ DD.$$

A scheme used by the Roman general Caesar simply shifted each letter:

$$A \to D,\ B \to E,\ C \to F,\ \cdots.$$

Cardinal Richelieu (1585–1642), by some accounts one of France's greatest statesmen, is reputed to have used an encryption method in which the actual message would be interspersed on a page with garbage, and the message would be recovered from the page by placing a "template" over the page that contained holes and slits. The message could then be read from the letters exposed by the holes.

Another simple method of encryption is to convert the individual letters in the message to integers, break the message into blocks (vectors) $v_1, \ldots, v_m$ of some fixed length $n$, and then apply an invertible $n \times n$ matrix $A$ with integer entries to obtain $Av_1, \ldots, Av_m$. The original message can then be recovered by applying the inverse $A^{-1}$ to each block.

Two large classes of encryption ciphers are the *substitution* and *affine* ciphers. In substitution ciphers, a permutation is applied to each letter of the alphabet. For example, the Caeser cipher described above was a substitution cipher. Another important substitution based cipher, one that affected the course of World War II, was the Enigma cipher that will be discussed in Chapter 4. In an affine cipher, the message is first digitized and then each digit or string is subjected to the transformation:$x \longrightarrow ax + b \bmod n$. For this to be invertible, $a$ must be invertible mod $n$. The affine cipher is a special case of the substitution cipher.

Schemes such as these have the feature that if you know how to encode a message, then you know immediately how to decode any message using the same scheme. Therefore it is extremely important that the method of encryption be kept secret.

R. Merkle, W. Diffie, and M. Hellman proposed a different kind of scheme, one in which anyone would know how to encrypt a message but only the recipient would know how to decrypt the message. Such a scheme would be useful to government departments, banks, and businesses that could publish the schemes to be used by anyone who wished to communicate with them, secure from any eavesdropper.

At first it might seem that Merkle, Diffie, and Hellman were proposing the impossible. But let us analyze this a little further. Suppose that we consider the encoding process as some sort of function $E$ mapping letters or words over the alphabet $a, b, c, \ldots$ to strings of zeros and ones—that is, words over the alphabet $\{0, 1\}$. The decoding process is then a function $D$ in the opposite direction. Since the decoding process must yield the original message, the functions $D$ and $E$ must compose to give the identity:

$$DE = I$$

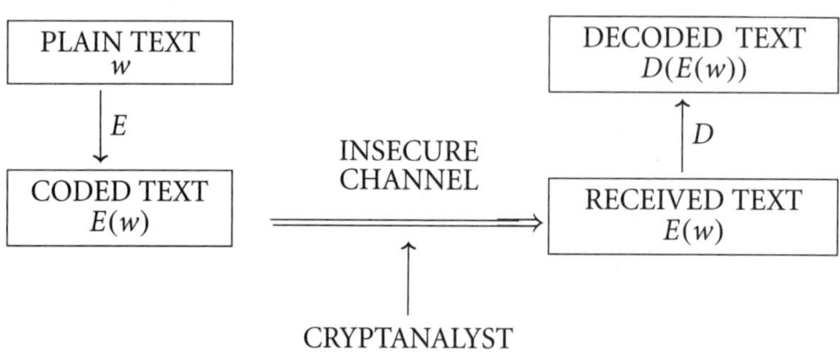


The idea put forward by Merkle, Diffie, and Hellman was that it would be possible to find suitable pairs of functions $(E, D)$ where knowledge of $E$ did not immediately reveal $D$ and that, although it might be possible to determine $D$ from $E$, the time and the cost would be too great for it to be practical. They labeled such functions $E$ as *trap door functions*.

Various pairs of functions $(E, D)$ satisfying the Merkle, Diffie, and Hellman requirements have been developed. We will present one such scheme that has attracted great attention and has been widely implemented. This scheme is called the *RSA system* after its inventors Rivest, Shamir, and Adleman (see [RSA]). Although Rivest, Shamir, and Adleman announced their technique in 1977, it appears that the British Security Services were aware of the RSA scheme some years earlier.

The "ingredients," as you might say, in the RSA system are as follows:

$p, q$ — distinct large primes (probably more than 150 decimal digits)

$$n = pq$$

$$\varphi(n) = (p - 1)(q - 1) \qquad \text{(by Corollary 1.10.4)}$$

$$d = \text{large positive integer with } (d, \varphi(n)) = 1$$

$$e \in \mathbb{N} \text{ such that } de \equiv 1 \bmod \varphi(n).$$

Note that such an integer $e$ will always exist on account of Theorem 1.7.1(iii). For this choice,

$$n = \text{the } \textit{modulus}$$

$$e = \text{the } \textit{encryption exponent}$$

$$d = \text{the } \textit{decryption exponent}$$

$$\{n, e\} = \text{the } \textit{public encryption key}$$

Any agency wishing to use this scheme would make the values of $n$ and $e$ publicly available. Anyone wishing to send a message only needs to know the values of $n$ and $e$.

Once the values of $n$, $e$, and $d$ have been selected, we can establish our encoding and decoding procedure.

### Encoding Procedure

(1) Digitize the message into a decimal word.

(2) Divide the decimal word into blocks of some suitable length $L$ so that each block $b$ in the message represents a number between 0 and $n - 1$.

(3) Encode each block $b$:

$$b \to b^e \bmod n$$

where $b^e \bmod n$ is the unique element in $[b^e]_n \cap \{0, 1, \ldots, n - 1\}$.

### Decoding Procedure

(1) Decode each received word $w$:

$$w \to w^d \bmod n$$

where $w^d \bmod n$ is the unique element in $[w^d] \cap \{0, 1, \ldots, n - 1\}$.

(2) Dedigitize $w^d \bmod n$.

The key question now is, if we start with a block $b$, encode it as $w = b^e$ and decode $w$ as $w^d$, do we recover $b$?

**Theorem 1.12.1** *Let $n$, $e$, and $d$ be as noted earlier and $b \in \{0, 1, 2, \ldots, n - 1\}$. Then $b$ is the unique integer in*

$$S = \{0, 1, 2, \ldots, n - 1\} \cap [(b^e)^d]_n.$$

**Proof.** By Lemma 1.6.1, there can be at most one integer in $S$. Consequently, all that we have to do is show that $(b^e)^d \equiv b \bmod n$. First let us suppose that $(b, n) = 1$.

By the choice of $e$ and $d$, we have $ed \equiv 1 \bmod \varphi(n)$ so that there exists $k \in \mathbb{N}$ with

$$ed = k\varphi(n) + 1.$$

Hence,

$$(b^e)^d = b^{ed} = b^{k\varphi(n)+1}$$

$$= (b^{\varphi(n)})^k b \equiv 1^k b \bmod n \quad \text{by Euler's Theorem}$$

$$\equiv b \bmod n,$$

as required. Now suppose that $(b, n) \neq 1$. That is possible since, after all, $n = pq$ is composite. If $b = 0$ then $(b^e)^d = 0$. So suppose that $b \neq 0$. Then either $p \mid b$ or $q \mid b$, but not both, since $n = pq$ and $1 \leq b < n$. Without loss of generality, it suffices to assume that $p \mid b$ and that $(b, q) = 1$. By Euler's Theorem applied to $q$, we have

$$b^{\varphi(q)} \equiv 1 \bmod q$$

so that, since $\varphi(n) = \varphi(p)\varphi(q)$,

$$b^{\varphi(n)} = (b^{\varphi(q)})^{\varphi(p)} \equiv 1^{\varphi(p)} \equiv 1 \bmod q.$$

Therefore,

$$b^{ed} = b^{k\varphi(n)+1} = (b^{\varphi(n)})^k b \equiv b \bmod q.$$

Since $p \mid b$ we have

$$b^{ed} \equiv 0 \equiv b \bmod p.$$

Thus, both $p$ and $q$ divide $b^{ed} - b$. However $(p, q) = 1$. Consequently, $n = pq$ also divides $b^{ed} - b$. In other words,

$$b^{ed} \equiv b \bmod n$$

in this case (where $p \mid b$) also.  $\square$

**Corollary 1.12.2** *The RSA procedure provides unique encoding and decoding for integers $b \in \{0, 1, 2, \ldots, n - 1\}$.*

**Scheme:**

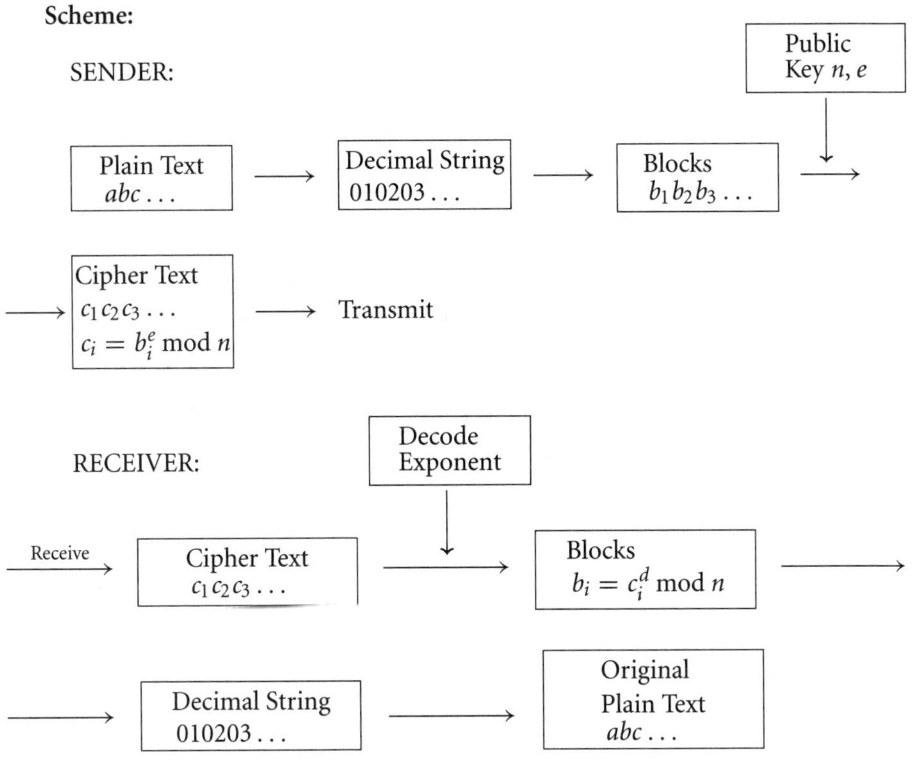


### A Decimal Representation of Alphabet, Numbers and Punctuation

| | | | | |
|---|---|---|---|---|
| $a = 01$ | $b = 02$ | $c = 03$ | $d = 04$ | $e = 05$ |
| $f = 06$ | $g = 07$ | $h = 08$ | $i = 09$ | $j = 10$ |
| $k = 11$ | $l = 12$ | $m = 13$ | $n = 14$ | $o = 15$ |
| $p = 16$ | $q = 17$ | $r = 18$ | $s = 19$ | $t = 20$ |
| $u = 21$ | $v = 22$ | $w = 23$ | $x = 24$ | $y = 25$ |
| $z = 26$ | $A = 27$ | $B = 28$ | $C = 29$ | $D = 30$ |
| $E = 31$ | $F = 32$ | $G = 33$ | $H = 34$ | $I = 35$ |
| $J = 36$ | $K = 37$ | $L = 38$ | $M = 39$ | $N = 40$ |
| $O = 41$ | $P = 42$ | $Q = 43$ | $R = 44$ | $S = 45$ |
| $T = 46$ | $U = 47$ | $V = 48$ | $W = 49$ | $X = 50$ |
| $Y = 51$ | $Z = 52$ | $0 = 53$ | $1 = 54$ | $2 = 55$ |
| $3 = 56$ | $4 = 57$ | $5 = 58$ | $6 = 59$ | $7 = 60$ |
| $8 = 61$ | $9 = 62$ | $\ = 63$ | $. = 64$ | $, = 65$ |
| $; = 66$ | $: = 67$ | $? = 68$ | $! = 69$ | $" = 70$ |
| $" = 71$ | $' = 72$ | | | |

**Example 1.12.3** Let $p = 113$, $q = 173$ so that $n = 19549$ and $\varphi(n) = 112 \times 172 = 19264$. Let $e = 31$. Then $d = 11807$.

Plain Text:     Timid     b o n e b r e a k e r     ?

Decimal String:   46   09   13   09   04   63   02   15   14   05   02   18   05   01   11   05   18   68

Blocks:     04609   01309   00463   00215   001405   00218   00501   01105   01868

$b_i^{31} \bmod 19549$

Cipher Text:     12774   16971   12161   09354   05666   06946   08265   13817   18250

$c_i^{11807} \bmod 19549$

Blocks Recovered:     04609   01309   00463   00215   001405   00218   00501   01105   01868

Plain Text Recovered:     Timid bonebreaker?

A minor procedural problem arises when we look at specific examples, such as Example 1.12.3. Since the modulus is $n = 19264$, which has 5 digits, if we break the message up into blocks of length 4, then all the resulting numbers represented by the blocks $b_i$ will certainly be less than $n$. However, when we calculate $b^e \bmod n$, we may obtain a number with 1, 2, 3, 4, or 5 digits. For instance $4609^{31} \equiv 12774 \bmod 19264$. So we require some device to separate the values of $b_1^e$, $b_1^e$, and so on. One possibility is to break up the message into blocks one less than the length of $n$ and add a zero at the beginning to produce a block of the same length as $n$ but still representing a number less than $n$. For example, replace 4609 by 04609. Then the encoded and decoded message can be treated uniformly in blocks of the same length (5 in this case).

Elegant as the ideas behind the RSA encryption scheme may be, their implementation depends critically on the ability to perform exponential calculations efficiently involving two large integers. Calculating $x^m$ as

$$x^m = \overbrace{(\ldots((x \cdot x) \cdot x)\ldots)}^{(m-1)-\text{multiplications}}$$

with $(m - 1)$ multiplications is too inefficient. One technique that is used is known as *exponentiation by squaring* or *square and multiply*. This technique dramatically reduces the number of calculations involved. It takes advantage of the representation of the exponent in base 2 arithmetic. For example, if

$$m = a_k 2^k + a_{k-1} 2^{k-1} + \ldots + a_0 \ (a_k \neq 0, a_i \in \{0, 1\}),$$

then we can write

$$x^k = ((\ldots (x^{a_k})^2 x^{a_{k-1}})^2 x^{a_{k-2}})^2 \cdots x^{a_1})^2 x^{a_0}$$

where the number of multiplications (alternating between squaring and multiplication by $x$) to be carried out is of the order of log $m$, as opposed to $m - 1$. (Of course all the calculations are actually carried out relative to the modulus $n$ adopted for the RSA encryption scheme.) The extent of the gain is easily appreciated by considering one or two examples, even small ones.

If $m = 15 = 2^3 + 2^2 + 2^1 + 2^0$, then we can write

$$x^{15} = (((x)^2 x)^2 x)^2 x$$

where the latter requires 6 multiplications as opposed to 14.

If $m = 65 = 2^6 + 1$, then

$$x^{65} = ((((((x)^2)^2)^2)^2)^2)^2 x$$

requiring 7 multiplications instead of 64.

Finally, if $m = 1000 = 2^9 + 2^8 + 2^7 + 2^6 + 2^5 + 2^3$, then

$$x^{1000} = (\ldots (x)^2 x)^2 x)^2 x)^2 x)^2)^2 x)^2)^2)^2$$

requiring just 14 multiplications versus 999 by direct computation.

The question that remains is: How secure is the RSA system? The task confronting a cryptanalyst is to discover the value of $d$ given the information $n$ and $e$. To do this requires a knowledge of $\varphi(n)$. But, $pq = n$ and $p + q = n - \varphi(n) + 1$. So a knowledge of $n$ and $\varphi(n)$ would make it possible to solve these equations to find the factorization of $n$. Thus, the task facing the cryptanalyst is to find the factorization of $n$ as a product of two primes.

When $n$ has more than 100 digits, this is considered a hard problem. However, ever faster computers combined with distributed or parallel computing and new factorization techniques are advancing constantly the frontiers of what is "manageable" in this regard.

To test and to demonstrate to the scientific community the security of their system, Rivest, Shamir, and Adelman published a short message that had been encoded using their method and a modulus $n$ (a product of two prime numbers) with 129 digits. This number became known in the literature as RSA 129.

$$RSA129 = 11438162575788886766923577997614661201021829672124236256256184293$$
$$5706935245733897830597123563958705058989075147599290026879543541$$
$$= 3490529510847650949147849619903898133417764638493387843990820577$$
$$\cdot \ 32769132993266709549961988190834461413177642967992942539798288533$$

The authors of this scheme estimated that using the conventional methods of the day, the number of calculations required to factor RSA 129 and break the code would take 40 quadrillion years. The challenge, with a nominal \$100 reward, was published in Martin Gardner's column in the Scientific American in 1977. What they did not reckon with were the advances that would be made in the development of both factoring algorithms and computing power. Using distributed computing techniques involving approximately 1700 computers around the world, in a massive cooperative attack, D. Atkins, M. Graff, A. Lenstra, and P. Leyland were able to announce on April 26, 1994, that they had factored RSA 129 and decoded the message as "The Magic Words are Squeamish Ossifrage". Despite this impressive achievement, the encryption community still believes that the advantage lies with the designers of this system because the complexity of factoring increases dramatically as the length of the modulus $n$ increases. Boneh [Bon, page 212] asserts that "Two decades of research into inverting the RSA function produced some insightful attacks, but no devastating attack has ever been found". However, the growing size of the keys and the computations involved may result in the advantage in utility passing to rival encryption schemes such as elliptic curve cryptography.

The exercises in this section illustrate one problem that can occur in RSA encryption if one wants to use a common modulus. There are various other pitfalls. For example, the number of digits in the decryption exponent should be at least as big as one quarter of the number of digits in $n$. For further good reading, see Patterson [Pat] and Stinson [Sti]. The topic of factorization of large integers, which is so crucial to RSA encryption, is discussed briefly in Section 7.3, to which an interested reader could skip at this point for a description of Pollard's $p - 1$ method.

**Exercises 1.12**

1. Choose distinct primes $p, q$ ($\geq 11$) and $d < n = pq$ such that $(d, \varphi(n)) = 1$.

   (i)  Find $e = d^{-1} \bmod \varphi(n)$.
   (ii) Pick a message with at least 10 characters. Now encode and decode your message using the RSA encryption scheme.

*2. Let $e_1, e_2, m, n \in \mathbb{N}$ where $(e_1, e_2) = 1 = (m, n)$ and $e_1, e_2 > 1$. Let

$$a = m^{e_1} \bmod n, \quad b = m^{e_2} \bmod n. \tag{1.14}$$

   Assuming that $a, b, e_1, e_2$ and $n$ are known, show how to recover the value of $m$. (Hint: Apply the Euclidean algorithm to $e_1, e_2$.)

   In the exercises to follow, let $p, q, \; n = pq, \; e$ and $d$ be chosen for an RSA encryption scheme as described earlier.

*3. (i)   Show that $e$ and $d$ are both odd.
    (ii)  Show that there exists $k > 0$ with $ed - 1 = 2^k y$ where $y$ is odd.
    (iii) Let $1 \leq b \leq n - 1$ and $(b, n) = 1$. Show that the order of $b$ in $\mathbb{Z}_n$ is of the form $2^j z$ where $j \leq k, \; z \mid y$.
    (iv)  Show that $a = b^{2^{j-1}z}$ satisfies $a^2 \equiv 1 \bmod n$.

*4. Let $1 < a < n - 1$ be such that

    (i)  $a^2 \equiv 1 \bmod n$
    (ii) $a \not\equiv \pm 1 \bmod n$.

    Show that either $(a + 1, n) = p$ or $(a + 1, n) = q$.

5. Exercises 3 and 4 show how a subscriber (with knowledge of one pair of values $e, d$) to a shared RSA encryption protocol using a common modulus can factorize the modulus $n$. Choose values of $p, q, \; n = pq, \; e$ and $d$ and, using only the values of $n, e,$ and $d$, a value of $b$ with $1 \leq b \leq n - 1$ and the method of Exercises 3 and 4, recover the values of $p$ and $q$. Note: Not every choice of $b$ is guaranteed to be successful.

*6. Let $P$ and $P_1$ be participants in a network in which internal communication among members uses RSA encryption with individual encryption and a decryption exponent, but with a common modulus. Let communication with $P$ use $(n, e, d)$ and with $P_1$ use $(n, e_1, d_1)$. The following steps show how $P_1$ may be able to decode $P$'s messages without needing to know the factorization of $n$. Let

$$f = (e, e_1 d_1 - 1), \quad g = (e_1 d_1 - 1)/f.$$

Assuming that $(e, g) = 1$, establish the following:

(i)  The Euclidean algorithm can be used to find $s, t \in \mathbb{Z}$ with

$$gs + et = 1.$$

(ii)  $(f, \varphi(n)) = 1$.
(iii)  $\varphi(n)$ divides $g$.
(iv)  $et \equiv 1 \bmod \varphi(n)$.
(v)  For $(m, n) = 1$, $(m^e)^t \equiv m \bmod n$.

7. Determine how many multiplications are required in the exponentiation by squaring method applied to $x^{500}$.

# 2

## Rings and Fields

In Chapter 1, we introduced $\mathbb{Z}_n$ and developed some of its properties as an algebraic structure. The important feature of $\mathbb{Z}_n$ as an algebraic structure is the fact that it possesses two algebraic binary operations (addition and multiplication) that have certain properties individually and that interact with each other through the distributive law.

Other algebraic structures share some of the properties of $\mathbb{Z}_n$. For example, the set of all $n \times n$ matrices of real numbers is an algebraic structure with two binary operations (again, addition and multiplication) and the set of all real valued functions of a real variable has three familiar binary operations (this time, addition, multiplication and composition).

In this chapter we begin a study of algebraic structures with two binary operations of addition and multiplication.

## 2.1 Basic Properties

Let $R$ be a nonempty set. Then a *binary operation* $*$ on $R$ is a rule that associates with each pair of elements $a, b \in R$ an element $a * b$. If $a * b$ is also an element of $R$, for all $a, b \in R$, then we say that $R$ is *closed* under $*$. The standard examples of binary operations are things like addition or multiplication of numbers, matrices, and so forth.

A simple example of a situation in which there is a reasonable operation but the set is not closed under that operation is the set $\mathbb{N}$ of natural numbers under division. We have $3, 4 \in \mathbb{N}$ but $\frac{3}{4} \notin \mathbb{N}$. In other words, $\mathbb{N}$ is not closed under division.

Let $R$ be a set with two binary operations which we will denote by $+$ and $\cdot$ (although we will usually indicate $\cdot$ simply by juxtaposition). Here are some of the properties that $(R, +, \cdot)$ might have: For all $a, b, c \in R$ (except in (ix))

$$
\begin{array}{ll}
\text{(0)} & R \text{ is closed with respect to } + \text{ and } \cdot \\
\text{(i)} & a + (b + c) = (a + b) + c \\
\text{(ii)} & a + b = b + a \\
\text{(iii)} & \text{there exists an element } 0_R \in R \text{ with } a + 0_R = a \\
\text{(iv)} & \text{there exists an element } -a \in R \text{ with } a + (-a) = 0_R \\
\text{(v)} & a(bc) = (ab)c \\
\text{(vi)} & a(b + c) = ab + ac \\
 & (b + c)a = ba + ca \\
\text{(vii)} & ab = ba \\
\text{(viii)} & \text{there exists an element } 1_R \in R \text{ with } 1_R \neq 0_R \text{ and} \\
 & a1_R = a = 1_R a \\
\text{(ix)} & \text{for all } a \in R \text{ with } a \neq 0, \text{ there exists an element } a^{-1} \in R \\
 & \text{with } aa^{-1} = a^{-1}a = 1_R.
\end{array}
$$

FIELD

Essentially, we have just listed all the rules that we usually apply in the familiar arithmetic of real or rational numbers. We adopt some of the same terminology. The element $0_R$ is called the *zero* and the element $1_R$ is called the *identity* of $R$.

Some of these laws have their own names:

| | |
|---|---|
| (0) | Closure |
| (i), (v) | Associative |
| (vi) | Distributive |
| (ii), (vii) | Commutative |
| (iii), (viii) | Existence of identities |
| (iv) | Existence of additive inverse |
| (ix) | Existence of multiplicative inverse |

Certain groupings of these properties show up with such frequency that they have been accorded special names:

| Properties | Name |
|---|---|
| (0)-(vi) | Ring |
| (0)-(vii) | Commutative ring |
| (0)-(vi), (viii) | Ring with identity |
| (0)-(viii) | Commutative ring with identity |
| (0)-(ix) | Field |

The simplest of these structures is a ring and the richest (in terms of having the most properties) is that of a field. Note that although we allow a ring to consist of a single element (i.e., $R = \{0_R\}$), we insist that $0_R \neq 1_R$ in a field, so that any field has at least two elements.

**Example 2.1.1**  The following are examples of rings:

(1)  $\mathbb{Z}, \mathbb{Q}, \mathbb{R}$.

(2)  The set $2\mathbb{Z}$ of all even integers.

(3)  $\mathbb{Z}_n$.

(4)  $M_n(R)$, the set of all $n \times n$ matrices over a ring $R$ with the usual addition and multiplication of matrices.

(5)  $\mathbb{Q}[x]$, the set of all polynomials with rational coefficients.

(6)  $\mathbb{R}[x]$, the set of all polynomials with real coefficients.

With all these rules that look so familiar to us since we have been using them in arithmetic without even thinking about them, it is tempting to assume that all the rules of arithmetic will always apply. However, we are already aware of certain examples when the normal rules do not apply. For instance, the multiplication of matrices is not commutative, the set of even integers does not have a multiplicative identity, and in section 1.7 we saw that products of nonzero elements in $\mathbb{Z}_n$ can be zero (something that is impossible in the arithmetic with which we are most familiar). All this warns us that we must be extremely careful in developing the properties of rings and must take nothing for granted. In the next result we gather together some basic observations.

**Lemma 2.1.2**  *Let* $(R, +, \cdot)$ *be a ring.*

    (i)  *The zero element* $0_R$ *is unique.*
    (ii)  *If* $R$ *has an identity* $1_R$, *then it is unique.*
    (iii)  *For all* $a \in R$, $0_R \cdot a = 0_R = a \cdot 0_R$.
    (iv)  *For all* $a \in R$, $-a$ *is unique.*
    (v)  *If* $a^{-1}$ *exists, then it is unique.*
    (vi)  *For all* $a, b \in R$,

$$(-a)b = a(-b) = -(ab)$$

$$-(-a) = a, \quad (-a)(-b) = ab.$$

    (vii)  $-0_R = 0_R$.

**Proof.**  (i) Let $0_R$ and $w$ be elements of $R$ such that $a + 0_R = a$ and $a + w = a$, for all $a \in R$. Then,

$$
\begin{aligned}
w &= w + 0_R & \text{by property of } 0_R \\
&= 0_R + w & \text{by commutativity of } + \\
&= 0_R & \text{by property of } w.
\end{aligned}
$$

Thus, $0_R$ is unique.

(ii) Exercise.

(iii) Let $a \in A$. Then,

$$
\begin{aligned}
0_R \cdot a &= (0_R + 0_R) \cdot a & \text{by property of } 0_R \\
&= 0_R \cdot a + 0_R \cdot a & \text{by distributivity}
\end{aligned}
$$

so that

$$
0_R \cdot a + (-0_R \cdot a) = 0_R \cdot a + 0_R \cdot a + (-0_R \cdot a)
$$

and

$$
0_R = 0_R \cdot a + 0_R = 0_R \cdot a.
$$

Similarly, we can show that $a \cdot 0_R = 0_R$.

(iv) Let $-a$ and $x$ be elements of $R$ such that $a + (-a) = 0_R$, $a + x = 0_R$. Then

$$
\begin{aligned}
x &= 0_R + x & \text{by property of } 0_R \\
&= a + (-a) + x & \text{by property of } -a \\
&= a + x + (-a) & \text{by commutativity} \\
&= 0_R + (-a) & \text{by property of } x \\
&= -a & \text{by property of } 0_R.
\end{aligned}
$$

Thus, the additive inverse of each element is unique.

(v) Exercise.

(vi) Let $a, b \in R$. Then we have

$$
\begin{aligned}
ab + (-a)b &= (a + (-a))b & \text{by distributivity} \\
&= 0_R \cdot b & \text{by property of } -a \\
&= 0_R & \text{by part (iii).}
\end{aligned}
$$

By part (iv) it now follows that $(-a)b = -(ab)$.

The verification of the remaining equalities is left as an exercise.

(vii) Exercise.  $\square$

Part (iii) of Lemma 2.1.2 is interesting, because there is nothing in the axioms for a ring that specifically relates $0_R$ to multiplication and so it might be expected that this would have to be part of the axioms. However, this is not the case, because it is a consequence of the listed axioms.

To simplify the notation, it is customary to write $a - b$ in place of $a + (-b)$. Henceforth, unless it is important to use $0_R$ and $1_R$ to avoid confusion, we will simply write 0 and 1 and expect that the meaning will be clear from the context.

Let $R$ be a ring with identity and $a \in R$. If $a^{-1}$ exists, then $a$ is said to be *invertible* or a *unit* and $a^{-1}$ is called the *inverse* of $a$. We denote the set of all invertible elements in $R$ by $R^*$. Since all nonzero elements in a field $F$ are invertible, it follows that in a field $F$, $F^*$ just consists of all the nonzero elements.

If $a$ and $b$ are elements of a commutative ring $R$ such that $a \neq 0_R$, $b \neq 0_R$, but $ab = 0_R$, then $a$ and $b$ are said to be *zero divisors*. Typically the presence of zero divisors complicates calculations, so it is nicer to work in contexts that are free of them. By an *integral domain* we mean a commutative ring with identity and no zero divisors. For example, $\mathbb{Z}$, $\mathbb{Q}$, and $\mathbb{R}$ are integral domains but $\mathbb{Z}_6$ is not. There is a connection between zero divisors and the cancellation laws.

**Lemma 2.1.3** *Let $R$ be a commutative ring with identity. Then the following are equivalent:*

 (i)  *$R$ is an integral domain.*
 (ii)  *$a, b, x \in R$, $x \neq 0$ and $ax = bx \Rightarrow a = b$.*

**Proof.** (i) *implies* (ii). We have

$$
\begin{aligned}
x \neq 0,\ ax = bx \Longrightarrow\ & (a - b)x = 0 \\
\Longrightarrow\ & a - b = 0 \text{ or } x = 0 \quad \text{by hypothesis} \\
\Longrightarrow\ & a - b = 0 \\
\Longrightarrow\ & a = b.
\end{aligned}
$$

(ii) *implies* (i). We argue by contradiction. Let $a, b \in R$ be such that $a \neq 0$, $b \neq 0$, and $ab = 0$. Then,

$$
\begin{aligned}
ab = 0 \Longrightarrow\ & ab = 0 = a \cdot 0 \\
\Longrightarrow\ & b = 0 \qquad\qquad \text{by hypothesis}
\end{aligned}
$$

which is a contradiction. Therefore, $R$ is an integral domain. $\square$

Fields are important examples of integral domains.

**Lemma 2.1.4** *Every field is an integral domain.*

**Proof.** Let $F$ be a field, $a, b, x \in F$, and $x \neq 0$. Since $F$ is a field, the element $x$ has a multiplicative inverse $x^{-1}$. Consequently,

$$
\begin{aligned}
ax = bx &\Longrightarrow (ax)x^{-1} = (bx)x^{-1} \\
&\Longrightarrow a(xx^{-1}) = b(xx^{-1}) \\
&\Longrightarrow a \cdot 1 \quad\;\; = b \cdot 1 \\
&\Longrightarrow a = b.
\end{aligned}
$$

It now follows from Lemma 2.1.3 that $F$ is an integral domain. $\quad\square$

Our principal goal in this chapter is to begin the study of fields. The only examples of fields that are ready to hand for us at the moment are $\mathbb{Q}$, $\mathbb{R}$ and $\mathbb{C}$ so it is interesting to discover that some of the rings $\mathbb{Z}_n$ are also fields.

**Theorem 2.1.5** $\mathbb{Z}_n$ *is a field if and only if $n$ is a prime.*

**Proof.** By Theorem 1.6.6 we already know that $\mathbb{Z}_n$ is a commutative ring with identity. Consequently,

$$
\begin{aligned}
\mathbb{Z}_n \text{ is a field} &\Longleftrightarrow \text{for all } a \in \mathbb{Z}_n \setminus \{0\}, a \text{ has an inverse} \\
&\Longleftrightarrow \text{for all } a \in \mathbb{Z}_n \setminus \{0\}, (a, n) = 1 \text{ (Theorem 1.7.1)} \\
&\Longleftrightarrow n \text{ is a prime.} \quad\square
\end{aligned}
$$

We shall see shortly how to construct many other examples of finite fields.

The following is a useful way of obtaining new examples of rings from examples that we already know about. Let $S$ and $T$ be rings and let $R = S \times T$ with addition and multiplication defined as follows:

$$
(a, b) + (c, d) = (a + c, b + d), \quad (a, b)(c, d) = (ac, bd).
$$

We describe these operations by saying that addition and multiplication are defined *componentwise*. Then $R$ is a ring (see exercises) called the *direct product* of $S$ and $T$.

We have seen that if a ring $R$ has an identity, then $1^2 = 1$ and $0^2 = 0$. Any element $x \in R$ such that $x^2 = x$ is called an *idempotent*. Thus, 0 and 1 are examples of idempotents. In general, a ring may have many idempotents.

**Exercises 2.1** In Exercises 1 through 6, let $R$ be a ring and $a, b \in R$.

1. Show that if $R$ has an identity then it is unique.

2. Show that $a(-b) = -(ab)$, $(-a)(-b) = ab$, $-(-a) = a$.

3. Show that if $m \in \mathbb{N}$, then $-(ma) = m(-a)$.

4. Show that if $R$ is a ring with identity, $a \in R$, and $a$ has an inverse, then it is unique.

5. Show that if $m \in \mathbb{N}$ and $a$ has an inverse, then $(a^m)^{-1} = (a^{-1})^m$ and $(-a)^{-1} = -(a^{-1})$.

6. Let $R$ be a ring with identity. Show that the set $U$ of invertible elements is closed under multiplication but not under addition.

7. Show that $(\mathbb{Z}\setminus\{0\}, \gcd, \cdot)$ is not a ring.

8. Let $R$ denote the set of all real valued functions of a real variable. Show that

   (i)  $(R, +, \cdot)$ is a ring
   (ii) $(R, +, \circ)$ (where $\circ$ denotes the composition of functions) is not a ring.

9. Show that $\mathbb{Z}[\sqrt{2}] = \{a + b\sqrt{2} \mid a, b \in \mathbb{Z}\}$ is a ring (with the usual addition and multiplication).

10. Show that $\mathbb{Q}[\sqrt{2}] = \{a + b\sqrt{2} \mid a, b \in \mathbb{Q}\}$ is a field.

11. Show that $G = \{a + bi \mid a, b \in \mathbb{Z}\}$ is a ring (where $i^2 = -1$). This ring is known as the *ring of Gaussian integers*. Which elements of $G$ are invertible?

12. Find an example of elements $a$, $b$ in a ring $R$ with $a \neq 0$ such that the equation $ax = b$ has more than one solution.

13. Let $R$ be a commutative ring with identity. Show that $R$ is a field if and only if every equation of the form $ax = b$ with $a, b \in R$ and $a \neq 0$ has a unique solution.

14. Find all idempotents in $\mathbb{Z}_{12}$.

15. Find all roots of the equation $x^2 - 1 = 0$ in (i) $\mathbb{Z}_7$, (ii) $\mathbb{Z}_8$, and (iii) $\mathbb{Z}_9$.

16. Let $F$ be a field. Show that the only solutions to the equation $x^2 = x$ in $F$ are $x = 0$ and $x = 1$.

17. Let $M_2(\mathbb{Z}_3)$ denote the set of all $2 \times 2$ matrices over $\mathbb{Z}_3$. Let $A \in M_2(\mathbb{Z}_3)$.

   (i)   What is the identity of $M_2(\mathbb{Z}_3)$?
   (ii)  How many elements are there in $M_2(\mathbb{Z}_3)$?
   (iii) Describe $-A$.
   (iv)  Is $M_2(\mathbb{Z}_3)$ commutative?
   (v)   Give examples of an invertible element and a noninvertible element.

18. Let $M_2(\mathbb{Z})$ denote the set of all $2 \times 2$ matrices over $\mathbb{Z}$. Which elements are invertible?

19. Let $S$, $T$ be rings with identity and $R = S \times T$ be the direct product of $S$ and $T$.

    (i)   Show that $R$ is a ring.
    (ii)  What is the zero of $R$?
    (iii) What is the identity of $R$?
    (iv)  If $S$ and $T$ are fields, does it follow that $R$ is a field? (Prove or give a counterexample.)

20. Let $S = \mathbb{Z} \times \mathbb{Z}$ with addition and multiplication defined by

$$(a, b) + (c, d) = (a + c, \, b + d), \quad (a, b)(c, d) = (a + bc, \, bd).$$

    Is $S$ a ring?

21. Let $R$ be a ring such that $x^2 = x$ for all $x \in R$. Show that $R$ is commutative.

22. Show that, for all $n \in \mathbb{N}$, every nonzero element of $\mathbb{Z}_n$ is either a unit or a zero divisor. Does the same assertion hold in $\mathbb{Z}$?

## 2.2  Subrings and Subfields

It will be convenient to be able to refer to multiples and powers of elements and so we introduce the following notation. For every element $a$ in a ring $R$, we define

$$0 \cdot a \; = 0_R \qquad\qquad 1 \cdot a \; = a$$

$$\underset{\text{integer}}{\uparrow} \qquad \underset{\text{ring element}}{\uparrow} \qquad \underset{\text{integer}}{\uparrow}$$

and for all $m \in \mathbb{N}$

$$(m + 1)a = ma + a, \quad (-m)a = -(ma).$$

Equivalently,

$$ma = \underbrace{a + a + \cdots + a}_{m \text{ terms}}, \quad (-m)a = \underbrace{(-a) + (-a) + \cdots + (-a)}_{m \text{ terms}}.$$

In this way we have assigned a meaning to $ma$ for all $m \in \mathbb{Z}$, $a \in R$ in such a way that the following rules are obeyed: For all $a, b \in R$, $m, n \in \mathbb{Z}$,

$$(m + n)a = ma + na$$

$$m(a + b) = ma + mb.$$

Next, for all $a \in R$, $m \in \mathbb{N}$ we define

$$a^1 = a, \; a^{m+1} = a^m \cdot a.$$

If $R$ has an identity then we extend this notation to

$$a^0 = 1$$

and if $a$ has an inverse then we define

$$a^{-m} = (a^{-1})^m.$$

If $a$ has an inverse then we have assigned a meaning to $a^m$ for all $m \in \mathbb{Z}$. Moreover,

$$a^m a^n = a^{m+n} \quad \text{for all } m, n \in \mathbb{Z}$$
$$(a^m)^n = a^{mn} \quad \text{for all } m, n \in \mathbb{Z}.$$

From Theorem 2.1.5 we know that $\mathbb{Z}_5$ is a field. Because $\mathbb{Z}_5$ is finite, there are things that happen in $\mathbb{Z}_5$ that do not happen in familiar examples of fields like $\mathbb{Q}$ and $\mathbb{R}$. For instance, in $\mathbb{Z}_5$, we have

$$5 \cdot 1 = 1 + 1 + 1 + 1 + 1 = 5 = 0.$$

Thus, as must happen in any finite field, there exist multiples of the identity that are zero, something that is impossible in $\mathbb{Q}$ or $\mathbb{R}$.

Let $R$ be a ring with identity. If there exists $n \in \mathbb{N}$ such that $n \cdot 1 = 0$, then the least such positive integer is called the *characteristic* of $R$. We will denote the characteristic of $R$ by *char*$(R)$. If there is no such integer, then $R$ is said to have characteristic 0. The rings $\mathbb{Z}$, $\mathbb{Q}$, and $\mathbb{R}$ have characteristic 0. For any $n \in \mathbb{N}$, $\mathbb{Z}_n$ has characteristic $n$. Note that for any nontrivial ring $R$ with identity—that is, any ring in which $1 \neq 0$—$R$ cannot have characteristic equal to 1 since that would imply that $1 = 1 \cdot 1 = 0$, which is a contradiction. Since the ring with just one element is rather uninteresting, we always assume that if $R$ is a ring with identity, then its characteristic is either 0 or finite and greater than one. The next two results present some of the basic properties of the characteristic of a ring.

**Lemma 2.2.1** *Let $R$ be a finite ring with identity $1_R$. Then $R$ has nonzero characteristic.*

**Proof.** Since $R$ is finite, the set

$$\{m1_R \mid m \in \mathbb{N}\}$$

must also be finite. Hence, there must exist $i, j \in \mathbb{N}$ with $1 \le i < j$ such that $i1_R = j1_R$. Then

$$(j - i)1_R = j1_R - i1_R = 0.$$

Therefore, the characteristic of $R$ lies between 1 and $j - i$. $\quad\square$

**Lemma 2.2.2** *Let $R$ be a ring of characteristic $n \ne 0$.*

  (i)  *$na = 0$, for all $a \in R$.*
  (ii)  *If $R$ is a field, then $n$ is a prime.*

**Proof.** (i) For all $a \in R$, we have

$$na = \underbrace{a + a + \cdots + a}_{n \text{ times}} = 1_R\, a + 1_R\, a + \cdots + 1_R\, a$$
$$= (1_R + \cdots + 1_R)a = (n1_R)a = 0_R \cdot a = 0_R.$$

(ii) Suppose that $n$ is not a prime, say $n = rs$ where $1 < r, s < n$. Then

$$0_R = n1_R = (rs)1_R \quad = \underbrace{1_R + \cdots + 1_R}_{rs \text{ times}} = \underbrace{(1_R + 1_R + \cdots + 1_R)}_{r \text{ times}} \underbrace{(1_R + 1_R + \cdots + 1_R)}_{s \text{ times}}$$
$$= (r1_R)(s1_R).$$

By the minimality of $n$, we must have $(r1_R) \ne 0$ and $(s1_R) \ne 0$, which, since $R$ is an integral domain, means that $(r1_R)(s1_R) \ne 0$. Thus we have arrived at a contradiction and $n$ must be a prime. $\quad\square$

Let $S$ be a ring (respectively, a field) and $R \subseteq S$. Then $R$ is a *subring* (respectively, *subfield*) if $R$ is a ring (respectively, a field) in its own right with respect to the *same operations as in $S$*.

**Example 2.2.3**

$$
\begin{array}{lll}
2\mathbb{Z} & \text{is a subring of} & \mathbb{Z} \\
\mathbb{Z} & \text{is a subring of} & \mathbb{Q} \\
\mathbb{Q} & \text{is a subfield of} & \mathbb{R} \\
\mathbb{R} & \text{is a subfield of} & \mathbb{C}
\end{array}
$$

**Lemma 2.2.4**  *Let S be a ring and $\emptyset \neq R \subseteq S$.*

   (i)  *R is a subring if and only if R is closed under $+$, $-$, and multiplication.*
   (ii)  *If S is a field then R is a subfield if and only if R is a subring, $1_S \in R$, and also*

$$a \in R,\ a \neq 0 \implies a^{-1} \in R,$$

*where $a^{-1}$ denotes the inverse of $a$ in $S$.*

**Proof.** Note that the properties of associativity, distributivity, and the commutativity of addition (and multiplication in part (ii)) are all automatically inherited by $R$ from $S$. Therefore it only remains to verify the ring conditions (iii) and (iv) to show that $R$ is a ring in part (i), and then to verify the field conditions (viii) and (ix) to show that $R$ is a field in part (ii). We leave this as an exercise.  $\square$

Subrings are usually quite plentiful and easy to find. One way to generate subrings is the following. For any ring $R$ and any $m \in \mathbb{N}$, we define the subset $mR$ of $R$ as follows:

$$mR = \{ma \mid a \in R\}.$$

This is always a subring of $R$ and it turns out that all the subrings of $\mathbb{Z}$ and $\mathbb{Z}_n$ are of this form (see the exercises).

There is a very interesting difference in the way that subrings behave in regard to zeros and identities. Let $R$ be a subring of a ring $S$ and let their respective zeros be $0_R$ and $0_S$. Then $0_R = 0_S$ (see the exercises). Now not all rings have identities, but let us assume that $S$ has an identity $1_S$. This does not force $R$ to have an identity. However, what is surprising is that even if $R$ does have an identity, $1_R$ say, then it is possible for this to be different from $1_S$—that is, it is possible to have $1_R \neq 1_S$. For example, let $S$ be the ring of all $2 \times 2$ matrices of real numbers and let $R$ be the set of real matrices of the form

$$\begin{bmatrix} 0 & 0 \\ 0 & a \end{bmatrix}$$

where $a \in \mathbb{R}$. Then $R$ is a subring of $S$, both $R$ and $S$ have identities, but

$$1_R = \begin{bmatrix} 0 & 0 \\ 0 & 1 \end{bmatrix} \neq \begin{bmatrix} 1 & 0 \\ 0 & 1 \end{bmatrix} = 1_S.$$

What is even more surprising is that $R$ is a field, even although $S$ itself is not a field! However, this is only an issue in very unusual circumstances since, generally speaking, we are only interested in subfields of fields and in this case there is only one identity involved (see the exercises).

Let $R$ and $S$ be rings and $\varphi : R \to S$ be a mapping such that

$$\varphi(a + b) = \varphi(a) + \varphi(b)$$
$$\varphi(ab) = \varphi(a)\varphi(b).$$

Then $\varphi$ is a *homomorphism*. If $\varphi$ is a homomorphism and a bijection, then $\varphi$ is an *isomorphism* and the rings $R$ and $S$ are said to be *isomorphic*. If $R = S$ then $\varphi$ is called an *automorphism*. If $\varphi : R \to S$ is an *isomorphism*, then so also is $\varphi^{-1} : S \to R$ (see the exercises). Note that although the symbol $\varphi$ is used to denote the Euler $\varphi$-function, it is also used, as here, to denote more general mappings. This should not result in any confusion because the meaning is usually clear from the context.

**Example 2.2.5** Let $S$ denote the set of all rational scalar multiples of the identity $n \times n$ matrix over $\mathbb{Q}$:

$$S = \{cI \mid c \in \mathbb{Q}\}.$$

Then $S$ is a ring and the mapping

$$\varphi : c \longrightarrow cI$$

is an isomorphism of $\mathbb{Q}$ onto $S$.

This example gives an isomorphism $\varphi : \mathbb{Q} \to S$, where we know both $\mathbb{Q}$ and $S$ really well. The real benefits of this idea become more apparent when we understand the structure of $S$ "well" and are able to show that some new ring $R$ is actually isomorphic to $S$, one that we already know.

**Theorem 2.2.6**  (i) *Let $F$ be a field of prime characteristic $p$. Let $1_F$ be the identity of $F$ and*

$$P = \{m1_F \mid m \in \mathbb{Z}\}.$$

*Then $P$ is a subfield of $F$ and the mapping*

$$\varphi : m \longrightarrow m1_F \quad (m \in \mathbb{Z}_p)$$

*is an isomorphism of $\mathbb{Z}_p$ onto $P$.*

   (ii) *Let $F$ be a field of characteristic $0$. Let $1_F$ be the identity of $F$ and*

$$P = \{(m1_F)(n1_F)^{-1} \mid m, n \in \mathbb{Z}\backslash\{0\}\} \cup \{0\}.$$

*Then P is a subfield of F and the mapping*

$$\varphi : \begin{cases} m/n & \to & (m1_F)(n1_F)^{-1} \ \text{ for } \ m/n \in \mathbb{Q}\backslash\{0\} \\ 0 & \to & 0_F \end{cases}$$

*is an isomorphism of $\mathbb{Q}$ onto P.*

**Proof.** (i) Clearly $P$ is closed under $+$, $-$, and multiplication so that $P$ is a subring. For any $m \in \mathbb{Z}$, let $q, r \in \mathbb{Z}$ be such that

$$m = pq + r \qquad 0 \le r < p.$$

Then

$$m1_F = q(p1_F) + r1_F = q \cdot 0_F + r1_F = r1_F$$

so that

$$P = \{0 \cdot 1_F, 1 \cdot 1_F, \ldots, (p-1) \cdot 1_F\}. \tag{2.1}$$

By the definition of the characteristic we know that $0 \cdot 1_F$ is the only element in this list equal to zero. It follows that these elements must also be distinct because, if $0 \le n < m < p$ and $m1_F = n1_F$, then

$$0_F = m \cdot 1_F - n \cdot 1_F = (m - n) \cdot 1_F$$

where $0 < m - n < p$, which is a contradiction. Thus the elements of $P$ are exactly those listed in (2.1). From this it is clear that $\varphi$ is well defined and, indeed, a bijection.

Now let $a \in P\backslash\{0_F\}$, say $a = m1_F$ where $1 \le m \le p - 1$. Since $p$ is a prime, it follows that $(m, p) = 1$. Consequently, from Corollary 1.3.8 we know that there exist $x, y \in \mathbb{Z}$ with $mx + py = 1$, so that

$$\begin{aligned} (m1_F)(x1_F) &= (1 - py)1_F \\ &= 1 \cdot 1_F - y(p1_F) \\ &= 1_F - 0_F \\ &= 1_F. \end{aligned}$$

Therefore, $x1_F = (m1_F)^{-1}$ where $x1_F \in P$. Hence, $P$ is a subfield.

Now let $\mathbb{Z}_p = \{0, 1, 2, \ldots, p - 1\}$ and consider the mapping $\varphi$. To see that $\varphi$ respects addition, we consider two cases. Let $m, n \in \mathbb{Z}_p$.

*Case (1).*  As integers, $m + n < p$. Then

$$\varphi(m + n) = (m + n)1_F = m1_F + n1_F = \varphi(m) + \varphi(n).$$

*Case (2).* As integers, $m + n \geq p$. Let $m + n = p + r$ where $0 \leq r < p$. Then $m + n = r$ (in $\mathbb{Z}_p$) so that

$$\varphi(m + n) = \varphi(r) = r 1_F$$

$$\begin{aligned}
\varphi(m) + \varphi(n) = m 1_F + n 1_F &= (m + n) 1_F \\
&= (p + r) 1_F = p 1_F + r 1_F \\
&= 0_F + r 1_F \qquad \text{since } F \text{ has characteristic } p \\
&= r 1_F \\
&= \varphi(m + n).
\end{aligned}$$

Therefore $\varphi$ respects addition. A similar argument will show that $\varphi(mn) = \varphi(m)\varphi(n)$. Thus $\varphi$ is an isomorphism.

(ii) An exercise. $\square$

The subfield of $F$ described in Theorem 2.2.6 is called the *prime subfield* or the *base subfield* of $F$. An important feature of fields of characteristic $p$ is that the arithmetic of sums raised to powers of $p$ is especially simple.

**Lemma 2.2.7** *Let $F$ be a field of characteristic $p$, a prime. Let $n \in \mathbb{N}$ and $a, b \in F$. Then*

$$(a \pm b)^{p^n} = a^{p^n} \pm b^{p^n}.$$

**Proof.** An exercise. (Hint: show that $p$ divides all the coefficients in expansion of $(a \pm b)^{p^n}$ except for the first and last.) $\square$

**Exercises 2.2**

1. What are the characteristics of the following rings?

    (i) $M_n(\mathbb{Z})$.
    (ii) $M_n(\mathbb{Z}_n)$.
    (iii) $\mathbb{Z} \times \mathbb{Z}$.
    (iv) $\mathbb{Z}_n \times \mathbb{Z}_n$.
    (v) $\mathbb{Z} \times \mathbb{Z}_n$.
    (vi) $\mathbb{Z}_4 \times \mathbb{Z}_6$.

2. Let $F$ be a field of characteristic $p$, a prime. Show that $(a \pm b)^p = a^p \pm b^p$, for all $a, b \in F$.

3. Let $R$ be a ring, $a \in R$ and $m \in \mathbb{N}$. Show that the following are subrings of $R$:

    (i) $\{mx \mid x \in R\}$.
    (ii) $\{x \in R \mid ax = 0\}$.

    (iii) $\{x \in R \mid xa = 0\}$.
    (iv) $\{x \in R \mid xy = yx \ \text{for all} \ y \in R\}$.

The subring in (iv) is called the *center* of R.

4. Let $R$ be any subring of $\mathbb{Z}$, $R \neq \{0\}$. Show that there exists $m \in \mathbb{N}$ with $R = m\mathbb{Z}$. (Hint: Show that $R$ contains a smallest positive integer.)

5. Let $R$ be any subring of $\mathbb{Z}_n$ $(n \in \mathbb{N})$. Show that there exists $m \in \mathbb{N}$ with $R = m\mathbb{Z}_n$.

6. Let $R$ be a subring of $S$. Show that $0_R = 0_S$.

7. Let $F$ be a subring of a field $G$. Show that, if $F$ is a field, then $F$ and $G$ share the same zero, identity, and characteristic.

8. Let $R = \{0, 5, 10\}$ in $\mathbb{Z}_{15}$. Show that $R$ is a subfield. What is the identity of $R$?

9. Find subrings $F$ and $G$ of $\mathbb{Z}_6$ that are isomorphic to $\mathbb{Z}_2$ and $\mathbb{Z}_3$, respectively. (Describe the isomorphisms explicitly.) What are the identities of $F$ and $G$?

10. Find an example of a ring $R$ and an element $a \in R$ such that the subrings described in parts (ii) and (iii) of Exercise 3 are distinct.

11. Let $R$ be a ring and $A$ be a subset of $R$. Let $S$ be the intersection of all subrings of $R$ that contain $A$. Show that $S$ is a subring of $R$.

12. Let $F$ be a field of characteristic $p$, a prime. Show that

$$G = \{a \in F \mid a^p = a\}$$

is a subfield of $F$.

13. Let $R$ and $S$ be rings and $\varphi : R \to S$ be an isomorphism. Show that $\varphi^{-1} : S \to R$ is also an isomorphism.

14. Let $R$ and $S$ be rings and $\varphi : R \to S$ be an isomorphism. Let $a \in R$. Prove the following:

    (i)   $R$ is commutative $\iff$ $S$ is commutative.
    (ii)  $R$ has an identity $\iff$ $S$ has an identity. When both have an identity, $\varphi(1_R) = 1_S$.
    (iii) $\varphi(a) = 0 \iff a = 0$.
    (iv) $a$ is a zero divisor $\iff$ $\varphi(a)$ is a zero divisor.
    (v)  $a$ is invertible $\iff$ $\varphi(a)$ is invertible.
    (vi) $b = a^{-1} \iff \varphi(b) = (\varphi(a))^{-1}$.

15. Let $R$ and $S$ be rings and $\varphi : R \to S$ be an isomorphism. Prove the following:

    (i) $R$ is an integral domain if and only if $S$ is an integral domain.
    (ii) $R$ is a field if and only if $S$ is a field.

16. Prove that the following pairs of rings are not isomorphic:

    (i) $\mathbb{Z}_6$ and $\mathbb{Z}_8$.
    (ii) $\mathbb{Z}_3 \times \mathbb{Z}_3$ and $\mathbb{Z}_9$.
    (iii) $\mathbb{Z}_{81}$ and $M_2(\mathbb{Z}_3)$.

17. Show that if $m \mid n$ and $\frac{n}{m}$ is not a prime then $m\mathbb{Z}_n$ has zero divisors.

*18. Let $m, n \in \mathbb{N}$ be such that $m \mid n$ and $\frac{n}{m} = p$ is a prime. Show that there exists $a \in \mathbb{N}$ with $1 \leq a < p$ and $(ma)^2 = ma$ in $\mathbb{Z}_n$ if and only if $(m, p) = 1$.

*19. Let $m, n,$ and $p$ be as in Exercise 18. Let $a \in \mathbb{N}$ be such that $1 < a < p$ and $(ma)^2 = ma$ in $\mathbb{Z}_n$. Show that $ma$ is the identity for $m\mathbb{Z}_n$.

*20. With $m, n,$ and $p$ as in Exercise 19, show that $m\mathbb{Z}_n$ is a field.

*21. Let $F$ be a finite field of characteristic $p$. Show that the mapping

$$\varphi : a \to a^p \quad (a \in F)$$

is an isomorphism of $F$ onto itself.

22. Show that any subring $R$ $(R \neq \{0\})$ of a finite field must be a subfield.

23. Show that, for any prime $p$,

$$\mathbb{Q}(\sqrt{p}) = \{a + b\sqrt{p} \mid a, b \in \mathbb{Q}\}$$

is a subfield of $\mathbb{R}$.

24. Show that

$$\varphi : a + b\sqrt{p} \to a - b\sqrt{p}$$

is an automorphism of $\mathbb{Q}(\sqrt{p})$.

*25. Let $R$ denote the set of matrices of the form

$$\begin{bmatrix} a & -b \\ b & a \end{bmatrix}$$

where $a, b \in \mathbb{R}$.

   (i)  Show that $R$ is a field with respect to matrix addition and multiplication.

  (ii)  Show that $R$ is isomorphic to $(\mathbb{C}, +, \cdot)$.


## 2.3  Review of Vector Spaces

It is assumed here that you have been exposed to the basic theory of vector spaces, perhaps in a standard introductory course to linear algebra. This section is only intended to provide a quick review of the basic ideas and to use them to obtain some important facts about finite fields.

A *vector space* over a field $F$ is a set of elements $V$ together with two operations, namely $+$ (addition) and $\cdot$ (scalar multiplication), satisfying the following axioms for all $x, y, z \in V$ and all $a, b \in F$:

(1)  $x + y \in V$.

(2)  $x + y = y + x$.

(3)  $x + (y + z) = (x + y) + z$.

(4)  There exists an element $0 \in V$ such that $x + 0 = x$.

(5)  There exists an element $-x$ such that $x + (-x) = 0$.

(6)  $a \cdot x \in V$.

(7)  For the identity $1 \in F$, $1 \cdot x = x$.

(8)  $a \cdot (x + y) = a \cdot x + a \cdot y$.

(9)  $(a + b) \cdot x = a \cdot x + b \cdot x$.

(10)  $(ab) \cdot x = a \cdot (bx)$.

The elements of $F$ are called *scalars* and the elements of $V$ are called *vectors*.


**Example 2.3.1**  Some familiar examples of vector spaces are the following:

| Field | Vectors | Dimension |
| --- | --- | --- |
| $\mathbb{Q}$ | $\mathbb{Q}^n$ | $n$ |
| $\mathbb{Q}$ | $\mathbb{R}^n$ | Infinite |
| $\mathbb{R}$ | $\mathbb{R}^n$ | $n$ |
| $\mathbb{Q}$ | $\mathbb{C}^n$ | Infinite |
| $\mathbb{R}$ | $\mathbb{C}^n$ | $2n$ |
| $\mathbb{C}$ | $\mathbb{C}^n$ | $n$ |
| $F$ | $F^n$ | $n$ |
| $\mathbb{Q}$ | $M_n(\mathbb{Q})$ | $n^2$ |
| $\mathbb{R}$ | $M_n(\mathbb{R})$ | $n^2$ |

Note that when we consider the vector spaces $\mathbb{Q}^n$ and $\mathbb{R}^n$ over $\mathbb{Q}$, the operations of vector addition and scalar multiplication are the same and the scalars are the same, only the underlying sets of vectors are different. In general, if $V$ and $W$ are vector spaces over the *same* field $F$, then we will say that $V$ is a *subspace* of $W$ if $V \subseteq W$. So, the vector space $\mathbb{Q}^n$ over $\mathbb{Q}$ is a subspace of the vector space $\mathbb{R}^n$ over $\mathbb{Q}$, but it is not a subspace of the vector space $\mathbb{R}^n$ over $\mathbb{R}$ because the field of scalars is different.

Let $V$ be a vector space over a field $F$ and $\mathcal{B} \subseteq V$. Then $\mathcal{B}$ is said to be (*linearly*) *dependent* if there exist $a_1, \ldots, a_n \in F$, not all of which are zero, and $v_1, \ldots, v_n \in \mathcal{B}$ with

$$a_1 v_1 + a_2 v_2 + \cdots + a_n v_n = 0.$$

If $\mathcal{B}$ is *not* dependent, then it is *linearly independent*. Put another way, the set $\mathcal{B}$ is linearly independent if whenever $a_1 v_1 + a_2 v_2 + \cdots + a_n v_n = 0$ for some scalars $a_i \in F$ and vectors $v_i \in \mathcal{B}$, then necessarily $a_1 = a_2 = \cdots = a_n = 0$.

The set $\mathcal{B}$ of vectors in $V$ is said to *span* $V$ if for all $v \in V$ there exist $c_1, \ldots, c_n \in F$ and $v_1, \ldots, v_n \in \mathcal{B}$ with

$$v = c_1 v_1 + c_2 v_2 + \cdots + c_n v_n.$$

**Lemma 2.3.2** *Let $V$ be a vector space over a field $F$ and $\mathcal{B} \subseteq V$. Then the following conditions are equivalent:*

(i) *$\mathcal{B}$ is linearly independent and spans $V$.*
(ii) *For every $v \in V$ there exist unique scalars $c_1, c_2, \ldots, c_n \in F$ and vectors $v_1, \ldots, v_n \in \mathcal{B}$ with*

$$v = c_1 v_1 + c_2 v_2 + \cdots + c_n v_n.$$

If $\mathcal{B}$ is a set of vectors in a vector space $V$ satisfying the conditions of Lemma 2.3.2, then $\mathcal{B}$ is a *basis* for $V$.

**Theorem 2.3.3** *Every vector space has a basis.*

**Theorem 2.3.4** *Let $V$ be a vector space over a field $F$. Let $\mathcal{B}_1$ and $\mathcal{B}_2$ be two bases for $V$. Then there exists a bijection from $\mathcal{B}_1$ to $\mathcal{B}_2$.*

Let $V$ be a vector space over a field $F$. Let $\mathcal{B}$ be a basis for $V$. If $\mathcal{B}$ is finite then $V$ is said to be *finite dimensional* and $|\mathcal{B}|$ is called the *dimension* of $V$. (By Theorem 2.3.4, the dimension of a vector space does not depend on any particular basis!)

Let $V$ and $W$ be vector spaces over the same field $F$. A mapping $\varphi :$ $V \to W$ is an *isomorphism* if it satisfies the following conditions:

(i)  $\varphi$ is a bijection.
(ii)  $\varphi(u + v) = \varphi(u) + \varphi(v)$ for all $u, v \in V$.
(iii)  $\varphi(au) = a\varphi(u)$ for all $a \in F, v \in V$.

If there exists an isomorphism $\varphi : V \to W$, then $V$ and $W$ are *isomorphic*.

**Lemma 2.3.5**  *Let $V$ be a vector space of dimension $n$ over a field $F$. Then $V$ is isomorphic to $F^n$.*

We can use Lemma 2.3.5 to tell us something very important about the relative sizes of finite fields and their subfields.

**Theorem 2.3.6**  *Let $F$ be a finite field and $E$ be a subfield of $F$. Then there exists $n \in \mathbb{N}$ such that $|F| = |E|^n$.*

**Proof.**  We consider $F$ as a vector space over the field $E$. Let $\mathcal{B}$ be a basis for $F$ over $E$. Since $F$ is finite, so also is $\mathcal{B}$. Let

$$\mathcal{B} = \{v_1, \ldots, v_n\}.$$

Then every element of $F$ can be written uniquely in the form

$$c_1 v_1 + c_2 v_2 + \cdots + c_n v_n \qquad \text{with } c_i \in E.$$

Therefore, the number of distinct elements in $F$ is exactly equal to the number of sequences of the form

$$(c_1, c_2, \ldots, c_n) \qquad \text{with } c_i \in E.$$

The number of choices for each $c_i$ is $|E|$, so that the number of such sequences is $|E|^n$. Thus $|F| = |E|^n$.  $\square$

In the previous section we saw that every finite field of charateristic $p$ contains a prime subfield isomorphic to $\mathbb{Z}_p$. We can now build on that to obtain an important restriction on the size of any finite field.

**Corollary 2.3.7**  *Let $F$ be a finite field of characteristic $p$. Then there exists $n \in \mathbb{N}$ with $|F| = p^n$.*

**Proof.**  In Theorem 2.2.6 we saw that $F$ contains a subfield $P$ with $p$ elements. Applying Theorem 2.3.6, we then find that there exists $n \in \mathbb{N}$ with

$$|F| = |P|^n = p^n.  \square$$

Essentially the same argument will give us more precise information regarding the relationship between the size of a finite field and the size of any subfield.

**Corollary 2.3.8** *Let F be a finite field of characteristic p and $|F| = p^n$ ($n \in \mathbb{N}$). Let E be a subfield of F. Then $|E| = p^m$ for some integer m such that $m \mid n$.*

**Proof.** Clearly $E$ must also have characteristic $p$, since $F$ and $E$ share the same identity. Thus, $|E| = p^m$ for some $m \in \mathbb{N}$. By Theorem 2.3.6 there exists $k \in \mathbb{N}$ with $|F| = |E|^k$ so that

$$p^n = |F| = |E|^k = (p^m)^k = p^{mk}.$$

Consequently, $n = mk$ and $m \mid n$. $\quad\square$

The next result gives us a picture of the additive structure of a finite field.

**Corollary 2.3.9** *Let F be a field with $p^n$ elements. Then as a vector space over $\mathbb{Z}_p$, F is isomorphic to $\mathbb{Z}_p^n$.*

**Proof.** Clearly $F$ has characteristic $p$ and prime subfield (isomorphic to) $\mathbb{Z}_p$. By Theorem 2.3.3, there exists a basis $\mathcal{B}$ for $F$ as a vector space over $\mathbb{Z}_p$. Since $F$ is finite, $\mathcal{B}$ must also be finite so that the dimension of $\mathcal{B}$ is $|\mathcal{B}| = m$, say. By Lemma 2.3.5, $F$ is isomorphic (as a vector space) to $\mathbb{Z}_p^m$. Consequently,

$$p^m = |\mathbb{Z}_p^m| = |F| = p^n$$

so that $m = n$ and the result follows. $\quad\square$

Of course, one of the first applications in any linear algebra course is the solution of systems of linear equations. All the standard arguments used in that context work equally well over an arbitrary field. In particular, the definitions of matrices and determinants are just as meaningful over an arbitrary field. We conclude our review with one final observation in that regard.

**Theorem 2.3.10** *Let A be an $n \times n$ matrix over a field F. Then A is invertible if and only if $\det(A) \neq 0$.*

One major question now confronting us is: Where can we find fields with $p^n$ elements? The answer to this question lies in the study of polynomials, which we will commence in the next section.

**Exercises 2.3**

1. Are the vectors $u = (1, -1, 2, 0)$, $v = (-3, 1, 1, -1)$, and $w = (1, 1, 1, 1)$ independent

   (i) in $\mathbb{R}^4$
   (ii) in $\mathbb{Z}_3^4$?

2. Do the vectors $u = (1, -1, 2, -2)$, $v = (-2, 1, -1, 2)$, $w = (2, -2, 1, -1)$, and $x = (-1, 2, -2, 1)$ span

   (i) $\mathbb{R}^4$
   (ii) $\mathbb{Z}_3^4$?

3. Show that

$$V = \{(x, y, u, v) \in \mathbb{R}^4 \mid x + 2y - u + 3v = 0\}$$

   is a subspace of $\mathbb{R}^4$. Find a basis for $V$.

4. Extend $\{(1, 1, 1, 1),\ (1, 0, 1, 0)\}$ to a basis for $\mathbb{R}^4$.

5. Let $V$ be a vector space over a field $F$ of dimension $n$. Let $A$ be an independent subset with $m < n$ elements. Prove that there exists a basis $B$ for $V$ with $A \subseteq B$.

6. Let $F$ be a field. Find a basis for $M_n(F)$ over $F$. What is the dimension of $M_n(F)$?

7. Let

$$A = \begin{bmatrix} 1 & 1 & 1 & -1 \\ 1 & 1 & 2 & 1 \\ 1 & 0 & 1 & 3 \\ 1 & 0 & 0 & 1 \end{bmatrix}$$

   over $\mathbb{Z}_3$.

   (i) Find the row rank of $A$.
   (ii) Find a basis for the row space.
   (iii) What is the dimension of the column space?

8. Let $A$ be the matrix in Exercise 7. Solve the system of equations $AX = 0$ (in $\mathbb{Z}_3$).

9. Let

$$A = \begin{bmatrix} 1 & 0 & 1 \\ 1 & 1 & 0 \\ 0 & 1 & 1 \end{bmatrix}$$

   over $\mathbb{Z}_5$. Find $A^{-1}$.

10. Let $F$ be either of the subrings of $\mathbb{Z}_6$ from Exercise 2.2.9 that are fields. Define addition and "scalar" multiplication as follows: For any $u, v \in \mathbb{Z}_6$, $a \in F$,

$$u + v = \text{sum of } u \text{ and } v \text{ in } \mathbb{Z}_6$$

$$av = \text{product of } a \text{ and } v \text{ in } \mathbb{Z}_6.$$

Is $\mathbb{Z}_6$ a vector space over $F$ with respect to these definitions?

11. Find a basis for the vector space of polynomials with rational coefficients as a vector space over $\mathbb{Q}$.

12. Let $F$ be a field and $n > 0$. Is

$$V = \{A \in M_n(F) \mid \det A = 0\}$$

a subspace of $M_n(F)$?

13. (i) Show that $\sqrt{p} \notin \mathbb{Q}$, for any prime $p$. (Hint: Try $p = 2$ first.)
    (ii) Show that $\sqrt{6} \notin \mathbb{Q}$.

14. Show that $1, \sqrt{2}$, and $\sqrt{3}$ are independent when $\mathbb{R}$ is viewed as a vector space over $\mathbb{Q}$.

15. What is the dimension of $\mathbb{Q}(\sqrt{p})$ when viewed as a vector space over $\mathbb{Q}$? (See Exercise 2.2.24.)

## 2.4 Polynomials

We now know a little bit about the size of finite fields (they must have $p^n$ elements), but the only examples that we have so far are the fields of the form $\mathbb{Z}_p$ with exactly $p$ elements. So, we might ask ourselves, are there fields with $2^2 = 4$, $2^3 = 8$, $3^2 = 9$ (and so forth) elements and, if so, what do they look like? It turns out that the key to progress on these questions is the study of polynomials.

Let $R$ be a ring. Then a *polynomial over $R$* is an expression of the form

$$a(x) = a_0 + a_1 x + a_2 x^2 + \cdots + a_n x^n$$

where each $a_i \in R$. We must be careful here because there is a real trap. We are conditioned from years of working with calculus and with real numbers to assume that *polynomial* means a special kind of function. Here we are going to give it a different interpretation. What is the real meaning of $x$? It is not an "unknown" to enable us to define a function; we are not thinking of $a(x)$ as a function. For example, consider the polynomials

$$x^3 + x^2, \quad x^2 + x$$

over $\mathbb{Z}_3$. We want to consider these as *distinct* polynomials. However, if we wish to do so we can use these expressions to define functions from $\mathbb{Z}_3$ to $\mathbb{Z}_3$ so that, say,

$$f(a) = a^2 + a \quad \text{for all} \quad a \in \mathbb{Z}_3$$

$$g(a) = a^3 + a^2 \quad \text{for all} \quad a \in \mathbb{Z}_3.$$

But then it is easily checked that these two functions are the *same*. This is not a problem that can occur when dealing with polynomials over the rational or real numbers (see Exercise 2.5.10). It can only occur over finite fields.

So what do we really mean by $x^3 + x^2$ and $x^2 + x$ and in what sense can we consider them to be distinct?

There is more than one approach that can be used to clarify the distinction between polynomials and functions. Our formal definition of a *polynomial over a ring R* will be a sequence $(a_0, a_1, \ldots)$ of elements from $R$, at most a finite number of which are nonzero. We define addition and multiplication of polynomials over $R$ by

$$(a_0, a_1, a_2, \ldots, a_n, \ldots) + (b_0, b_1, \ldots, b_n, \ldots)$$
$$= (a_0 + b_0, a_1 + b_1, a_2 + b_2, \ldots, a_n + b_n, \ldots)$$

$$(a_0, a_1, a_2, \ldots, a_n, \ldots) \cdot (b_0, b_1, b_2, \ldots, b_n, \ldots)$$
$$= (a_0 b_0, a_0 b_1 + a_1 b_0, a_0 b_2 + a_1 b_1 + a_2 b_0, \ldots).$$

Clearly, the result of addition or multiplication is another sequence with, at most, a finite number of nonzero elements.

These rules of addition and multiplication exactly parallel our customary understanding of addition and multiplication of polynomials as expressions in a *variable*:

$$(a_0 + a_1 x + a_2 x^2 + \cdots) + (b_0 + b_1 x + b_2 x^2 + \cdots)$$
$$= (a_0 + b_0) + (a_1 + b_1)x + (a_2 + b_2)x^2 + \cdots$$

and

$$(a_0 + a_1 x + a_2 x^2 + \cdots)(b_0 + b_1 x + b_2 x^2 + \cdots)$$
$$= a_0 b_0 + (a_0 b_1 + a_1 b_0)x + (a_0 b_2 + a_1 b_1 + a_2 b_0)x^2 + \cdots.$$

Since the familiar notation for polynomials is more convenient in calculations, we revert back to that notation while keeping in mind that a polynomial is really a sequence (and not a function). We have the following correspondence:

$$a_0 + a_1 x + a_2 x^2 + \cdots + a_n x^n \leftrightarrow (a_0, a_1, a_2, \ldots, a_n, 0, 0, \ldots)$$
$$0 \leftrightarrow (0, 0, 0, \ldots)$$
$$a \leftrightarrow (a, 0, 0, \ldots)$$
$$x \leftrightarrow (0, 1, 0, \ldots).$$

For any commutative ring $R$ with an identity, we denote by $R[x]$ the set of polynomials (in $x$) over $R$—that is, with coefficients in $R$. With respect to the addition and multiplication in $R[x]$ introduced above, $R[x]$ is itself a commutative ring with identity and is called the *ring of polynomials over R*.

For the polynomial

$$f(x) = a_n x^n + a_{n-1} x^{n-1} + \cdots + a_0$$

with $a_n \neq 0$, we call $a_i$ the *ith coefficient* and $a_n$ the *leading coefficient*. We call $n$ the *degree* of $f(x)$ and we write $\deg(f(x))$ for the degree of $f(x)$. If $\deg(f(x)) = 0$, then $f(x) = a_0 \in F$ is called a *constant*. A polynomial of degree 1 is called *linear*. In other words, we adopt all the language concerning polynomials that is familiar to us. Note that no degree has been assigned to the zero polynomial. In some circumstances it is convenient to assign a special value, such as minus infinity or $-1$, to the degree of the zero polynomial and accompany it with some special rules as to how to combine it with other degrees. Here we will simply treat the zero polynomial either as a constant with degree zero or as a special case. If $a_n = 1$ then $f(x)$ is *monic*. We identify $x^0$ with 1.

**Lemma 2.4.1** *Let R be an integral domain and*

$$f(x) = a_m x^m + \cdots + a_0 \quad (a_m \neq 0)$$
$$g(x) = b_n x^n + \cdots + b_0 \quad (b_n \neq 0)$$

*be elements of $R[x]$. Then*

$$\deg(f(x)g(x)) = m + n$$

*and the leading coefficient of $f(x)g(x)$ is $a_m b_n$.*

**Proof.** An exercise. $\square$

The next result reveals that $\mathbb{Z}$ and $F[x]$ have something in common. It is something that we will take advantage of shortly.

**Corollary 2.4.2** *Let R be an integral domain. Then $R[x]$ is also an integral domain.*

**Proof.** Let $f(x), g(x) \in R[x] \setminus \{0\}$. If $f(x)$ and $g(x)$ are both constants—that is, $f(x), g(x) \in R$—then $f(x)g(x) \neq 0$ since $R$ is an integral domain. If either $\deg(f(x)) \geq 1$ or $\deg(g(x)) \geq 1$, then, by Lemma 2.4.1, $\deg(f(x)g(x)) = \deg(f(x)) + \deg(g(x)) \geq 1$ so that $f(x)g(x) \neq 0$. The result follows. $\square$

For the most part we will be interested in the ring $F[x]$ of polynomials over a field, but also in the ring $Z[x]$, which is why the results pertaining to integral domains will be important. However, *henceforth in this chapter*, if not otherwise mentioned, *F will denote a field.*

**Theorem 2.4.3 (Division Algorithm for Polynomials)** *Let R be a ring with identity and let $f(x), g(x) \in R[x]$ where $g(x) \neq 0$ and the leading coefficient of $g(x)$ is invertible in R. Then there exist unique polynomials $q(x)$ and $r(x)$ in $R[x]$ such that*

$$f(x) = g(x)q(x) + r(x) \tag{2.2}$$

*where either $r(x) = 0$ or $\deg(r(x)) < \deg(g(x))$.*

**Note.** If $R$ is, in fact, a field, then the hypothesis that the leading coefficient of $g(x)$ be invertible is automatically satisfied.

**Proof.** We begin by showing the existence of polynomials $q(x)$ and $r(x)$ satisfying (2.2).

If $\deg(g(x)) = 0$, then $g(x) = a$ is a nonzero constant in $R$. By hypothesis, $a$ has an inverse and

$$f(x) = aa^{-1}f(x) = g(x)(a^{-1}f(x)) + 0$$

where $q(x) = a^{-1}f(x)$, $r(x) = 0$ satisfy (2.2).

So now suppose that $\deg(g(x)) \geq 1$. Let $P$ denote the set of all polynomials of the form

$$f(x) - g(x)q(x) \qquad (q(x) \in R[x]).$$

The degrees of all the polynomials in $P$ gives us a set of nonnegative integers. By the well-ordering principle, this set has a smallest element, $k$ say. Let $r(x) = f(x) - g(x)q(x)$ be such a polynomial of degree $k$. We now want to show that $\deg(r(x)) < \deg(g(x))$.

Suppose that $k = \deg(r(x)) \geq \deg(g(x))$. Let $\deg(g(x)) = m$ and let

$$g(x) = a_m x^m + \cdots + a_0 \quad (a_m \neq 0)$$
$$r(x) = b_k x^k + \cdots + b_0 \quad (b_k \neq 0).$$

Then $a_m$ has an inverse in $R$. So let

$$r_1(x) = r(x) - b_k a_m^{-1} x^{k-m} g(x). \tag{2.3}$$

Since the terms in $x^k$ from $r(x)$ and $b_k a_m^{-1} x^{k-m} g(x)$ will cancel, it follows that $\deg(r_1(x)) < \deg(r(x))$. Also,

$$\begin{aligned}
r_1(x) &= r(x) - b_k a_m^{-1} x^{k-m} g(x) \\
&= f(x) - g(x)q(x) - b_k a_m^{-1} x^{k-m} g(x) \\
&= f(x) - g(x)[q(x) + b_k a_m^{-1} x^{k-m}] \\
&\in P.
\end{aligned}$$

This contradicts the choice of $r(x)$ as a polynomial with least degree in $P$. Therefore, we must have $\deg(r(x)) < \deg(g(x))$, and so $q(x)$ and $r(x)$ do indeed satisfy (2.2).

It remains for us to establish the uniqueness of the polynomials $q(x)$ and $r(x)$. Let $q_1(x)$ and $r_1(x)$ also be polynomials satisfying (2.2)—that is,

$$f(x) = g(x)q_1(x) + r_1(x) \quad \text{and} \quad \deg(r_1(x)) < \deg(g(x)).$$

Then

$$\begin{aligned}
0 &= f(x) - f(x) \\
&= g(x)q(x) + r(x) - g(x)q_1(x) - r_1(x) \\
&= g(x)(q(x) - q_1(x)) + r(x) - r_1(x)
\end{aligned}$$

so that

$$g(x)(q(x) - q_1(x)) = r_1(x) - r(x). \tag{2.4}$$

If $q(x) - q_1(x) \neq 0$, then, since the leading coefficient of $g(x)$ is invertible, the degree of the left-hand side of (2.4) is at least as great as the degree of $g(x)$ whereas $\deg(r_1(x) - r(x)) < \deg(g(x))$. This is a contradiction. Therefore, $q(x) - q_1(x) = 0$ so that $r_1(x) - r(x) = 0$. Hence $q(x) = q_1(x)$ and $r(x) = r_1(x)$, and we have established uniqueness. $\quad\square$

The division algorithm has many useful applications concerning roots and factors. Let $f(x), g(x) \in F[x]$. If there exists a polynomial $h(x) \in F[x]$ such that

$$f(x) = g(x)h(x)$$

then $g(x)$ is a *divisor* or *factor* of $f(x)$ (in $F[x]$) and we say that $g(x)$ *divides* $f(x)$ and write $g(x) \mid f(x)$.

For $f(x), g(x), d(x) \in F[x]$, we say that $d(x)$ is a *greatest common divisor* (*gcd*) of $f(x)$ and $g(x)$ if

(i)  $d(x) \mid f(x), g(x)$
(ii) if $c(x) \mid f(x), g(x)$, then $c(x) \mid d(x)$.

Now, for every $a \in F^*$ we have $aa^{-1} = 1$. Hence,

$$\left. \begin{array}{l} f(x) = d(x)d'(x) \\ g(x) = d(x)d''(x) \end{array} \right\} \implies \begin{array}{l} f(x) = ad(x)(a^{-1}d'(x)) \\ g(x) = ad(x)(a^{-1}d''(x)) \end{array}$$

Thus, if $d(x)$ is a greatest common divisor of $f(x)$ and $g(x)$, then any nonzero scalar multiple $ad(x)$ ($a \neq 0$) is also a greatest common divisor. Thus, a greatest common divisor is *not* unique. For example, if

$$f(x) = (x-2)(x-3) \text{ and } g(x) = (x-2)(2x+3) \text{ in } \mathbb{Q}[x],$$

then clearly $x - 2 \mid f(x)$ and $g(x)$. But we also have

$$f(x) = (5x-10)\left(\frac{1}{5}x - \frac{3}{5}\right) \text{ and } g(x) = (5x-10)\left(\frac{2}{5}x + \frac{3}{5}\right)$$

so that $5x - 10 = 5(x-2)$ also divides $f(x)$ and $g(x)$. However, if we impose one further condition then we obtain uniqueness.

**Theorem 2.4.4 (Euclidean Algorithm for Polynomials)** *Let $F$ be a field and $f(x), g(x)$ be nonzero polynomials in $F[x]$. Then $f(x)$ and $g(x)$ have a unique monic greatest common divisor $d(x)$. Moreover, there exist polynomials $\lambda(x)$, $\mu(x) \in F[x]$ such that*

$$d(x) = \lambda(x)f(x) + \mu(x)g(x). \tag{2.5}$$

**Proof.** The proof of the existence of a greatest common divisor

$$d(x) = d_0 + d_1 x + \cdots d_n x^n \quad (d_n \neq 0)$$

for $f(x)$ and $g(x)$ follows from the division algorithm for polynomials (Theorem 2.4.3) exactly as the proof of the corresponding result (Theorem 1.3.4) for integers follows from the division algorithm for integers (Lemma 1.3.2). If $d_n \neq 1$, then $d_n^{-1}d(x)$ is also a greatest common divisor and is monic. So let us assume that $d(x)$ is a monic greatest common divisor.

Now let $c(x)$ be another monic greatest common divisor. Then $d(x) \mid c(x)$ and $c(x) \mid d(x)$ so that there exist polynomials $a(x), b(x)$ with

$$c(x) = a(x)d(x), \quad d(x) = b(x)c(x).$$

Hence, $\deg(c(x)) \geq \deg(d(x))$ and $\deg(d(x)) \geq \deg(c(x))$, which implies that $\deg(c(x)) = \deg(d(x))$. Consequently,

$$\deg(a(x)) = \deg(b(x)) = 0$$

so that $a(x)$ and $b(x)$ are constants. Since $c(x)$ and $d(x)$ are monic, $a(x) = b(x) = 1$. Thus, $c(x) = d(x)$ and there is only one monic greatest common divisor of $f(x)$ and $g(x)$, and we have established the required uniqueness.

The proof of the existence of polynomials $\lambda(x)$ and $\mu(x)$ to satisfy (2.5) follows exactly as for the corresponding part of Theorem 1.3.4, expressing the greatest common divisor of two integers $a, b$ in the form $ax + by$, where $x, y \in \mathbb{Z}$.  $\square$

Henceforth, by *the* greatest common divisor of two polynomials $f(x)$ and $g(x)$ in $F[x]$, we mean the unique *monic* greatest common divisor of $f(x)$ and $g(x)$.

When it comes to calculating the greatest common divisor of two polynomials, the *extended Euclidean algorithm* works exactly as before.

**Example 2.4.5** Find the gcd of $f(x) = x^5 + 4x^3 + 4x^2 + 3$ and $g(x) = x^2 + 3x + 2$ where $f(x), g(x) \in \mathbb{Z}_5[x]$.

We apply the division algorithm by performing a long division. (Note that the coefficients are always calculated in $\mathbb{Z}_5$.)

$$
\begin{array}{r}
x^3 \phantom{+} +2x^2 \phantom{+} +x + 2 \\
\hline
x^2 + 3x + 2 \,\big)\, x^5 + 0.x^4 + 4.x^3 + 4.x^2 + 0.x + 3 \\
x^5 + 3x^4 + 2x^3 \\
\hline
2x^4 + 2x^3 + 4x^2 \\
2x^4 + x^3 + 4x^2 \\
\hline
x^3 + 0\cdot x^2 + 0\cdot x \\
x^3 + 3x^2 + 2x \\
\hline
2x^2 + 3x + 3 \\
2x^2 + x + 4 \\
\hline
2x + 4
\end{array}
$$

Thus,

$$f(x) = g(x)(x^3 + 2x^2 + x + 2) + (2x + 4).$$

Now we divide $g(x)$ by $2x + 4$:

$$
\begin{array}{r}
3x+3 \\ \hline
2x+4\,)\,\overline{x^2+3x+2} \\
x^2+2x \\ \hline
x+2 \\
x+2 \\ \hline
0
\end{array}
$$

Then $g(x) = (2x + 4)(3x + 3)$ so that

$$f(x) = g(x)(x^3 + 2x^2 + x + 2) + (2x + 4)$$
$$g(x) = (2x + 4)(3x + 3)$$

and we find that $2x + 4$ is a greatest common divisor for $f(x)$ and $g(x)$. However, this is not a monic polynomial, so we do not refer to it as *the* greatest common divisor. We have $2x + 4 = 2(x + 2)$ so that $x + 2$ is a monic greatest common divisor and $x + 2 = \gcd(f(x), g(x))$. Therefore, remembering that our coefficients are drawn from $\mathbb{Z}_5$, we have

$$
\begin{aligned}
\gcd(f(x), g(x)) &= x + 2 \\
&= 3 \times 2 \times (x + 2) \\
&= 3 \times (2x + 4) \\
&= 3 \times (f(x) - (x^3 + 2x^2 + x + 2)g(x)) \\
&= 3f(x) - 3(x^3 + 2x^2 + x + 2)g(x) \\
&= 3f(x) + 2(x^3 + 2x^2 + x + 2)g(x).
\end{aligned}
$$

Alternatively, the extended Euclidean algorithm then leads to the following table:

| $i$ | $q_i$ | $r_i$ | $x_i$ | $y_i$ |
|---|---|---|---|---|
| 0 | — | $x^5 + 4x^3 + 4x^2 + 3$ | 1 | 0 |
| 1 | — | $x^2 + 3x + 2$ | 0 | 1 |
| 2 | $x^3 + 2x^2 + x + 2$ | $2x + 4$ | 1 | $-(x^3 + 2x^2 + x + 2)$ |
| 3 | $3x + 3$ | 0 | | |

From the table we conclude that

$$2x + 4 = 1 \cdot f(x) - (x^3 + 2x^2 + x + 2)g(x).$$

However, $2x + 4$ is not a monic polynomial so we will obtain *the* greatest common divisor if we multiply this equation by 3 to obtain

$$x + 2 = 3(2x + 4) = 3 \cdot f(x) + (2x^3 + 4x^2 + 2x + 4)g(x).$$

**Exercises 2.4**

1. Compute the sum and product of the following polynomials:

    (i) $x^3 + 2x^2 + 2$ and $2x^2 + x + 1$ in $\mathbb{Z}_3[x]$.
    (ii) $2x^3 + 4x^2 + 3$ and $x^3 + 5x^2 + 4$ in $\mathbb{Z}_6[x]$.

2. Show that the following pairs of polynomials induce the same mappings:

    (i) $x^4 + x^3 + 1, x^2 + x + 1$ on $\mathbb{Z}_3$.
    (ii) $x^5 - x, 0$ on $\mathbb{Z}_5$.
    (iii) $x^p - x, 0$ on $\mathbb{Z}_p$.
    (iv) $x^2 + 1, x^4 + 1$ on $\mathbb{Z}_4$.

3. Show that $6x + 1$ is invertible in $\mathbb{Z}_{12}[x]$.

4. Find a nonconstant invertible polynomial in $\mathbb{Z}_4[x]$.

5. Show that if $n \in \mathbb{N}$ and $n$ is not a prime, then $\mathbb{Z}_n[x]$ is not an integral domain.

6. Let $F$ be a field and $f(x) \in F[x]$. Show that $f(x)$ is invertible in $F[x]$ if and only if $f(x)$ is a nonzero constant polynomial.

7. Show that in $\mathbb{Z}_3[x]$,

$$(x + 1)^3 = x^3 + 1.$$

8. Show that in $\mathbb{Z}_p[x]$, where $p$ is a prime,

$$(x + 1)^p = x^p + 1.$$

9. Show that, if $n > 1$ and $n$ is not a prime, then in $\mathbb{Z}_n[x]$,

$$(x + 1)^n \neq x^n + 1.$$

10. Show that if $p$ is a prime and $f_i(x) \in \mathbb{Z}_p[x]$, $i = 1, 2, \ldots, n$, then

$$(f_1(x) + \cdots + f_n(x))^p = f_1(x)^p + f_2(x)^p + \cdots + f_n(x)^p.$$

11. Let $f(x) = a_n x^n + a_{n-1} x^{n-1} + \cdots + a_0 \in \mathbb{Z}_p[x]$ where $p$ is a prime. Show that

$$(f(x))^p = f(x^p).$$

12. Let $f(x) = x^4 + 2x^3 + 1$ and $g(x) = x^2 + x + 1$. Find the quotient and remainder when $f(x)$ is divided by $g(x)$ as polynomials in

    (i) $\mathbb{Q}[x]$
    (ii) $\mathbb{Z}_3[x]$
    (iii) $\mathbb{Z}_5[x]$.

13. With $f(x)$ and $g(x)$ as in Exercise 12, find $\gcd(f(x), g(x))$ in

    (i) $\mathbb{Q}[x]$
    (ii) $\mathbb{Z}_3[x]$
    (iii) $\mathbb{Z}_5[x]$.

*14. With $f(x)$ and $g(x)$ as in Exercise 13, find polynomials $\lambda(x), \mu(x)$ for which

$$\gcd(f(x), g(x)) = \lambda(x)f(x) + \mu(x)g(x)$$

in

    (i) $\mathbb{Q}[x]$
    (ii) $\mathbb{Z}_3[x]$
    (iii) $\mathbb{Z}_5[x]$.

15. Show that if $R$ and $S$ are isomorphic rings, then $R[x]$ is isomorphic to $S[x]$.

## 2.5 Polynomial Evaluation and Interpolation

For any ring $R$, $f(x) \in R[x]$ and $a \in R$, we can *evaluate* the polynomial $f(x)$ at $a$ by substituting the element $a$ for every occurrence of $x$ in $f(x)$ and calculating the resulting expression in $R$. We denote the result by $f(a)$. For example, if $f(x) = 2x^2 + x + 1 \in \mathbb{Q}[x]$ and $a = 3$, then

$$f(3) = 2 \cdot 9 + 3 + 1 = 22$$

whereas if $f(x) = 2x^2 + x + 1 \in \mathbb{Z}_5[x]$ and $a = 3$, then

$$f(3) = 2 \cdot 9 + 3 + 1 = 22 \equiv 2 \bmod 5.$$

In other words, *associated* with each polynomial $f(x) \in R[x]$ there is a function

$$f(x) : a \longrightarrow f(a) \quad (a \in R)$$

that *evaluates* the polynomial at each point $a \in R$. As we saw in the previous section, *different* polynomials may yield the *same* function.

If $f(x) \in R[x]$ and $a \in R$ is such that $f(a) = 0$, then $a$ is a *zero* or *root of* $f(x)$ *in R.*

**Lemma 2.5.1** *Let $R$ be an integral domain, $a, a_1, \ldots, a_m$ be distinct elements in $R$, and $f(x) \in R[x]$. Then*

    (i) *$a$ is a zero for $f(x) \iff x - a$ is a factor of $f(x)$*
    (ii) *$a_1, \ldots, a_m$ are zeros for $f(x) \iff (x - a_1)(x - a_2) \cdots (x - a_m)$ is a factor of $f(x)$.*

**Proof.** (i) First suppose that $f(a) = 0$. Now divide $f(x)$ by $x - a$. By the division algorithm for polynomials, there exist polynomials $q(x)$ and $r(x)$ in $R[x]$ such that

$$f(x) = (x - a)q(x) + r(x)$$

where either $r(x) = 0$ or $\deg(r(x)) < \deg(x - a) = 1$. Consequently, $r(x) \in R$—say, $r(x) = c \in R$—and

$$f(x) = (x - a)q(x) + c.$$

Evaluating $f(x)$ at $a$, we find that

$$0 = f(a) = (a - a)q(a) + c$$
$$= 0 \cdot q(a) + c$$
$$= c.$$

Thus,

$$f(x) = (x - a)q(x)$$

and $x - a$ is a factor of $f(x)$.

Conversely, now suppose that $x - a$ divides $f(x)$—say, $f(x) = (x - a)q(x)$ for some $q(x) \in R[x]$. Then

$$f(a) = (a - a)q(a) = 0 \cdot q(a) = 0$$

so that $a$ is a root of $f(x)$.

(ii) We argue by induction. We know from the first part that the claim holds when $m = 1$. So assume that $m > 1$. Let $a_1, \ldots, a_m$ be zeros for $f(x)$. By the first part, there exists a polynomial $q(x) \in R[x]$ such that

$$f(x) = (x - a_1)q(x)$$

and, necessarily, $\deg(q(x)) < \deg(f(x))$. Then, for $2 \le i \le m$,

$$0 = f(a_i) = (a_i - a_1)q(a_i).$$

Since the $a_i$ are all distinct, we have $a_i - a_1 \neq 0$ provided that $i \neq 1$. Since $R$ is an integral domain, we must have $q(a) = 0$ for $a = a_2, a_3, \ldots, a_k$. By the induction hypothesis, this implies that

$$q(x) = (x - a_2) \cdots (x - a_m)q_2(x)$$

for some polynomial $q_2(x)$. Thus,

$$f(x) = (x - a_1)(x - a_2) \cdots (x - a_m)q_2(x)$$

as required. Conversely, it is clear that if $(x - a_1)(x - a_2) \cdots (x - a_m)$ divides $f(x)$, then the elements $a_1, a_2 \cdots a_m$ are all zeros of $f(x)$.  $\square$

**Lemma 2.5.2** *Let $F$ be a field and $f(x) \in F[x]$ be a nonzero polynomial of degree $n$. Then $f(x)$ can have, at most, $n$ distinct zeros in $F$.*

**Proof.** Let $f(x)$ have distinct zeros $a_1, a_2 \cdots a_m$. By Lemma 2.5.1, it follows that there exists a polynomial $q(x) \in F[x]$ such that

$$f(x) = (x - a_1)(x - a_2) \cdots (x - a_m)q(x)$$

from which it follows that the degree of $f(x)$ is at least $m$. In other words, $m \leq n$ as required.  $\square$

Exercise 2 in this section shows that Lemma 2.5.2 is not necessarily true if $F$ is not a field.

For any complex number $z = a + ib$, the number

$$\bar{z} = a - ib$$

is the *complex conjugate* of $z$. Note that $\bar{z} = z$ if and only if $z \in \mathbb{R}$. There is a bond between $z$ and $\bar{z}$ that is of particular interest to us.

**Lemma 2.5.3** *The mapping*

$$\varphi : a + ib \to a - ib \quad (a + ib \in \mathbb{C})$$

*is an automorphism of $\mathbb{C}$. Moreover, for any complex number $z$, we have that $\varphi(z) = z$ if and only if $z$ is a real number.*

**Proof.** Clearly $\phi$ is a bijection. Let $a + ib, c + id \in \mathbb{C}$. Then

$$
\begin{aligned}
\varphi((a + ib) + (c + id)) &= \varphi(a + c + i(b + d)) \\
&= a + c - i(b + d) \\
&= (a - ib) + (c - id) \\
&= \varphi(a + ib) + \varphi(c + id)
\end{aligned}
$$

and

$$
\begin{aligned}
\varphi((a + ib) \cdot (c + id)) &= \varphi(ac - bd + i(ad + bc)) \\
&= ac - bd - i(ad + bc) \\
&= (a - ib) \cdot (c - id) \\
&= \varphi(a + ib) \cdot \varphi(c + id).
\end{aligned}
$$

Therefore, $\phi$ respects addition and multiplication and so $\phi$ is an isomorphism. The final claim in the Lemma is clear. $\square$

This will enable us to establish the important fact that the complex roots of a polynomial with real coefficients occur in conjugate pairs.

**Theorem 2.5.4** *Let $f(x) \in \mathbb{R}[x]$ and $z \in \mathbb{C}$ be a root of $f(x)$. Then $\bar{z}$ is also a root of $f(x)$.*

**Proof.** Let $f(x) = a_0 + a_1 x + \cdots + a_n x^n$ where $a_i \in \mathbb{R}$. Let $\varphi : \mathbb{C} \to \mathbb{C}$ be defined as in Lemma 2.5.3. Note that $\varphi(a) = a$ for all $a \in \mathbb{R}$. Then

$$
0 = f(z) = a_0 + a_1 z + \cdots + a_n z^n
$$

so that

$$
\begin{aligned}
0 &= \varphi(0) \\
&= \varphi(a_0) + \varphi(a_1) \varphi(z) + \cdots + \varphi(a_n) \varphi(z)^n \\
&= a_0 + a_1 \bar{z} + \cdots + a_n \bar{z}^n \\
&= f(\bar{z}).
\end{aligned}
$$

Thus, $\bar{z}$ is also a root of $f(x)$, as required. $\square$

Of course, Theorem 2.5.4 is telling us something new only if $z \notin \mathbb{R}$. If $z \in \mathbb{R}$, then $\varphi(z) = \bar{z} = z$ and Theorem 2.5.4 yields no information. One simple consequence of Theorem 2.5.4 is as follows.

**Corollary 2.5.5** *Let $f(x) \in \mathbb{R}[x]$ and $\deg(f(x)) = 3$. Then either $f(x)$ has three real roots (and so factors completely over $\mathbb{R}$) or else $f(x)$ has one real root and two distinct complex roots of the form $z, \bar{z}$.*

In the previous discussion, we looked at how a polynomial can assume the value zero. From time to time, we are interested in constructing polynomials that assume certain assigned values, not necessarily just zero, for certain values of the variable $x$. For instance, we may be interested in finding, or at least in knowing of the existence of, a polynomial that will yield the values $y_1, y_2, \ldots, y_n$ when $x$ assumes the distinct values $x_1, x_2, \ldots, x_n$, respectively. Another way of describing this would be to say that we want to find a polynomial $f(x) \in F[x]$ for which the graph $\{(x, f(x)) \mid x \in F\}$ passes through the points $(x_i, y_i), 1 \le i \le n$. This process is called *interpolation* or *polynomial interpolation*. We describe two ways in which this can be accomplished: Lagrange interpolation and Newton interpolation.

**Lagrange Interpolation.** Let

$$f(x) = \frac{(x - x_2)(x - x_3) \cdots (x - x_n)}{(x_1 - x_2)(x_1 - x_3) \cdots (x_1 - x_n)} \cdot y_1$$
$$+ \frac{(x - x_1)(x - x_3) \cdots (x - x_n)}{(x_2 - x_1)(x_2 - x_3) \cdots (x_2 - x_n)} \cdot y_2 + \cdots.$$

We can write this a bit more compactly if we introduce the polynomials

$$L_i(x) = \prod_{j \ne i} (x - x_j)$$

and then the Lagrange formula can be written as

$$f(x) = \sum_i \frac{L_i(x)}{L_i(x_i)} y_i.$$

**Newton Interpolation.** In this case, we define the polynomial that we want inductively. Let

$$f_1(x) = y_1$$

$$f_2(x) = f_1(x) + (y_2 - y_1)\frac{(x - x_1)}{(x_2 - x_1)}$$

$$\cdots = \cdots$$

$$f_n(x) = f_{n-1}(x) + (y_n - f_{n-1}(x_n))\frac{(x - x_1) \cdots (x - x_{n-1})}{(x_n - x_1) \cdots (x_n - x_{n-1})}.$$

We leave it as an exercise for you to verify that these polynomials do indeed assume the values required at the appropriate places.

**Example 2.5.6** Using both the Lagrange and Newton interpolation methods, find a polynomial $f(x) \in \mathbb{Z}_5$ that has the values $1, 0, 3$ at the points $0, 1, 3$, respectively.

Lagrange's method yields:

$$
f(x) = \frac{(x-1)(x-2)}{(0-1)(0-2)} \cdot 1 + \frac{(x-0)(x-2)}{(1-0)(1-2)} \cdot 0 + \frac{(x-0)(x-1)}{(2-0)(2-1)} \cdot 3
$$

$$
= 3(x^2 - 3x + 2) + 3 \cdot 3x(x-1)
$$

$$
= 2x^2 + 2x + 1.
$$

Newton's method yields:

$$
f_1(x) = 1
$$

$$
f_2(x) = f_1(x) + (y_2 - f_1(1))\frac{(x-0)}{(1-0)}
$$

$$
= 1 + (0-1)x = 1 - x
$$

$$
f_3(x) = f_2(x) + (y_3 - f_2(2))\frac{(x-0)(x-1)}{(2-0)(2-1)}
$$

$$
= 1 - x + 4 \cdot 3x(x-1)
$$

$$
= 2x^2 + 2x + 1.
$$

As we will see in a moment, the fact that the methods produce the same polynomial in this example is no accident.

One attraction of Newton's method over Lagrange's method is that if you already have a polynomial that has the desired values $y_1, \ldots, y_n$ at $n$ points $x_1, \ldots, x_n$ and want a polynomial that has these values at these points and a value of $y_{n+1}$ at an additional point $x_{n+1}$, then with Newton's method it is only necessary to calculate one additional term whereas with Lagrange's method it is necessary to recalculate the whole polynomial. Note that with both the Newton and Lagrange methods applied to $n$ values, we obtain a polynomial of degree at most $n - 1$. This bound on the degree has a very nice consequence.

**Theorem 2.5.7** *Let $x_i, y_i, 1 \le i \le n$ be elements of a field $F$ where the $x_i$ are distinct. Then there exists a unique polynomial $f(x) \in F[x]$ of degree $n - 1$ such that $y_i = f(x_i), 1 \le i \le n$.*

**Proof.** We can use either the Lagrange or Newton form of interpolation to guarantee the existence of a polynomial with the desired properties. Now suppose that that $f(x), g(x) \in F[x]$ both satisfy the requirements—that is,

$$f(x_i) = g(x_i) = y_i \quad \text{for all } 1 \le i \le n.$$

Let $h(x) = f(x) - g(x)$. Then,

$$h(x_i) = f(x_i) - g(x_i) = y_i - y_i = 0.$$

By Lemma 2.5.1, it follows that $(x - x_i)$ divides $h(x)$ for all $1 \le i \le n$. Hence,

$$(x - x_1) \cdots (x - x_n) \mid h(x).$$

However, $(x - x_1) \cdots (x - x_n)$ is a polynomial of degree $n$, so that the only way that it can divide $h(x)$, which is at most of degree $n - 1$, is if $h(x) = 0$. Thus, $f(x) - g(x) = 0$ and $f(x) = g(x)$. Thus there is a unique polynomial with the required properties.  $\square$

**Exercises 2.5**  Throughout the exercises we abide by our convention that $F$ denotes a field.

1. Let $a \in F$ and $f(x) \in F[x]$. Show that $f(a)$ is the remainder when $f(x)$ is divided by $x - a$.

2. Find all zeros of $x^2 + 2x$ in $\mathbb{Z}_{12}$. Why does this not contradict Lemma 2.5.2?

3. Let $p$ be a prime number. Show that there exist infinitely many $f(x) \in \mathbb{Z}_p[x]$ such that $f(a) = 0$ for all $a \in \mathbb{Z}_p$.

4. Show that for every function $g : \mathbb{Z}_2 \to \mathbb{Z}_2$, there exists a polynomial $f(x) \in \mathbb{Z}_2[x]$ such that $g(a) = f(a)$ for all $a \in \mathbb{Z}_2$.

5. Repeat Exercise 4 but with an arbitrary finite field $F$ in place of $\mathbb{Z}_2$.

6. Using both Lagrange and Newton interpolation, find a polynomial $f(x) \in \mathbb{Z}_7[x]$ that assumes the values $2, 1, 0$ when $x = 0, 1, 2$, respectively.

7. Repeat the previous exercise over the field $\mathbb{Q}$.

8. Let $f(x), g(x) \in F[x]$ both assume the values $y_1, y_2, \ldots, y_m$ at the distinct points $x_1, x_2, \ldots, x_m$, respectively. Show that there exists a polynomial $q(x) \in F[x]$ such that

$$f(x) = g(x) + (x - x_1) \cdots (x - x_m) q(x).$$

9. Let $F$ be an infinite field and $f(x) \in F[x]$ be such that $f(a) = 0$ for all $a \in F$. Show that $f(x) = 0$.

10. Let $F$ be an infinite field and $f(x), g(x) \in F[x]$ be such that $f(a) = g(a)$ for all $a \in F$. Show that $f(x) = g(x)$.

11. Let $x_1, \ldots, x_n$ be distinct elements in $F$. Prove that the $n$ *Lagrange Polynomials* $L_i(x)$ constitute a basis in $F[x]$ for the subspace of polynomials of degree, at most, $n - 1$.

## 2.6 Irreducible Polynomials

We now come to a concept for polynomials that strongly resembles the concept of a prime number. Throughout this section, $F$ denotes a field. Since $f(x) = a^{-1}(a f(x))$ for every $f(x) \in F[x]$, $a \in F^*$, we see that every polynomial can be factored into two polynomials. The crucial point to observe in this factorization is that $af(x)$ has the same degree as $f(x)$.

Let $R$ be an integral domain and $f(x) \in R[x]$ be a nonconstant polynomial—that is, $\deg(f(x)) \geq 1$. Then $f(x)$ is said to be *reducible* if there exist polynomials $g(x), h(x) \in R[x]$, *neither of which is a constant* such that

$$f(x) = g(x)h(x).$$

A polynomial that is not reducible is said to be *irreducible*. In other words, a polynomial $f(x) \in R[x]$ with $\deg(f(x)) \geq 1$ is said to be *irreducible* if it satisfies the following condition:

$$f(x) = g(x)h(x), \text{ where } g(x), h(x) \in R[x],$$
$$\implies \text{either } g(x) \text{ or } h(x) \text{ is a constant.}$$

For example, the polynomial $x^2 + 1$ is irreducible over $\mathbb{Z}$. To see this, suppose that

$$x^2 + 1 = (ax + b)(cx + d)$$

where $a, b, c, d \in \mathbb{Z}$. Then, equating coefficients, we must have $ac = 1$ so that $a, c = \pm 1$. Without loss of generality, we may assume that $a = 1$. Hence, substituting $x = -b$ we obtain $b^2 + 1 = 0$, which is clearly impossible. On the other hand,

$$x^4 + x^3 + x + 1 = (x + 1)(x^3 + 1)$$

in $\mathbb{Z}[x]$, so this polynomial is reducible.

**Lemma 2.6.1** *Let $f(x), g(x) \in F[x]$ and $f(x)$ be monic and irreducible. Then,*

$$\gcd\left(f(x), g(x)\right) = 1 \quad \text{or} \quad f(x).$$

**Proof.** Let $d(x) = \gcd\left(f(x), g(x)\right)$ and $h(x) \in F[x]$ be such that

$$f(x) = d(x)h(x).$$

By the hypothesis on $f(x)$, either $d(x)$ or $h(x)$ must be a constant. If $d(x)$ is a constant then $d(x) = 1$, since $d(x)$ is monic. If $h(x)$ is a constant, then $f(x)$ and $d(x)$ have the same degree and are both monic. Consequently, $h(x) = 1$ so that $d(x) = f(x)$.  $\square$

Clearly, all polynomials of degree 1 are irreducible. For degrees 2 and 3, there is a very useful test. Suppose that $f(x) \in F[x]$ is reducible of degree 2 or 3. This means that there exist polynomials $g(x), h(x) \in F[x]$, neither of which is a constant, such that

$$f(x) = g(x)h(x).$$

Then,

$$\deg\left(f(x)\right) = \deg\left(g(x)\right) + \deg\left(h(x)\right).$$

Consequently, since $\deg(f(x)) = 2$ or 3,

$$\text{either} \quad \deg\left(g(x)\right) = 1 \quad \text{or} \quad \deg\left(h(x)\right) = 1.$$

Suppose that $\deg(g(x)) = 1$. Then, for some $a, b \in F$, with $a \neq 0$,

$$g(x) = ax + b = a(x + a^{-1}b) = a(x - c)$$

where $c = -a^{-1}b$. Thus we find that $f(x)$ is divisible by a polynomial of the form $x - c$ so that, by Lemma 2.5.1, $f(x)$ has a root $c$ in $F$. We have, therefore, established the direct part of the next result.

**Lemma 2.6.2** *Let $f(x) \in F[x]$ and $\deg(f(x)) = 2$ or 3. Then $f(x)$ is reducible if and only if $f(x)$ has a zero in $F$. Equivalently, $f(x)$ is irreducible if and only if $f(x)$ has no zeros in $F$.*

**Proof.** It only remains to show that if $f(x)$ has a zero in $F$, then it is reducible, but this follows immediately from Lemma 2.5.1. $\square$

**Example 2.6.3** The following polynomials are irreducible over $\mathbb{Z}_2$:

$$1 + x + x^2, \quad 1 + x + x^3, \quad 1 + x^2 + x^3.$$

Consider $f(x) = 1 + x^2 + x^3$. We have

$$f(0) = 1, \quad f(1) = 3$$

so that $f(x)$ has no zero in $\mathbb{Z}_2$. By Lemma 2.6.2, $f(x)$ is irreducible. Similar arguments apply to the other polynomials.

**Example 2.6.4** The following polynomials are irreducible over $\mathbb{Z}_3$:

$$1 + x^2, \ 1 + x + 2x^2, \ 1 + 2x + 2x^2.$$

The existence of irreducible polynomials of various degrees is extremely important to the study of finite fields.

**Theorem 2.6.5** *Let $F$ be a finite field and $n \in \mathbb{N}$. Then there exists an irreducible monic polynomial in $F[x]$ of degree $n$.*

**Proof.** The number of monic polynomials of degree $n$ in $F[x]$—that is, polynomials of the form $a_0 + a_1 x + \cdots + a_{n-1} x^{n-1} + x^n$—is simply $|F|^n$, since each of the coefficients $a_0, \ldots, a_{n-1}$ can be chosen arbitrarily. Any reducible monic polynomial $f(x) \in F[x]$ of degree $n$ can be written as a product of two or more irreducible polynomials of degree less than $n$. So the basic idea is to calculate inductively the number of irreducible monic polynomials of each degree and from that calculate the number of reducible monic polynomials of degree $n$ with the outcome being less than $|F|^n$. For illustrative purposes here, we will just deal with the smallest case when $n = 2$. A proof of the general claim using a different approach can be found in Corollary 2.12.10. The total number of monic polynomials of degree 2 is just $|F|^2$. Any reducible monic polynomial $f(x)$ of degree 2 must factor as $(x + a)(x + b)$ for some $a, b \in F$. The number of ways in which this can happen is

$$\binom{|F|}{2} \quad \text{if} \quad a \neq b$$

$$|F| \quad \text{if} \quad a = b.$$

Therefore, the number of reducible monic polynomials of degree 2 is

$$\binom{|F|}{2} + |F|$$

so that the number of irreducible monic polynomials of degree 2 is

$$|F|^2 - \left(\binom{|F|}{2} + |F|\right) = \binom{|F|}{2} > 0.$$

Thus, there do exist irreducible polynomials of degree 2 over any finite field.  $\square$

**Theorem 2.6.6** *Let $f(x) \in F[x]$. Then $f(x)$ can be written as a product of irreducible polynomials. If*

$$f(x) = g_1(x) \cdots g_r(x) = h_1(x) \cdots h_s(x)$$

*where the polynomials $g_i(x)$, $h_i(x)$ are irreducible, then $r = s$ and the order of the factors may be rearranged so that for all $i$, $1 \le i \le r$, there exists $a_i \in F^*$ with $g_i(x) = a_i h_i(x)$.*

**Proof.** The proof is parallel to that of the corresponding result for integers (Theorem 1.4.2). See the exercises for more details.  $\square$

The following fact is very useful when factorizing polynomials with integer coefficients.

**Proposition 2.6.7** *Let $f(x) \in \mathbb{Z}[x]$. If $f(x)$ is reducible in $\mathbb{Q}[x]$, then it is reducible in $\mathbb{Z}[x]$.*

**Proof.** Let $f(x) = g_1(x) h_1(x)$ where $g_1(x), h_1(x) \in \mathbb{Q}[x]$ and $\deg(g_1(x))$, $\deg(h_1(x)) \ge 1$. Let $a$ (respectively, $b$) be the product of all the denominators of the coefficients in $g_1(x)$ (respectively, $h_1(x)$). Let $g(x) = a \, g_1(x)$ and $h(x) = b \, h_1(x)$. Then

$$f(x) = \frac{1}{ab} g(x) h(x)$$

where $g(x), h(x) \in \mathbb{Z}[x]$. Let $p$ be a prime that divides $ab$ and let

$$g(x) = a_m x^m + \cdots + a_0$$
$$h(x) = b_n x^n + \cdots + b_0.$$

Since $f(x) \in \mathbb{Z}[x]$, it follows that $p$ divides every coefficient of $g(x)h(x)$. In other words, if we consider $g(x)$ and $h(x)$ as polynomials in $\mathbb{Z}_p[x]$, we have

$g(x)h(x) = 0$. But by Corollary 2.4.2, $\mathbb{Z}_p[x]$ is an integral domain. Therefore, either $g(x) = 0$ or $h(x) = 0$ in $\mathbb{Z}_p[x]$. Thus, either $p$ divides all the coefficients of $g(x)$ or else $p$ divides all the coefficients of $h(x)$. Without loss of generality, we may assume the former and may write

$$f(x) = \frac{1}{(ab)/p}\left(\frac{1}{p}g(x)\right)h(x)$$

where $\frac{1}{p}g(x), h(x) \in \mathbb{Z}[x]$. Repeating this argument sufficiently often, we will finally divide the whole of $ab$ into $g(x)h(x)$ and express $f(x)$ as a product of two polynomials with integer coefficients. $\quad\square$

One general application of the preceding result is the following famous criterion for irreducibility in $\mathbb{Q}[x]$.

**Theorem 2.6.8 (Eisenstein's Criterion)** *Let $p$ be a prime and $f(x) = a_n x^n + a_{n-1}x^{n-1} + \cdots + a_0 \in \mathbb{Z}[x]$ be such that*

    (i)   $p \nmid a_n$
    (ii)  $p \mid a_{n-1}, \ldots, a_0$
    (iii)  $p^2 \nmid a_0$.

*Then, $f(x)$ is irreducible in $\mathbb{Q}[x]$.*

**Proof.** Let $f(x) = g(x)h(x)$ where

$$g(x) = b_s x^s + b_{s-1}x^{s-1} + \cdots + b_0 \in \mathbb{Q}[x]$$
$$h(x) = c_t x^t + c_{t-1}x^{t-1} + \cdots + c_0 \in \mathbb{Q}[x]$$

and $s, t \geq 1$. By Proposition 2.6.7, we may assume that $g(x), h(x) \in \mathbb{Z}[x]$.

Since $a_0 = b_0 c_0$ and $p^2$ does not divide $a_0$, it follows that $p$ does not divide both $b_0$ and $c_0$. So we may assume that $p \mid c_0$ and $p$ does not divide $b_0$. Now $a_n = b_s c_t$ so that $p$ does not divide $c_t$. Let $k$ be the smallest integer such that

$$p \mid c_0, \; p \mid c_1, \ldots, p \mid c_{k-1}, \; p \nmid c_k.$$

From the preceding remarks, we know that $0 < k \leq t$. Then

$$a_k = b_0 c_k + b_1 c_{k-1} + \cdots + b_k c_0 \equiv b_0 c_k \bmod p$$

so that $p$ does not divide $a_k$. By the hypothesis of the theorem, this means that $k = n$, which implies that $t = n$ and $s = 0$, which is a contradiction. $\quad\square$

The factorization of a polynomial into a product of irreducible polynomials depends very much on the field of coefficients. For example,

$$x^{10} + 4x^7 - 3x^6 + 5x^2 + 1$$

has the following factorizations into a product of irreducible polynomials:

$$(x+1)(x^9 - x^8 + x^7 + 3x^6 - 6x^5 + 6x^4 - 6x^3 + 6x^2 - x + 1) \text{ in } \mathbb{Z}[x] \text{ or } \mathbb{Q}[x]$$

$$(x+1)(x^9 + 2x^8 + x^7 + 2x + 1) \text{ in } \mathbb{Z}_3[x]$$

$$(x+1)(x+3)(x^8 + x^7 + 3x^6 + 4x^5 + 2x^4 + 4x^2 + 4x + 2) \text{ in } \mathbb{Z}_5[x]$$

$$(x+1)(x+3)^2(x+5)^2(x^5 + 4x^4 + 4x^3 + 2x^2 + 4x + 1) \text{ in } \mathbb{Z}_7[x]$$

$$(x+1)(x+2)(x^3 + 5x^2 + 8x + 2)(x^4 + x^3 + 4x^2 + 2x + 7) \text{ in} \mathbb{Z}_{11}[x].$$

**Exercises 2.6**

1. Show that the polynomials $1 + x + x^2, 1 + x + x^3, 1 + x^2 + x^3 \in \mathbb{Z}_2[x]$ are irreducible.

2. Show that the polynomials listed in Exercise 1 are the only irreducible polynomials in $\mathbb{Z}_2[x]$ of degrees 2 and 3.

3. Show that the following polynomials in $\mathbb{Z}_3[x]$ are irreducible:

$$\text{(i) } 1 + x^2, \qquad \text{(ii) } 2 + x + x^2, \qquad \text{(iii) } 2 + 2x + x^2,$$

$$\text{(iv) } 2 + 2x^2, \quad \text{(v) } 1 + 2x + 2x^2, \quad \text{(vi) } 1 + x + 2x^2.$$

4. Show that the monic polynomials listed in Exercise 3 are the only irreducible monic polynomials in $\mathbb{Z}_3[x]$ of degree 2.

5. Show that the following are irreducible in $\mathbb{Z}_2[x]$:

(i) $x^4 + x + 1$.
(ii) $x^4 + x^3 + 1$.
(iii) $x^4 + x^3 + x^2 + x + 1$.

6. Show that the number of irreducible quadratic monic polynomials in $\mathbb{Z}_p[x]$ is $\frac{1}{2}p(p-1)$. (Hint: First find the number of reducible quadratic monic polynomials.)

7. Show that the number of irreducible cubic monic polynomials in $\mathbb{Z}_p[x]$ is $\frac{1}{3}p(p^2 - 1)$.

8. (i) Find all reducible polynomials in $\mathbb{Z}_2[x]$ of degree 4.
   (ii) Find all irreducible polynomials in $\mathbb{Z}_2[x]$ of degree 4.

9. Factorize the following into products of irreducible polynomials:

(i) $x^3 + 2x^2 + x + 1$ in $\mathbb{Z}_3[x]$.
(ii) $x^3 + x + 2$ in $\mathbb{Z}_5[x]$.
(iii) $x^3 + 3x^2 + 3$ in $\mathbb{Z}_7[x]$.

10. Factorize the following into products of irreducible polynomials:

    (i) $x^4 + x^3 + x + 1$ in $\mathbb{Z}_2[x]$.
    (ii) $x^4 + 1$ in $\mathbb{Z}_3[x]$.
    (iii) $x^4 + 1$ in $\mathbb{Z}_5[x]$.

11. Show that if $f(x) = a_n x^n + a_{n-1} x^{n-1} + \cdots + a_1 x + a_0 \in F[x]$ is irreducible, then so also is $g(x) = a_0 x^n + a_1 x^{n-1} + \cdots + a_n$.

12. Let $f(x) = ax^2 + bx + c \in \mathbb{R}[x]$, $a \neq 0$. Give necessary and sufficient conditions in terms of the coefficients $a$, $b$, and $c$ for $f(x)$ to be irreducible over $\mathbb{R}$.

13. Let $f(x) \in F[x]$ be a nonconstant polynomial. Show, by induction on $\deg(f(x))$, that $f(x)$ can be written as a product of irreducible polynomials.

14. Let $p(x), q(x), r(x) \in F[x]$. Show that if $p(x) \mid q(x)r(x)$ and $(p(x), q(x)) = 1$, then $p(x) \mid r(x)$.

15. Show that if $p(x), q_i(x) \in F[x]$, $p(x)$ is irreducible and $p(x) \mid q_1(x) \cdots q_n(x)$, then $p(x) \mid q_i(x)$ for some $i$.

16. Use Exercises 13, 14, and 15 to complete the proof of Theorem 2.6.6.

17. Are the following polynomials reducible in $\mathbb{Q}[x]$?

    (i) $2x^4 - 6x^3 + 9x + 15$.
    (ii) $x^5 + 10x^2 + 50x + 20$.
    (iii) $x^3 - 2x + 6$.

18. Let $p$ be a prime.

    (i) Show that $x^p - 1 = (x - 1)(x^{p-1} + x^{p-2} + \cdots + x + 1)$.
    (ii) Show that $x^{p-1} + x^{p-2} + \cdots + x + 1$ is irreducible in $\mathbb{Z}[x]$. (Hint: Try substituting $x = y + 1$ in $\frac{x^p - 1}{x - 1}$.)

The next two exercises develop some properties of the *formal derivative* $f'(x)$ of a polynomial $f(x) = A_0 + a_1 x + a_2 x^2 + \cdots + a_n x^n \in F[x]$:

$$f'(x) = a_1 + 2a_2 x + 3a_3 x^2 + \cdots + na_n x^{n-1}.$$

19. Let $f(x) = \sum_{i=0}^{m} a_i x^i$, $g(x) = \sum_{j=0}^{n} b_j x^j \in F[x]$ and $a \in F$, where $F$ is a field. Establish the following:

    (i) $(af(x))' = af'(x)$.
    (ii) $(f(x) + g(x))' = f'(x) + g'(x)$.
    (iii) $(f(x)x^k)' = f'(x)x^k + kf(x)x^{k-1}$.
    (iv) $(f(x)g(x))' = f'(x)g(x) + f(x)g'(x)$.
    (v) $(g(x)^n)' = ng(x)^{n-1}g'(x)$      $(n \in \mathbb{N})$.
    (vi) $f(g(x))' = f'(g(x))g'(x)$.

20. Let $f(x) = \sum\limits_{i=0}^{n} a_i x^i \in F[x]$. Show that $(f(x), f'(x)) \neq 1$ if and only if there exists an irreducible polynomial $p(x)$ such that $p(x)^2$ divides $f(x)$.

21. Let $F$ be a subfield of a field $G$. Let $f(x), g(x) \in F[x]$ and $h(x) \in G[x]$ be such that $f(x) = g(x)h(x)$. Let

$$f(x) = \sum_{i=0}^{\ell} a_i x^i, \qquad g(x) = \sum_{i=0}^{m} b_i x^i, \qquad h(x) = \sum_{i=0}^{n} c_i x^i.$$

Show that $c_i \in F$ for all $i$—that is, $h(x) \in F[x]$.

## 2.7 Construction of Fields

In chapter 1 we saw how, for each prime $p$, we could construct a field $\mathbb{Z}_p$ from $\mathbb{Z}$ by using the congruence relation $\equiv_p$. We have also seen how $F[x]$, like $\mathbb{Z}$, is an integral domain and has polynomials (we called them *irreducible*) with properties resembling those of prime numbers. We are now going to copy within the context of $F[x]$ what we did with $\mathbb{Z}$ to construct the field $\mathbb{Z}_p$. We are now going to perform the same kind of construction on $F[x]$. Throughout this section, $F$ denotes an arbitrary field (which need not be finite).

Let $f(x) \in F[x]$ be a nonconstant polynomial. We can use $f(x)$ to define a relation $\equiv_{f(x)}$, or simply $\equiv$, on $F[x]$ as follows:

$$a(x) \equiv b(x)(\bmod f(x)) \iff f(x) \mid a(x) - b(x).$$

Equivalently,

$$a(x) \equiv b(x) \iff a(x) = b(x) + k(x)f(x) \ \text{ for some } \ k(x) \in F[x].$$

**Theorem 2.7.1**  (i) $\equiv$ *is an equivalence relation on $F[x]$.*

 (ii) *$a(x) \equiv c(x)$ and $b(x) \equiv d(x)$ then $a(x) + b(x) \equiv c(x) + d(x)$ and $a(x)b(x) \equiv c(x)d(x)$.*

 (iii) *The operations $+$ and $\cdot$ on the set $F[x]/f(x)$ of $\equiv$ classes defined by*

$$[a(x)] + [b(x)] = [a(x) + b(x)], [a(x)][b(x)] = [a(x)b(x)]$$

 *are well defined.*

 (iv) *With respect to these operations, $F[x]/f(x)$ is a commutative ring with 1.*

**Proof.**  The proof will be established in the exercises. The arguments are almost identical to those used in Theorem 1.5.3, Lemma 1.6.5, and Theorem 1.6.6, with the appropriate substitution of *polynomial* for *integer* and so on.  $\square$

We refer to the elements of $F[x]/f(x)$ as the *congruence classes* mod $f(x)$. Note that

$$[0] = \{k(x)f(x) \mid k(x) \in F[x]\}$$

is the zero of $F[x]/f(x)$ and that

$$[1] = \{1 + k(x)f(x) \mid k(x) \in F[x]\}$$

is the identity of $F[x]/f(x)$. Note that the set

$$\overline{F} = \{[a] \mid a \in F\}$$

of all the elements of $F[x]/f(x)$ of the form $[a]$ is a field that is clearly isomorphic to $F$ under the isomorphism $a \to [a]$. It is customary to identify $F$ and $\overline{F}$ by considering the elements $a$ and $[a]$ to be the "same". In this way we can consider any polynomial

$$g(x) = a_0 + a_1 x + \cdots + a_m x^m \in F[x]$$

also as a polynomial in $\overline{F}[x]$—namely,

$$g(x) = [a_0] + [a_1]x + \cdots + [a_m]x^m.$$

It is easy to see that $F[x]/f(x)$ will not always be a field. If we let $f(x) = (x-1)^2 \in \mathbb{Z}_p[x]$, then $x - 1$ is not congruent to 0 modulo $f(x)$ so that in $\mathbb{Z}_p[x]/f(x)$, we have

$$[x - 1] \neq [0]$$

but

$$[x - 1][x - 1] = [(x - 1)^2] = [f(x)] = [0].$$

Thus, $F[x]/f(x)$ is not an integral domain in this case.

In chapter 1 we saw how useful it was to have a nice set of representatives of the classes modulo $n$—namely $0, 1, 2, \ldots, n - 1$. Many calculations were greatly simplified by reducing to one set of representatives. It turns out that a similar device will help us analyze $F[x]/f(x)$.

**Theorem 2.7.2** *Let $F$ be a field, $f(x) \in F[x]$, $n = \deg(f(x)) \geq 1$, and $\equiv$ be as above. Let $g(x) \in F[x]$ and $g(x) = f(x)q(x) + r(x)$ where $\deg(r(x)) < \deg(f(x))$.*

    (i) $g(x) \equiv r(x)$.
    (ii) *Every $\equiv$ class contains a unique polynomial of degree less than $n$.*

(iii) *If $F$ is finite, then*

$$|F[x]/f(x)| = |F|^n.$$

**Proof.** (i) Let $g(x) \in F[x]$ and

$$g(x) = q(x)f(x) + r(x) \quad \text{where} \quad \deg(r(x)) < n.$$

Then

$$f(x) \mid g(x) - r(x)$$

and $g(x) \equiv r(x)$.

(ii) By the division algorithm there always exist polynomials $q(x), r(x)$ that satisfy the hypothesis of the theorem. Thus, every $\equiv$ class contains a polynomial of degree less than $n$. Now suppose that

$$g(x) \equiv h(x) \quad \text{where} \quad \deg(h(x)) < n.$$

Then, $h(x) \equiv r(x)$ so that $f(x) \mid (h(x) - r(x))$ whereas $\deg(h(x) - r(x)) < n$. This can only happen if $h(x) - r(x) = 0$. Therefore $h(x) = r(x)$ and each $\equiv$ class contains exactly one polynomial of degree less than $n$.

(iii) Now suppose that $F$ is finite. Every $\equiv$-class contains a unique polynomial of degree less than $n$. Hence,

$$|F[x]/f(x)| = \text{number of} \equiv \text{classes}$$
$$= |\{a_0 + a_1 x + \cdots + a_{n-1}x^{n-1} \mid a_i \in F\}|$$
$$= |F|^n$$

where the last equality follows from the fact that we have $|F|$ independent choices for each of the coefficients $a_i$. $\quad\square$

Just as in chapter 1 where we identified $\mathbb{Z}_n$ with $\{0, 1, 2, \ldots, n - 1\}$, we can now identify $F[x]/f(x)$ with the set $\{a_0 + a_1 x + \cdots + a_{n-1}x^{n-1} \mid a_i \in F\}$ of all polynomials over $F$ of degree, at most, $n - 1$.

**Example 2.7.3** Let $f(x) = x^2 + x + 1 \in \mathbb{Z}_2[x]$. By Theorem 2.7.2, we have

$$\mathbb{Z}_2[x]/(x^2 + x + 1) = \{0, 1, x, 1 + x\}$$

where the elements add and multiply as for elements of $\mathbb{Z}_2[x]$, except that we use the equality $x^2 = x + 1$ to reduce any polynomial of degree 2 or more to a linear polynomial. Thus,

$$x(1 + x) = x + x^2 = x + x + 1 = 1$$

for example, so that $x$ and $x + 1$ are both invertible elements. Since 1 is also an invertible element, it follows that $\mathbb{Z}_2[x]/(x^2 + x + 1)$ is a field with 4 elements. In the same way,

$$\mathbb{Z}_2[x]/(x^2 + 1) = \{0, 1, x, 1 + x\}$$

but in this case

$$(1 + x)^2 = 1 + 2x + x^2 = 1 + x^2 = 1 + 1 = 0$$

so that $\mathbb{Z}_2[x]/(x^2 + 1)$ has zero divisors and is not an integral domain and therefore is not a field. The next result characterizes the circumstances in which we will obtain a field.

**Theorem 2.7.4** *Let $f(x)$, $F$, and $\equiv$ be as above and $G = F[x]/f(x)$.*

    (i)  *$G$ is a field if and only if $f(x)$ is irreducible in $F[x]$.*
    (ii)  *$G$ contains a root of $f(x)$—namely, $[x]$.*

**Proof.** By Theorem 2.7.1, $G$ is a commutative ring with 1.
    (i) Let $G$ be a field. Suppose $f(x)$ is reducible—say,

$$f(x) = g(x)h(x) \quad \text{where} \quad g(x), h(x) \quad \text{are not constants.}$$

Then $\deg(g(x)) < \deg(f(x))$ and $\deg(h(x)) < \deg(f(x))$ so that $[g(x)] \neq 0 \neq [h(x)]$ but

$$[g(x)][h(x)] = [f(x)] = [0].$$

Since a field has no zero divisors, this is a contradiction. Hence, $f(x)$ is irreducible.

    Now suppose that $f(x)$ is irreducible. Since we know that $G$ is a commutative ring with identity, it only remains to show that every nonzero element in $G$ has an inverse. Let $g(x) \in F[x]$, $[g(x)] \neq 0$. Then $f(x)$ does not divide $g(x)$. Consequently, $(f(x), g(x)) = 1$ so that by Theorem 2.4.4 there exist $\lambda(x), \mu(x) \in F[x]$ with

$$\lambda(x) f(x) + \mu(x) g(x) = 1.$$

Hence, $\mu(x)g(x) \equiv 1$ or $[\mu(x)][g(x)] = [1]$ and $[g(x)]$ has an inverse—namely, $[\mu(x)]$. Therefore, $G$ is a field.

(ii) Let $f(x) = a_0 + a_1 x + \cdots + a_n x^n$ and consider the element $[x]$ in $G$. Then, substituting $[x]$ into the polynomial $f(x)$, we obtain

$$
\begin{aligned}
f([x]) &= [a_0] + [a_1][x] + [a_2][x]^2 + \cdots + [a_n][x]^n \\
&= [a_0] + [a_1 x] + [a_2 x^2] + \cdots + [a_n x^n] \\
&= [a_0 + a_1 x + a_2 x^2 + \cdots + a_n x^n] \\
&= [f(x)] \\
&= [0].
\end{aligned}
$$

Thus, $[x]$ is a root of the polynomial $f(x)$ in $G$.  $\square$

Let $f(x) \in F[x]$. Then $f(x)$ has an irreducible factor, say $g(x)$. We have seen in Theorem 2.7.4 that there is then a field $G$ containing $F$ which contains a root of $g(x)$, which will necessarily also be a root of $f(x)$. Let us call that root $a$ (in Theorem 2.7.4, $a = [x]$). By Lemma 2.5.1(i), $f(x)$ then has a linear factor $x - a$. We can repeat that process to obtain a complete factorization of $f(x)$ into a product of linear factors.

**Theorem 2.7.5** *Let $f(x) \in F[x]$. Then there exists a field $K$ containing $F$ as a subfield such that $f(x)$ factors into a product of linear factors in $K[x]$.*

**Proof.** We argue by induction on the degree of $f(x)$. If the degree of $f(x)$ is one, then $f(x)$ is a linear polynomial and the claim holds. So suppose that the degree of $f(x)$ is $n$ and that the claim holds for all fields and for all polynomials of degree less than $n$. First assume that $f(x)$ is irreducible. By Theorem 2.7.4(i), there exists a field $G$ containing $F$ as a subfield and containing a root, $a$ say, of $f(x)$. Then, by Lemma 2.5.1(i), we can factor $f(x)$ in $G[x]$ as $f(x) = (x - a)g(x)$, for some polynomial $g(x)$ in $G[x]$. Since the degree of $g(x)$ is $n - 1$, it follows from the induction hypothesis that there exists a field $K$ containing the field $G$ as a subfield such that $g(x)$, and therefore also $f(x)$, factors into a product of linear factors in $K[x]$. Thus the claim holds in this case.

Now let us assume that $f(x)$ is reducible as $f(x) = g(x)h(x)$, where both $g(x)$ and $h(x)$ have smaller degrees than $f(x)$. Then, by the induction hypothesis, there exists a field $G$ containing $F$ such that $g(x)$ factors into a product of linear factors in $G[x]$. Again by the induction hypothesis, there exists a field $K$ containing $G$ such that $h(x)$ factors into linear factors in $K[x]$. Thus both $g(x)$ and $h(x)$, and therefore also $f(x)$, factor into linear factors in $K[x]$ and the claim holds.  $\square$

We are now in a position to establish the existence of fields of size $p^n$.

**Theorem 2.7.6** *For every prime p and every positive integer n, there exists a unique field with $p^n$ elements.*

**Proof.** Let $f(x) = x^{p^n} - x \in \mathbb{Z}_p[x]$. By Theorem 2.7.5, there exists a field $G$, say, such that $f(x)$ factors into a product of linear factors in $G[x]$:

$$f(x) = (x - a_1)(x - a_2) \cdots (x - a_{p^n})$$

where $a_1, a_2, \ldots, a_n \in G$. If there existed a repeated root, then (see Exercise 2.6.20) $f(x)$ and $f'(x)$ would have a nonconstant common factor. But

$$f'(x) = p^n x^{p^n - 1} - 1 = -1.$$

So that is impossible and $f(x)$ must have $p^n$ distinct roots. By Lemma 2.2.7, these roots form a subfield of $G$ with $p^n$ elements. This establishes the existence of a field with $p^n$ elements.

The assertion that this field is "unique" simply means that any two fields with $p^n$ elements will be isomorphic. From Corollary 2.3.9, we know that any two fields of the same size are isomorphic as vector spaces and therefore have the isomorphic additive structures. From Theorem 2.10.6 (to come) we will see that they have isomorphic multiplicative structures. But the real challenge is to show that there is an isomorphism that simultaneously respects both the additive and multiplicative structure. We will see how to achieve this in Section 2.13. $\square$

Finite fields are often referred to as *Galois fields* after the brilliant young French mathematician Évariste Galois in recognition of his pioneering work in this area. Galois died tragically in a duel at the age of 20. Since there is a unique field (to within isomorphism) of size $p^n$ for each prime $p$ and positive integer $n$, we can refer to *the* Galois field with $p^n$ elements. We will denote it by $\mathrm{GF}(p^n)$.

**Exercises 2.7** In Exercises 1 through 3, let $F$ be a field and $f(x) \in F[x]$. Let $\equiv$ be the relation on $F[x]$ defined as at the beginning of this section.

1. Show that $\equiv$ is an equivalence relation in $F[x]$.

2. Show that if $a(x) \equiv c(x)$ and $b(x) \equiv d(x)$, then
   $a(x) + b(x) \equiv c(x) + d(x)$ and $a(x)b(x) \equiv c(x)d(x)$.

3. Show that $F[x]/f(x)$ is a commutative ring with identity with respect to the operations

   $$[a(x)] + [b(x)] = [a(x) + b(x)], [a(x)][b(x)] = [a(x)b(x)].$$

4. Find all elements in $\mathbb{Z}_5[x]/(x^2 + 1)$ that are *idempotent*—that is, such that $[a(x)]^2 = [a(x)]$.

5. Find all the invertible elements in $\mathbb{Z}_5[x]/(x^2 + 1)$. (Hint: Consider the arbitrary element $[ax + b]$ and the cases (i) $a \neq 0, b = 0$, (ii) $a = 0,\ b \neq 0$, (iii) $a, b \neq 0$.)

6. Which of the following are fields?

   (i) $\mathbb{Z}_5[x]/(x^2 + x + 1)$.
   (ii) $\mathbb{Z}_5[x]/(x^2 + 2x + 3)$.
   (iii) $\mathbb{Z}_5[x]/(x^2 + x + 4)$.

7. For which of the following numbers are there fields of that size?

   (i) 11.
   (ii) 22.
   (iii) 27.
   (iv) 32.
   (v) 48.
   (vi) 121.

8. What is the characteristic of the following fields?

   (i) GF(16).
   (ii) GF(49).
   (iii) GF(81).
   (iv) GF(625).

9. Which of the following pairs of rings/fields are isomorphic?

   (i) $\mathbb{Z}_5, \text{GF}(5)$.
   (ii) $\mathbb{Z}_8, \text{GF}(8)$.
   (iii) $\mathbb{Z}_{17}, \text{GF}(17)$.
   (iv) $\mathbb{Z}_{27}, \text{GF}(27)$.

*10. Let $F$ be a subfield of a field $G$, $m(x) \in F[x]$ be an irreducible monic polynomial of degree $n$, and let $\alpha \in G$ be a root of $m(x)$.

   (i) Show that

$$F(\alpha) = \{a_0 + a_1\alpha + \cdots + a_{n-1}\alpha^{n-1} \mid a_i \in F\}$$

   is a subfield of $G$. (Hint for finding inverses: If $f(x) = a_0 + a_1\alpha + \cdots + a_{n-1}\alpha^{n-1} \in F(\alpha)$, then the greatest common divisor of $a_0 + a_1 x + \cdots a_{n-1}x^{n-1}$ and $m(x)$ is 1.)
   (ii) By taking $F = \mathbb{Q}$, $G = \mathbb{C}$, $m(x) = x^3 - 2$, show that it is possible to have $F(\alpha) \neq F(\beta)$ for distinct roots $\alpha, \beta$ of $m(x)$.

11. Let $F$ be a field.

   (i) Formulate a theorem for polynomials in $F[x]$ modeled on the Chinese remainder theorem for integers.
   (ii) Prove the theorem from part (i).

## 2.8 Extension Fields

If $F$ is a subfield of a field $G$ then we also say that $G$ is an *extension field* of $F$. In the previous section we saw one method of constructing extension fields by means of the construction $F[x]/f(x)$. We will now consider some examples, focusing in particular on extensions of $\mathbb{Z}_p$.

In any discussions involving polynomials to date, we have used the variable $x$. Returning to our original definition of a polynomial (as an infinite sequence with only a finite number of nonzero terms) we saw that the "variable" $x$ really stands for the sequence

$$(0, 1, 0, 0, 0, \ldots).$$

Clearly, any symbol other than $x$—for example, $u$, $v$, $w$, $y$, or $z$—would serve our purpose equally well. To avoid confusion as we begin to construct new fields and then consider polynomials over these new fields, we will adopt other variables such as $u, v, w, y, z$ as the primary variable to be used in the construction process and will refer to whichever variable is used as the *construction variable*.

As a first illustration we begin with a familiar example using an unfamiliar variable. Let us work with the variable $i$. Then the polynomial $i^2 + 1 \in \mathbb{R}[i]$. Since the square of any real number is either positive or zero, it is clear that there is no real number $i$ with $i^2 = -1$. Thus, $i^2 + 1$ is irreducible over the field $\mathbb{R}$. We can therefore construct the field $\mathbb{R}[i]/(i^2 + 1)$. As mentioned in the previous section, we may identify the elements of this field with the polynomials in $i$ of, degree less than 2, which is the degree of $i^2 + 1$. Thus,

$$\mathbb{R}[i]/(i^2 + 1) = \{a + bi | a, b \in \mathbb{R}\}.$$

In this field, $i^2 = -1$ so that $i$ is a root of $x^2 + 1$. Of course, what we have done is simply to reconstruct the complex numbers using the theory of the previous section.

**Example 2.8.1** Suppose we wish to construct a field with $9 = 3^2$ elements. Then we are looking for a field with a finite characteristic that must be 3. If we start with $\mathbb{Z}_3$ and can find an irreducible polynomial $f(y) \in \mathbb{Z}_3[y]$ of degree 2, then $F = \mathbb{Z}_3[y]/f(y)$ will have $3^2 = 9$ elements.

From section 2.6, we know that $f(y) = 1 + y^2$ is an irreducible polynomial in $\mathbb{Z}_3[y]$. By Theorem 2.7.2 and the remarks following it, we can identify the elements of $F$ with the polynomials in $y$ of degree less than 2:

$$F = \{0, 1, 2, y, 1 + y, 2 + y, 2y, 1 + 2y, 2 + 2y\}.$$

Of course, just listing the elements of $F$ does not tell us anything until we describe the addition and multiplication. Addition is almost trivial. We add the elements as polynomials and then reduce the coefficients modulo 3:

$$1 + (2 + y) = 3 + y = 0 + y = y$$
$$(2 + y) + (1 + 2y) = 3 + 3y = 0$$
$$(1 + y) + (1 + 2y) = 2 + 3y = 2$$

and so forth.

To compute products in $F$ we need to note that the zero in $F = \mathbb{Z}_3[y]/f(y)$ is the class

$$[0] = [f(y)].$$

In other words, in $F$,

$$1 + y^2 = 0$$

so that $y^2 = -1$ or, equivalently,

$$y^2 = 2.$$

Thus, to compute products we simply combine the elements as polynomials and then replace every occurrence of $y^2$ by 2:

$$(1 + y)(1 + 2y) = 1 + 3y + 2y^2 = 1 + 2y^2 = 1 + 2 \cdot 2 = 5 = 2$$
$$y(2 + 2y) = 2y + 2y^2 = 2y + 2 \cdot 2 = 4 + 2y = 1 + 2y.$$

A complete list of all products in $F$ can be given in a table like the following, in which we have left the completion of the table as an exercise.

MULTIPLICATION TABLE FOR GF(9)

|         | 0 | 1     | 2     | $y$  | $1+y$  | $2+y$  | $2y$ | $1+2y$ | $2+2y$ |
|---------|---|-------|-------|------|--------|--------|------|--------|--------|
| 0       | 0 | 0     | 0     | 0    | 0      | 0      | 0    | 0      | 0      |
| 1       | 0 | 1     | 2     | $y$  | $1+y$  | $2+y$  | $2y$ | $1+2y$ | $2+2y$ |
| 2       | 0 | 2     | 1     | $2y$ | $2+2y$ | $1+2y$ | $y$  | $2+y$  | $1+y$  |
| $y$     | 0 | $y$   | $2y$  |      |        |        |      |        |        |
| $1+y$   | 0 | $1+y$ | $2+2y$ |     |        |        |      |        |        |
| $2+y$   | 0 | $2+y$ | $1+2y$ |     |        |        |      |        |        |
| $2y$    | 0 | $2y$  | $y$   | 4    | $1+2y$ |        |      |        |        |
| $1+2y$  | 0 | $1+2y$ | $2+y$ |     |        |        |      |        |        |
| $2+2y$  | 0 | $2+2y$ | $1+y$ |     |        |        |      |        |        |

**Remark 2.8.2**

   (i) Since $|F| = 9$, we can write $F = \mathrm{GF}(9)$.

   (ii) The constants 0, 1, and 2 constitute a subfield of $F$ isomorphic to $\mathbb{Z}_3$. Thus, $F$ is an "extension" of $\mathbb{Z}_3$.

  (iii) We have seen that the polynomial $f(x) = 1 + x^2$ has no root in $\mathbb{Z}_3$. However, it does have a root in $\mathrm{GF}(9)$ since, if we substitute the element $y$ in $\mathrm{GF}(9)$ for $x$, then we obtain zero:

$$f(y) = 1 + y^2 = 0.$$

  (iv) We have also seen that the polynomial $g(x) = 1 + x^2$ is irreducible over $\mathbb{Z}_3$. This is no longer the case if we consider $g(x)$ as a polynomial over $\mathrm{GF}(9)$ since

$$(x + y)(x - y) = x^2 - y^2 = x^2 - 2 = x^2 + 1.$$

Thus, by extending the field over which we are working we can find new roots and factorizations.

The idea of a polynomial having no root in one field yet having a root in a larger field is a familiar one. For example, the polynomial $x^2 - 2$ has no root in the field $\mathbb{Q}$ of rational numbers and is irreducible in $\mathbb{Q}[x]$, but does have a root $\sqrt{2} \in \mathbb{R}$ and factors in $\mathbb{R}[x]$ as $(x - \sqrt{2})(x + \sqrt{2})$.

**Example 2.8.3** We have choices in how we construct $\mathrm{GF}(p^n)$. For example,

$$x^2 + 1, x^2 + x + 2, x^2 + 2x + 2$$

are all irreducible over $\mathbb{Z}_3$. Any one of these can be used to construct a copy of $\mathrm{GF}(9)$. In our first example we used $x^2 + 1$. Whichever polynomial we use, we will get a field with 9 elements. By Theorem 2.7.6, we must obtain the "same" field in each case. Let us now construct $\mathrm{GF}(9)$ using $x^2 + x + 2$. To avoid confusion between the two examples, we will use $u$ as the construction variable in this example. The elements are then all linear polynomials

$$a + bu \qquad a, b \in \mathbb{Z}_3.$$

Addition is as before. For multiplication we use the relation $u^2 = 2u + 1$:

| $\cdot$ | 0 | 1 | 2 | $u$ | $2u$ | $1+u$ | $2+u$ | $1+2u$ | $2+2u$ |
|---|---|---|---|---|---|---|---|---|---|
| 0 | 0 | 0 | 0 | 0 | 0 | 0 | 0 | 0 | 0 |
| 1 | 0 | 1 | 2 | $u$ | $2u$ | $1+u$ | $2+u$ | $1+2u$ | $2+2u$ |
| 2 | 0 | 2 | | | | | | | |
| $u$ | 0 | $u$ | $2u$ | $1+2u$ | $2+u$ | 1 | $1+u$ | $2+2u$ | 2 |
| $2u$ | 0 | $2u$ | | | | | | | |
| $1+u$ | 0 | $1+u$ | | | | | | | |
| $2+u$ | 0 | $2+u$ | | | | | | | |
| $1+2u$ | 0 | $1+2u$ | | | | | | | |
| $2+2u$ | 0 | $2+2u$ | | | | | | | |

This table "looks" different from the previous table since we now have

$$u^2 = 1 + 2u$$

whereas (using $x^2 + 1$) we previously had

$$y^2 = 2.$$

To demonstrate that

$$\mathbb{Z}_3[u]/(u^2 + u + 2) \qquad \text{and} \qquad \mathbb{Z}_3[y]/(y^2 + 1)$$

are really the "same" field, we must produce an isomorphism from one to the other.

It is not immediately obvious how to do this, so we will return to this question in section 2.13.

**Example 2.8.4** By the method of Lemma 2.6.2, we can verify that the polynomial $f(y) = 1 + y + y^3 \in \mathbb{Z}_5[y]$ is irreducible over $\mathbb{Z}_5$. We can therefore use $f(y)$ to construct the field $\mathbb{Z}_5[y]/f(y)$. This field has $5^3 = 125$ elements and is therefore GF(125). Obtained in this way, the elements of GF(125) are the polynomials of the form

$$a_0 + a_1 y + a_2 y^2 \qquad \text{where} \qquad a_0, a_1, a_2 \in \mathbb{Z}_5.$$

As before, addition is performed by combining coefficients as elements of $\mathbb{Z}_5$. For example,

$$(1 + 4y + 2y^2) + (3 + 3y + 4y^2) = 4 + 7y + 6y^6$$
$$= 4 + 2y + y^2.$$

To calculate products, we take advantage of the fact that

$$[0] = [1 + y + y^3]$$

so that $y^3 = -1 - y = 4 + 4y$. Hence, for example,

$$\begin{aligned}
(1 + 2y^2)(3y + 4y^2) &= 3y + 4y^2 + 6y^3 + 8y^4 \\
&= 3y + 4y^2 + (4 + 4y) + 3y(4 + 4y) \\
&= 3y + 4y^2 + 4 + 4y + 12y + 12y^2 \\
&= 4 + 4y + y^2.
\end{aligned}$$

**Remark 2.8.5**

(i) The constants $0, 1, 2, 3, 4$ form a subfield of GF(125) isomorphic to $\mathbb{Z}_5$. Thus, $F$ is an extension of $\mathbb{Z}_5$.

(ii) The polynomial $f(x) = 1 + x + x^3$ has no root in $\mathbb{Z}_5$. However,

$$f(y) = 1 + y + y^3 = 0$$

in GF(125). Thus, $f(x)$ does have a root in this extension field of $\mathbb{Z}_5$.

(iii) The polynomial $f(x) = 1 + x + x^3$ is irreducible over $\mathbb{Z}_5$. However, since $y$ is a root of $f(x)$ as a polynomial over GF(125), $f(x)$ must have a factor $x - y$. Thus we can divide $f(x)$ by $x - y$:

$$
\require{enclose}
\begin{array}{r}
x^2 \;\;+yx \qquad +(1 + y^2) \\[2pt]
\hline
x - y\,\overline{\smash{)}\,x^3 + 0.x^2 + x + 1} \\[2pt]
\underline{x^3 - yx^2} \qquad\qquad\quad \\[2pt]
yx^2 \;\;+x \qquad\quad \\[2pt]
\underline{yx^2 \;\;-y^2 x} \qquad\quad \\[2pt]
(1 + y^2)x + 1 \\[2pt]
\underline{(1 + y^2)x - y(1 + y^2)} \\[2pt]
1 + y + y^3 = 0
\end{array}
$$

Thus, as polynomials over GF(125), we have

$$x^3 + x + 1 = (x - y)(x^2 + yx + (1 + y^2)).$$

**Example 2.8.6** Consider $f(z) = z^2 + 2 \in \mathbb{Q}[z]$. Since $z^2 + 2$ has no zeros in $\mathbb{Q}$ (see Exercise 13 in section 2.3), it is irreducible in $\mathbb{Q}[z]$. Let $F = \mathbb{Q}[z]/f(z)$. As before, we can identify $F$ with the set of polynomials of degree less than 2, which is the degree of $f(z)$:

$$F = \{a + bz \mid a, b \in \mathbb{Q}\}.$$

Addition in $F$ is just the usual addition of polynomials. To calculate products, we note that

$$[0] = [z^2 + 2]$$

so that, in $F$, $z^2 = -2$. Hence,

$$(a + bz)(c + dz) = ac + (ad + bc)z + bdz^2$$
$$= (ac - 2bd) + (ad + bc)z.$$

Thus, $z$ behaves just like $\sqrt{-2}$ and we can think of $F$ as

$$\{a + b\sqrt{-2} \mid a, b \in \mathbb{Q}\}.$$

This field is often denoted by $\mathbb{Q}(\sqrt{-2})$.

**Remark 2.8.7**

    (i)  The constants $\{a \mid a \in \mathbb{Q}\}$ form a subfield isomorphic to $\mathbb{Q}$.

   (ii)  The polynomial $f(x) = x^2 + 2$ has no root in $\mathbb{Q}$. However,

$$f(z) = z^2 + 2 = -2 + 2 = 0.$$

        Thus $f(x)$ does have a root in $F$.

  (iii)  The polynomial $f(x) = x^2 + 2$ is irreducible in $\mathbb{Q}[x]$. However, in $F[x]$ we have

$$x^2 + 2 = (x + z)(x - z).$$

We have seen in Theorem 2.7.5 how, for any field F and polynomial $f(x)$ in $F[x]$, there must exist a larger field over which $f(x)$ factors completely into linear factors. This leads naturally to the question of the existence of fields over which every polynomial factors into linear factors. We already know one field with this property.

**Theorem 2.8.8 (The Fundamental Theorem of Algebra)** *Let $f(x) \in \mathbb{C}[x]$ be a monic polynomial and* $\deg f(x)) = n$*. Then there exist $\alpha_1, \alpha_2, \ldots, \alpha_n \in \mathbb{C}$ with*

$$f(x) = (x - \alpha_1)(x - \alpha_2) \cdots (x - \alpha_n).$$

**Proof.** In order to prove this, it is necessary to have a precise description of the real numbers. It is also necessary to employ some deep analysis. See Brown and Churchill [BC], page 166.   $\square$

Let $F$ be a field such that every $f(x) \in F[x]$ factors into a product of linear factors over $F$. This is equivalent to saying that $F$ contains a complete set of roots for $f(x)$. Then $F$ is said to be *algebraically closed*. A field $K$ containing $F$ as a subfield is said to be an *algebraic closure* of $F$ if $K$ is algebraically closed and every element of $K$ is a root of some polynomial in $F[x]$. If $K_1$ and $K_2$ are both algebraic closures of $F$, then $K_1$ and $K_2$ are isomorphic. Thus we may speak of *the* algebraic closure of $F$. By Theorem 2.8.8, $\mathbb{C}$ is clearly the algebraic closure of $\mathbb{R}$. In fact, algebraically closed fields are reasonably plentiful.

**Theorem 2.8.9** *Every finite field $F$ has an algebraic closure and for any two algebraic closures $K_1, K_2$ of $F$ there exists an isomorphism from $K_1$ to $K_2$ that restricts to the identity mapping on $F$.*

**Proof.** For a proof of this and even more general results concerning algebraic closures, see [McCa], Chapter 1. $\quad\square$

In particular, it follows from Theorem 2.8.9 that each of the fields $Z_p$ is contained in an algebraically closed field that, since it shares the same identity with $Z_p$, will be of characteristic $p$ and must be infinite (see the exercises).

Since $\mathbb{C}$ is algebraically closed, every polynomial of the form $x^n - z (z \in \mathbb{C})$ has roots in $\mathbb{C}$. We conclude this section with some observations concerning the roots of complex numbers. The polar representation of complex numbers is particularly helpful here in providing a simple way to present the $n$th roots of a complex number. Let $z = r(\cos\theta + i\sin\theta) \in \mathbb{C}$, and suppose that we wish to describe the roots of $x^n - z$, that is, we wish to describe the $n$th roots of $z$. Let $x = s(\cos\alpha + i\sin\alpha)$ be a solution. Then

$$r\left(\cos\theta + i\sin\theta\right) = x^n = s^n\left(\cos n\alpha + i\sin n\alpha\right).$$

Hence,

$$s^n = r \quad \text{and} \quad n\alpha = \theta + 2k\pi$$

or

$$s^n = r^{1/n} \quad \text{and} \quad \alpha = \frac{1}{n}\theta + \frac{2k\pi}{n}.$$

By taking $k = 0, 1, 2, \ldots, n-1$, we obtain $n$ distinct solutions (corresponding to $n$ points equally spaced around a circle of radius $r^{1/n}$). But we know from the general theory that a polynomial of degree $n$ can have at most $n$ distinct roots. Hence, the distinct $n$th roots of $z = r(\cos\theta + i\sin\theta)$ are

$$x_i = r^{1/n}\left(\cos\left(\frac{\theta + 2k\pi}{n}\right) + i\sin\left(\frac{\theta + 2k\pi}{n}\right)\right) \qquad 0 \le k \le n-1.$$

Thus we have the following proposition.

**Proposition 2.8.10** *Let $\alpha \in \mathbb{C}, \alpha \neq 0$, and $n \in \mathbb{N}$. Then the equation $x^n = \alpha$ has $n$ distinct solutions in $\mathbb{C}$.* $\quad\square$

In particular, from Proposition 2.8.10, we see that every nonzero complex number has two distinct sqare roots, three distinct cube roots, four distinct quartic roots and so on. The roots of $x^n - 1 \in \mathbb{C}[x]$ are called the $n$th *roots of unity*.

For any $n$th root of unity $\alpha$, it is evident that every power of $\alpha$ is also an $n$th root of unity. From the discussion preceding the Proposition, we know that $x^n - 1$ has $n$ distinct roots. If $\alpha$ is such that the powers $1, \alpha, \alpha^2, \ldots, \alpha^{n-1}$ are all distinct, then these must be all the $n$th roots of unity and, when that is the case, $\alpha$ is called a *primitive $n$th root of unity*. We can then write

$$x^n - 1 = \prod_{0 \leq k < n} (x - \alpha^k).$$

Let

$$\alpha = \cos\left(\frac{2\pi}{n}\right) + i\sin\left(\frac{2\pi}{n}\right).$$

Then $\alpha$ is clearly a primitive $n$th root of unity. Moreover, for every positive integer $k$, we have

$$\alpha^k = \cos\left(\frac{2k\pi}{n}\right) + i\sin\left(\frac{2k\pi}{n}\right)$$

so that $\alpha^k$ is a primitive $n$ root of unity if and only if $k$ and $n$ are relatively prime. Thus the number of primitive $n$th roots of unity is simply $\varphi(n)$. The polynomial

$$\Phi_n(x) = \prod_{1 \leq k < n, (k,n)=1} (x - \alpha^k)$$

where the factors run over all the primitive $n$th roots of unity, has degree $\varphi(n)$ and is known as the $n$th *cyclotomic polynomial*.

More generally, let $k$ be an integer with $0 \leq k \leq n - 1$ and let $d = (k, n)$. Then

$$(\alpha^k)^{n/d} = (\alpha^n)^{k/d} = 1$$

Hence, $\alpha^k$ is a root of $x^{n/d} - 1$ and the powers

$$1 = (\alpha^k)^0, (\alpha^k)^1 (\alpha^k)^2 \ldots (\alpha^k)^{\frac{n}{d}-1}$$

are the $\frac{n}{d}$ distinct roots of $x^{n/d} - 1$. In other words, $\alpha^k$ is a primitive $\frac{n}{d}$th root of unity. Thus, in the factorization of $x^n - 1$ above, if we group together the factors involving the primitive $\frac{n}{d}$th roots of unity, as we let d range over the divisors of $n$, then we obtain:

$$x^n - 1 = \prod_{d|n} \Phi_{n/d}(x) = \prod_{d|n} \Phi_d(x)$$

since, as $d$ runs through all possible divisors of $n$, so also does $n/d$. Here are some examples:

$$\Phi_1(x) = x - 1$$
$$\Phi_2(x) = x + 1$$
$$\Phi_4(x) = (x - i)(x + i) = x^2 + 1.$$

while
$$x^4 - 1 = \Phi_1(x)\Phi_2(x)\Phi_4(x) = (x - 1)(x + 1)(x^2 + 1)$$

Note that each of $\Phi_1(x)$, $\Phi_2(x)$ and $\Phi_4(x)$ is a polynomial in $\mathbb{Z}[x]$, that is, they all have integer coefficients. This might look like just a coincidence due to the fact that the integers 1, 2, and 4 are so small. However, it turns out that this is always the case.

**Theorem 2.8.11** *For all positive integers $n$, $\Phi_n(x)$ is a polynomial with integer coefficients.*

**Proof.** We argue by induction. We have already seen that the claim is valid for $n = 1$, so suppose that the claim holds for all $\Phi_m(x)$ for $m < n$. We have

$$\Phi_n(x) = \frac{x^n - 1}{\prod_{d|n, d<n} \Phi_d(x)}.$$

A simple application of the division algorithm will now yield the desired result. By the induction assumption, each polynomial factor in the denominator has integer coefficients. Let the proper divisors of $n$ be $d_1 = 1, d_2, \ldots, d_k$. Then we can simply apply the division algorithm successively to $x^n - 1$, $(x^n - 1)/\Phi_{d_1}(x), \ldots$ with successive divisors equal to $\Phi_{d_1}(x), \Phi_{d_2}(x) \ldots$ At each step, we are dividing one monic polynomial with integer coefficients by another monic polynomial with integer coefficients and so, at each step, we must obtain a monic polynomial with integer coefficients. Therefore, after all the divisions have been completed, we are left with a monic polynomial with integer coefficients, namely $\Phi_n(x)$. $\square$

The cyclotomic polynomials are also irreducible in $\mathbb{Z}[x]$, but a proof of this fact is beyond our reach at this point.

**Exercises 2.8** In Exercises 1 through 5, let GF(9) be constructed as in Example 2.8.1

1. Find a basis for GF(9) as a vector space over $\mathbb{Z}_3$. What is the dimension of GF(9) over $\mathbb{Z}_3$?

2. Show that every element of GF(9) is a root of the polynomial $x^9 - x \in \mathbb{Z}_3[x]$.

3. Factorize $x^9 - x \in \mathbb{Z}_3[x]$ into a product of irreducible polynomials. Determine the roots in GF(9) of each irreducible factor.

4. Factorize $x^4 - 1 \in \mathrm{GF}(9)[x]$ into a product of irreducibles (in $\mathrm{GF}(9)[x]$).

5. Show that the polynomial $x^3 - x + 1$ is irreducible over GF(9).

6. Establish the following:

   (i) $\mathbb{Z}_3[y]/(1 + y^2)$ contains a root of $x^2 + x + 2$.
   (ii) $\mathbb{Z}_3[y]/(2 + y + y^2)$ contains a root of $x^2 + 1$.

7. What are the dimensions of the following vector spaces?

   (i) GF(64) over GF(4).
   (ii) GF(81) over $\mathbb{Z}_3$.

8. Construct GF(4). (Describe the elements and describe the operations of addition and multiplication.)

9. Construct GF(8). (Describe the elements and describe the operations of addition and multiplication.)

10. Use the irreducible polynomial $1 + x + x^4 \in \mathbb{Z}_2[x]$ to construct GF(16). (Describe the elements and describe the operations of addition and multiplication.)

11. Construct GF(27). (Describe the elements and the operations of addition and multiplication.)

12. Show that an algebraically closed field must be infinite.

*13. Does $\mathbb{Z}_8$ contain a subring isomorphic to GF(4)?

14. A subring $I$ of a ring $R$ is called an *ideal* if it satisfies the additional condition

$$a \in I, \ x \in R \implies ax, xa \in I.$$

For an ideal $I$ of a ring $R$, define a relation $\equiv$ on $R$ by

$$a \equiv b \iff a - b \in I.$$

Establish the following:

   (i) $\equiv$ is an equivalence relation on $R$.
   (ii) If $a \equiv b$ and $c \equiv d$, then $a + c \equiv b + d$ and $ac \equiv bd$.

Now let the set of $\equiv$ classes be denoted by $R/I$ and show

(iii)  $R/I$ is a ring.

## 2.9 Multiplicative Structure of Finite Fields

From the examples of the previous section, we have seen how to calculate products in any finite field (provided that we know the base field and the irreducible polynomial used to construct the finite field). We will see that, viewed from the correct perspective, the multiplicative structure of a finite field is extremely simple. We must first develop some preliminary ideas.

**Lemma 2.9.1**  *Let $p$ be a prime number, $n \in \mathbb{N}$, and let $\alpha \in \mathrm{GF}(p^n)$.*

(i)  *If $\alpha \neq 0$, then $\alpha^{p^n-1} = 1$.*
(ii)  *$\alpha$ is a root of $x^{p^n} - x$.*

**Proof.**  The technique used here is the same as we used in Euler's Theorem.

(i) Let the distinct nonzero elements of $\mathrm{GF}(p^n)$ be $a_1, a_2, \ldots, a_{p^n-1}$. By the cancellation law (since $\alpha \neq 0$), we find that the elements

$$\alpha a_1, \alpha a_2, \ldots, \alpha a_{p^n-1}$$

are all distinct and nonzero. However, there are only $p^n - 1$ nonzero elements in $\mathrm{GF}(p^n)$. Hence,

$$\{a_1, a_2, \ldots, a_{p^n-1}\} = \{\alpha a_1, \alpha a_2, \ldots, \alpha a_{p^n-1}\}$$

so that

$$a_1 a_2 \cdots a_{p^n-1} = (\alpha a_1)(\alpha a_2) \cdots (\alpha a_{p^n-1})$$

$$= a_1 a_2 \cdots a_{p^n-1} \alpha^{p^n-1}.$$

Multiplying by $(a_1 a_2 \cdots a_{p^n-1})^{-1}$ we find that

$$\alpha^{p^n-1} = 1$$

as required.

(ii) If $\alpha \neq 0$, then $\alpha^{p^n-1} = 1$ by part (i). Hence $\alpha^{p^n} = \alpha$. If $\alpha = 0$, then trivially $\alpha^{p^n} = \alpha$. Thus, in all cases, $\alpha$ is a root of $x^{p^n} - x$.  $\square$

Let $\alpha \in \mathrm{GF}(p^n)^*$. Then we see from Lemma 2.9.1 that there exists $t \in \mathbb{N}$ with $\alpha^t = 1$—namely, $t = p^n - 1$. Hence we can define the *order* of $\alpha$, written ord $(\alpha)$, to be the smallest positive integer $t$ such that $\alpha^t = 1$. Some basic properties of ord $(\alpha)$ are provided next.

**Lemma 2.9.2** *Let $\alpha \in \mathrm{GF}(p^n)^*$ have order $t$.*

(i) $1, \alpha, \alpha^2, \ldots, \alpha^{t-1}$ *are distinct.*
(ii) $\alpha^m = 1$ *if and only if $t \mid m$.*
(iii) $t \mid p^n - 1$.

**Proof.** (i) Let $0 \le i \le j < t$. Then

$$\alpha^i = \alpha^j \Rightarrow \alpha^{j-i} = 1 \quad \text{where } 0 \le j - i < t$$
$$\Rightarrow j - i = 0$$
$$\Rightarrow i = j.$$

(ii) Let $\alpha^m = 1$ and $m = tq + r$ where $0 \le r < t$. Then

$$1 = \alpha^m = \alpha^{tq+r} = (\alpha^t)^q \alpha^r = 1 \cdot \alpha^r = \alpha^r.$$

However, $t$ is the smallest positive integer with $\alpha^t = 1$. Hence, $r = 0$ and $t \mid m$. Conversely, if $m = tq \ (q \in \mathbb{N})$, then

$$\alpha^m = (\alpha^t)^q = 1^q = 1.$$

(iii) By Lemma 2.9.1, $\alpha^{p^n-1} = 1$. By part (ii), this implies that $t \mid p^n - 1$. $\square$

Let $F$ be a finite field. An element $\alpha \in F^*$ is a *generator* of $F^*$ or a *primitive element* of $F$ if

$$F^* = \{\alpha^i \mid i \ge 0\}.$$

In $\mathbb{Z}_2$, it is clear that 1 is primitive. In $\mathbb{Z}_3$,

$$1^2 = 1, 2^1 = 2, 2^2 = 4 = 1$$

so that 2 is primitive and 1 is not primitive.
In $\mathbb{Z}_7$,

$$2^1 = 2, 2^2 = 4, 2^3 = 8 = 1$$
$$3^1 = 3, 3^2 = 9 = 2, 3^3 = 6, 3^4 = 18 = 4$$
$$3^5 = 12 = 5, 3^6 = 15 = 1$$

so that 2 is not a primitive element in $\mathbb{Z}_7$, but 3 is.

**Example 2.9.3** Consider GF(9) as constructed in Example 2.8.1. The most obvious element to test first is $y$ (the construction variable). However,

$$y^1 = y, y^2 = -1, y^3 = -y, y^4 = 1$$

so that $y$ is not primitive. On the other hand,

$$(1+y)^1 = 1+y \qquad\qquad (1+y)^5 = 2+2y$$
$$(1+y)^2 = 2y \qquad\qquad (1+y)^6 = y$$
$$(1+y)^3 = 1+2y \qquad\qquad (1+y)^7 = 2+y$$
$$(1+y)^4 = 2 \qquad\qquad (1+y)^8 = 1.$$

Thus, $1+y$ is a primitive element. Note that $1+2y, 2+2y$, and $2+y$ are also primitive elements in this example, so that primitive elements are not, in general, unique.

In Example 2.8.1 we used the irreducible polynomial $1 + y^2$ to construct a copy of GF(9). In section 2.6, we saw that $2 + u + u^2 = 2(1 + 2u + 2u^2)$ is also an irreducible polynomial of degree 2 over $\mathbb{Z}_3$, and in Example 2.8.3 we used $g(u) = 2 + u + u^2$ to construct another copy of GF(9). When we change the polynomial used to construct a field in this way, the role of the construction variable may also change. In the previous example, we saw that $y$ is not a primitive element in $\mathbb{Z}_3[y]/(1 + y^2)$. In $\mathbb{Z}_3[u]/(2 + u + u^2)$, the construction variable $u$ is a primitive element (see the exercises).

## Exercises 2.9

1. For each of the following fields, determine the order of each nonzero element and also which elements are primitive:

    (i) $\mathbb{Z}_7$.
    (ii) $\mathbb{Z}_{13}$.

2. Find the multiplicative order of each element in GF(9)* and identify the primitive elements. (Use the construction of GF(9) in Example 2.8.3.)

3. (i) Show that the order of each element of GF($2^n$)*, $n \geq 1$, is odd.
   (ii) Show that the order of each nonidentity element of GF(257)* is even.

4. (i) Show that every nonidentity element of GF($2^n$)* is primitive for $n = 2, 3, 5, 7$.
   (ii) Show that GF($2^4$) contains nonidentity elements that are not primitive.

5. Let $F$ and $G$ be finite fields and $\varphi : F \to G$ be an isomorphism. Let $a \in F^*$. Show that ord $(a) =$ ord $(\varphi(a))$.

## 2.10 Primitive Elements

Let us focus a little more closely on the existence and properties of primitive elements.

**Lemma 2.10.1** *Let $\alpha \in \mathrm{GF}(p^n)$.*

   (i) *$\alpha$ is primitive if and only if $\mathrm{ord}\,(\alpha) = p^n - 1$.*
   (ii) *If $\alpha$ is primitive then*

$$\mathrm{GF}(p^n)^* = \{1, \alpha, \alpha^2, \ldots, \alpha^{p^n - 2}\}.$$

**Proof.** (i) First assume that $\alpha$ is primitive. Note that the powers of $\alpha$ start repeating at $p^n - 1$ since

$$\alpha^{p^n - 1} = 1, \quad \alpha^{p^n} = \alpha, \quad \alpha^{p^n + 1} = \alpha^2, \ldots.$$

Therefore, if $\{\alpha^i \mid i \geq 0\}$ contains $p^n - 1$ distinct elements, then they must be $1, \alpha, \alpha^2, \ldots, \alpha^{p^n - 2}$. Hence,

$$\mathrm{ord}\,(\alpha) \geq p^n - 1$$

whereas, by Lemma 2.9.2, $\mathrm{ord}\,(\alpha)$ divides $p^n - 1$. Hence, $\mathrm{ord}\,(\alpha) = p^n - 1$.

Conversely, suppose that $\mathrm{ord}\,(\alpha) = p^n - 1$. Then, by Lemma 2.9.2, the elements

$$1, \alpha, \alpha^2, \ldots, \alpha^{p^n - 2}$$

are distinct. However, $\mathrm{GF}(p^n)$ only has $p^n - 1$ nonzero elements. Therefore, these are all the nonzero elements. Hence, $\alpha$ is primitive.

(ii) This follows immediately from the proof of (i).  $\square$

This result and those of the previous section simplify the task of finding the order of nonzero elements and of finding primitive elements. By Lemma 2.10.1, we are looking for elements of order $p^n - 1$ and, by Lemma 2.9.2, the order of any element $\alpha$ divides $p^n - 1$. In other words, $\alpha$ is primitive if and only if $\alpha^m \neq 1$ for all integers $m \mid p^n - 1$ with $m < p^n - 1$.

**Example 2.10.2** $|\mathrm{GF}(8)^*| = 7$. Therefore, the order $t$ of any $\alpha \in \mathrm{GF}(8)^*$ divides 7 so that $t = 1$ or 7. Hence, if $\alpha \in \mathrm{GF}(8)$ and $\alpha \neq 0, 1$, then $\alpha$ is primitive.

**Example 2.10.3** If $\mathrm{GF}(9) = \mathbb{Z}_3[u]/(u^2 + u + 2)$, then the order of any nonzero element divides 8. Since $u^2 = 2u + 1$ and $u^4 = 2$, it follows that $u$ has order 8 and, therefore, that $u$ is primitive.

In the next two results, by way of preparation for the main theorem, we describe how to determine the order of certain combinations of elements.

**Lemma 2.10.4** *Let F be a finite field, $\alpha \in F^*$, $t = \mathrm{ord}\,(\alpha)$, and $m \in \mathbb{N}$.*

   (i) *If $m \mid t$, then $\mathrm{ord}\,(\alpha^m) = t/m$.*
   (ii) *If $(t, m) = 1$, then $\mathrm{ord}\,(\alpha^m) = t$.*

**Proof.** Let $k = \mathrm{ord}\,(\alpha^m)$.
   (i) Since

$$(\alpha^m)^{t/m} = \alpha^t = 1 \tag{2.6}$$

and

$$\alpha^{mk} = (\alpha^m)^k = 1 \tag{2.7}$$

it follows from Lemma 2.9.2 (ii) that (2.6) implies that

$$k \ \text{ divides } \ t/m \tag{2.8}$$

and from (2.7) that

$$t \mid mk. \tag{2.9}$$

From (2.9) we have that $t/m$ divides $k$. Combined with (2.8) this shows that $k = t/m$.

   (ii) Again, from (2.7) we have that $t \mid mk$. However, $(t, m) = 1$ so that we must have $t \mid k$. On the other hand,

$$(\alpha^m)^t = (\alpha^t)^m = 1^m = 1$$

so that, by Lemma 2.9.2 (ii), $k \mid t$. Hence, $k = t$. $\quad\square$

**Lemma 2.10.5** *Let F be a finite field, $\alpha, \beta \in F^*$, $s = \mathrm{ord}\,(\alpha)$, $t = \mathrm{ord}\,(\beta)$, and $(s, t) = 1$. Then $\mathrm{ord}\,(\alpha\beta) = st$.*

**Proof.** Let $m = \mathrm{ord}\,(\alpha\beta)$. Then

$$(\alpha\beta)^{st} = (\alpha^s)^t(\beta^t)^s = 1 \cdot 1 = 1.$$

Therefore, $m \mid st$. Let us say $m = s't'$ where $s' \mid s$ and $t' \mid t$. Also,

$$1 = (1)^{t/t'} = ((\alpha\beta)^m)^{t/t'}$$
$$= (\alpha\beta)^{s't} = \alpha^{s't}(\beta^t)^{s'}$$
$$= \alpha^{s't}$$

so that, by Lemma 2.9.2 (ii), $s \mid s't$, but $(s, t) = 1$. Therefore, $s \mid s'$ and it follows that $s' = s$. Similarly, $t' = t$ and we have

$$m = s't' = st. \quad \square$$

We are now ready for the main result concerning the existence of primitive elements.

**Theorem 2.10.6** *For all primes $p$ and all $n \in \mathbb{N}$, there is a primitive element in* $\mathrm{GF}(p^n)$.

**Proof.** By Lemma 2.9.2, we know that the order of every nonzero element divides $p^n - 1$. Let $t$ be the largest order of any element of $\mathrm{GF}(p^n)^*$ and let $\alpha$ be one of the elements with order $t$. If $t = p^n - 1$, then by Lemma 2.9.2, $\alpha$ is a primitive element. So suppose that $t < p^n - 1$. The polynomial $x^t - 1$ has, at most, $t$ roots. Since $\mathrm{GF}(p^n)^*$ has $p^n - 1$ elements, it follows that there exists an element $\beta \in \mathrm{GF}(p^n)^*$ that is not a root of $x^t - 1$. This means that $m = \mathrm{ord}\,(\beta)$ does not divide $t$.

Consequently, there exists a prime number $q$ and an integer $d \in \mathbb{N}$ such that $q^d \mid m$ but $q^d \nmid t$. Let $e, 0 \le e < d$, be the largest power of $q$ that divides $t$ and let

$$\alpha' = \alpha^{q^e}.$$

Then, by Lemma 2.10.4 (i),

$$\mathrm{ord}\,(\alpha') = (\mathrm{ord}\,\alpha)/q^e = t/q^e.$$

We also have

$$\gcd(t/q^e, q) = 1.$$

Now let

$$\beta' = \beta^{m/q^d}.$$

By Lemma 2.10.4 (i),

$$\mathrm{ord}\,(\beta') = m/(m/q^d) = q^d.$$

Therefore,

$$(\operatorname{ord}(\alpha'), \operatorname{ord}(\beta')) = (t/q^e, q^d) = 1$$

so that, by Lemma 2.10.5,

$$\operatorname{ord}(\alpha'\beta') = (t/q^e)q^d$$
$$= t \cdot q^{d-e}$$
$$> t$$

which contradicts the choice of $t$. Hence, $t = p^n - 1$, as required. $\square$

The existence of primitive elements in finite fields means that we have two ways of representing the elements of any finite field of the form $G = F[x]/p(x)$, where $p(x)$ is an irreducible polynomial of degree $n$. The first is as polynomials in $F[x]$ of degree, at most, $n - 1$. Each such polynomial is completely determined by the $n$-tuple of its coefficients:

$$a_0 + a_1 x + \cdots + a_{n-1} x^{n-1} \iff (a_0 a_1 a_2 \cdots a_{n-1}).$$

In addition, if $\alpha$ is a primitive element in $G$, then every nonzero element is a power of $\alpha$. If we know every element of $G$ in both forms, then we can use the vector form when adding elements, and the power form when multiplying elements.

**Example 2.10.7** Let $\mathrm{GF}(9) = \mathbb{Z}_3[y]/(y^2 + 1)$. We have seen that $\alpha = 1 + y$ is a primitive element in $F$. So we have

$$\alpha = 1 + y \leftrightarrow (11)$$
$$\alpha^2 = 1 + 2y + y^2 = 2y \leftrightarrow (02)$$
$$\alpha^3 = 1 + y^3 = 1 + 2y \leftrightarrow (12)$$
$$\alpha^4 = (2y)^2 = y^2 = 2 \leftrightarrow (20)$$
$$\alpha^5 = \alpha^4 \alpha = 2 + 2y \leftrightarrow (22)$$
$$\alpha^6 = \alpha^4 \alpha^2 = 4y = y \leftrightarrow (01)$$
$$\alpha^7 = \alpha^4 \alpha^3 = 2(1 + 2y) = 2 + y \leftrightarrow (21)$$
$$\alpha^8 = (\alpha^4)^2 = 1 \leftrightarrow (10).$$

In summary, the elements of GF(9) and the orders of the nonzero elements are as follows:

| Element | | | Order |
|---|---|---|---|
| 0 | 0 | (00) | |
| 1 | 1 | (10) | 1 |
| $\alpha$ | $1+y$ | (11) | 8 |
| $\alpha^2$ | $2y$ | (02) | 4 |
| $\alpha^3$ | $1+2y$ | (12) | 8 |
| $\alpha^4$ | 2 | (20) | 2 |
| $\alpha^5$ | $2+2y$ | (22) | 8 |
| $\alpha^6$ | $y$ | (01) | 4 |
| $\alpha^7$ | $2+y$ | (21) | 8 |

To illustrate the use of this "double bookkeeping," we have (using $y=\alpha^6$, $\alpha^8=1$)

$$(1+y^3)(y+y^4)^5 = (1+\alpha^{18})(\alpha^6+\alpha^{24})^5$$
$$= (1+\alpha^2)(\alpha^6+1)^5$$
$$= ((10)+(02))((01)+(10))^5$$
$$= (12)(11)^5$$
$$= \alpha^3\alpha^5$$
$$= \alpha^8$$
$$= 1.$$

**Example 2.10.8** If GF(16) $= \mathbb{Z}_2[y]/(y^4+y+1)$, then $\alpha = y$ is a primitive element and we have the following:

| | | | | | | |
|---|---|---|---|---|---|---|
| 0 | 0 | (0000) | | $\alpha^8$ | $1+y^2$ | (1010) |
| $\alpha$ | $y$ | (0100) | | $\alpha^9$ | $y+y^3$ | (0101) |
| $\alpha^2$ | $y^2$ | (0010) | | $\alpha^{10}$ | $1+y+y^2$ | (1110) |
| $\alpha^3$ | $y^3$ | (0001) | | $\alpha^{11}$ | $y+y^2+y^3$ | (0111) |
| $\alpha^4$ | $1+y$ | (1100) | | $\alpha^{12}$ | $1+y+y^2+y^3$ | (1111) |
| $\alpha^5$ | $y+y^2$ | (0110) | | $\alpha^{13}$ | $1+y^2+y^3$ | (1011) |
| $\alpha^6$ | $y^2+y^3$ | (0011) | | $\alpha^{14}$ | $1+y^3$ | (1001) |
| $\alpha^7$ | $1+y+y^3$ | (1101) | | $\alpha^{15}$ | 1 | (1000) |

**Exercises 2.10**

1. With GF(9) as described in Example 2.10.7, show that

$$((2 + y)^5 + (1 + 2y)^4)^6 = 2.$$

2. With GF(16) as described in Example 2.10.8, simplify $((1 + y)^4 + (y + y^2)^5)^6$.

3. Let $\alpha$ be a primitive element in GF($p^n$) and $m \in \mathbb{N}$. Let $d = (p^n - 1, m)$. Show that ord $(\alpha^m) = (p^n - 1)/d$.

4. Let $\alpha$ be a primitive element in GF($p^n$) and $m \in \mathbb{N}$.

    (i)  Show that $\alpha^m$ is primitive if and only if $(m, p^n - 1) = 1$.
    (ii)  Show that the number of primitive elements in GF($p^n$) is $\varphi(p^n - 1)$.

5. Find the number of primitive elements in each of the following:

    (i)  GF(8).
    (ii)  GF(9).
    (iii)  GF(25).
    (iv)  GF(32).
    (v)  GF(81).

6. Find all the primitive elements in GF(16) $= \mathbb{Z}_2[y]/(y^4 + y + 1)$.

7. (i)  Show that there exists an element in GF($p^n$)* of every possible order.
    (ii)  Find elements of each possible order in GF(9)* (as described in Example 2.10.7).

*8. (i)  Let $m \in \mathbb{N}$ and $a \in \mathbb{Z}_m^*$. Show that there exist $(a, m)$ distinct solutions to the equation $ax = 0$ in $\mathbb{Z}_m$.
    (ii)  Let $p$ be prime, $a \in \mathbb{Z}_p^*$, and $a = \alpha^s \neq 1$ where $\alpha$ is a primitive root of $\mathbb{Z}_p$. Show that the number of distinct solutions mod $(p - 1)$ to $a^x \equiv 1 \bmod p$ is $(s, p - 1)$.

## 2.11 Subfield Structure of Finite Fields

We have seen in section 2.2 how every field of characteristic $p$ will contain a subfield isomorphic to $\mathbb{Z}_p$. We shall now see how to identify all the subfields of any finite field GF($p^n$). We will achieve this by a careful examination of the factorization of the polynomial $x^{p^n} - x$. For this we take advantage of the following standard piece of algebra: For every positive integer $k$,

$$a^k - 1 = (a - 1)(a^{k-1} + a^{k-2} + \cdots + a + 1).$$

This is equally valid whether $a$ is an integer, a field element, or a variable.

**Lemma 2.11.1** *Let $p$ be a prime number and $m, n \in \mathbb{N}$ be such that $m \mid n$.*

  (i) $p^m - 1$ *divides* $p^n - 1$.
  (ii) $x^{p^m-1} - 1$ *divides* $x^{p^n-1} - 1$.

**Proof.** (i) Let $k \in \mathbb{N}$ be such that $n = mk$. We have

$$p^n - 1 = p^{mk} - 1$$
$$= (p^m)^k - 1$$
$$= (p^m - 1)((p^m)^{k-1} + (p^m)^{k-2} + \cdots + p^m + 1).$$

  (ii) From part (i) we know that there exists $t \in \mathbb{N}$ with $p^n - 1 = (p^m - 1)t$. Then,

$$x^{p^n-1} - 1 = (x^{(p^m-1)})^t - 1$$
$$= (x^{p^m-1} - 1)(x^{(p^m-1)(t-1)} + \cdots + x^{p^m-1} + 1)$$

and we have established (ii). $\square$

We are now ready to characterize all the subfields of $G = \mathrm{GF}(p^n)$.

**Theorem 2.11.2** *Let $p$ be a prime number and $n \in \mathbb{N}$. Then, $G = \mathrm{GF}(p^n)$ contains a subfield with $k$ elements if and only if $k = p^m$ where $m \mid n$. Moreover, if there exists a subfield $F$ with $p^m$ elements (where $m \mid n$), then*

$$F = \{\alpha \in \mathrm{GF}(p^n) \mid \alpha \text{ is a root of } x^{p^m} - x\}$$

*and so it is unique.*

**Proof.** If $\mathrm{GF}(p^m)$ is a subfield of $\mathrm{GF}(p^n)$, then we know from Corollary 2.3.8 that $m \mid n$.

  Now suppose that $m \mid n$ and let

$$F = \{\alpha \in \mathrm{GF}(p^n) \mid \alpha \text{ is a root of } x^{p^m} - x\}.$$

First we show that $F$ is a subfield of $\mathrm{GF}(p^n)$. Clearly, $0, 1 \in F$. Let $a, b \in F$, then, by Lemma 2.2.7,

$$(a \pm b)^{p^m} = a^{p^m} \pm b^{p^m} = a \pm b$$

and

$$(ab)^{p^m} = (a^{p^m})(b^{p^m}) = ab$$

and, for $a \in F$, $a \neq 0$, we have

$$(a^{-1})^{p^m} = (a^{p^m})^{-1} = a^{-1}.$$

Thus, $F$ is a subfield of $GF(p^n)$.

Next, we wish to determine the size of $F$.

By Lemma 2.9.1, we know that every element of $GF(p^n)$ is a root of $x^{p^n} - x$. In other words, the polynomial $x^{p^n} - x$ has $p^n$ distinct roots in $GF(p^n)$. Now

$$x^{p^n} - x = x(x^{p^n-1} - 1)$$

where $x$ has just one root—namely, 0. Consequently, $x^{p^n-1} - 1$ has $p^n - 1$ distinct roots in $GF(p^n)$—namely, all the nonzero elements.

Since $m \mid n$, by Lemma 2.11.1, there exists a polynomial $g(x)$ with

$$x^{p^n-1} - 1 = (x^{p^m-1} - 1)g(x).$$

Since $x^{p^n-1} - 1$ has $p^n - 1$ roots in $GF(p^n)$, the polynomial $x^{p^m-1} - 1$ must have $p^m - 1$ distinct roots. Hence, $x^{p^m} - x$ will have $p^m$ roots in $GF(p^n)$. Thus, $|F| = p^m$ so that $F = GF(p^m)$ as required.

Now let $K$ be a subfield of $G$ with $p^m$ elements and let $F$ be as in the statement of the theorem. By Lemma 2.9.1, every element of $K$ is a root of the polynomial

$$x^{p^m} - x$$

and therefore belongs to $F$. Thus, $K \subseteq F$ and, since $|F| = p^m = |K|$, we must have $K = F$. Thus, $F$ is the unique subfield of $G$ with $p^m$ elements. $\quad\square$

For small values of $n$ it is possible to illustrate the containment relation between subfields of $GF(p^n)$. For example, if $n = 18$ then the divisors of 18 are $1, 2, 3, 6, 9, 18$, so that we have the following arrangement of subfields:

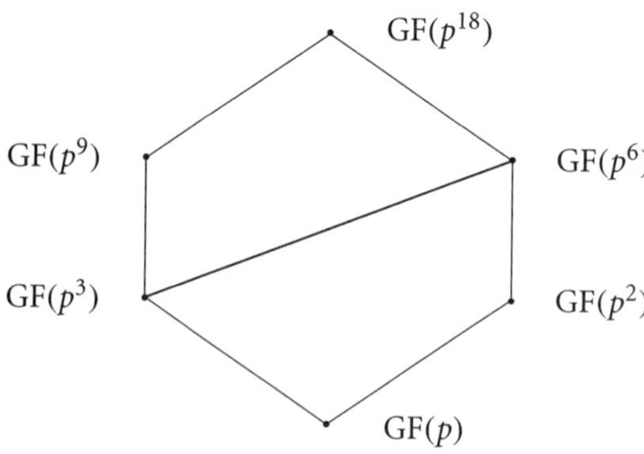

In the next lemma, we provide an alternative way to identify a subfield of a specific size in $\mathrm{GF}(p^n)$.

**Lemma 2.11.3** *Let $m, n \in \mathbb{N}$, $m \mid n$, $p$ be a prime, and $\alpha$ be a primitive element in $\mathrm{GF}(p^n)$. Let $\beta = \alpha^q$ where $q = \frac{p^n - 1}{p^m - 1}$. Then,*

$$G = \{\beta^k \mid 1 \le k \le p^m - 1\} \cup \{0\}$$

*is a subfield of $\mathrm{GF}(p^n)$ with $p^m$ elements.*

**Proof.** Since $\alpha$ is a primitive element, it follows that the elements $\alpha, \alpha^2, \alpha^3, \ldots, \alpha^{p^n-1} = 1$ are all distinct and therefore the elements $\beta, \beta^2, \beta^3, \ldots, \beta^{p^m-1}$ are also distinct. Thus, $G$ contains $p^m$ distinct elements. However, since $\beta^{p^m-1} = \alpha^{p^n-1} = 1$, we see that all the elements of $G$ are roots of the polynomial $x^{p^m} - x$, which implies that $G$ is just the field constructed in Theorem 2.11.2.  $\square$

### Exercises 2.11

1. Let $m, n \in \mathbb{N}$ and $m \mid n$. Let $F$ and $G$ be subfields of $\mathrm{GF}(p^n)$ with $|F| = |G| = p^m$. Show that $F = G$.

2. By Theorem 2.11.2, $\mathrm{GF}(16)$ contains a subfield with four elements.

    (i) Describe $\mathrm{GF}(16)$ with the help of the irreducible polynomial $x^4 + x + 1 \in \mathbb{Z}_2[x]$.
    (ii) Find the subfield of $\mathrm{GF}(16)$, as constructed in (i), with four elements.

3. Draw a diagram representing the mutual containments of the subfields of $\mathrm{GF}(p^{24})$ along the lines of the diagram in section 2.11.

4. Let $F = \mathbb{Z}_2[z]/(z^2 + z + 1), f(y) = y^2 + y + z \in F[y]$, and $G = F[y]/f(y)$. Show that $f(y)$ is irreducible in $F[y]$ and find the subfield of $G$ that is isomorphic to $\mathrm{GF}(4)$.

5. Let $n \in \mathbb{N}$. Show that there exists a nonconstant polynomial $f(x) \in \mathbb{Z}_p[x]$ that has no root in $\mathrm{GF}(p^n)$.

6. Let $F = \mathrm{GF}(p^n)$ and $q = p^n$. Let $F = \{a_1, a_2, \ldots, a_q\}$. For each $t \in F$, $t \neq 0$, let $A_t$ be the $q \times q$ matrix with $(i, j)$th entry

$$ta_i + a_j \qquad (1 \le i, j \le q).$$

    (i) Show that $A_t$ is a Latin Square for all $t \in F^*$.
    (ii) Show that, for $t \neq u$, $A_t$ and $A_u$ are orthogonal (see Exercise 16 in section 1.7).

## 2.12 Minimal Polynomials

Let $F = \mathrm{GF}(p^n)$ be a finite field of characteristic $p$ and $\alpha \in F^*$. From Lemma 2.9.1 (i), we know that $\alpha$ is a root of the monic polynomial $x^{p^n-1} - 1 \in \mathbb{Z}_p[x]$. More generally, if $F$ is a subfield of a field $G$ and $\alpha \in G$ is such that $f(\alpha) = 0$ for some $f(x) \in F[x]$, then we say that $\alpha$ is *algebraic* over $F$. By the preceding remark, every element of $\mathrm{GF}(p^n)$ is algebraic over $\mathbb{Z}_p$. Also $\sqrt{2}$, as a root of $x^2 - 2$, is algebraic over $\mathbb{Q}$. If $\alpha \in G$ is algebraic over $F$ and $m(x) \in F[x]$ is a monic polynomial of the smallest possible degree for which $\alpha$ is a root ($m(\alpha) = 0$), then we will call $m(x)$ a *minimal polynomial for $\alpha$ (with respect to $F$)*. If we wish to emphasize that $m(x)$ is a minimal polynomial for the element $\alpha$, then we write $m_\alpha(x)$. Note that since $m_\alpha(x)$ is a monic polynomial and $m_\alpha(\alpha) = 0$, we must have $\deg(m_\alpha(x)) \geq 1$.

**Theorem 2.12.1** *Let $F$ be a subfield of $G$, $\alpha \in G$, $m(x) \in F[x]$ be a minimal polynomial for $\alpha$ over $F$ and $f(x) \in F[x]$.*

> (i) $f(\alpha) = 0 \implies m(x) \mid f(x)$.
> (ii) $F = \mathbb{Z}_p, G = \mathrm{GF}(p^n) \implies m(x) \mid x^{p^n-1} - 1$.
> (iii) $m(x)$ *is irreducible.*
> (iv) $m(x)$ *is unique.*

**Proof.** (i) By the division algorithm, there exist polynomials $q(x), r(x)$ with

$$f(x) = q(x)m(x) + r(x) \quad \text{where} \ \deg(r(x)) < \deg(m(x)).$$

Then,

$$0 = f(\alpha) = q(\alpha)m(\alpha) + r(\alpha) = q(\alpha) \cdot 0 + r(\alpha)$$
$$= r(\alpha).$$

Since $\deg(r(x)) < \deg(m(x))$ we must have $r(x) = 0$. Thus, $m(x) \mid f(x)$ .

(ii) The second claim follows from part (i) and Lemma 2.9.1 (i).

(iii) Suppose that $m(x) = g(x)h(x)$ for some nonconstant polynomials $g(x), h(x) \in F[x]$. Then,

$$0 = m(\alpha) = g(\alpha)h(\alpha)$$

so that either $g(\alpha) = 0$ or $h(\alpha) = 0$. If $g(\alpha) = 0$, then $m(x) \mid g(x)$ and $\deg(g(x)) = m(x)$. Likewise, if $h(\alpha) = 0$, then $\deg(h(x)) = m(x)$. Thus, $m(x)$ is irreducible.

(iv) Let $g(x)$ also be a minimal polynomial for $\alpha$. Then $g(\alpha) = 0$ and, by (i), $m(x) \mid g(x)$. Switching the roles of $m(x)$ and $g(x)$ we also have $g(x) \mid m(x)$. Since both are monic polynomials, it follows that $m(x) = g(x)$ and (iv) holds. $\square$

In light of Theorem 2.12.1, we can now refer to $m(x)$ as *the* minimal polynomial for $\alpha$ over $F$. Some minimal polynomials are pretty obvious. For example, if $\alpha \in \mathbb{Z}_p$, then the minimal polynomial for $\alpha$ is simply $x - \alpha$:

$$m_0(x) = x, \; m_1(x) = x - 1, \; \ldots \; , \; m_{p-1}(x) = x - p + 1.$$

Indeed, for any field $G$, the minimal polynomial for $\alpha \in G$ over $G$ is just $x - \alpha$.

The next lemma, although not very methodical, can be quite helpful in finding minimal polynomials.

**Lemma 2.12.2** *Let $F$ be a subfield of the field $G$, $\alpha \in G$, and $g(x) \in F[x]$ be monic and irreducible. If $\alpha$ is a root of $g(x)$, then $m_\alpha(x) = g(x)$.*

**Proof.** Since $g(\alpha) = 0$, it follows from Theorem 2.12.1, that $m_\alpha(x)$ divides $g(x)$. Since $g(x)$ is irreducible, we must have $\deg(m_\alpha(x)) = \deg(g(x))$ and, since both are monic, we must have $g(x) = m_\alpha(x)$. $\quad\square$

Lemma 2.12.2 is quite useful when dealing with simple situations since it means that if you know, by whatever means, an irreducible monic polynomial $g(x)$ for which $\alpha$ is a root, then you have found the minimal polynomial. For example, we know that $x^2 - 2 \in \mathbb{Q}[x]$ is monic, irreducible, and has $\sqrt{2}$ as a root. Hence, $m_{\sqrt{2}}(x) = x^2 - 2$ over $\mathbb{Q}$.

**Example 2.12.3** Find the minimal polynomial for $\alpha = \sqrt{2} + \sqrt{3} \in \mathbb{R}$ over $\mathbb{Q}$. One approach for such numbers that can be quite simple is to compute some simple expressions for $\alpha$ and determine whether some combination will eliminate the square roots. We have

$$\alpha^2 = 5 + 2\sqrt{6}, \quad (\alpha^2 - 5)^2 = 24$$

so that $\alpha$ is a root of $(x^2 - 5)^2 - 24$, or

$$x^4 - 10x^2 + 1.$$

Using Proposition 2.6.7, it is not too hard to see that this polynomial is irreducible over $\mathbb{Q}$. Hence, $m_\alpha(x) = x^4 - 10x^2 + 1$ over $\mathbb{Q}$. Unfortunately, this method is a bit haphazard.

**Example 2.12.4** A more methodical method that works over an arbitrary field of arbitrary characteristic is to test arbitrary monic polynomials of degree $1, 2, 3$, and so on, using linear algebra to solve for the coefficients. So, for the same $\alpha$ as in the previous example, we might first test a general monic linear polynomial $a_0 + x$. Substituting $\alpha$, we obtain

$$a_0 + (\sqrt{2} + \sqrt{3}) = a_0 + \sqrt{2} + \sqrt{3}.$$

Now we take advantage of the fact that $1, \sqrt{2}, \sqrt{3}$ are independent vectors in the vector space $\mathbb{R}$ over $\mathbb{Q}$ (a fact that can be verified easily). Thus, the previous expression reduces to zero if and only if we have $a_0 = 1 = 1 = 0$. Thus, $\alpha$ is not a root of any linear polynomial—a fact that must have been obvious from the beginning (since otherwise we could have solved to find $\alpha \in \mathbb{Q}$), but checking it formally illustrates the method. We now repeat the checking process with arbitrary polynomials of degree 2 and 3 with similarly negative results. Now consider an arbitrary monic polynomial of degree 4:

$$a_0 + a_1 x + a_2 x^2 + a_3 x^3 + x^4.$$

If we substitute $\alpha$ into this polynomial and equate the coefficients of the independent vectors $1, \sqrt{2}, \sqrt{3}, \sqrt{6}$ to zero, then we obtain the following system of linear equations in the variables $a_0, \ldots, a_3$:

$$a_0 + 0a_1 + 5a_2 + 0a_3 + 49 = 0$$

$$0a_0 + a_1 + 0a_2 + 11a_3 + 0 = 0$$

$$0a_0 + a_1 + 0a_2 + 9a_3 + 0 = 0$$

$$0a_0 + 0a_1 + 2a_2 + 0a_3 + 20 = 0.$$

Solving this system of linear equations, we obtain

$$a_0 = 1$$

$$a_1 = a_3 = 0$$

$$a_2 = -10.$$

Thus, we obtain the polynomial $x^4 - 10x^2 + 1$ exactly as before. Because there was no polynomial of lesser degree with $\alpha$ as a root, we know immediately that this is the minimal polynomial for $\alpha$.

The next three lemmas exhibit some fascinating and theoretically important information regarding the manner in which the roots of a minimal polynomial over a finite field are related.

**Lemma 2.12.5** *Let F be a finite field of characteristic p and $\alpha \in F^*$. Let $\alpha$ be a root of $f(x) \in \mathbb{Z}_p[x]$. Then $\alpha^p$ is also a root of $f(x)$.*

**Proof.** Let

$$f(x) = a_0 + a_1 x + \cdots + a_m x^m \quad (a_i \in \mathbb{Z}_p).$$

By Fermat's Theorem, $a_i^p = a_i$. Hence,

$$f(\alpha^p) = a_0 + a_1 \alpha^p + \cdots + a_m (\alpha^p)^m$$

$$= a_0^p + (a_1 \alpha)^p + \cdots + (a_m \alpha^m)^p$$

$$= (a_0 + a_1\alpha + \cdots + a_m\alpha^m)^p \quad \text{by Lemma 2.2.7}$$

$$= 0^p = 0$$

and $\alpha^p$ is a root of $f(x)$.  $\square$

In particular, Lemma 2.12.5 tells us that since $\alpha$ is a root of $m_\alpha(x)$, so also is $\alpha^p$ and, therefore, $(\alpha^p)^p = \alpha^{p^2} = \alpha^{p^3}$ as well, and so forth.

Let $\alpha \in \mathrm{GF}(p^n)^*$. By Lemma 2.9.1 (ii) we know that $\alpha^{p^n} = \alpha$. Let $k$ be the smallest positive integer such that $\alpha^{p^k} = \alpha$. Then $\alpha, \alpha^p, \alpha^{p^2}, \ldots, \alpha^{p^{k-1}}$ are all distinct (see the exercises) and we call them the *conjugates* (or *conjugate roots*) of $\alpha$ and write

$$C(\alpha) = \left\{ \alpha, \alpha^p, \ldots, \alpha^{p^{k-1}} \right\}.$$

**Lemma 2.12.6**  *For all $i \in \mathbb{N}$,*

$$C(\alpha) = C(\alpha^{p^i}).$$

**Proof.**  Since $(\alpha^{p^{k-1}})^p = \alpha^{p^k} = \alpha$, it is clear that

$$C(\alpha) = \left\{ \alpha^{p^t} \mid t \in \mathbb{N} \right\} = \left\{ \alpha, \alpha^p, \alpha^{p^2}, \ldots, \alpha^{p^{k-1}} \right\}.$$

Therefore,

$$C(\alpha^{p^i}) = \left\{ \alpha^{p^i}, \alpha^{p^{i+1}}, \ldots \right\} \subseteq C(\alpha)$$

so that there must be an integer $j$ with $0 \le j \le k - 1$ such that

$$\alpha^{p^i} = \alpha^{p^j}.$$

Then,

$$\left( \alpha^{p^i} \right)^{p^{k-j}} = (\alpha^{p^j})^{p^{k-j}} = \alpha^{p^k} = \alpha$$

from which it follows that $\alpha \in C(\alpha^{p^i})$. Therefore,

$$C(\alpha) \subseteq C\left( \alpha^{p^i} \right)$$

and equality prevails.  $\square$

One immediate consequence of Lemma 2.12.6 is that "being a conjugate of" is a symmetric relation: If $\beta$ is a conjugate of $\alpha$, then $\alpha$ is a conjugate of $\beta$. Thus, we can speak of $\alpha$ and $\beta$ as "being conjugates".

**Lemma 2.12.7** *Let $\alpha \in \mathrm{GF}(p^n)$ and $C(\alpha)$ be as above and $m(x)$ be the minimal polynomial for $\alpha$ over $\mathbb{Z}_p$. Then*

$$(x - \alpha)(x - \alpha^p) \cdots (x - \alpha^{p^{k-1}}) \mid m_\alpha(x)$$

*as polynomials in $\mathrm{GF}(p^n)[x]$.*

**Proof.** This follows immediately from Lemma 2.12.5 and the remarks preceding Lemma 2.12.6.  $\square$

Lemma 2.12.7 states that the polynomial

$$c(x) = \prod_{\beta \in C(\alpha)} (x - \beta) = (x - \alpha)(x - \alpha^p) \cdots (x - \alpha^{p^{k-1}})$$

divides $m_\alpha(x)$ in $\mathrm{GF}(p^n)[x]$. This tells us something about $m_\alpha(x)$, but it is a little unsatisfactory since $m_\alpha(x) \in \mathbb{Z}_p[x]$, but we have no information so far concerning the coefficients of this polynomial $c(x)$. However, we are now in a position to describe $m_\alpha(x)$.

**Theorem 2.12.8** *Let $\alpha \in \mathrm{GF}(p^n)^*$ and $m_\alpha(x)$ be the minimal polynomial for $\alpha$ over $\mathbb{Z}_p$. Then,*

$$m_\alpha(x) = \prod_{\beta \in C(\alpha)} (x - \beta).$$

*Moreover,*

$$x^{p^n - 1} - 1 = \prod_{\alpha} m_\alpha(x)$$

*where the product runs over all distinct minimal polynomials $m_\alpha(x)$ for $\alpha \neq 0$.*

**Proof.** Let

$$c(x) = \prod_{\beta \in C(\alpha)} (x - \beta)$$

$$= a_0 + a_1 x + \cdots + a_{k-1} x^{k-1} + x^k \qquad (a_i \in \mathrm{GF}(p^n)).$$

Our main task is to show that $c(x)$ is indeed a polynomial over $\mathbb{Z}_p$. Toward that end we have

$$(c(x))^p = \prod_{\beta \in C(\alpha)} (x - \beta)^p = \prod_{\beta \in C(\alpha)} (x^p - \beta^p)$$

$$= \prod_{\beta \in C(\alpha)} (x^p - \beta) \qquad \text{by Lemma 2.12.6}$$

$$= c(x^p)$$

$$= a_0 + a_1 x^p + \cdots + a_{k-1} x^{p(k-1)} + x^{pk}. \qquad (2.10)$$

On the other hand, we also have

$$(c(x))^p = (a_0 + a_1 x + \cdots + x^k)^p$$

$$= a_0^p + a_1^p x^p + \cdots + a_{k-1}^p x^{(k-1)p} + x^{kp}. \qquad (2.11)$$

Equating the coefficients in (2.10) and (2.11) we find that

$$a_i^p = a_i \quad (i = 0, 1, \ldots, k)$$

which means that $a_i$ is a root of $x^p - x$ for $i = 0, 1, \ldots, k$. By Theorem 2.11.2, this means that $a_i \in \mathrm{GF}(p) = \mathbb{Z}_p$ for all $i$. Thus, $c(x) \in \mathbb{Z}_p[x]$. By Lemma 2.12.7, we already know that $c(x) \mid m_\alpha(x)$. Since $c(x)$ and $m_\alpha(x)$ are both monic polynomials, it follows from the minimality of $m_\alpha(x)$ that $c(x) = m_\alpha(x)$ as required.

With regard to the final claim, first note that every nonzero element $\alpha$ of $\mathrm{GF}(p^n)$ is a root of $x^{p^n-1} - 1$ so that $x - \alpha$ divides $x^{p^n-1} - 1$ in $\mathrm{GF}(p^n)[x]$. Thus, in $\mathrm{GF}(p^n)[x]$, we have

$$x^{p^n-1} - 1 = \prod_\alpha (x - \alpha)$$

where the product runs over all nonzero elements $\alpha$ in $\mathrm{GF}(p^n)$. If we group together the factors containing elements in the same conjugacy class, then we see from the first part of the theorem that each group of factors becomes the minimal polynomial corresponding to that class of conjugates, and therefore we can write

$$x^{p^n-1} - 1 = \prod_\alpha m_\alpha(x)$$

as required. $\quad \square$

**Corollary 2.12.9** *Let $\alpha, \beta \in \mathrm{GF}(p^n)^*$ and $\alpha$ and $\beta$ be conjugates. Then $m_\alpha(x) = m_\beta(x)$.*

We finally arrive at the long anticipated verification of the existence of irreducible polynomials of all degrees over the integers modulo $p$.

**Corollary 2.12.10** *For every prime $p$ and every positive integer $n$, there exists an irreducible polynomial of degree $n$ over $\mathbb{Z}_p$.*

**Proof.** By Theorem 2.10.6, we know that there exists a primitive element in $\mathrm{GF}(p^n)$.

Let $\alpha$ be such an element, let $m_\alpha(x)$ be the minimal polynomial for $\alpha$ over $\mathbb{Z}_p$, and let the degree of $m_\alpha(x)$ be $m$. By the division algorithm, for any positive integer $k$, we can write

$$x^k = q_k(x)m_\alpha(x) + r_k(x).$$

where $q_k(x), r_k(x) \in \mathbb{Z}_p[x]$, $r_k(x)$ is unique and either $r_k(x) = 0$ or the degree of $r_k(x)$ is less than the degree of $m_\alpha(x) = m$. Substituting $\alpha$ into this equation, we find that $\alpha^k = r_k(\alpha)$. It follows that the number of distinct powers of $\alpha$ (which we know to be $p^n - 1$) must be at most the number of nonzero polynomials of degree less than $m$, that is, $p^m - 1$. Thus, $n \leq m$. On the other hand, for any two polynomials $r(x), s(x)$ of degree less than $m$

$$r(\alpha) = s(\alpha) \Rightarrow r(\alpha) - s(\alpha) = 0 \Rightarrow r(x) = s(x)$$

since the degree of $r(x) - s(x)$ is less than $m$, the degree of $m_\alpha(x)$. Thus the number of elements in $\mathrm{GF}(p^n)$, that is, $p^n$ is at least as big as the number of polynomials of degree less then $m$, which is $p^m$. Thus $m \leq n$. Therefore we have equality, $m = n$ and $m_\alpha(x)$ is a polynomial of degree $n$. However, by Theorem 2.12.1(iii), minimal polynomials are always irreducible and so $m_\alpha(x)$ is an irreducible polynomial of degree $n$. $\square$

**Example 2.12.11** Consider $\mathrm{GF}(9) = \mathbb{Z}_3[z]/(1 + z^2)$. Here, $p = 3$.

    (i) For $\alpha \in \mathbb{Z}_3$, $m_\alpha(x) = x - \alpha$.

    (ii) $\alpha = z$. We have

$$z^3 = z^2 \cdot z = 2z$$

$$z^{3^2} = (z^3)^3 = (2z)^3 = 8 \cdot z^3 = 8 \cdot 2z = z$$

so that $C(z) = \{z, 2z\}$ and

$$m_z(x) = (x - z)(x - 2z) = x^2 - (3z)x + 2z^2$$

$$= x^2 + 4 = x^2 + 1.$$

By Corollary 2.12.9, we also have $m_{2z} = x^2 + 1$.

(iii) $\alpha = 1 + z$. We have

$$(1 + z)^3 = 1^3 + z^3 = 1 + z^2 \cdot z = 1 + 2z$$

$$(1 + z)^{3^2} = ((1 + z)^3)^3 = (1 + 2z)^3 = 1 + 8z^3$$

$$= 1 + 16z = 1 + z$$

so that $C(1 + z) = \{1 + z, 1 + 2z\}$ and

$$m_{1+z}(x) = m_{1+2z}(x) = (x - (1 + z))(x - (1 + 2z))$$

$$= x^2 - 2x + (1 + 2z^2)$$

$$= x^2 - 2x + 5$$

$$= x^2 - 2x + 2$$

$$= x^2 + x + 2.$$

(iv) $\alpha = 2 + z$. We have

$$(2 + z)^3 = 2^3 + z^3 = 2 + 2z$$

$$(2 + z)^{3^2} = (2 + 2z)^3 = 2^3 + 2^3 z^3 = 2 + 2.2z = 2 + z$$

so that $C(\alpha) = \{2 + z, 2 + 2z\}$ and

$$m_{2+z}(x) = m_{2+2z}(x) = (x - (2 + z))(x - (2 + 2z))$$

$$= x^2 - x + (4 + 2z^2)$$

$$= x^2 + 2x + 8$$

$$= x^2 + 2x + 2.$$

Illustrating the final claim of the Theorem 2.12.8, we then have

$$x^8 - 1 = (x - 1)(x - 2)(x^2 + 1)(x^2 + x + 2)(x^2 + 2x + 2).$$

**Example 2.12.12** Consider $\mathrm{GF}(16) = \mathbb{Z}_2[z]/(1 + z + z^4)$. Here, $p = 2$ and $\alpha = z$ is a primitive element.

(i) Standard calculations show that the conjugate elements are grouped as

$$C(z) = \{z, z^2, z^4, z^8\}$$
$$C(z^3) = \{z^3, z^6, z^9, z^{12}\}$$
$$C(z^5) = \{z^5, z^{10}\}$$
$$C(z^7) = \{z^7, z^{11}, z^{13}, z^{14}\}.$$

(ii) The corresponding minimal polynomials can then be calculated as

$$m_z(x) = 1 + x + x^4$$
$$m_{z^3}(x) = 1 + x + x^2 + x^3 + x^4$$
$$m_{z^5}(x) = 1 + x + x^2$$
$$m_{z^7}(x) = 1 + x^3 + x^4.$$

(iii) Combining the minimal polynomials we get

$$x^{15} - 1 = (1 + x)(1 + x + x^2)(1 + x + x^4)(1 + x^3 + x^4)$$
$$\times (1 + x + x^2 + x^3 + x^4).$$

### Exercises 2.12

1. Find minimal polynomials in $\mathbb{Q}[x]$ for

    (i) $\sqrt{2}$
    (ii) $1 + \sqrt{2}$.

2. (i) Show that $\sqrt{3} \notin \mathbb{Q}(\sqrt{2})$.
    (ii) Find the minimal polynomial for $\sqrt{3}$ over $\mathbb{Q}(\sqrt{2})$.
    (iii) Find the minimal polynomial for $\sqrt{6}$ over $\mathbb{Q}(\sqrt{2})$.

3. (i) Show that $\sqrt[4]{2} \notin \mathbb{Q}(\sqrt{2})$.
    (ii) Find the minimal polynomial for $\sqrt[4]{2}$ over $\mathbb{Q}(\sqrt{2})$.

4. (i) Produce a table for $\text{GF}(8) = \mathbb{Z}_2[y]/(y^3 + y + 1)$ in the style of
    Example 2.10.7 using the primitive element $\alpha = y$.
    (ii) Determine the conjugacy classes of all elements.
    (iii) Determine the minimal polynomial for each element of $\text{GF}(8)$
    over $\mathbb{Z}_2$.

5. (i) Determine the conjugacy classes in $\text{GF}(9)$ constructed as
    $\mathbb{Z}_3[y]/(y^2 + y + 2)$.
    (ii) Determine the minimal polynomial for each element of $\text{GF}(9)$
    over $\mathbb{Z}_3$.

6.  (i)  Determine the conjugacy classes in GF(16) as described in Example
    2.10.8.
    (ii)  Determine the minimal polynomials for
    (a)  $y + y^2$
    (b)  $y$.

7.  Let $\alpha \in$ GF($p^n$). Show that $\alpha$ is a root of a nonzero polynomial $f(x) \in$
    $\mathbb{Z}_p[x]$ with $\deg(f(x)) \le n$. (Hint: Consider GF($p^n$) as a vector space
    over $\mathbb{Z}_p$.)

8.  Let $\alpha \in$ GF($p^n$)* and $t \in \mathbb{N}$ be the smallest integer such that $\alpha^{p^t} = \alpha$.
    Show that the elements

$$\alpha, \alpha^p, \alpha^{p^2}, \ldots, \alpha^{p^{t-1}}$$

are all distinct.

## 2.13 Isomorphisms between Fields

The goal of this section is to demystify the construction of isomorphisms
between (small) finite fields. Suppose that $f(x)$ and $g(x)$ are both irreducible
polynomials of degree $n$ over $\mathbb{Z}_p$. Then we can use either $f(x)$ or $g(x)$ to construct GF($p^n$) as $\mathbb{Z}_p[x]/f(x)$ or $\mathbb{Z}_p[x]/g(x)$. This means that $\mathbb{Z}_p[x]/f(x)$ and
$\mathbb{Z}_p[x]/g(x)$ must be isomorphic. But how can we find such an isomorphism?
We begin with some simple general observations.

**Lemma 2.13.1**  *Let F and G be fields and $\varphi : F \to G$ be an isomorphism.*

(i)  $\varphi(0_F) = 0_G$.
(ii)  $\varphi(1_F) = 1_G$.
(iii)  $\text{char}(F) = \text{char}(G)$.

**Proof.**  (i) We have

$$\varphi(0_F) = \varphi(0_F + 0_F)$$
$$= \varphi(0_F) + \varphi(0_F)$$

so that

$$0_G = \varphi(0_F) - \varphi(0_F)$$
$$= \varphi(0_F) + \varphi(0_F) - \varphi(0_F)$$
$$= \varphi(0_F).$$

(ii) Exercise.

(iii) Let $m = \mathrm{char}(F)$ and $n = \mathrm{char}(G)$. Then,

$$\varphi(0_F) = 0_G \qquad = n \cdot 1_G$$
$$= n\varphi(1_F) = \varphi(n \cdot 1_F).$$

Since $\varphi$ is injective, it follows that

$$n \cdot 1_F = 0_F.$$

Hence, $m \leq n$. Since $\varphi^{-1} : G \to F$ is also an isomorphism, we also have $n \leq m$. Thus, $m = n$. $\quad\square$

Now let us be a little more specific and consider isomorphisms between finite fields of the form constructed in section 2.7.

**Lemma 2.13.2** *Let $F$ and $G$ be fields of characteristic $p$, and $\varphi : F \to G$ be an isomorphism. Let $h(x) \in \mathbb{Z}_p[x]$ and $\alpha \in F$.*

(i) $\varphi(a) = a$ *for all $a \in \mathbb{Z}_p$.*
(ii) $\alpha$ *is a root of $h(x)$ if and only if $\varphi(\alpha)$ is a root of $h(x)$.*

**Proof.** (i) Let $a \in \mathbb{Z}_p$. Then

$$\varphi(a) = \varphi(a \cdot 1) = a\varphi(1) = a \cdot 1 = a.$$

(ii) Let $h(x) = a_0 + a_1 x + \cdots + a_n x^n$. Then, since $\varphi$ is an isomorphism,

$$h(\alpha) = 0 \iff a_0 + a_1\alpha + \cdots + a_n\alpha^n = 0$$
$$\iff \varphi(a_0 + a_1\alpha + \cdots + a_n\alpha^n) = \varphi(0) = 0$$
$$\iff \varphi(a_0) + \varphi(a_1)\varphi(\alpha) + \cdots + \varphi(a_n)\varphi(\alpha)^n = 0$$
$$\iff a_0 + a_1\varphi(\alpha) + \cdots + a_n\varphi(\alpha)^n = 0$$
$$\iff h(\varphi(\alpha)) = 0$$

from which the claim follows. $\quad\square$

Lemma 2.13.2 provides the key to the construction of isomorphisms between finite fields of the same characteristic. To see how this works, let

$$F = \mathbb{Z}_p[y]/f(y), \qquad G = \mathbb{Z}_p[z]/g(z)$$

where $f(x)$ and $g(x)$ are both irreducible polynomials in $\mathbb{Z}_p[x]$ of degree $n$. Then $F$ and $G$ are two descriptions of $\mathrm{GF}(p^n)$ and thus are isomorphic. To find such an isomorphism $\varphi$, first recall that the elements of $F$ are represented as polynomials in $\mathbb{Z}_p[y]$ of degree, at most, $n - 1$ that is, they are of the form

$$\alpha = a_0 + a_1 y + a_2 y^2 + \cdots + a_{n-1} y^{n-1}.$$

Since $\varphi$ is an isomorphism and since $\varphi(a) = a$ for all $a \in \mathbb{Z}_p$ by Lemma 2.13.2 (i), we must have

$$\varphi(\alpha) = \varphi(a_0) + \varphi(a_1)\varphi(y) + \varphi(a_2)\varphi(y)^2 + \cdots + \varphi(a_{n-1})(\varphi(y))^{n-1}$$
$$= a_0 + a_1\varphi(y) + a_2\varphi(y)^2 + \cdots + a_{n-1}(\varphi(y))^{n-1}.$$

Thus, "all" that we have to do is to define $\varphi(y)$ suitably and this is where Lemma 2.13.2 (ii) comes into play. Somewhat surprisingly, no matter what value we assign to $\varphi(y)$, our mapping will respect addition. To see this, let $\alpha = a_0 + a_1 y + \cdots + a_{n-1} y^{n-1}$ and $\beta = b_0 + b_1 y + \cdots + b_{n-1} y^{n-1}$. Then

$$\varphi(\alpha + \beta) = \varphi((a_0 + b_0) + (a_1 + b_1)y + \cdots + (a_{n-1} + b_{n-1})y^{n-1})$$
$$= (a_0 + b_0) + (a_1 + b_1)\varphi(y) + \cdots + (a_{n-1} + b_{n-1})(\varphi(y))^{n-1}$$
$$= a_0 + a_1\varphi(y) + \cdots + a_{n-1}(\varphi(y))^{n-1} + b_0 + b_1\varphi(y)$$
$$+ \cdots + b_{n-1}(\varphi(y))^{n-1}$$
$$= \varphi(\alpha) + \varphi(\beta).$$

So the whole problem reduces to that of assigning a value to $\varphi(y)$ that ensures that $\varphi$ respects multiplication. This leads us to the next theorem.

**Theorem 2.13.3** *Let $f(x), g(x) \in \mathbb{Z}_p[x]$ be irreducible monic polynomials of degree $n$. Let $\beta \in \mathbb{Z}_p[z]/g(z)$. Then the mapping*

$$\varphi: a_0 + a_1 y + a_2 y^2 + \cdots + a_{n-1} y^{n-1}$$
$$\longrightarrow a_0 + a_1\beta + a_2\beta^2 + \cdots + a_{n-1}\beta^{n-1} \qquad (2.12)$$

*is an isomorphism of $F = \mathbb{Z}_p[y]/f(y)$ to $G = \mathbb{Z}_p[z]/g(z)$ if and only if $\beta$ is a root of $f(x)$.*

**Proof.** First suppose that $\varphi$ is an isomorphism. By Theorem 2.7.4 (ii), we know that $y$ is a root of $f(x)$ and so, by Lemma 2.13.2, it follows that $\beta = \varphi(y)$ must also be a root of $f(x)$.

Now suppose, conversely, that $\beta = \varphi(y)$ is a root of $f(x)$. From the discussion preceding the theorem we know that $\varphi$, as defined in (2.12), respects addition. So it remains to show that $\varphi$ is a bijection and respects multiplication. We leave the verification of the fact that $\varphi$ is a bijection to the exercises, but will now show that $\varphi$ respects multiplication. Let $u, v \in F$ and apply the division algorithm to the polynomial $uv$ in $y$:

$$uv = f(y)q(y) + r(y)$$

where $r(y) = 0$ or $\deg r(y) < \deg(f(y))$. Then $uv = r(y)$ in $F$ and

$$\varphi(uv) = \varphi(r(y)) = r(\beta)$$

whereas

$$\begin{aligned}
\varphi(u)\varphi(v) &= u(\beta)v(\beta) \\
&= f(\beta)q(\beta) + r(\beta) \\
&= r(\beta)
\end{aligned}$$

since $\beta$ is a root of $f(\beta)$. Thus, $\varphi$ respects multiplication and the proof is complete. $\quad\square$

We illustrate the procedure described here by returning to Example 2.8.1.

**Example 2.13.4** Let $f(x) = 1 + x^2$, $g(x) = 2 + x + x^2 \in \mathbb{Z}_3[x]$, and set

$$F = \mathbb{Z}_3[y]/f(y), \qquad G = \mathbb{Z}_3[u]/g(u).$$

Now $y$ is a root of $1 + x^2$ so that under any isomorphism $\varphi : F \to G$, $\varphi(y)$ must also be a root of $1 + x^2$. It is not hard to discover that the roots of $1 + x^2$ in $G$ are $\pm(u + 2)$. Thus we obtain two isomorphisms $\varphi_1, \varphi_2 : F \to G$ defined by

$$\varphi_1(a_0 + a_1 y) = a_0 + a_1(u + 2)$$

and

$$\varphi_2(a_0 + a_1 y) = a_0 - a_1(u + 2).$$

To illustrate the action of $\varphi_1$ on products, consider

$$(1 + y)(2 + y) = 2 + 3y + y^2 = 2 + 2 = 1$$

so that

$$\varphi_1((1 + y)(2 + y)) = \varphi_1(1) = 1.$$

On the other hand,

$$\varphi_1(1+y)\varphi_1(2+y) = (1+u+2)(2+u+2)$$
$$= u(u+1)$$
$$= u^2 + u$$
$$= 1.$$

Thus,

$$\varphi_1((1+y)(2+y)) = \varphi_1(1+y)\varphi_1(2+y).$$

An interesting special case of the previous discussion occurs when we take $F = G$. In that case we obtain an isomorphism of $F$ to itself—in other words, an *automorphism* of $F$. For example, suppose that we take $F$ to be as in Example 2.13.4 and $G = F$. Then the roots of $f(x)$ in $F$ are $y$ (which we already knew) and $2y$. Consequently, we will obtain a nontrivial automorphism of $F$ if we define $\varphi(y) = 2y$ that is,

$$\varphi(a_0 + a_1 y) = a_0 + 2a_1 y.$$

We are now finally in a position to establish the uniqueness of the Galois field with $p^n$ elements.

**Corollary 2.13.5** *For each prime number $p$ and each $n \in \mathbb{N}$ there is a unique field with $p^n$ elements.*

**Proof.** We have seen that for each prime number $p$ and each $n \in \mathbb{N}$ there exists a field with $p^n$ elements. It remains to establish uniqueness. To this end, let $F$ and $G$ be fields with $p^n$ elements. Then necessarily $F$ and $G$ both have characteristic $p$ and the prime subfield of each is (isomorphic to) $\mathbb{Z}_p$. Let $\alpha$ be a primitive element in $F$ with minimal polynomial $m_\alpha$. By the proof of Corollary 2.12.10, $\deg(m_\alpha) = n$. By Theorem 2.12.8, $m_\alpha \mid x^{p^n-1} - 1$. However, $G$ has $p^n$ elements, all of which are roots of $x^{p^n} - x = 0$, and therefore $x^{p^n} - x = x(x^{p^n-1} - 1)$ has a complete set of roots in $G$. But, $m_\alpha(x)$ is an irreducible factor of $x^{p^n-1} - 1$ in $\mathbb{Z}_p[x]$. Therefore, $G$ contains a complete set of roots for $m_\alpha(x)$. Let $\beta$ be any of the roots of $m_\alpha(x)$ in $G$. Since $\deg(m_\alpha) = n$, it follows that

$$1, \alpha, \alpha^2, \ldots, \alpha^{n-1}$$

are independent as vectors in $F$ as a vector space over $\mathbb{Z}_p$. Therefore these elements form a basis for $F$ (which has dimension $n$). Now define $\theta : F \longrightarrow \mathbb{Z}_p[y]/m_\alpha(y)$ by

$$\theta : a_0 + a_1\alpha + \cdots + a_{n-1}\alpha^{n-1} \longrightarrow a_0 + a_1 y + \cdots + a_{n-1}y^{n-1}.$$

The same argument as in the proof of Theorem 2.13.3 will now show that $\theta$ is an isomorphism. Similarly $G$ is isomorphic to $\mathbb{Z}_p[z]/g(z)$ for some irreducible polynomial $g(z) \in \mathbb{Z}_p[z]$ of degree $n$. But, as we saw earlier, $G$ contains a root $\beta$ of $m_\alpha(x)$. Consequently, by Theorem 2.13.3, the fields $\mathbb{Z}_p[y]/m_\alpha(y)$ and $\mathbb{Z}_p[z]/g(z)$ are also isomorphic. So we have

$$F \cong \mathbb{Z}_p[y]/m_\alpha(y) \cong \mathbb{Z}_p[z]/g(z) \cong G.$$

Therefore, the fields $F$ and $G$ are also isomorphic, as required. $\qquad\square$

**Exercises 2.13**

1. Construct an isomorphism from

$$F = \mathbb{Z}_2[y]/(1 + y + y^3) \quad \text{to} \quad G = \mathbb{Z}_2[z]/(1 + z^2 + z^3).$$

2. Construct an isomorphism from

$$F = \mathbb{Z}_3[y]/(2 + y + y^2) \quad \text{to} \quad G = \mathbb{Z}_3[z]/(2 + 2z + z^2).$$

3. Construct a nonidentity automorphism of $F = \mathbb{Z}_5[y]/(1 + y + y^2)$.

4. Show that the mapping $\varphi$ constructed in the converse part of Theorem 2.13.3 is a bijection.

5. The elements $y$ and $1 + y$ are both primitive in $\mathrm{GF}(9) = \mathbb{Z}_3[y]/(2 + y + y^2)$ (see text). Show that there does not exist an automorphism $\varphi$ of $\mathrm{GF}(9)$ with $\varphi(y) = 1 + y$.

6. Let $F$ and $G$ be finite fields of characteristic $p$, and $\varphi : F \to G$ be an isomorphism. Let $\alpha \in F$. Show that

   (i) $\alpha$ is primitive if and only if $\varphi(\alpha)$ is primitive
   (ii) $m_\alpha(x) = m_{\varphi(\alpha)}(x)$.

7. Let $F$ be a subfield of a field $G$ and $\alpha, \beta \in G$ have the same minimal polynomial $m(x) \in F[x]$ of degree $n$. Define $\varphi : F(\alpha) \to F(\beta)$ (see Exercise 10 in section 2.7 for the definition of $F(\alpha), F(\beta)$) by

$$\varphi(a_0 + a_1\alpha + \cdots + a_{n-1}\alpha^{n-1}) = a_0 + a_1\beta + \cdots + a_{n-1}\beta^{n-1}.$$

   Show that $\varphi$ is an isomorphism of $F(\alpha)$ to $F(\beta)$.

8. In the notation of Example 2.12.10, let $\alpha = z$, $\beta = 2z$. Show that the mapping

$$\varphi : a + b\alpha \longrightarrow a + b\beta \qquad (a, b \in \mathbb{Z}_3)$$

is an automorphism of GF(9). (Hint: Recall Exercise 10 in section 2.6.)

9. Repeat Exercise 8 with $\alpha = 1 + z$ and $\beta = 1 + 2z$.

10. Let $\alpha$ be a primitive element in $GF(p^n)$. Show that every conjugate $\beta$ of $\alpha$ is also primitive.

11. Let $F$ be a subfield of the finite field $G$, $\alpha \in G \backslash F$, and $m_\alpha(x)$ be the minimal polynomial for $\alpha$ over $F$. Show that the fields $F(\alpha)$ and $F[y]/m_\alpha(y)$ are isomorphic.

**12. The addition and multiplication tables for the hexadecimal field $H$ are provided here. This system is commonly used in communication systems because its 16 elements can be represented by packets of 4 0/1 bits more efficiently than can the decimal system. Identify this field by giving

   (i)  a polynomial description of the field together with
   (ii)  the corresponding polynomial form for each element in the table and
   (iii)  the description of each element in the table as a power of a primitive element/polynomial.

### (a) Addition Table

| + | 0 | 1 | 2 | 3 | 4 | 5 | 6 | 7 | 8 | 9 | A | B | C | D | E | F |
|---|---|---|---|---|---|---|---|---|---|---|---|---|---|---|---|---|
| 0 | 0 | 1 | 2 | 3 | 4 | 5 | 6 | 7 | 8 | 9 | A | B | C | D | E | F |
| 1 | 1 | 0 | 3 | 2 | 5 | 4 | 7 | 6 | 9 | 8 | B | A | D | C | F | E |
| 2 | 2 | 3 | 0 | 1 | 6 | 7 | 4 | 5 | A | B | 8 | 9 | E | F | C | D |
| 3 | 3 | 2 | 1 | 0 | 7 | 6 | 5 | 4 | B | A | 9 | 8 | F | E | D | C |
| 4 | 4 | 5 | 6 | 7 | 0 | 1 | 2 | 3 | C | D | E | F | 8 | 9 | A | B |
| 5 | 5 | 4 | 7 | 6 | 1 | 0 | 3 | 2 | D | C | F | E | 9 | 8 | B | A |
| 6 | 6 | 7 | 4 | 5 | 2 | 3 | 0 | 1 | E | F | C | D | A | B | 8 | 9 |
| 7 | 7 | 6 | 5 | 4 | 3 | 2 | 1 | 0 | F | E | D | C | B | A | 9 | 8 |
| 8 | 8 | 9 | A | B | C | D | E | F | 0 | 1 | 2 | 3 | 4 | 5 | 6 | 7 |
| 9 | 9 | 8 | B | A | D | C | F | E | 1 | 0 | 3 | 2 | 5 | 4 | 7 | 6 |
| A | A | B | 8 | 9 | E | F | C | D | 2 | 3 | 0 | 1 | 6 | 7 | 4 | 5 |
| B | B | A | 9 | 8 | F | E | D | C | 3 | 2 | 1 | 0 | 7 | 6 | 5 | 4 |
| C | C | D | E | F | 8 | 9 | A | B | 4 | 5 | 6 | 7 | 0 | 1 | 2 | 3 |
| D | D | C | F | E | 9 | 8 | B | A | 5 | 4 | 7 | 6 | 1 | 0 | 3 | 2 |
| E | E | F | C | D | A | B | 8 | 9 | 6 | 7 | 4 | 5 | 2 | 3 | 0 | 1 |
| F | F | E | D | C | B | A | 9 | 8 | 7 | 6 | 5 | 4 | 3 | 2 | 1 | 0 |

(b) Multiplication Table

| × | 0 | 1 | 2 | 3 | 4 | 5 | 6 | 7 | 8 | 9 | A | B | C | D | E | F |
|---|---|---|---|---|---|---|---|---|---|---|---|---|---|---|---|---|
| 0 | 0 | 0 | 0 | 0 | 0 | 0 | 0 | 0 | 0 | 0 | 0 | 0 | 0 | 0 | 0 | 0 |
| 1 | 0 | 1 | 2 | 3 | 4 | 5 | 6 | 7 | 8 | 9 | A | B | C | D | E | F |
| 2 | 0 | 2 | 4 | 6 | 8 | A | C | E | 3 | 1 | 7 | 5 | B | 9 | F | D |
| 3 | 0 | 3 | 6 | 5 | C | F | A | 9 | B | 8 | D | E | 7 | 4 | 1 | 2 |
| 4 | 0 | 4 | 8 | C | 3 | 7 | B | F | 6 | 2 | E | A | 5 | 1 | D | 9 |
| 5 | 0 | 5 | A | F | 7 | 2 | D | 8 | E | B | 4 | 1 | 9 | C | 3 | 6 |
| 6 | 0 | 6 | C | A | B | D | 7 | 1 | 5 | 3 | 9 | F | E | 8 | 2 | 4 |
| 7 | 0 | 7 | E | 9 | F | 8 | 1 | 6 | D | A | 3 | 4 | 2 | 5 | C | B |
| 8 | 0 | 8 | 3 | B | 6 | E | 5 | D | C | 4 | F | 7 | A | 2 | 9 | 1 |
| 9 | 0 | 9 | 1 | 8 | 2 | B | 3 | A | 4 | D | 5 | C | 6 | F | 7 | E |
| A | 0 | A | 7 | D | E | 4 | 9 | 3 | F | 5 | 8 | 2 | 1 | B | 6 | C |
| B | 0 | B | 5 | E | A | 1 | F | 4 | 7 | C | 2 | 9 | D | 6 | 8 | 3 |
| C | 0 | C | B | 7 | 5 | 9 | E | 2 | A | 6 | 1 | D | F | 3 | 4 | 8 |
| D | 0 | D | 9 | 4 | 1 | C | 8 | 5 | 2 | F | B | 6 | 3 | E | A | 7 |
| E | 0 | E | F | 1 | D | 3 | 2 | C | 9 | 7 | 6 | 8 | 4 | A | B | 5 |
| F | 0 | F | D | 2 | 9 | 6 | 4 | B | 1 | E | C | 3 | 8 | 7 | 5 | A |

## 2.14 Error Correcting Codes

In electronic communications such as with satellites, space stations, Mars landers, or cellular phones, the signals can be weak or there may be interference as a result of background noise, resulting in the loss or corruption of some of the transmitted data. In data storage on CDs, DVDs, hard drives, and so on, data can also become corrupted for various reasons such as scratches or magnetic interference. The exponential growth in data storage and digital communication would not have been possible without the development of techniques to deal with such problems. If you are chatting with a friend at a noisy party and don't catch a comment ("Was that 'If you join our firm we can offer you a handsome salary' or 'If you join our firm we can't offer you a handsome salary'?"), the solution is easy. You just say, "Pardon, but I didn't catch that. Could you repeat it please". This approach, known as the *automatic repeat request*, can be a simple and effective solution. If an error or ambiguity is detected, then you ask for the message to be repeated. Bar codes like the UPC are designed with this approach in mind. It is designed to detect (most) errors in the scanning process and then, if an error is detected, the bar code should simply be read again.

But what if it is impractical to request that every message containing an error be resent (for example, from a Mars lander), or if the source of data has been damaged and resending or rereading the data will not improve the situation (as would be the case with a scratch on a CD)? In these situations, the only solution is to accept that there will be some errors and try to devise means to correct them. This is where error correcting codes play a critical role. The basic idea is to introduce a sufficient amount of redundancy (a bit like the check digits in bar codes) to make it possible to reconstruct the original message. Of course, it should be appreciated that error correction can't work miracles and can only be effective in the presence of a relatively small number of errors. If you toss a CD in the fire for a couple of minutes, it is not likely that you will be able to recover much of the data afterward. On the other hand, it can do some very impressive things. For instance, if you happen to scratch a CD with a ring or scratch it when you put it into its sleeve, then you might hope that the error correcting features of the CD would fix that well enough that there would be no telltale "clicks".

In general, a *code* just means a collection of words that might be used in some encoding scheme. For example, at one extreme, one might take all words over some alphabet. If we chose the English alphabet then the code would include all the words in the usual English dictionary, but also "words" such as $abcd \ldots z$, $zzzyyywww$, $yllis$, and so forth. If we chose the binary alphabet $\{0, 1\}$, then the code would include all binary strings. However, generally speaking one wants to work with a subset of all the words over the chosen alphabet rather than the set of all words, for reasons that will become apparent later. Here we just provide a brief glimpse at one particular approach using one particular class of codes known as *linear block codes*. In these codes, the digitized data are divided into uniform blocks of convenient size that are then treated as points in a vector space over a finite field. These kinds of codes are used in CDs and DVDs, and were used by NASA on the *Mariner* spacecraft and on the *Voyager* spacecraft to transmit data and color photographs from Jupiter, Saturn, Uranus, and Neptune.

The following somewhat trivial example illustrates the idea. Suppose that we wish to transmit messages involving just four numbers: 0, 1, 2, and 3. Here are two different ways in which we might encode the letters of our messages:

|       | 0      | 1      | 2      | 3      |
|-------|--------|--------|--------|--------|
| $C_1$ | 00     | 01     | 10     | 11     |
| $C_2$ | 000000 | 000111 | 111000 | 111111 |

If the number 2 is transmitted in $C_1$ (as 10) and an error is introduced so that either 00 or 11 is received, then the receiver has little choice but to assume that 00 (that is, 0) or 11 (that is, 3) was the original message. However, if $C_2$ is used and a single error is introduced, then it is still possible to identify the original message. In addition, even if two errors are introduced, then the recipient will be aware that errors have been introduced.

Note that when we speak of the code $C_1$ (respectively, $C_2$) we mean the set of words used in the code. In the previous example,

$$C_1 = \{00, 01, 10, 11\}$$

$$C_2 = \{000000, 000111, 111000, 111111\}.$$

We are interested in properties of these sets of words. The actual original encoding ($2 \to 111000$, and so on) is not so important.

These codes are both binary since their letters are drawn from $\mathbb{Z}_2 = \{0, 1\}$. More generally, by a *code $C$ over a field $F$ of length $n$*, we mean a subset $C$ of $F^n$. Then the elements of $C$ are $n$-tuples of the form $(a_1 a_2 a_3 \cdots a_n)$, $a_i \in F$.

It is clear that the feature of the code $C_2$ used here that makes it possible to recognize that some errors have occurred is the fact that the code words differ from each other in more positions than do the code words in code $C_1$. A useful measure of this feature is given by the following: For any $a = (a_1 a_2 \cdots a_n)$, $b = (b_1 b_2 \cdots b_n) \in F^n$, the *(Hamming) distance* between $a$ and $b$ is

$$d(a, b) = |\{i \mid a_i \neq b_i\}|.$$

The *(Hamming) distance of a code $C \subseteq F^n$* is

$$d(C) = \min\{d(a, b) \mid a, b \in C, a \neq b\}.$$

For example,

$$d((1011011), (0111100)) = 5$$

whereas

$$d(C_1) = 1, \quad d(C_2) = 3.$$

The Hamming distance has properties resembling the familiar properties of distance between two real numbers as measured by the absolute value of their difference.

**Lemma 2.14.1** *For any $a, b, c \in F^n$,*

(i) $d(a, b) = 0 \iff a = b$
(ii) $d(a, b) = d(b, a)$
(iii) $d(a, c) \leq d(a, b) + d(b, c).$

**Proof.** Parts (i) and (ii) are obvious.

(iii) Let $a = (a_i)$, $b = (b_i)$ and $c = (c_i)$. Then

$$a_i \neq c_i \implies \text{either } a_i \neq b_i \text{ or } b_i \neq c_i.$$

Therefore,

$$|\{i \mid a_i \neq c_i\}| \leq |\{i \mid a_i \neq b_i\}| + \{\{i \mid b_i \neq c_i\}|$$

or

$$d(a, c) \leq d(a, b) + d(b, c). \quad \square$$

Consider what happens with codes of small distance. Let $C$ be a code.

$\mathbf{d}(C) = 1$. This means that there are two code words $u, v$ with $d(u, v) = 1$. It only takes one error in the right place to convert the code word $u$ into the code word $v$ and vice versa. So any recipient of the messages $u, v$, from a channel where it is possible to expect one error, would have no idea whether the original message that was sent was $u$ or $v$. So such a code would not even detect some quite small errors.

$\mathbf{d}(C) = 2$. Now there is at least one pair of code words (say, $u, v$) that differ from each other in just two positions, but nothing smaller than that. It requires at least two errors in the appropriate positions to convert $u$ to $v$ or vice versa. This means that any single error in a code word will not be sufficient to convert it to a new code word. If a single error is introduced into a code word, then the new word will not be a code word and will therefore be detectable. So we say that a code of distance two can *detect* single-errors and is called a *single-error detecting* code.

$\mathbf{d}(C) = 3$. For this kind of code, it takes at least three errors in the right positions to change one code word to a second code word. If the channel of communication introduces one or two errors into a code word, the received word will still not be a code word and thus will be recognized as not being a code word. Therefore, this code can *detect* two errors and would be called a *two-error detecting* code. However, in this case we can go a step further. If a single error is introduced into a code word $u$ to produce a noncode word $w$, then it will be recognizable as not being a code word. But now if we ask, as the recipients of the word $w$, what is the word that was most likely sent, we can see that it would only have taken one error in the transmission of the word $u$ to have produced the word $w$. However, to obtain $w$ from any other code word $v$ would have required at least two errors in $v$ in the right positions. So, if we feel that it is unlikely that there would be more than one error in the received word, then we may feel that it is a safe bet that $u$, the nearest neighbor, so to speak, was the original word transmitted. In such a case, the code $C$ is said to be *single-error correcting*.

Being able to "correct" one, two, three, … errors may not seem like much, but it is amazing how by combining codes and adopting techniques such as interleaving words it is possible to spin this out to an ability to correct dozens, even hundreds of errors in a block of transmitted words.

The *nearest-neighbor* decoding principle consists of decoding any received word to the nearest code word, if such exists.

**Theorem 2.14.2** *Let $C \subseteq F$ be a code with Hamming distance $\delta \geq 2e+1$ where $e \in \mathbb{N}$. Let $c$ be a transmitted code word that is received as $w$ with, at most, $e$ errors. Then $c$ is the unique code word satisfying $d(c, w) \leq e$.*

**Proof.** Exercise. $\square$

A code $C \subseteq F^n$ is said to be *linear* if it is a (vector) subspace of $F^n$ (over the field $F$). The big attraction in dealing with linear codes is that it is much easier to determine the distance of the code than it is for an arbitrary code.

**Lemma 2.14.3** *Let $F$ be a finite field and $C$ be a linear code in $F^n$. Then,*

$$d(C) = \min\{d(w, 0) \mid w \in C,\ w \neq 0\}.$$

**Proof.** By the definition of Hamming distance, for any $u, v \in C$,

$$d(u, v) = d(u - v, v - v) = d(w, 0)$$

where

$$w = u - v \in C \quad \text{and} \quad 0 = v - v \in C$$

since $C$ is linear. Hence,

$$\{d(u, v) \mid u, v \in C\} = \{d(w, 0) \mid w \in C\}$$

and

$$d(C) = \min\{d(u, v) \mid u, v \in C\}$$
$$= \min\{d(w, 0) \mid w \in C\}. \quad \square$$

There are three important numbers associated with any linear code $C \subseteq F^n$:

    (i)  $n$, the length of code words.
    (ii)  $\dim C$.
    (iii)  $d(C)$.

In the search for good codes it is desirable to make both $d(C)$ and the ratio $(\dim C)/n$ as large as possible. The larger the value of $d(C)$, the greater will be the error-correcting capabilities. The larger the value of $(\dim C)/n$, the more efficiently the code uses space. However, $(\dim C)/n$ tends to decrease as

$d(C)$ increases. This makes the task of finding good codes both challenging and interesting.

A linear code $C \subseteq F^n$ is *cyclic* if

$$(a_0 a_1 \cdots a_{n-1}) \in C \implies (a_{n-1} a_0 \cdots a_{n-2}) \in C.$$

For example,

$$C = \{(000), (110), (011), (101)\}$$

is a cyclic code in $\mathbb{Z}_2^3$. We will describe a method for constructing cyclic codes that will also give us information about the dimension of the code that we are constructing.

Let $p$ be a prime and $n > 1$. Let

$$R = \mathbb{Z}_p[x]/(x^n - 1).$$

Since $x^n - 1$ is reducible, $R$ is not a field. Nonetheless, as before, we can identify the elements of $R$ with the set of all polynomials

$$a_0 + a_1 x + a_2 x^2 + \cdots a_{n-1} x^{n-1} \quad (a_i \in F)$$

and the arithmetic of $R$ is considerably simplified by the equality $x^n = 1$.

Let $g(x) \in \mathbb{Z}_p[x]$ and $g(x) \mid x^n - 1$ in $\mathbb{Z}_p[x]$. Define

$$I_{g(x)} = \{k(x)g(x) \in R \mid k(x) \in \mathbb{Z}_p[x]\}$$
$$C_{I_{g(x)}} = \{(a_0 a_1 \cdots a_{n-1}) \mid a_0 + a_1 x + \cdots + a_{n-1} x^{n-1} \in I_{g(x)}\}.$$

Note that $I_{g(x)}$ is a subset of $R$ and that the polynomial products defining $I_{g(x)}$ are calculated in $R$, that is, modulo $x^n - 1$.

An important basic property of $I_{g(x)}$ is that it is closed under multiplication by elements of $\mathbb{Z}_p[x]$ since we have, for any $h(x) \in \mathbb{Z}_p[x]$,

$$\begin{aligned}
h(x)I_{g(x)} &= \{h(x)k(x)g(x) \in R \mid k(x) \in \mathbb{Z}_p[x]\} \\
&\subseteq \{k^*(x)g(x) \in R \mid k^*(x) \in \mathbb{Z}_p[x]\} \\
&= I_{g(x)}.
\end{aligned}$$

This leads to the following important method for generating cyclic codes.

**Lemma 2.14.4** $C_{I_{g(x)}}$ *is a cyclic subspace of* $\mathbb{Z}_p^n$.

**Proof.** The verification that $C_{I_{g(x)}}$ is a subspace is left to you as an exercise. To see that $C_{I_{g(x)}}$ is cyclic, we have (taking full advantage of the simplification $x^n = 1$)

$$
\begin{aligned}
(a_0 \cdots a_{n-1}) \in C_{I_{g(x)}} &\implies a_0 + a_1 x + \cdots + a_{n-1} x^{n-1} \in I_{g(x)} \\
&\implies x(a_0 + a_1 x + \cdots + a_{n-1} x^{n-1}) \in I_{g(x)} \\
&\implies a_0 x + a_1 x^2 + \cdots + a_{n-2} x^{n-1} + a_{n-1} x^n \in I_{g(x)} \\
&\implies a_0 x + a_1 x^2 + \cdots + a_{n-2} x^{n-1} + a_{n-1} \in I_{g(x)} \\
&\implies a_{n-1} + a_0 x + \cdots + a_{n-2} x^{n-1} \in I_{g(x)} \\
&\implies (a_{n-1} a_0 \cdots a_{n-2}) \in C_{I_{g(x)}}
\end{aligned}
$$

so that $C_{I_{g(x)}}$ is indeed cyclic.   $\square$

The next observation gives us a convenient test for membership in a cyclic subspace of the form $C_{I_{g(x)}}$ from which we can deduce some useful information regarding the dimension of $C_{I_{g(x)}}$.

**Lemma 2.14.5** *Let* $g(x) \in \mathbb{Z}_p[x]$ *divide* $x^n - 1$ *and* $v = (a_0 a_1 \ldots a_{n-1})$ *and* $h(x) = a_0 + a_1 x + a_2 x^2 + \ldots + a_{n-1} x^{n-1}$. *Then*

$$
v \in C_{I_{g(x)}} \iff g(x) \mid h(x) \ \text{in} \ \mathbb{Z}_p[x].
$$

*Moreover,* $\dim(C_{I_{g(x)}}) = n - \deg(g(x))$.

**Proof.** By hypothesis, there exists $s(x) \in \mathbb{Z}_p[x]$ such that $x^n - 1 = g(x)s(x)$. First assume that $v \in C_{I_{g(x)}}$. By the definition of $C_{I_{g(x)}}$, $h(x) \in I_{g(x)}$. In other words, there exists $k(x) \in \mathbb{Z}_p[x]$ such that

$$
h(x) = k(x)g(x) \mod x^n - 1.
$$

This means that there exists $t(x) \in \mathbb{Z}_p[x]$ with

$$
\begin{aligned}
h(x) &= k(x)g(x) + t(x)(x^n - 1) \\
&= k(x)g(x) + t(x)g(x)s(x) \\
&= g(x)(k(x) + t(x)s(x))
\end{aligned}
$$

where $g(x)$ divides $h(x)$ in $\mathbb{Z}_p[x]$. Conversely, suppose that $g(x)$ divides $h(x)$ in $\mathbb{Z}_p[x]$. Then there exists $k(x) \in \mathbb{Z}_p[x]$ with $h(x) = k(x)g(x) \in I_{g(x)}$ so that $v \in C_{I_{g(x)}}$.

For the final claim, let $\deg(g(x)) = m$. We will establish the equivalent claim that $I_{g(x)}$ has dimension $n - \deg(g(x))$. Let $h(x) \in I_{g(x)}$. By the first part of the lemma, there exists $k(x) \in \mathbb{Z}_p[x]$ with $h(x) = k(x)g(x)$ in $\mathbb{Z}_p[x]$ where, necessarily, $\deg(k(x)) \leq n - m - 1$. Let

$$k(x) = k_0 + k_1 x + k_2 x^2 + \cdots + k_{n-m-1} x^{n-m-1}.$$

Then

$$h(x) = k(x)g(x) = k_0 g(x) + k_1 x g(x)$$
$$+ k_2 x^2 g(x) + \cdots + k_{n-m-1} x^{n-m-1} g(x)$$

so that $h(x)$ is a linear combination (over $\mathbb{Z}_p$) of the polynomials

$$g(x), xg(x), x^2 g(x), \ldots, x^{n-m-1} g(x).$$

Since these polynomials have degrees $m, m+1, m+2, \cdots, n-1$, respectively, it is clear that they are independent. Hence, they form a basis for $I_{g(x)}$ and we have that $\dim(I_{g(x)}) = n - 1 - \deg(g(x))$, as required. $\square$

It is clear that in this context, the factorizations of $x^n - 1$ are of considerable importance. See the Table of Factorizations of $x^n - 1$ in $\mathbb{Z}_2[x]$ for a few such factorizations over $\mathbb{Z}_2$.

Table of Factorizations of $x^n - 1$ in $\mathbb{Z}_2[x]$

| | |
|---|---|
| $x - 1$ | $1 + x$ |
| $x^2 - 1$ | $(1 + x)^2$ |
| $x^3 - 1$ | $(1 + x)(1 + x + x^2)$ |
| $x^4 - 1$ | $(1 + x)^4$ |
| $x^5 - 1$ | $(1 + x)(1 + x + x^2 + x^3 + x^4)$ |
| $x^6 - 1$ | $(1 + x)^2(1 + x + x^2)^2$ |
| $x^7 - 1$ | $(1 + x)(1 + x + x^3)(1 + x^2 + x^3)$ |
| $x^8 - 1$ | $(1 + x)^8$ |
| $x^9 - 1$ | $(1 + x)(1 + x + x^2)(1 + x^3 + x^6)$ |
| $x^{10} - 1$ | $(1 + x)^2(1 + x + x^2 + x^3 + x^4)^2$ |
| $x^{11} - 1$ | $(1 + x)(1 + x + \cdots + x^{10})$ |
| $x^{12} - 1$ | $(1 + x)^4(1 + x + x^2)^4$ |
| $x^{13} - 1$ | $(1 + x)(1 + x + \cdots + x^{12})$ |
| $x^{14} - 1$ | $(1 + x)^2(1 + x + x^3)^2(1 + x^2 + x^3)^2$ |
| $x^{15} - 1$ | $(1 + x)(1 + x + x^2)(1 + x + x^2 + x^3 + x^4)(1 + x + x^4)(1 + x^3 + x^4)$ |
| $x^{16} - 1$ | $(1 + x)^{16}$ |

**Example 2.14.6** Let $g(x) = (1 + x)(1 + x + x^2) = 1 + x^3$. From the table, we see that $g(x) \mid x^6 - 1$. Here, $g(x)$ corresponds to the vector

$(100100) \in \mathbb{Z}_2^6$ and

$$C_{I_{g(x)}} = \{(000000), (100100), (010010), (001001),$$
$$(110110), (011011), (101101), (111111)\}.$$

By Lemma 2.14.3, we see that $d(C_{I_{g(x)}}) = 2$, and it is easy to see that the dimension of $C_{I_{g(x)}}$ is 3.

If we choose our parameters carefully, we can get some important information regarding $d(C_{I_{g(x)}})$.

**Theorem 2.14.7** *Let $p$ be a prime, $k \in \mathbb{N}$, $g(x) \in \mathbb{Z}_p[x]$, $g(x) \mid x^n - 1$, $g(x) \mid x^{p^k-1} - 1$, and $\alpha$ be an element of order $n$ in $\mathrm{GF}(p^k)^*$. Let the roots of $g(x)$ in $\mathrm{GF}(p^k)$ contain a run of $m$ consecutive powers of $\alpha$:*

$$\alpha^{l+1}, \alpha^{l+2}, \ldots, \alpha^{l+m}.$$

*Then $d(C_{I_{g(x)}}) \geq m + 1$.*

**Note.** By Theorem 2.12.8, $x^{p^k-1} - 1$ has a complete set of roots in $GF(p^k)$ and, since $g(x)$ divides $x^{p^k-1} - 1$, $g(x)$ also has a complete set of roots in $GF(p^k)$. Since every element of $GF(p^k)^*$ has order dividing $p^k - 1$, it follows that $n$ must divide $p^k - 1$. Furthermore, for any divisor $n$ of $p^k - 1$ and any primitive element $\beta \in GF(p^k)$, the element $\alpha = \beta^{(p^k-1)/n}$ will have order $n$. Consequently such elements exist and are easy to find.

**Proof.** For the sake of simplicity, we will just take $l = 0$ and $m = 3$. The proof of the general case uses induction and is a bit more complicated. However, this case illustrates quite well the essential ideas of the general case. Under the assumptions $l = 0, m = 3$, we wish to establish that $d(C_{I_{g(x)}}) \geq 4$. By Lemma 2.14.3, it will suffice to show that every *nonzero* element of $C_{I_{g(x)}}$ has at least four nonzero components. So, let $C = C_{I_{g(x)}}$ and consider any nonzero element $v = (a_0 a_1 \ldots a_{n-1}) \in C$. We want to show that at least four of the $a_i$ are nonzero. We will argue by contradiction. Suppose that no more than three of the $a_i$ are nonzero and that they are among the three components $a_r, a_s, a_t$ (at least one of which is nonzero), where $r < s < t$.

Let

$$h(x) = a_0 + a_1 x + \cdots + a_{n-1} x^{n-1}.$$

Then, since $g(\alpha^i) = 0$, for $i = 1, 2, 3$, we have

$$v \in C \implies h(x) \in I_{g(x)}$$
$$\implies g(x) \mid h(x) \quad \text{(by Lemma 2.14.5)}$$
$$\implies h(\alpha^i) = 0, \quad i = 1, 2, 3$$
$$\implies \sum_{j=0}^{n-1} a_j(\alpha^i)^j = 0, \quad i = 1, 2, 3.$$

Taking advantage of the fact that all the $a_i$ other than $a_r$, $a_s$, and $a_t$ are zero, this reduces to a system of three equations:

$$a_r \alpha^r + a_s \alpha^s + a_t \alpha^t = 0.$$
$$a_r \alpha^{2r} + a_s \alpha^{2s} + a_t \alpha^{2t} = 0. \qquad (2.13)$$
$$a_r \alpha^{3r} + a_s \alpha^{3s} + a_t \alpha^{3t} = 0.$$

Now consider this as a system of three equations in the three unknowns $a_r, a_s, a_t$. This system has a determinant of coefficients

$$\begin{vmatrix} \alpha^r & \alpha^s & \alpha^t \\ \alpha^{2r} & \alpha^{2s} & \alpha^{2t} \\ \alpha^{3r} & \alpha^{3s} & \alpha^{3t} \end{vmatrix} = \alpha^{r+s+t} \begin{vmatrix} 1 & 1 & 1 \\ \alpha^r & \alpha^s & \alpha^t \\ \alpha^{2r} & \alpha^{2s} & \alpha^{2t} \end{vmatrix}$$

$$= \alpha^{r+s+t} \begin{vmatrix} 1 & 1 & 1 \\ 0 & \alpha^s - \alpha^r & \alpha^t - \alpha^r \\ 0 & \alpha^{2s} - \alpha^{r+s} & \alpha^{2t} - \alpha^{r+t} \end{vmatrix}$$

$$= \alpha^{r+s+t} \begin{vmatrix} \alpha^s - \alpha^r & \alpha^t - \alpha^r \\ \alpha^{2s} - \alpha^{r+s} & \alpha^{2t} - \alpha^{r+t} \end{vmatrix}$$

$$= \alpha^{r+s+t}(\alpha^s - \alpha^r)(\alpha^t - \alpha^r) \begin{vmatrix} 1 & 1 \\ \alpha^s & \alpha^t \end{vmatrix}$$

$$= \alpha^{r+s+t}(\alpha^s - \alpha^r)(\alpha^t - \alpha^r)(\alpha^t - \alpha^s).$$

However, $\alpha$ is of order $n$ in $\mathrm{GF}(p^k)$ and therefore the elements $\alpha^r, \alpha^s$, and $\alpha^t$ are distinct. Thus, the system of equations (2.13) has a nonzero determinant of coefficients and therefore has a unique solution $a_r = a_s = a_t = 0$, which contradicts our assumption. Hence, at least four of the components $a_i$ must be nonzero and therefore $d(C) \geq 4$, as required.  $\square$

Note that the information obtained in Theorem 2.14.7 regarding $d(C_{I_{g(x)}})$ can depend on the choice of the element $\alpha$ (see exercises). Note also that Theorem 2.14.7 guarantees a *minimum* distance for the corresponding code. The actual distance of the code might be larger than the guaranteed minimum. Linear codes of the sort constructed in Theorem 2.14.7 were first developed by R. C. Bose and D. Ray-Chaudhuri and, independently, by A. Hocquenghem. As a result, they are referred to as *BCH codes.* One important refinement of the BCH codes was introduced by I. Reed and G. Solomon. In *Reed-Solomon codes,* it is not required that the polynomial $g(x)$ be over the base field (for example, $\mathbb{Z}_p$). This makes it possible to choose exactly the roots that we want $g(x)$ to have and no more. So, for instance, for any element $\alpha$ in $\mathrm{GF}(p^k)$ of order $n$, we can define

$$g(x) = (x - \alpha^{l+1})(x - \alpha^{l+2}) \cdots (x - \alpha^{l+m})$$

so that $g(x)$ is guaranteed to have a run of $m$ powers of $\alpha$ as roots, and thus the corresponding code generated by $g(x)$ will have distance of at least $m + 1$. Of course, the coefficients of code elements are then elements of $\mathrm{GF}(p^k)$, rather than the base field or the prime subfield, but this is easily handled computationally. The big gain here is that, by Lemma 2.14.5, as the degree of $g(x)$ decreases, the dimension of the code space increases. Thus, we can achieve a larger code space with the same distance, or a larger distance with the same dimension for the code space. Thus there is an interesting trade-off between the code distance and the code dimension.

**Example 2.14.8** Let $\mathrm{GF}(16) = \mathbb{Z}_2[z]/(z^4 + z + 1)$. See Example 2.12.12 for background details. Then $\alpha = z$ is a primitive element. Let

$$g(x) = (x^4 + x^3 + x^2 + x + 1)(x^4 + x^3 + 1) \in \mathbb{Z}_2[x].$$

Then $g(x)$ divides $x^{15} - 1$. Moreover (from Example 2.12.12), we see that,

$$
\begin{aligned}
x^4 + x^3 + x^2 + x + 1 \quad &\text{has roots} \quad \alpha^3, \alpha^6, \alpha^9, \alpha^{12} \\
x^4 + x^3 + 1 \quad &\text{has roots} \quad \alpha^7, \alpha^{11}, \alpha^{13}, \alpha^{14}.
\end{aligned}
$$

It follows that $g(x)$ has the run

$$\alpha^{11}, \alpha^{12}, \alpha^{13}, \alpha^{14}$$

among its roots. Therefore,

$$d(C_{I_{g(x)}}) \geq 5.$$

The results of this section introduce some of the fundamental concepts in error correcting codes. Hopefully, this will inspire the reader to explore this fascinating and important topic further. A nice introduction to algebraically based error correcting codes can be found in [Van].

**Exercises 2.14**

1. Let $C = \{(00000), (10101), (11101), (00011)\}$. Find $d(C)$.

2. For any $\alpha = (a_1 a_2 \cdots a_n) \in \mathbb{Z}_p{}^n$, let

$$w(\alpha) = |\{i \mid a_i \neq 0\}|.$$

   Then $w(\alpha)$ is called the *weight* of $\alpha$. Show that if $C \subseteq \mathbb{Z}_p{}^n$ is a linear code, then

$$d(C) = \min\{w(\alpha) \mid \alpha \in C, \alpha \neq 0\}.$$

3. Let $C$ be the code in Exercise 2, with $n = 4$. List all elements in the sphere $S_0$ of radius 2 about $0 = (0000)$ where

$$S_0 = \{\alpha \in C \mid d(0, \alpha) \leq 2\}.$$

4. Construct a binary code $C \subseteq \mathbb{Z}_2{}^8$ with 8 elements and $d(C) = 4$.

5. Let $n \in \mathbb{N}$ and let $C$ be the binary code of length $n$ consisting of all elements of $\mathbb{Z}_2{}^n$ of even "weight"—that is, with an even number of nonzero components.

    (i) Show that $C$ is a cyclic code.
    (ii) What is the dimension of $C$?
    (iii) What is $d(C)$?
    (iv) Show that $C$ contains exactly half the elements of $\mathbb{Z}_2{}^n$.

6. If a message is transmitted using the code

$$C = \{(11000), (01101), (10110), (00011)\}$$

   decode the following received vectors using the nearest-neighbor decoding strategy:

    (i) $(10100)$.
    (ii) $(01010)$.

7. For each of the following polynomials

    (i) $f(x) = x^4 + x + 1$
    (ii) $g(x) = x^4 + x^3 + x^2 + x + 1$
    (iii) $h(x) = (x^4 + x + 1)(x^4 + x^3 + x^2 + x + 1)$

   in $\mathbb{Z}_2[x]$, and using the presentation of GF(16) in Example 2.12.11, find a lower bound for $d(C_{I_{g(x)}})$ based on Theorem 2.14.7.

8. In $\mathbb{Z}_2[x]$, $x^7 - 1 = (x+1)(x^3 + x^2 + 1)(x^3 + x + 1)$. Let
   $g(x) = (x+1)(x^3 + x + 1)$.

   (i) Find the code $C$ corresponding to $g(x)$.
   (ii) Find a primitive element $\alpha \in \mathrm{GF}(8)$.
   (iii) Which powers of $\alpha$ are roots of $g(x)$?
   (iv) What can you say about $d(C)$, from the information in (iii)?

9. Prove Theorem 2.14.2.

10. Prove that $C_{I_{g(x)}}$, as defined prior to Lemma 2.14.4, is a subspace of $\mathbb{Z}_p^n$.

11. Assume the following: The polynomials $f(x) = x^5 + x^3 + 1$,
    $g(x) = x^5 + x^4 + x^3 + x + 1 \in \mathbb{Z}_2[x]$ are irreducible so that, in
    particular, we may take $\mathbb{Z}_2[y]/f(y)$ as a representation of $\mathrm{GF}(32)$. Let
    $\alpha = y$. Then $\beta = \alpha^3$ is a root of $g(x)$.

    (i) Find $C(\alpha^3)$.
    (ii) What does Theorem 2.14.7 tell us about $d(C_{I_{g(x)}})$ based on (i)?
    (iii) Find $C(\beta)$ (in terms of $\beta$).
    (iv) What does Theorem 2.14.7 tell us about $d(C_{I_{g(x)}})$ based on (iii)?

# 3

# Groups and Permutations

Groups first appeared in the early part of the 19th century in the study of solutions to polynomial equations by Abel and Galois. Indeed, it appears that Galois was the first to use the term *group*. The examples that appeared in the work of Abel and Gauss are what we would now recognize as groups of permutations (of roots of polynomials). The abstract definition of a group emerged later in the century in the work of Cayley.

Group theory is, perhaps, at its most picturesque in the study of symmetry—a topic that we will be able to introduce here. We will also see its usefulness in various kinds of counting problems. Nowadays, however, group theory is not just a beautiful part of mathematics, but is also an essential tool in such diverse areas as quantum physics, crystallography, and cryptography.

## 3.1 Basic Properties

A *group* consists of a nonempty set $G$ together with a binary operation $*$ (that is, a pair $(G, *)$) with the following properties:

G1: For all $x, y \in G$, $x * y \in G$.
G2: For all $x, y, z \in G$, $x * (y * z) = (x * y) * z$.
G3: There exists an element $1 \in G$ with

$$1 * x = x * 1 = x \quad \text{for all } x \in G.$$

G4:For all $x \in G$, there exists an element $x' \in G$ with

$$x * x' = x' * x = 1.$$

Strictly speaking, it can be argued that axiom G1 is superfluous. If one assumes that the values of a binary operation are always, by definition, in the set itself, then it is not necessary to list this first axiom. However, it is none-the-less useful to include it, since we may inadvertently consider an operation such as subtraction on the set of positive integers where closure is lacking. It is important to appreciate that we have adopted the symbol $*$ to represent an unspecified operation. In particular situations, we would, naturally, replace $*$ with whatever specific symbol seemed most appropriate. For instance, if we were discussing addition of numbers, vectors or matrices, it would be most natural to use the symbol $+$ for addition. On the other hand, if we were discussing multiplication of numbers or matrices, then it would be most natural to adopt the symbol for multiplication. Except when it is really important to identify which operation $*$ is being used, we usually write just $xy$ instead of $x * y$. The expression $xy$ is referred to as the *juxtaposition* of $x$ and $y$. When there is no risk of confusion concerning the operation, it is usual to refer to "the group $G$", rather than the more formal "the group $(G, *)$".

Systems $(G, *)$ that have some but not all the properties G1 through G4 are also important especially in relation to the theory of automata in theoretical computing science. In particular, $(G, *)$ is a *semigroup* if it has properties G1 and G2, and is a *monoid* if it has properties G1 through G3. Some aspects of these systems will be explored in the exercises.

There are many properties that a group might have in addition to G1 through G4. One is

$$\text{G5. For all} \, x, y \in G, \; x * y = y * x.$$

A group satisfying G5 is called a *commutative group* or an *abelian group* (in honor of the Norwegian mathematician N. H. Abel).

**Example 3.1.1** Some examples of groups are the following:

(1) $(\mathbb{Q}, +)$, $(\mathbb{R}, +)$, $(\mathbb{C}, +)$

(2) $(V, +)$, for every vector space $V$

(3) $(\mathbb{Q}^+, \cdot)$, $(\mathbb{R}^+, \cdot)$, $(\mathbb{C}^*, \cdot)$

(4) $(\mathbb{Z}_n, +)$

(5) The set of invertible matrices over a field, with respect to matrix multiplication

(6) $\{1, -1, i, -i\}$ under multiplication

(7) $(F, +)$ for every field $F$

(8) $(F^*, \cdot)$ for every field $F$

At this point, it is a useful exercise for you to identify some algebraic structures that are not groups (such as real valued functions of a real variable under composition).

**Lemma 3.1.2** *Let G be a group and $x \in G$.*

(i) *$G$ has only one element satisfying G3.*
(ii) *$G$ has only one element $x'$ satisfying G4.*

**Proof.**  (i) Let 1 and $e$ both satisfy G3. Then

$$e = 1\,e \quad \text{by G3 applied to 1}$$
$$= 1 \quad\ \text{by G3 applied to } e.$$

(ii) The proof of (ii) is similar to that of (i) and is left to you as an exercise.  $\square$

Let $G$ be a group. We will denote the unique element satisfying G3 by 1, $1_G$, $e$, or $e_G$ as convenient and call it the *identity* of $G$. However, if we are using addition as the group operation (for example, in $(\mathbb{Q}, +)$), then it is more natural to refer to the unique element satisfying G3 as the *zero* and to denote it by 0. We denote the unique element satisfying G4 by $x^{-1}$, unless the group operation is addition, in which case we denote it by $-x$, and call it the *inverse* of $x$.

**Lemma 3.1.3** *Let G be a group and $x, y \in G$.*

(i) $(x^{-1})^{-1} = x$.
(ii) $(xy)^{-1} = y^{-1}x^{-1}$.

**Proof.**  (i) We leave this for you as an exercise.
(ii) We have

$$xy(y^{-1}x^{-1}) = x(yy^{-1})x^{-1} \quad \text{by G2}$$
$$= x\,1\,x^{-1} \quad\quad\ \text{by G4}$$
$$= xx^{-1} \quad\quad\ \ \text{by G3}$$
$$= 1 \quad\quad\quad\ \ \text{by G4}.$$

Similarly, $y^{-1}x^{-1}xy = 1$. Thus, $y^{-1}x^{-1}$ satisfies G4 (with respect to $xy$) so that $y^{-1}x^{-1}$ must be the inverse of $xy$.  $\square$

**Lemma 3.1.4** *Let $G$ be a group and $a, b, c \in G$.*
   *(i) The cancellation laws hold in $G$—that is,*

$$ac = bc \Rightarrow a = b,$$
$$ca = cb \Rightarrow a = b.$$

   *(ii) The equation $ax = b$ has a unique solution (for $x$).*

**Proof.** (i) We have

$$
\begin{aligned}
ac = bc &\Rightarrow (ac)c^{-1} = (bc)c^{-1} \\
&\Rightarrow a(cc^{-1}) = b(cc^{-1}) \\
&\Rightarrow a1 = b1 \\
&\Rightarrow a = b.
\end{aligned}
$$

This establishes the first implication in (i) and the second follows from a similar argument.

   (ii) We first show that the equation has a solution. Let $c = a^{-1}b$. Then

$$
\begin{aligned}
ac = a(a^{-1}b) &= (aa^{-1})b \\
= 1b \quad\quad &= b
\end{aligned}
$$

so that $c$ is a solution. Now suppose that $d$ is also a solution. Then

$$ac = b = ad$$

so that $ac = ad$ and, by part (i), $c = d$. Thus, the solution is unique. $\quad\square$

   It is useful when calculating in a group to be able to use exponents, so we define, for any element $x$ of a group $G$,

$$x^0 = 1, x^1 = x$$
$$x^{n+1} = (x^n)x \quad (n \geq 0)$$
$$x^{-n} = (x^{-1})^n \quad (n > 0).$$

Similarly, when the operation is addition, we find it useful to adopt the usual notation for multiples:

$$0 \cdot x = 0, 1 \cdot x = x$$
$$(n+1)x = nx + x \quad (n \geq 0)$$
$$(-n-1)x = (-n)x - x \quad (n > 0).$$

Typically we assume that any unspecified group operation is multiplication. However, every result could be reformulated in terms of $*$ or $+$ (even if $+$ is not a commutative operation).

**Lemma 3.1.5** *Let G be a group and $x \in G$. Then, for any $m, n \in \mathbb{Z}$,*

$$x^m x^n = x^{m+n}, \quad (x^m)^n = x^{mn}.$$

**Proof.** Exercise. $\square$

For any group $G$, we call $|G|$, that is, the number of elements in $G$, the *order* of $G$. If $x \in G$ and there exists a positive integer $m$ with $x^m = 1$, then we call the least such integer the *order* of $x$ and denote it by ord $(x)$ or $|x|$. If there is no such integer, then we say that $x$ has *infinite order*. Note that we always have $|1| = 1$ whenever 1 denotes the identity in a group.

**Example 3.1.6** (1) In $(\mathbb{Z}, +)$, every nonzero element has infinite order.
    (2) In $(\mathbb{Z}_{12}, +)$ we have

| $x$ | order of $x$ |
|:---:|:---:|
| 0 | 1 |
| 1,5,7,11 | 12 |
| 2,10 | 6 |
| 3,9 | 4 |
| 4,8 | 3 |
| 6 | 2 |

**Lemma 3.1.7** *Let G be a group and $a \in G$ be an element of order $m$. Let $n \in \mathbb{Z}$.*

(i) $|a^{-1}| = m$.
(ii) $a^n = 1 \iff m \mid n$.

**Proof.** (i) Exercise.
    (ii) If $m \mid n$, then there exists an integer $k$ with $n = km$. Hence,

$$a^n = a^{km} = (a^m)^k = 1^k = 1.$$

Conversely, suppose that $a^n = 1$. By the division algorithm, there exist integers $q, r$ with

$$n = qm + r \quad \text{with } 0 \le r < m.$$

Then

$$1 = a^n = a^{qm+r} = (a^m)^q \, a^r$$
$$= 1^q \, a^r = a^r.$$

By the minimality of $m$, we must have $r = 0$. Therefore, $n = qm$ and $m \mid n$ as desired. $\square$

If $G = \{e = a_1, a_2, \ldots, a_n\}$ is a group, then it is possible to construct a *multiplication table* (sometimes called a *Cayley table*) that displays the values of all products in $G$:

|       | $e$   | $a_2$ | $a_3$ | $\ldots$ | $a_n$ |
|-------|-------|-------|-------|----------|-------|
| $e$   | $e$   | $a_2$ | $a_3$ |          | $a_n$ |
| $a_2$ | $a_2$ |       |       |          |       |
| $a_3$ | $a_3$ |       |       |          |       |
| $\vdots$ |    |       |       |          |       |
| $a_n$ | $a_n$ |       |       |          |       |

The $(i, j)$th entry in the table is the element $a_i a_j$.

An important observation concerning the multiplication table of a group is that the elements in any one row (respectively, column) must all be distinct. This is a consequence of the cancellation law established in Lemma 3.1.4(i). By Lemma 3.1.4(ii), every element of $G$ must appear in each row and column. Thus, each row and each column is a rearrangement of the elements of $G$. In other words, every Cayley table is a latin square.

**Example 3.1.8** We will use the idea of a Cayley table to show that there is only one group of order 3 (apart from renaming the elements). Let $G$ be a group of order 3. We know that it must have an identity; let us call it $e$. Let the remaining elements be $a$ and $b$. When we look at the Cayley table, we find that there are just four positions to be filled:

|     | $e$ | $a$ | $b$ |
|-----|-----|-----|-----|
| $e$ | $e$ | $a$ | $b$ |
| $a$ | $a$ | $*$ | $*$ |
| $b$ | $b$ | $*$ | $*$ |

What options do we have in filling out the second row? We have two positions into which we must place $e$ and $b$ in some order. However, the third column already has a $b$. So we must place $b$ in the second column and $e$ in the third to obtain the following:

$$
\begin{array}{c|ccc}
 & e & a & b \\
\hline
e & e & a & b \\
a & a & b & e \\
b & b & * & *
\end{array}
$$

This leaves us only one option for the third row:

$$
\begin{array}{c|ccc}
 & e & a & b \\
\hline
e & e & a & b \\
a & a & b & e \\
b & b & e & a.
\end{array}
$$

Note that this is essentially the same table as we obtain with $(\mathbb{Z}_3, +)$ under the relabeling

$$0 \leftrightarrow e, \quad 1 \leftrightarrow a, \quad 2 \leftrightarrow b.$$

An interesting way of presenting a group is in terms of *generators* and *relations*. For example, if we write

$$G = \langle a, b \mid a^2 = b^3 = e, ab = b^{-1}a \rangle$$

then we mean that every element of $G$ can be expressed as a product of $a$'s and $b$'s, and their inverses, and that the multiplication is completely determined by the relations

$$a^2 = b^3 = e, \quad ab = b^{-1}a.$$

Notice that, since $a^2 = e$ we have $a^{-1} = a$, and since $b^3 = e$ we have $b^{-1} = b^2$, so that we could write $ab = b^{-1}a$ as $ab = b^2 a$. Thus, every element of $G$ can be written as a product involving just $a$ and $b$ without using inverses at all. Moreover, if we write down a product $x_1 x_2 x_3 \cdots x_n$ where each $x_i \in \{a, b\}$, then we can replace every occurrence of $ab$ by $b^2 a$. This means that every element of $G$ is of the form $b^r a^s$ for some $r, s \geq 0$. Since $a^2 = b^3 = e$, we can actually restrict $r$ and $s$ so that

$$0 \leq r \leq 2, \ 0 \leq s \leq 1.$$

Hence,

$$G = \{e, a, b, ba, b^2, b^2 a\}$$

and we may use the given rules to calculate the full multiplication table.

There is a very simple way of constructing new groups from "old" groups. Let $G_1, \ldots, G_n$ be groups and let

$$G = G_1 \times G_2 \times \cdots \times G_n$$

be endowed with the multiplication $*$ where

$$(a_1, a_2, \ldots, a_n) * (b_1, b_2, \ldots, b_n) = (a_1 b_1, a_2 b_2, \ldots, a_n b_n).$$

Then $(G, *)$ is a group (see the exercises) known as the (*external*) *direct product* of the groups $G_1, \ldots, G_n$. The basic properties of the direct product of groups are explored more fully in section 4.3. Of course, to be able to take advantage of this construction, we need to have a good supply of groups to start with. One important and interesting family of groups arises as discussed next.

Let $P$ be a set of points in $\mathbb{R}^1$, $\mathbb{R}^2$, or $\mathbb{R}^3$. Then a *rotational symmetry* of $P$ is a bijection of the set $P$ to itself that can be achieved by means of some rigid physical motion. Usually we want to take $P$ to be the set of points in some geometric shape like a triangle, square, . . . or cube, tetrahedron, and so on, or perhaps to be the set of vertices of such a shape.

It is a somewhat surprising fact that every rigid motion of an object in three dimensions that leaves the object occupying exactly the same space as it did before the motion (but with a different orientation) can be achieved by means of a single rotation about some axis through the object. The verification of this important fact unfortunately lies beyond the scope of this introductory text.

For any fixed figure $P$, we can combine such rotations by performing them successively one after the other. We can also invert them by reversing the rotation, and there is an "identity" that leaves everything fixed. In this way we obtain a *group of rotational symmetries.* Note that, if $R$ and $S$ are both rotational symmetries then we write $RS$ to represent the rotational symmetry that is obtained by first applying $S$ and then $R$. This is consistent with the manner in which we write the composition of two functions.

## Example 3.1.9

(1) A line segment has 2 rotational symmetries.

(2) An equilateral triangle has 6 rotational symmetries.

(3) A square has 8 rotational symmetries.

(4) A regular $n$-gon has $2n$ rotational symmetries.

We will denote the group of symmetries of an equilateral triangle by $D_3$, of a square by $D_4$, and of a regular $n$-gon by $D_n$.

Consider the rotational symmetries of a square:

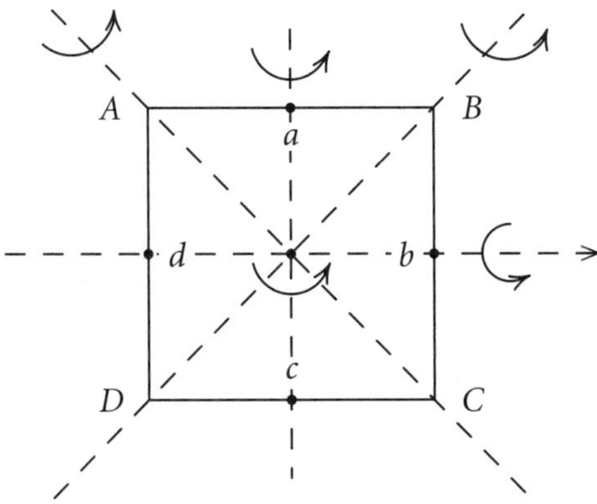

Let $R$ denote the clockwise rotation through $90°$ about an axis that is perpendicular to the plane of the square. Repeating this rotation gives us 4 distinct rotations, including the identity $I$:

$$I, R, R^2, R^3.$$

In addition to these rotations (in the plane of the square), it is clear that we can also flip the square over—that is, by rotating the square through $180°$ about one of the following axes:

$\rho_{A_1}$: the axis $A_1$ that runs from top left to bottom right
$\rho_{A_2}$: the axis $A_2$ that runs from top right to bottom left
$\rho_{A_3}$: the axis $A_3$ that runs from top middle to bottom middle
$\rho_{A_4}$: the axis $A_4$ that runs from left side middle to right side middle.

By considering all possible positions of the corners, we see that we have found all possible rotations and it is easily confirmed that all combinations of rotations will yield rotations so that we do indeed have a group of rotations

$$D_4 = \{I, R, R^2, R^3, \rho_{A_1}, \rho_{A_2}, \rho_{A_3}, \rho_{A_4}\}.$$

If we let $\rho = \rho_{A_1}$, it is easy to see that

$$\rho_{A_2} = R^2\rho, \quad \rho_{A_3} = R\rho, \quad \rho_{A_4} = R^3\rho$$

(always remembering that we are composing functions so that in computing $R\rho$, we apply $\rho$ first and then $R$). This allows us to list the elements of $D_4$ as

$$D_4 = \{I, R, R^2, R^3, \rho, R\rho, R^2\rho, R^3\rho\}$$

where every rotation is expressed as a product of powers of just two rotations—namely, $R$ and $\rho$—which is often useful. There is nothing special about the choice of $\rho = \rho_{A_1}$ here; any of the choices $\rho = \rho_{A_2}$, $\rho = \rho_{A_3}$, $\rho = \rho_{A_4}$ would work equally well. Note that $|D_4| = 8$.

In a similar manner we can see that

$$D_n = \{I, R, R^2, \ldots, R^{n-1}, \rho, R\rho, \ldots, R^{n-1}\rho\}$$

where $R$ denotes the rotation about the axis perpendicular to the regular $n$-gon through an angle of $360°/n$ and $\rho$ denotes a rotation about a suitable axis that bisects the $n$-gon (see Fig. 3.1). Thus $D_n$ has precisely $2n$ elements. It is for this reason that the notation for this group is not standard and some authors choose to write $D_{2n}$ (highlighting the size of the group) where we will write $D_n$, (highlighting the number of vertices of the underlying n-gon).

The simplest groups of rotations of three-dimensional objects arise in association with $n$-faced pyramids, $n > 3$, with bases forming regular polygons (see Fig. 3.2 for square-based and pentagonal-based pyramids).

Clearly, the only rotational symmetries are those about a vertical axis through the apex by multiples of $360°/n$. If $R$ denotes the rotation through

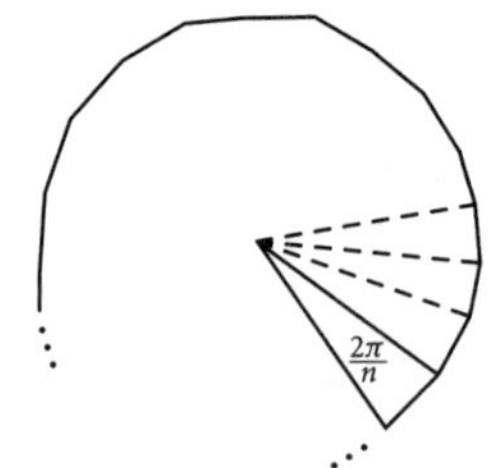

**Figure 3.1**

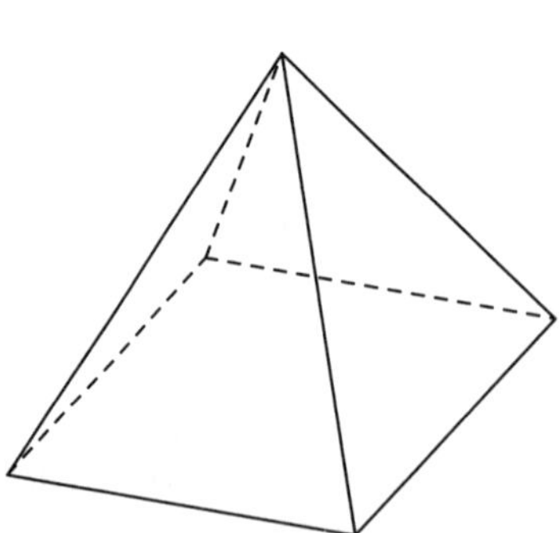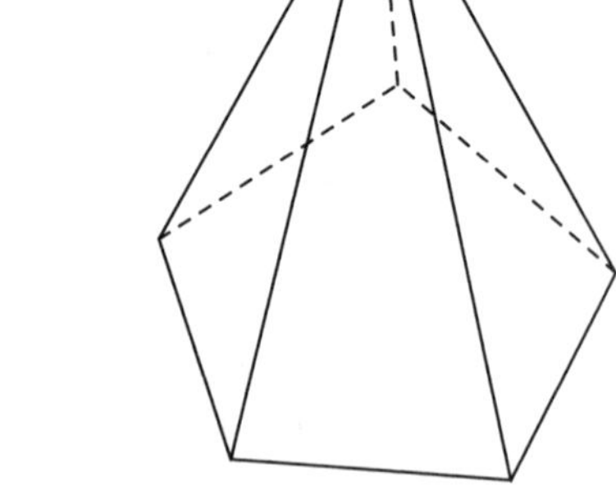

**Figure 3.2**

$360°/n$, then the group is

$$\{I, R, R^2, \ldots, R^{n-1}\}.$$

**Exercises 3.1**

1. Show that the set of invertible elements in $\mathbb{Z}_n$ is an Abelian group with respect to multiplication.

2. Are the following groups? ($\mathcal{P}(X)$ denotes the set of all subsets of a set $X$ and for any subsets $A$ and $B$ of $X$, $A * B = (A\backslash B) \cup (B\backslash A)$.)

    (i) $(\mathbb{Z}, -)$.
    (ii) $(\mathbb{Q}^*, \div)$.
    (iii) $(\mathcal{P}(X), \cap)$.
    (iv) $(\mathcal{P}(X), *)$.
    (v) $(\mathbb{Z}_n, \cdot)$.

3. Let $X$ be a nonempty set.

    (i) Is the set $\mathcal{T}_X$ of all mappings of $X$ to $X$ a group?
    (ii) Is the set of all injective mappings from $X$ to $X$ a group?
    (iii) Is the set of all surjective mappings from $X$ to $X$ a group?

4. Show that the set $U = \{1, -1, i, -i\}$ of complex numbers is a group with respect to multiplication.

5. Suppose that $G$ is a finite subset of $(\mathbb{R}^*, \cdot)$, which is a group. Describe the possibilities for $G$.

6. Write out the Cayley table for $(\mathbb{Z}_4, +)$.

7. Write out the Cayley table for the group $U$ of invertible elements in

    (i) $\mathbb{Z}_6$
    (ii) $\mathbb{Z}_8$
    (iii) $\mathbb{Z}_{10}$.

8. Let $G = \{e, a, b, c\}$. Using the fact that Cayley tables are latin squares, show that there are essentially (that is, to within just exchanging the roles of elements $a, b, c$) only two group structures on $G$ with $e$ as the identity. (These two groups must look like $(\mathbb{Z}_4, +)$ and $(\mathbb{Z}_2, +) \times (\mathbb{Z}_2, +)$. The group $(\mathbb{Z}_2, +) \times (\mathbb{Z}_2, +)$ is called the *Klein 4-group*.)

9. Show that the inverse of an element in a group is unique.

10. Show that for any element $x$ in a group $G$

    (i) $(x^{-1})^{-1} = x$
    (ii) $|x^{-1}| = |x|$.

11. Let $G$ be a group, $a, b \in G$, $|a| = m$, $|b| = n$. Establish the following:

   (i) $k \mid m \implies |a^k| = \frac{m}{k}$.

   (ii) $(k, m) = 1 \implies |a^k| = m$.

   (iii) $ab = ba$, $(m, n) = 1 \implies |ab| = mn$.

12. Let $G$ be a group. Show that there is exactly one element $x \in G$ such that $x^2 = x$.

13. Let $x$ and $y$ be elements of a group such that $(xy)^2 = x^2 y^2$. Show that $xy = yx$.

14. Let $G$ be a group such that $x^2 = e$ for all $x \in G$. Show that $G$ is abelian.

15. Let $(G, *)$ be the direct product of the groups $G_i$ where $G_i$ has identity $e_i$. Show that $(G, *)$ is a group where

   (i) $e_G = (e_1, \ldots, e_n)$

   (ii) $(a_1, \ldots, a_n)^{-1} = (a_1^{-1}, a_2^{-1}, \ldots, a_n^{-1})$

   (iii) $|G| = |G_1| \, |G_2| \, \cdots \, |G_n|$.

16. Describe every element of $D_3$ as a single rotation and as a product of powers of two "generating" rotations.

17. Describe every element of $D_n$ as a single rotation.

18. Construct a Cayley table for $D_3$.

19. Show that $D_n$ $(n \geq 3)$ is not abelian.

*20. Let $F$ be a field. Show that the set $\mathrm{Aut}(F)$ of all automorphisms of $F$ is a group with respect to composition of functions.

21. Let $P$ be the prime subfield of a field $F$. Show that $\alpha(a) = a$ for all $a \in P, \alpha \in \mathrm{Aut}(F)$.

22. Let $X$ be a nonempty set and $F$ denote the set of all sequences of the form $x_1 x_2 \cdots x_n$ $(x_i \in X)$, including the empty sequence $\emptyset$. Define the product of two words in $F$ to be the word obtained by joining the two sequences

$$(x_1 x_2 \cdots x_m) * (y_1 y_2 \cdots y_n) = x_1 x_2 \cdots x_m y_1 y_2 \cdots y_n.$$

   Show that $(F, *)$ is a monoid.

23. Show that the set $T_X$ of all mappings from a set $X$ to itself under composition of functions is a monoid.

24. Show that the set $C_{\mathbb{R}}$ of all continuous functions from $\mathbb{R}$ to $\mathbb{R}$ under composition of functions is a monoid.

25. Let $I$ be a nonempty set and $S = I \times I$ and define a multiplication on $S$ by $(i, j) * (k, \ell) = (i, \ell)$. Show that $S$ is a semigroup and is a monoid if and only if $|I| = 1$.

26. Let $S$ be a semigroup. Define the relations $\mathcal{L}$ and $\mathcal{R}$ on $S$ by

$$a \mathcal{L} b \iff Sa \cup \{a\} = Sb \cup \{b\} \quad \text{and} \quad a \mathcal{R} b \iff aS \cup \{a\} = bS \cup \{b\}.$$

   (i) Show that $\mathcal{L}$ and $\mathcal{R}$ are equivalence relations on $S$.
   (ii) Let $a, b, c, d, x, y \in S$ be such that

$$a \mathcal{L} b, b \mathcal{R} c, a = xb \quad \text{and} \quad c = by.$$

   Let $d = xby$. Show that $a \mathcal{R} d$ and $d \mathcal{L} c$.

27. Let $S$ be a monoid and call an element $a \in S$ *invertible* if there exists an element $x \in S$ with $ax = xa = 1$. Show that the set $U$ of all invertible elements of $S$ is a group.

## 3.2 Subgroups

Let $H$ be a subset of $G$. If $H$ is itself a group under the same operations as in $G$, then $H$ is a *subgroup* of $G$. For every group $G$, we always have the subgroups $\{e\}$ and $G$. We refer to $\{e\}$ as the *trivial* subgroup. If $H$ is a subgroup of $G$ and $H \neq G$, then $H$ is a *proper* subgroup.

Since not all subsets are subgroups, our first task is to develop a suitable test for subgroup status!

**Theorem 3.2.1** *Let $H$ be a nonempty subset of a group $G$. Then the following statements are equivalent:*

   (i) *$H$ is a subgroup of $G$.*
   (ii) *$a, b \in H \implies ab, a^{-1} \in H$.*
   (iii) *$a, b \in H \implies ab^{-1} \in H$.*

*Note:* Before proceeding to the proof, take note of the important precondition that $H \neq \emptyset$. The easiest way to establish that $H \neq \emptyset$ is usually to show that $e \in H$.

**Proof.** (i) *implies* (ii). Let $a, b \in H$. From the definition of a subgroup, we know immediately that $ab \in H$. We also know that $H$ must have an identity $e_H$. If $e$ denotes the identity of $G$, then

$$e \, e_H = e_H = e_H \, e_H$$

so that, by cancellation in $G$, we have $e_H = e$. Let $a^{-1}$ denote the inverse of $a$ in $G$. Since $H$ is a subgroup, there exists $x \in H$ with

$$ax = e = aa^{-1}.$$

Again, by cancellation in $G$, we have $a^{-1} = x \in H$.

(ii) *implies* (iii). Let $a, b \in H$. By (ii), we have $b^{-1} \in H$ and, therefore, by (ii) again we have $ab^{-1} \in H$.

(iii) *implies* (i). By hypothesis, $H \neq \emptyset$. So let $a, b \in H$. Then

$$e = aa^{-1} \in H \tag{3.1}$$

$$b^{-1} = eb^{-1} \in H \tag{3.2}$$

from which we have

$$ab = a(b^{-1})^{-1} \in H.$$

Thus, $H$ is closed under the operation of $G$. The multiplication must be associative, since it is associative in $G$. By (3.1) we know that $H$ contains an identity element and, by (3.2), each element of $H$ has an inverse in $H$. Therefore G1 through G4 are satisfied and $H$ is a group with respect to the operation in $G$. $\square$

As part of the proof of Theorem 3.2.1, we have shown that a group and a subgroup must share the same identity. For the sake of highlighting this fact, we state it separately.

**Corollary 3.2.2** *Let $H$ be a subgroup of $G$. Then $H$ and $G$ have the same identity.*

**Theorem 3.2.3** *Let $G$ be a group and $a \in G$. Then*

$$\langle a \rangle = \{a^n \mid n \in \mathbb{Z}\}$$

*is a subgroup of $G$.*

**Proof.** Since $a = a^1 \in G$, we know that $G \neq \emptyset$. For $a^m, a^n \in \langle a \rangle$ we have

$$a^m(a^n)^{-1} = a^m a^{-n} = a^{m-n} \in \langle a \rangle.$$

By Theorem 3.2.1, $\langle a \rangle$ is a subgroup. $\square$

We refer to the subgroup $\langle a \rangle$ introduced in Theorem 3.2.3 as the *(cyclic) subgroup of $G$ generated by $a$*. If $\langle a \rangle = G$, then we say that $G$ is a *cyclic* group *generated by $a$*. More generally, for any nonempty subset $A$ of $G$, we denote the set of all products of elements, or inverses of elements from $A$ by $\langle A \rangle$. Then $\langle A \rangle$ is a subgroup of $G$, indeed, the smallest subgroup of $G$ containing $A$ and we call it the *subgroup generated by $A$*. For the sake of completeness, we define the subgroup generated by the empty set $\emptyset$ to be $\{1\}$. Thus $\langle \emptyset \rangle = \langle 1 \rangle = \{1\}$.

**Example 3.2.4**  In $(\mathbb{Z}_{12}, +)$, we have

$$\langle 0 \rangle = \{0\}$$
$$\langle 1 \rangle = \langle 5 \rangle = \langle 7 \rangle = \langle 11 \rangle = \mathbb{Z}_{12}$$
$$\langle 2 \rangle = \langle 10 \rangle = \{0, 2, 4, 6, 8, 10\}$$
$$\langle 3 \rangle = \langle 9 \rangle = \{0, 3, 6, 9\}$$
$$\langle 4 \rangle = \langle 8 \rangle = \{0, 4, 8\}$$
$$\langle 6 \rangle = \{0, 6\}.$$

An important subgroup of any group $G$ is its *center:*

$$Z(G) = \{a \in G \mid ax = xa \quad \text{for all } x \in G\}.$$

We say that two elements $a$, $b$ in a group $G$ *commute* if $ab = ba$. Thus the center of a group $G$ consists of those elements that commute with every other element of the group.

**Example 3.2.5**  For every group $G$, the center $Z(G)$ of $G$ is a subgroup of $G$.

**Proof.**  Exercise.  $\square$

It is time to give a formal expression to the idea that two groups are "essentially the same". Let $(G, *)$ and $(H, \cdot)$ be groups. A mapping $\varphi : G \to H$ is a *homomorphism* if

(i) $\varphi(a * b) = \varphi(a) \cdot \varphi(b)$ for all $a, b \in G$.

If, in addition, we have

(ii) $\varphi$ is bijective

then we say that $\varphi$ is an *isomorphism* and that $(G, *)$ and $(H, \cdot)$ (or simply $G$ and $H$) are *isomorphic* and write $G \cong H$.

**Example 3.2.6**  Let

$$\theta : a \to e^a \ (a \in \mathbb{R})$$
$$\varphi : a \to \ln a \, (a \in \mathbb{R}^+)$$

where $e$ is the usual exponential constant $2.71\ldots$ from calculus. Then $\theta$ is an isomorphism of $(\mathbb{R},+)$ to $(\mathbb{R}^+,\cdot)$ and $\varphi$ is an isomorphism from $(\mathbb{R}^+,\cdot)$ to $(\mathbb{R},+)$.

**Theorem 3.2.7** *Let $G$ be a cyclic group with generator $a$.*

(i) *If there exist distinct integers $i$ and $j$ with $a^i = a^j$, then there exists $n \in \mathbb{N}$ and an isomorphism $\varphi : G \to (\mathbb{Z}_n,+)$ with $\varphi(a) = 1$.*

(ii) *Otherwise, $G \cong (\mathbb{Z},+)$.*

**Proof.** (i) Let $i$ and $j$ be distinct integers with $a^i = a^j$. Without loss of generality, we may assume that $i < j$. Then

$$e = a^i a^{-i} = a^j a^{-i} = a^{j-i}$$

where $j - i > 0$. Thus, we know that for some positive integer $k$, we have $a^k = e$.

Let $n$ be the smallest positive integer such that $a^n = e$. The preceding argument then shows that the elements $e, a, \ldots, a^{n-1}$ are all distinct. For any $m \in \mathbb{Z}$, let $q$ and $r \in \mathbb{Z}$ be such that

$$m = nq + r \qquad 0 \le r < n.$$

Then

$$a^m = (a^n)^q a^r = a^r.$$

Hence,

$$G = \{e = a^0, a, \ldots, a^{n-1}\}.$$

Now define $\varphi : G \to \mathbb{Z}_n$ by

$$\varphi(a^r) = r \qquad 0 \le r \le n - 1.$$

Clearly, $\varphi$ is a bijection. Also, for any $i, j \in \{0, 1, 2, \ldots, n - 1\}$, let $s, t \in \mathbb{Z}$ be such that

$$i + j = ns + t \qquad 0 \le t \le n - 1.$$

Then

$$\varphi(a^i a^j) = \varphi(a^{i+j})$$
$$= \varphi(a^t)$$
$$= t$$
$$\equiv_n i + j$$
$$= \varphi(a^i) + \varphi(a^j)$$

and $\varphi$ is an isomorphism.

(ii) Now suppose that for all distinct integers $i, j$ we have $a^i \neq a^j$. Define $\varphi : G \to \mathbb{Z}$ by

$$\varphi(a^i) = i \qquad (i \in \mathbb{Z}).$$

It follows immediately from the hypothesis in this case that $\varphi$ is a bijection. Also

$$\varphi(a^i a^j) = \varphi(a^{i+j}) = i + j$$
$$= \varphi(a^i) + \varphi(a^j).$$

Therefore $\varphi$ is an isomorphism. $\quad\square$

**Corollary 3.2.8** *Let $G$ be a group and $a \in G$. Then the order of $a$ equals the order of $\langle a \rangle$—that is, $|a| = |\langle a \rangle|$.*

**Proof.** If there exist distinct integers $i, j \in \mathbb{Z}$ with $i \neq j$ and $a^i = a^j$, then by Theorem 3.2.7, there exist $n \in \mathbb{N}$ and an isomorphism $\varphi : \langle a \rangle \to \mathbb{Z}_n$ such that $\varphi(a) = 1$. Then

$$|\langle a \rangle| = |\mathbb{Z}_n| = n$$

whereas

$$|a| = |\varphi(a)| = |1| = n.$$

Thus, $|a| = |\langle a \rangle|$.

If, on the other hand, $a^i \neq a^j$ whenever $i \neq j$, then clearly $|a|$ and $|\langle a \rangle|$ are both infinite. $\quad\square$

The following notation is useful when discussing the subgroups of $\mathbb{Z}$ or $\mathbb{Z}_n$. For any abelian group $(G, +)$ and any $m \in \mathbb{N}$, let

$$mG = \{mg \mid g \in G\}.$$

This will be explored further in the exercises.

### Exercises 3.2

1. Let $Q^+ = \{x \in \mathbb{Q} \mid x > 0\}$.

   (i) Is $(Q^+, +)$ a subgroup of $(Q, +)$?
   (ii) Is $(Q^+, \cdot)$ a subgroup of $(Q, +)$?
   (iii) Is $(\mathbb{Z}_7^*, \cdot)$ a subgroup of $(\mathbb{Z}_7, +)$?

2. Find the smallest subgroup of $(\mathbb{Z}_{16}, +)$ containing

   (i) 10
   (ii) 4 and 7.

3. Find all subgroups of

   (i) $D_3$
   (ii) $D_4$.

   Show that $D_4$ contains a subgroup that is not cyclic.

4. Find subgroups of $D_n$ of orders 2 and $n$.

5. Let $G$ be a group. Show that the center $Z(G)$ of $G$ is a subgroup.

6. Let $G$ be a group and $a \in G$. Show that $Z_G(a) = \{x \in G \mid ax = xa\}$ is a subgroup of $G$.

7. Find

   (i) $Z(D_3)$
   (ii) $Z(D_4)$
   (iii) $Z(D_{2n+1})$, $n \geq 2$
   (iv) $Z(D_{2n})$, $n \geq 2$.

*8. Let $Q_8$ be the subgroup of the group of $2 \times 2$ invertible matrices with complex entries generated by the matrices

$$A = \begin{bmatrix} 0 & 1 \\ -1 & 0 \end{bmatrix} \qquad B = \begin{bmatrix} 0 & i \\ i & 0 \end{bmatrix}.$$

   (i) Find the 8 elements of $Q_8$.
   (ii) Show that $Q_8$ is not isomorphic to $D_4$.
   (iii) Show that $Q_8$ is not commutative.
   (iv) Find all subgroups of $Q_8$.
   (v) Show that every proper subgroup of $Q_8$ is cyclic.

   $Q_8$ is known as the *quaternion group*.

*9. (Alternative definition of $Q_8$.) Let

$$Q_8 = \langle a, b \mid a^4 = e, \ b^2 = a^2, \ ba = a^{-1}b \rangle.$$

Show that every element in $Q_8$ can be written in the form $a^i b^j$ where $0 \le i \le 3$, $0 \le j \le 1$, and deduce that $|Q_8| = 8$.

(By completing the Cayley tables for $Q_8$ (as in Exercise 8) and as defined here, it becomes evident that the correspondence $a \leftrightarrow A$, $b \leftrightarrow B$ provides an isomorphism between the two descriptions.)

10. Let $G$ be a group and $H$ be a finite subset of $G$ such that

    (i) $H \ne \emptyset$
    (ii) $a, b \in H \implies ab \in H$.

    Show that $H$ is a subgroup of $G$.

11. Let $(G, +)$ be an abelian group and $m \in \mathbb{N}$. Show that $mG$ is a subgroup of $G$.

12. Let $(G, +)$ be an abelian group and $n \in \mathbb{N}$. Show that $G_n = \{a \in G \mid na = 0\}$ is a subgroup of $G$.

13. Show that every subgroup $A$ of $(\mathbb{Z}, +)$ is of the form $m\mathbb{Z}$ for some $m \in \mathbb{N}$.

14. Show that every subgroup $A$ of $(\mathbb{Z}_n, +)$ is of the form $m\mathbb{Z}_n$ for some $m \in \mathbb{N}$ such that $m \mid n$.

15. Show that if $H$ and $K$ are subgroups of a group $G$, then so also is $H \cap K$.

16. Let $\varphi : G \to H$ be an isomorphism where $G$ and $H$ are groups. Let $a \in G$.

    (i) Show that $\varphi(a^{-1}) = \varphi(a)^{-1}$.
    (ii) Show that $|\varphi(a)| = |a|$.

17. Let $G, H$ be groups and $\varphi : G \to H$ be a homomorphism.

    (i) Show that $\varphi(G) = \{\varphi(a) \mid a \in G\}$ is a subgroup of $H$.
    (ii) Show that $\ker(\varphi) = \{a \in G \mid \varphi(a) = e_H\}$ is a subgroup of $G$.

18. Show that $(\mathbb{Z}_2, +) \times (\mathbb{Z}_2, +)$ is not isomorphic to $(\mathbb{Z}_4, +)$.

19. Show that $(\mathbb{Z}_2, +) \times (\mathbb{Z}_3, +)$ is isomorphic to $(\mathbb{Z}_6, +)$.

20. Let $m, n \in \mathbb{N}$ be relatively prime. Show that $(\mathbb{Z}_m, +) \times (\mathbb{Z}_n, +)$ is isomorphic to $(\mathbb{Z}_{mn}, +)$.

21. Find an irregular $m$-gon with a group of rotations isomorphic to $(\mathbb{Z}_n, +)$.

22. Let $F$ be a field and $H$ a subgroup of $\mathrm{Aut}(F)$. Show that

$$K = \{a \in F \mid \alpha(a) = a \quad \text{for all } \alpha \in H\}$$

is a subfield of $F$.

23. Let $G$ be a group.

    (i) Show that, for any $a \in G$, the mapping

$$\varphi_a : x \rightarrow axa^{-1} \qquad (x \in G)$$

    is an isomorphism of $G$ to itself.

    (ii) Show that the mapping

$$\theta : a \rightarrow \varphi_a \qquad (a \in G)$$

    is a homomorphism of $G$ into the group of permutations $S_G$ of $G$.

    (iii) Show that the *kernel* of $\theta$—that is, $\{a \in G \mid \theta(a) = e\}$—is equal to $Z(G)$, the center of $G$.

## 3.3 Permutation Groups

By a *permutation* of a set $X$ we mean a bijection of $X$ onto $X$. Generally, we denote the set of all permutations of $X$ by $S_X$. However, if $X = \{1, 2, 3, \ldots, n\}$ then we write $S_X = S_n$.

Let $\alpha \in S_n$. Since $\alpha$ is defined on a finite set and has only a finite number of values, we can display them in an array with two rows:

$$\alpha = \begin{pmatrix} 1 & 2 & 3 & \cdots & n \\ \alpha(1) & \alpha(2) & \alpha(3) & & \alpha(n) \end{pmatrix}.$$

For example,

$$\alpha = \begin{pmatrix} 1 & 2 & 3 & 4 & 5 \\ 2 & 3 & 5 & 4 & 1 \end{pmatrix}$$

means that $\alpha \in S_5$ and

$$\alpha(1) = 2, \ \alpha(2) = 3, \ \alpha(3) = 5, \ \alpha(4) = 4, \ \alpha(5) = 1.$$

An element of the form

$$\begin{pmatrix} 1 & 2 & 3 & \cdots & n \\ a_1 & a_2 & a_3 & & a_n \end{pmatrix}$$

describes a unique permutation of $S_n$ if and only if every element of $\{1, 2, \ldots, n\}$ appears exactly once in the second row. The number of ways in which this can happen is precisely $n!$. Thus we have the following lemma.

**Lemma 3.3.1** $|S_n| = n!$.

Of special importance is the *identity* permutation on a set $X$ defined by

$$\epsilon(x) = x, \quad \text{for all } x \in X.$$

If $X = \{1, 2, \ldots, n\}$, then we write the identity permutation as

$$\epsilon = \begin{pmatrix} 1 & 2 & 3 & \cdots & n \\ 1 & 2 & 3 & & n \end{pmatrix}.$$

Since permutations are functions, we can take two permutations $\alpha$ and $\beta$ (of the *same* set $X$) and form their *composition* $\alpha \circ \beta$. This is defined as follows: For $x \in X$,

$$\alpha \circ \beta(x) = \alpha(\beta(x)).$$

Generally we write simply $\alpha\beta$ for $\alpha \circ \beta$.

In $S_5$, let

$$\alpha = \begin{pmatrix} 1 & 2 & 3 & 4 & 5 \\ 2 & 3 & 4 & 5 & 1 \end{pmatrix}, \quad \beta = \begin{pmatrix} 1 & 2 & 3 & 4 & 5 \\ 2 & 1 & 4 & 5 & 3 \end{pmatrix}.$$

Then,

$$\alpha\beta = \begin{pmatrix} 1 & 2 & 3 & 4 & 5 \\ 3 & 2 & 5 & 1 & 4 \end{pmatrix}$$

and $\alpha$ has an inverse

$$\alpha^{-1} = \begin{pmatrix} 1 & 2 & 3 & 4 & 5 \\ 5 & 1 & 2 & 3 & 4 \end{pmatrix}.$$

**Theorem 3.3.2** *For every nonempty set $X$, the set $S_X$ is a group.*

**Proof.** (i) For all $\alpha, \beta \in S_X$ it is straightforward to verify that $\alpha \circ \beta$ is also a bijection so that $\alpha \circ \beta \in S_X$. Thus, $S_X$ is closed under the binary operation $\circ$.

(ii) Since permutations are just special kinds of functions and we have seen in Lemma 1.1.7 that the composition of functions is associative, it follows that the associative law holds in $S_X$.

(iii) With $\epsilon$ denoting the identity permutation, we have

$$\alpha \circ \epsilon = \epsilon \circ \alpha = \alpha \quad \text{for all } \alpha \in S_X$$

so that $\epsilon$ is an identity for $S_X$.

(iv) It remains to show that every element has an inverse. Let $\alpha \in S_X$. For any $y \in X$ there exists a unique $x \in X$ with $\alpha(x) = y$ (since $\alpha$ is a bijection). So we can define a mapping $\beta$ as follows: For every $y \in X$

$$\beta(y) = x \quad \text{where} \quad \alpha(x) = y.$$

Then $\beta$ is a bijection and $\alpha \circ \beta = \beta \circ \alpha = \epsilon$. Thus, $\beta = \alpha^{-1}$.
We have now established that $S_X$ is a group. $\quad\square$

For any $a_1, a_2, \ldots, a_n \in X$, $n > 1$, we denote by $(a_1 a_2 \ldots a_n)$ the permutation $\alpha$ of $X$ such that

$$\alpha(a_i) = a_{i+1} \quad (1 \le i \le n - 1)$$
$$\alpha(a_n) = a_1$$
$$\alpha(x) = x \qquad x \in X \backslash \{a_1, a_2, \ldots, a_n\}.$$

We call $(a_1 a_2 \ldots a_n)$ an *n-cycle*. For example, $(2\,5)$ is a 2-cycle and $(1\,3\,5)$ is a 3-cycle. Another term for a 2-cycle is *transposition*. It is convenient to extend this notation to include $1$-*cycles*. By a $1$-*cycle* $(a)$, we will mean the permutation that maps the element $a$ to itself and all other elements to themselves. In other words, a 1-cycle is just another way of writing the identity permutation. This can be a useful device when writing a permutation as a product of cycles. Of course, we can still represent cycles in the previous notation if that is convenient. For instance, in $S_7$,

$$(1\,3\,5\,7) = \begin{pmatrix} 1 & 2 & 3 & 4 & 5 & 6 & 7 \\ 3 & 2 & 5 & 4 & 7 & 6 & 1 \end{pmatrix}.$$

Note that

$$(1\,3\,5\,7) = (3\,5\,7\,1) = (5\,7\,1\,3)$$
$$= (7\,1\,3\,5)$$

so that there is more than one way to write a cyclic permutation. We can compose cycles easily (but carefully, remembering that the one on the right acts first):

$$(1\,3\,5\,7)(2\,3\,4\,7) = \begin{pmatrix} 1 & 2 & 3 & 4 & 5 & 6 & 7 \\ 3 & 5 & 4 & 1 & 7 & 6 & 2 \end{pmatrix}.$$

The inverse of a cycle is also easily determined.

**Lemma 3.3.3** *For all $a_1, a_2, \ldots, a_n \in X$,*

$$(a_1 a_2 \ldots a_n)^{-1} = (a_n a_{n-1} \ldots a_2 a_1).$$

**Proof.** Exercise.  □

As a special case of Lemma 3.3.3, we have

$$(a_1 a_2)^{-1} = (a_2 a_1) = (a_1 a_2).$$

Thus, every transposition is equal to its own inverse.

Two cycles $(a_1 \ldots a_m)$ and $(b_1 \ldots b_n)$ are *disjoint* if $a_i \neq b_j$ for all $i, j$. In general, permutations do not commute. Even cycles need not commute:

$$(1\,2\,3)(2\,3) = (1\,2)$$
$$(2\,3)(1\,2\,3) = (1\,3).$$

However,

**Lemma 3.3.4** *Disjoint cycles commute.*

**Proof.** Let $\alpha = (a_1 a_2 \ldots a_m)$ and $\beta = (b_1 b_2 \ldots b_n)$ be disjoint cycles in $S_X$. Then

$$\alpha\beta(x) = \begin{cases} \alpha(x) & \text{if } x = a_i & \text{for some } i \\ \beta(x) & \text{if } x = b_j & \text{for some } j \\ x & \text{if } x \neq a_i, b_j & \text{for all } i, j \end{cases}$$

$$= \beta\alpha(x).$$

Thus, $\alpha\beta = \beta\alpha$.  □

The importance of disjoint cycles flows from the following theorem.

**Theorem 3.3.5** *Let $X$ be a finite set. Then every element of $S_X$ can be written as a product of disjoint cycles. Moreover, such a representation is unique to within the order in which the cycles are written.*

**Proof.** Let $\alpha \in S_X$. Pick $x \in X$ and consider $x, \alpha(x), \alpha^2(x), \ldots$. Since $X$ is finite, there exist positive integers $i, j$ with $i < j$ and $\alpha^i(x) = \alpha^j(x)$. Then,

$$x = \alpha^{-i}\alpha^i(x) = \alpha^{-i}\alpha^j(x) = \alpha^{j-i}(x)$$

so that the first element to repeat in the sequence $x, \alpha(x), \alpha^2(x), \ldots$ must be $x$ itself. Thus we can represent the action of $\alpha$ on the subset $\{x, \alpha(x), \alpha^2(x), \ldots\}$ by a cycle

$$(x \, \alpha(x) \, \alpha^2(x) \, \cdots).$$

Now pick $y \in X \setminus \{x, \alpha(x), \alpha^2(x), \ldots\}$ and repeat the argument. This leads to a description of $\alpha$ as a product of disjoint cycles:

$$(x\, \alpha(x)\, \alpha^2(x) \cdots)(y\, \alpha(y)\, \alpha^2(y) \cdots) \cdots .$$

Clearly, this representation is unique modulo the order of the cycles. $\qquad \square$

To illustrate the previous theorem, consider

$$\begin{pmatrix} 1 & 2 & 3 & 4 & 5 & 6 & 7 & 8 & 9 & 10 \\ 4 & 8 & 10 & 6 & 2 & 7 & 1 & 5 & 9 & 3 \end{pmatrix} = (1\,4\,6\,7)(2\,8\,5)(3\,10)(9)$$

$$= (1\,4\,6\,7)(2\,8\,5)(3\,10)$$

where, as is customary, we have omitted cycles of length 1.

With the help of cycles, we can list the elements of $S_n$ (for small values of $n$):

$$S_2 = \{\epsilon, (12)\}$$
$$S_3 = \{\epsilon, (12), (13), (23), (123), (132)\}.$$

Here is an amusing application of cycles. A standard pack of 52 playing cards is shuffled as follows. The pack is numbered 1, 2, 3, ..., 51, 52 from top to bottom. Now divide the pack into two piles consisting of the first 26 cards and the last 26 cards and "shuffle" them into the order 27, 1, 28, 2, 29, 3, ....

(i) Show that repeating this process a sufficient number of times will return the cards to their original order.

(ii) What is the minimal number of repetitions that will return the pack to its original order?

(iii) Can you describe another shuffle that would require a greater number of repetitions to return the pack of cards to its original order?

Answer: We are really discussing permutations in $S_{52}$. The shuffle that is described consists of one cycle:

$$\alpha = (1\ 2\ 4\ 8\ 16\ 32\ 11\ 22\ 44\ 35\ \ldots 29\ 5\ 10\ 20\ 40\ 27).$$

The order of $\alpha$ is 52. Hence, 52 repetitions of this shuffle will return the pack to its original order and that is the least number of repetitions to achieve this.

The "shuffle" corresponding to the permutation

$$\beta = (1\ 2\ 3 \ldots .49)(50\ 51\ 52)$$

has order $49 \times 3 = 714$. There are many possible examples.

**Exercises 3.3**

1. Evaluate each of the following in $S_6$ as products of disjoint cycles and in the form

$$\begin{pmatrix} 1 & 2 & 3 & 4 & 5 & 6 \\ a & b & c & d & e & f \end{pmatrix}.$$

   (i) $(1\,3\,5)(2\,3\,4\,5)$.
   (ii) $(1\,2\,3)(3\,4\,5)(2\,6\,3)$.
   (iii)

$$\begin{pmatrix} 1 & 2 & 3 & 4 & 5 & 6 \\ 3 & 6 & 5 & 1 & 2 & 4 \end{pmatrix}\begin{pmatrix} 1 & 2 & 3 & 4 & 5 & 6 \\ 6 & 5 & 4 & 3 & 2 & 1 \end{pmatrix}.$$

   (iv) $(2\,4\,3\,5\,1)^{-1}$.
   (v) $((2\,4\,5)(1\,3\,6))^{-1}$.

2. Find the order of the following elements of $S_{10}$:

   (i) $(1\,3\,5\,7)$.
   (ii) $(1\,3\,5)(2\,4\,6\,8)$.
   (iii) $(1\,2\,3\,4)(5\,6\,7\,8\,9\,10)$.
   (iv)

$$\begin{pmatrix} 1 & 2 & 3 & 4 & 5 & 6 & 7 & 8 & 9 & 10 \\ 5 & 6 & 9 & 8 & 4 & 10 & 1 & 7 & 3 & 2 \end{pmatrix}.$$

3. (i) Express

$$\alpha = \begin{pmatrix} 1 & 2 & 3 & 4 & 5 & 6 \\ 6 & 5 & 1 & 3 & 2 & 4 \end{pmatrix}$$

     as a product of transpositions.
  (ii) Express $\alpha = (1\,2\,3\,4\,5)$ as a product of 3-cycles.

4. What is the order of a $k$-cycle?

5. Show that the product of any two distinct elements in $S_3$ of order 2 yields an element of order 3.

6. Let $\alpha \in S_n$ and $\alpha = \alpha_1 \alpha_2 \cdots \alpha_m$ as a product of disjoint cycles. Show that

$$|\alpha| = \mathrm{lcm}\{\,|\alpha_i| \mid 1 \le i \le m\,\}.$$

7. What is the largest order of any element in

   (i) $S_5$
   (ii) $S_6$
   (iii) $S_{10}$?

8. Define $\varphi : \mathbb{Z}_{48} \to \mathbb{Z}_{48}$ by

$$\varphi(a) = 5a \quad (a \in \mathbb{Z}_{48}).$$

   Show that $\varphi$ is a permutation of $\mathbb{Z}_{48}$.

*9. Find $Z(S_2)$. Show that $Z(S_n) = \{\epsilon\}$ for all $n \geq 3$.

10. Find all subgroups of $S_3$.

11. (i) Show that $D_3 \cong S_3$.
    (ii) Show that $D_n$ is not isomorphic to $S_n$, for $n > 3$.

12. Let $GF(9) = \mathbb{Z}_3[y]/(y^2 + 1)$. Then

$$\varphi : a \to a^3 \quad (a \in GF(9))$$

   is a permutation of $GF(9)$ called the *Frobenius automorphism*. Write $\varphi$
   as a product of disjoint cycles.

*13. Let $F$ be a field. For all $a, b \in F$ with $a \neq 0$, let $t_{a,b}$ be the mapping of $F$
   into $F$ defined by

$$t_{a,b}(x) = ax + b \quad (x \in F).$$

   Show that $t_{a,b} \in S_F$ and that the set $H$ of all such permutations is a
   subgroup of $S_F$. (Note that the operation in $H$ is the composition of
   functions.)

*14. Let $R = \mathbb{Z}_p[x]/(x^n - 1)$ and $\alpha$ be the mapping defined by

$$\alpha(a(x)) = x \cdot a(x) \quad (a(x) \in R).$$

   Show that $\alpha$ is a permutation of $R$.

15. Let $G$ be a subgroup of $S_n$, let $A \subseteq \{1, 2, \ldots, n\}$ and let

$$S(A) = \{\alpha \in G \mid \alpha(a) = a, \ \forall \, a \in A\}.$$

   (i) For $G = S_5$ and $A = \{1, 3\}$, list the elements of $S(A)$.
   (ii) For any subgroup $G$ of $S_n$, show that $S(A)$ is a subgroup of $G$.

16. Let $G$ be the group of rotations of a rectangular box of dimensions $a \times b \times c$ where $a$, $b$, and $c$ are all different. Describe each element of $G$.

17. Repeat Exercise 16 assuming that $a = b \neq c$.

## 3.4 Matrix Groups

Throughout this section, let $F$ denote an arbitrary field. We will denote by $M_n(F)$ the set of all $n \times n$ matrices over $F$, with addition and multiplication in $F$, defined in the usual way: For $A = (a_{ij})$, $B = (b_{ij})$

$$A + B = (a_{ij} + b_{ij}), \quad AB = \left( \sum_{k=1}^{n} a_{ik} a_{kj} \right).$$

A third operation of scalar multiplication is also useful. For $a \in F$,

$$aA = (aa_{ij}).$$

We will assume that you have some familiarity with matrices, although perhaps only with matrices with rational or real entries. So we will begin by reviewing, without detailed verification, the basic ideas and results concerning matrices that we require. If you have only seen these results for matrices over the rationals or reals, you should have no difficulty verifying that they hold equally well for matrices over an arbitrary field.

Of special importance is the identity $n \times n$ matrix $I_n = (d_{ij})$ where

$$d_{ij} = \begin{cases} 1 & \text{if } i = j, \\ 0 & \text{otherwise.} \end{cases}$$

**Lemma 3.4.1** $M_n(F)$ *is a ring with identity.*

**Proof.** See any introductory text on linear algebra and matrices for a verification of the ring axioms, for example [Nic]. $\quad\square$

Of course, for $n > 1$, $M_n(F)$ is a noncommutative ring.

The following are called the *elementary row operations* on a matrix:

R1: Multiply a row by a nonzero element $a \in F^*$.
R2: Interchange two rows.
R3: Add a multiple of one row to another different row.

The *determinant* $\det(A)$ of an $n \times n$ matrix $A \in M_n(F)$ is defined inductively on $n$ as follows:

For $A \in M_1(F)$, say $A = (a)$, $\det(A) = a$.

For $A \in M_n(F)$, with $n > 1$, let $A_{ij}$ denote the matrix obtained from $A$ by deleting the $i$th row and $j$th column. Then

$$\det(A) = a_{11} \det(A_{11}) - a_{12} \det(A_{12}) + \cdots + (-1)^{n-1} a_{1n} \det(A_{1n}).$$

This is known as the Laplace expansion of the determinant across the first row. Thus, det is a function from $M_n(F)$ to $F$.

For instance, it is easily seen that $\det(I_n) = 1$ and that, more generally, if $A = (a_{ij})$ is a diagonal matrix (that is, $a_{ij} = 0$ if $i \neq j$), then

$$\det(A) = a_{11} a_{22} \cdots a_{nn}.$$

Another way to calculate the determinant of a matrix $A$ is to reduce $A$ to a simpler matrix using elementary row operations. The next result tells us, among other things, how to keep track of the value of a determinant as we apply elementary row operations.

**Lemma 3.4.2** *Let $A \in M_n(F)$, $a \in F$.*

   (i)  $\det(aA) = a^n \det(A)$.
  (ii)  *If $B$ is obtained from $A$ by multiplying a row of $A$ by $a \in F$, then* $\det(B) = a \det(A)$.
 (iii)  *If $B$ is obtained from $A$ by interchanging the $i$th and $j$th rows, where $i \neq j$, then $\det(B) = (-1) \det(A)$.*
 (iv)  *If $B$ is obtained from $A$ by adding a multiple of one row of $A$ to another, then $\det(B) = \det(A)$.*
  (v)  *For all $A, B \in M_n(F)$,*

$$\det(AB) = \det(A) \cdot \det(B).$$

**Proof.** See [HW], [Nic] or [Z] for a complete derivation of these results from the Laplace expansion. $\square$

A matrix $A \in M_n(F)$ is an *invertible matrix* if there exists a matrix $B \in M_n(F)$ with $AB = BA = I_n$.

The next result gives us several ways of determining whether a matrix is invertible.

**Lemma 3.4.3** *Let $A \in M_n(F)$. Then the following statements are equivalent:*

   (i)  *$A$ is invertible.*
  (ii)  *There exists a sequence of elementary row operations that, when applied to $A$, will reduce $A$ to $I_n$.*
 (iii)  *There exists a sequence of elementary row operations that, when applied to $I_n$, yield $A$.*
 (iv)  *The rows of $A$ are independent as vectors in $F^n$.*
  (v)  $\det(A) \neq 0$.

**Proof.**  See [HW], [Nic] or [Z] for a complete derivation of these results.  $\square$

If $A, X \in M_n(F)$ are such that $AX = XA = I_n$, then $X$ is unique and we say that $X$ is the *inverse* of $A$ and write $X = A^{-1}$. If $A$ and $B$ are both invertible matrices, then

$$(AB)^{-1} = B^{-1}A^{-1}.$$

There is a simple algorithm for calculating the inverse of an invertible $n \times n$ matrix $A$. First write $A$ and $I_n$ side by side to form an $n \times 2n$ matrix

$$B = [A \ \ I_n].$$

Now apply to $B$ any sequence of elementary row operations that will reduce $A$ to $I_n$. The resulting matrix is of the form

$$[I_n \ \ C]$$

where $C = A^{-1}$.

Our particular interest in this section lies in the set of invertible matrices: We define

$$GL_n(F) = \{A \in M_n(F) \mid A \ \text{is invertible}\}.$$

**Theorem 3.4.4**  $GL_n(F)$ *is a group with respect to matrix multiplication.*

**Proof.**  Clearly $I_n$ is invertible so that $I_n \in GL_n(F)$, and $GL_n(F)$ is nonempty and has an identity (namely, $I_n$). Let $A, B \in GL_n(F)$. Then by the discussion preceding the Theorem, we have that $AB$ is invertible with $(AB)^{-1} = B^{-1}A^{-1}$.

Thus, $GL_n(F)$ is closed under multiplication and we already know that matrix multiplication is associative. Finally, for $A \in GL_n(F)$, it is clear that the inverse of $A^{-1}$ is just $A$ so that $A^{-1} \in GL_n(F)$. Therefore, $GL_n(F)$ is a group.  $\square$

The group $GL_n(F)$ is called the *general linear* group (over $F$). There are many interesting groups associated with $GL_n(F)$. We consider one here and a few others in the exercises. Let

$$P_n(F) = \{A = (a_{ij}) \in GL_n(F) \mid \ \ (\text{i}) \ \ a_{ij} = 0 \text{ or } 1, \ \forall \, i, j,$$
$$(\text{ii}) \ A \text{ has exactly one nonzero entry}$$
$$\text{in each row and column}\}.$$

**Lemma 3.4.5**  *Let $A \in P_n(F)$.*

  (i)  *$A$ can be reduced to $I_n$ by a sequence of row interchanges.*
  (ii)  *$\det A = \pm 1$.*

**Proof.** (i) By the definition of $P_n(F)$, $A$ has exactly one nonzero entry in the first column in, say, the $i$th position. Then the first entry in the $i$th row is 1 and all other entries in the $i$th row are 0. So, if we interchange the first and $i$th rows, then we obtain a matrix of the form

$$\begin{bmatrix} 1 & 0 & \cdots & 0 \\ 0 & & & \\ \vdots & & B & \\ 0 & & & \end{bmatrix}$$

where $B \in P_{n-1}(F)$. Repeating the process with $B$, we obtain the desired result.

(ii) This follows immediately from part (i) and Lemma 3.4.2 (iii).  $\square$

**Theorem 3.4.6** (i) $P_n(F)$ *is a subgroup of* $GL_n(F)$.

(ii) $|P_n| = n!$.

**Proof.** (i) Clearly, $I_n \in P_n(F)$ and it is easily verified that $P_n(F)$ is closed under multiplication. Now let $A \in P_n(F)$. Then $A$ can be reduced to $I_n$ by applying a sequence of row exchanges. Applying these same row exchanges to the matrix

$$[A \;\; I_n]$$

converts $I_n$ to $A^{-1}$. Consequently, $A^{-1}$ can be obtained from $I_n$ by a sequence of row interchanges, and therefore $A^{-1} \in P_n(F)$. Thus, $P_n(F)$ is a subgroup of $GL_n(F)$.

(ii) If we wish to construct an element of $P_n$, then we can do it one column at a time. In the first column, we can place a 1 in any of the $n$ positions. This provides $n$ choices. After an entry of 1 has been placed in the first column, the remaining entries in the column must be 0 and the remaining entries in the row of that entry must also be 0. This means that we will have $n - 1$ choices for the second column, then $n - 2$ choices for the third column, and so on. Thus we have

$$n(n - 1)(n - 2) \cdots 3 \cdot 2 \cdot 1 = n!$$

choices in all.  $\square$

It turns out that $P_n(F)$ is not a completely new example of a group at all, but just a familiar group in a different guise.

**Theorem 3.4.7** *Let $n \geq 2$ and, for each $\alpha \in S_n$, define $A_\alpha = (a_{ij}) \in P_n(F)$ by*

$$a_{ij} = \begin{cases} 1 & \text{if } \alpha(j) = i, \\ 0 & \text{otherwise.} \end{cases}$$

*Then the mapping $\mu : \alpha \to A_\alpha$ is an isomorphism of $S_n$ onto $P_n(F)$.*

**Proof.** Let $\alpha, \beta \in S_n$ and $\alpha \neq \beta$. Let $A_\alpha = (a_{ij})$, $A_\beta = (b_{ij})$. Since $\alpha \neq \beta$, there exists an element $j$ $(1 \leq j \leq n)$ with $\alpha(j) \neq \beta(j)$. Let $i = \alpha(j)$. Then

$$a_{ij} = 1, \quad b_{ij} = 0$$

so that $A_\alpha \neq A_\beta$. Thus, $\mu$ is injective. Since $|S_n| = n! = |P_n|$, it follows from Lemma 1.1.6 that $\mu$ is a bijection.

Now let $A_\alpha A_\beta = C = (c_{ij})$. Then,

$$c_{ij} \neq 0 \iff \sum_k a_{ik} b_{kj} \neq 0$$

$$\iff \exists\, k \ \text{(necessarily unique) with } a_{ik} \neq 0 \neq b_{kj}$$

$$\iff \exists\, k \ \text{with } \beta(j) = k, \ \alpha(k) = i$$

$$\iff \alpha\beta(j) = i$$

$$\iff (i, j)^{\text{th}} \ \text{entry of } A_{\alpha\beta} \text{ is nonzero.}$$

We also see from the previous calculations that if $c_{ij} \neq 0$, then there exists some $k$ with $c_{ij} = a_{ik} b_{kj} = 1 \cdot 1 = 1$. Thus,

$$C = A_{\alpha\beta}$$

and

$$\mu(\alpha)\mu(\beta) = A_\alpha A_\beta = C = A_{\alpha\beta} = \mu(\alpha\beta).$$

Hence, $\mu$ is an isomorphism. $\quad \square$

**Exercises 3.4**

1. Let $F = \mathbb{Z}_3[y]/(1 + y^2)$. Find the inverse of

$$A = \begin{bmatrix} y & 1 + 2y & 2 + 2y \\ 2y & 2 + y & 2 + y \\ 2 & 1 + y & 2 \end{bmatrix}$$

in $GL_n(F)$.

2. Let $M$ be the set of matrices $A$ of the form

$$A = \begin{bmatrix} 1 & a \\ 0 & b \end{bmatrix}$$

where $b \neq 0$. Show that $M$ is a subgroup of $GL_n(F)$.

3. Show that

$$SL_n(F) = \{A \in GL_n(F) \mid \det A = 1\}$$

is a subgroup of $GL_n(F)$. $SL_n(F)$ is called the *special linear group*.

4. A matrix $A \in GL_n(F)$ is *orthogonal* if $A^T = A^{-1}$ where, if $A$ is the $n \times n$ matrix $(a_{ij})$, then $A^T$ denotes the *transpose* of $A$, that is, the matrix $(a_{ji})$. Show that

$$O_n(F) = \{A \in GL_n(F) \mid A \text{ is orthogonal}\}$$

is a subgroup of $GL_n(F)$. $O_n(F)$ is called the *orthogonal group*.

5. Let

$$SO_n(F) = \{A \in O_n(F) \mid \det A = 1\}.$$

Show that $SO_n(F)$ is a subgroup of $O_n(F)$. $SO_n(F)$ is called the *special orthogonal group*.

6. Show that

$$D_n(F) = \{A \in GL_n(F) \mid A \text{ is a diagonal matrix}\}$$

is a subgroup of $GL_n(F)$.

7. Show that

$$UT_n(F) = \{A \in GL_n(F) \mid A \text{ is upper triangular}\}$$

is a subgroup of $GL_n(F)$.

8. Show that the set

$$SUT_n(F) = \{A \in UT_n(F) \mid \det A = 1\}$$

is a subgroup of $UT_n(F)$.

9. Show that

$$SUT'_n(F) = \{A \in SUT_n(F) \mid a_{ii} = 1 \text{ for } 1 \leq i \leq n\}$$

is a subgroup of $SUT_n(F)$.

10. Let $A, B \in SUT_n(F)$. Show that

$$A^{-1}B^{-1}AB \in SUT_n'(F).$$

11. Show that $SO_2(\mathbb{R})$ consists of all matrices of the form

$$\begin{bmatrix} \cos\theta & \sin\theta \\ -\sin\theta & \cos\theta \end{bmatrix}$$

with $0 \le \theta < 2\pi$.

## 3.5 Even and Odd Permutations

It turns out that there is a very interesting and important division of permutations into two classes that is not at all obvious at first sight.

Take $F = \mathbb{Q}$ in Theorem 3.4.7. Then by Lemma 3.4.5 (ii), we see that for all $\alpha \in S_n$,

$$\det(\mu(\alpha)) = \pm 1.$$

So let us define $\alpha$ to be

$$\begin{aligned} even \quad &\text{if} \quad \det(\mu(\alpha)) = 1, \\ odd \quad &\text{if} \quad \det(\mu(\alpha)) = -1. \end{aligned}$$

and the *sign* of $\alpha$ to be

$$\mathrm{sgn}(\alpha) = \det(\mu(\alpha)) = \begin{cases} 1 & \text{if } \alpha \text{ even} \\ -1 & \text{if } \alpha \text{ odd} \end{cases}$$

We refer to $\mathrm{sgn}(\alpha)$, or to whether $\alpha$ is even or odd, as the *parity* of $\alpha$. The parity of permutations is a key ingredient in an alternative definition of the determinant of a square matrix that runs as follows: for any $n \times n$ matrix $A = (a_{ij})$ over a field,

$$\det(A) = \sum_{\sigma \in S_n} \mathrm{sgn}(\sigma) a_{1\sigma(1)} a_{2\sigma(2)} \cdots a_{n\sigma(n)}.$$

Of course, in order to adopt this as the *definition* of the determinant function, one must first show that the parity of a permutation is well-defined. To see that the two definitions of the determinant are indeed equivalent, we refer the interested reader to [HW], [Nic] or [Z].

This is all very well and good, but is there an easier or more direct way of determining whether a permutation is even or odd than calculating first $\mu(\alpha)$

and then $\det(\mu(\alpha))$? Yes there is, as we shall see in Theorem 3.5.3. We require a few preliminary observations.

**Lemma 3.5.1**  (i) $\epsilon$ *is even.*
  (ii) *Every transposition is odd.*

**Proof.**  (i) We have $\mu(\epsilon) = I_n$ so that

$$\det(\mu(\epsilon)) = \det I_n = 1.$$

(ii) Let $\alpha = (ij) \in S_n$ be a transposition. Then

$$\mu(\alpha) = \begin{array}{c} \\ \\ i^{\text{th}}\text{ row} \\ \\ j^{\text{th}}\text{ row} \\ \\ \\ \end{array} \left[ \begin{array}{cccccc} 1 & 0 & \cdots & \cdot & \cdot & 0 \\ 0 & 1 & & & & \\ \cdot & & & & & \\ \cdot & & & 0 & 1 & \\ \cdot & & & & & \\ \cdot & & & 1 & 0 & \\ \cdot & & & & & \\ 0 & & & 0 & 0 & 1 \end{array} \right] \cdot$$

$$\phantom{\mu(\alpha) = } \qquad\qquad i^{\text{th}}\text{ col} \quad j^{\text{th}}\text{ col}$$

Hence, $\mu(\alpha)$ can be reduced to $I_n$ by simply interchanging rows $i$ and $j$. Therefore,

$$\det(\mu(\alpha)) = -1,$$

so that $\alpha$ is odd.  $\square$

In the light of Lemma 3.5.1, we will be able to determine the parity of a permutation if we can just express it as a product of transpositions.

**Lemma 3.5.2** *Let $X$ be a finite set. Then every permutation in $S_X$ can be written as a product of transpositions.*

**Proof.**  Let $\alpha \in S_X$. By Theorem 3.3.5, we can write $\alpha$ as a product of disjoint cycles—say,

$$\alpha = \alpha_1 \alpha_2 \cdots \alpha_k$$

where the $\alpha_i$ are disjoint cycles. Therefore, it suffices to show that each $\alpha_i$ can be written as a product of 2-cycles. Let

$$\alpha_i = (a_1 a_2 \cdots a_m).$$

Then

$$\alpha_i = (a_1 a_m)(a_1 a_{m-1}) \cdots (a_1 a_3)(a_1 a_2)$$

as required.  $\square$

The description of a permutation as a product of 2-cycles is not unique:

$$(1\,2\,3\,4\,5) = (15)(14)(13)(12)$$
$$= (21)(25)(24)(23)$$
$$\epsilon = (12)(12) = (12)(34)(34)(12).$$

**Theorem 3.5.3** *Let $\alpha \in S_n$. Then $\alpha$ is even if and only if $\alpha$ can be written as a product of an even number of transpositions.*

**Proof.**  By Lemma 3.5.2, $\alpha$ can be expressed as a product of transpositions—say,

$$\alpha = \alpha_1 \alpha_2 \cdots \alpha_k$$

where each $\alpha_i$ is a transposition. Then

$$\mu(\alpha) = \mu(\alpha_1 \alpha_2 \cdots \alpha_k)$$
$$= \mu(\alpha_1)\mu(\alpha_2) \cdots \mu(\alpha_k)$$

so that

$$\det(\mu(\alpha)) = \det(\mu(\alpha_1)) \cdot \det(\mu(\alpha_2)) \cdots \det(\mu(\alpha_k))$$
$$= (-1) \cdot (-1) \cdots (-1)$$
$$= (-1)^k.$$

Thus, $\alpha$ is even if and only if $k$ is even, and the claim follows.  $\square$

Note that it follows from the proof of Theorem 3.5.3 that if a permutation $\alpha \in S_n$ can be written as a product of an even (odd) number of transpositions, then every expression for $\alpha$ as a product of transpositions must have an even (odd) number of transpositions.

We denote by $A_n$ the set of all even permutations in $S_n$. In light of the next result, $A_n$ is a subgroup of $S_n$ that is known as the *alternating subgroup* of $S_n$.

**Theorem 3.5.4** *$A_n$ is a subgroup of $S_n$ and $|A_n| = \frac{1}{2}|S_n|$.*

**Proof.** We leave the verification that $A_n$ is a subgroup for you as an exercise. To see that $A_n$ contains exactly half the elements, let $B_n = S_n \backslash A_n$. In other words, $B_n$ consists of all the odd elements. Let $\tau$ be any transposition and let

$$\varphi : \alpha \to \alpha\tau \quad (\alpha \in S_n).$$

Then

$$\varphi(\alpha) = \varphi(\beta) \implies \alpha\tau = \beta\tau$$
$$\implies \alpha = \beta$$

so that $\varphi$ is one-to-one. Since $S_n$ is finite, it follows that $\varphi$ is a permutation of $S_n$. Now

$$\alpha \in A_n \implies \exists \text{ transpositions } \tau_1, \ldots, \tau_{2k} \text{ with } \alpha = \tau_1 \ldots \tau_{2k}$$
$$\implies \varphi(\alpha) = \alpha\tau = \tau_1 \tau_2 \ldots \tau_{2k}\tau \in B_n.$$

Similarly,

$$\alpha \in B_n \implies \exists \text{ transpositions } \tau_1, \ldots, \tau_{2k+1} \text{ with } \alpha = \tau_1 \ldots \tau_{2k+1}$$
$$\implies \varphi(\alpha) = \alpha\tau = \tau_1 \ldots \tau_{2k+1}\tau \in A_n.$$

Therefore, $\varphi$ maps $A_n$ into $B_n$ and $B_n$ into $A_n$. Since $\varphi$ is one-to-one, we must have $|A_n| = |B_n|$ so that $|A_n| = \frac{1}{2}|S_n|$. $\quad \square$

Since $S_2 = \{\epsilon, (12)\}$, we have $A_2 = \{\epsilon\}$. The situation in $S_3$ is a little more interesting:

$$S_3 = \{\epsilon, (12), (13), (23), (123), (132)\}$$

where

$$(123) = (13)(12)$$
$$(132) = (12)(13).$$

Thus,

$$A_3 = \{\epsilon, (123), (132)\}.$$

**Exercises 3.5**

1. Show that a $k$-cycle ($k > 1$) is even if and only if $k$ is odd.

2. List the elements of

    (i) $S_4$,
    (ii) $A_4$,

    as products of disjoint cycles.

3. Find the center of $A_3, A_4$.

4. Prove that $A_n$ is a subgroup of $S_n$.

5. Let $\alpha \in S_n$, $\beta \in A_n$. Show that $\alpha^{-1}\beta\alpha \in A_n$.

6. Let $\alpha, \beta \in S_n$. Show that $\alpha^{-1}\beta^{-1}\alpha\beta \in A_n$.

7. Let $a, b, c, d$ be distinct integers with $1 \le a, b, c, d \le n$. Show that

    (i) $(ab)(ac) = (acb)$
    (ii) $(ab)(cd) = (acb)(acd)$.
    (iii) Show that, for $n \ge 3$, $A_n$ consists precisely of those permutations that can be written as a product of 3-cycles.

8. Let $H$ be a subgroup of $S_n$. Show that either $H \subseteq A_n$ or else $|H \cap A_n| = \frac{1}{2}|H|$.

9. Let $n \ge 2$, $1 \le a < b \le n$, and $\alpha, \beta \in S_n$ be the elements

$$\alpha = (1\ \ 2), \qquad \beta = (1\ \ 2\ \ 3\ \cdots\ n).$$

    Establish the following:

    (i) $(a\ b) = (a\ a+1)(a+1\ a+2)\cdots(b-2\ b-1)(b-1\ b)$
    $(b-1\ b-2)(b-2\ b-3)\cdots(a+2\ a+1)(a+1\ a)$.
    (ii) $\beta^i\alpha\beta^{-i} = (1+i\ \ 2+i)$     (for $i = 1, 2, \ldots, n-2$),
    $\beta^{n-1}\alpha\beta^{-(n-1)} = (1\ n)$.
    (iii) Every element of $S_n$ can be expressed as a product of the permutations $\alpha$ and $\beta$.

## 3.6 Cayley's Theorem

It is tempting to think of groups of permutations just as nice examples of groups. However, the next result tells us that, in fact, every group is isomorphic to a group of permutations. Thus, the study of groups of permutations is at the very heart of group theory.

**Theorem 3.6.1 (Cayley's Theorem)** *Let $G$ be a group. For each $a \in G$, define a mapping*

$$\rho_a : x \to ax \quad (x \in G).$$

(i) $\rho_a \in S_G$.
(ii) $H = \{\rho_a \mid a \in G\}$ *is a subgroup of* $S_G$.
(iii) *The mapping*

$$\chi : a \to \rho_a \quad (a \in G)$$

*is an isomorphism of $G$ onto $H$.*

**Proof.** (i) For any $x, y \in G$

$$\rho_a(x) = \rho_a(y) \Longrightarrow ax = ay$$
$$\Longrightarrow x = y$$

so that $\rho_a$ is injective. Also, for any $x \in G$,

$$\rho_a(a^{-1}x) = a(a^{-1}x) = aa^{-1}x = x$$

which implies that $\rho_a$ is surjective. Thus, $\rho_a$ is a bijection of $G$ onto itself—that is, $\rho_a \in S_G$.

(ii) Clearly, $H \neq \emptyset$. For any $\rho_a, \rho_b \in S_G$, we have

$$\begin{aligned}
(\rho_a\rho_b)(x) &= \rho_a(\rho_b(x)) \qquad \text{for all } x \in G \\
&= \rho_a(bx) \\
&= a(bx) \\
&= (ab)x \\
&= \rho_{ab}(x)
\end{aligned} \tag{3.3}$$

so that $\rho_a\rho_b = \rho_{ab} \in H$ and $H$ is closed. Also, it is straightforward to verify that $(\rho_a)^{-1} = \rho_{a^{-1}} \in H$. Thus $H$ is a subgroup of $S_G$.

(iii) Let $a, b \in G$. Then

$$\begin{aligned}
\chi(a) = \chi(b) &\Longrightarrow \rho_a = \rho_b \\
&\Longrightarrow \rho_a(1) = \rho_b(1) \\
&\Longrightarrow a \cdot 1 = b \cdot 1 \\
&\Longrightarrow a = b.
\end{aligned}$$

Thus, $\chi$ is injective. Clearly, $\chi$ is surjective so that $\chi$ is a bijection. Finally, for all $a, b \in G$,

$$\chi(ab) = \rho_{ab}$$
$$= \rho_a \rho_b \qquad \text{by (3.3)}$$
$$= \chi(a)\chi(b).$$

Therefore, $\chi$ is an isomorphism as required. $\quad\square$

Historically, the first groups to be studied were permutation groups. The idea of axiomatizing groups came later. Cayley's Theorem shows that our group axioms capture exactly the concept of a group of permutations.

To illustrate Cayley's Theorem: consider $\mathbb{Z}_5{}^* = \{1, 2, 3, 4\}$. Then

$$\rho_1 = \begin{pmatrix} 1 & 2 & 3 & 4 \\ 1 & 2 & 3 & 4 \end{pmatrix} = \epsilon$$

$$\rho_2 = \begin{pmatrix} 1 & 2 & 3 & 4 \\ 2 & 4 & 1 & 3 \end{pmatrix}$$

$$\rho_3 = \begin{pmatrix} 1 & 2 & 3 & 4 \\ 3 & 1 & 4 & 2 \end{pmatrix}$$

and so forth. In $(\mathbb{Z}_5, +)$, on the other hand,

$$\rho_0 = \begin{pmatrix} 0 & 1 & 2 & 3 & 4 \\ 0 & 1 & 2 & 3 & 4 \end{pmatrix} = \epsilon$$

$$\rho_1 = \begin{pmatrix} 0 & 1 & 2 & 3 & 4 \\ 1 & 2 & 3 & 4 & 0 \end{pmatrix}$$

and so on.

### Exercises 3.6

1.  (i) Describe the Cayley representation $\chi$ of $(\mathbb{Z}_4, +)$.
    (ii) Give an example of a permutation $\alpha \in S_{\mathbb{Z}_4} \setminus \chi(\mathbb{Z}_4)$.
    (iii) Find $\chi(\mathbb{Z}_4) \cap A_4$.

2.  (i) Describe the Cayley representation $\chi$ of $(\mathbb{Z}_7{}^*, \cdot)$.
    (ii) Give an example of a permutation $\alpha \in S_{\mathbb{Z}_7{}^*} \setminus \chi(\mathbb{Z}_7{}^*)$.
    (iii) Find $\chi(\mathbb{Z}_7{}^*) \cap A_6$.

3.  (i) Describe the Cayley representation $\chi$ of $S_3$.
    (ii) Give an example of a permutation $\alpha \in S_{S_3} \setminus \chi(S_3)$.
    (iii) Find $\chi(S_3) \cap A_6$.

*4. Let $\chi : G \to S_G$ be the Cayley representation of a finite group $G$, let $a \in G$ and $\chi(a) = \rho_a$. Let $\rho_a$ be written as a product of disjoint cycles $\rho_a = \alpha_1 \alpha_2 \cdots \alpha_m$. Establish the following:

(i) For all $i$, $1 \le i \le m$, the length of $\alpha_i$ is equal to the order of $a$.
(ii) $m = |G|/|a|$, the index of $\langle \alpha \rangle$ in $G$.

*5. Let $\chi : G \to S_G$ be the Cayley representation of a finite group $G$. Show that $\chi(G)$ contains only even permutations if and only if one of the following conditions prevails:

(a) Every element of $G$ has odd order.
(b) $G$ is not cyclic and 4 divides $|G|$.

(Hint: Use Exercise 4.)

## 3.7 Lagrange's Theorem

Let $H$ be a subgroup of a group $G$. Following a pattern that by now may seem familiar, we define a relation $\equiv_H$ on $G$ by

$$a \equiv_H b \iff ab^{-1} \in H.$$

Equivalently,

$$a \equiv_H b \iff ab^{-1} = h, \quad \text{for some } h \in H$$
$$\iff a = hb, \quad \text{for some } h \in H.$$

Then, for all $a, b, c \in G$, we see that

$$aa^{-1} = e \in H \implies a \equiv_H a$$

$$a \equiv_H b \implies ab^{-1} \in H$$
$$\implies ba^{-1} = (ab^{-1})^{-1} \in H$$
$$\implies b \equiv_H a$$

and

$$a \equiv_H b,\ b \equiv_H c \implies ab^{-1},\ bc^{-1} \in H$$
$$\implies ac^{-1} = (ab^{-1})(bc^{-1}) \in H$$
$$\implies a \equiv_H c.$$

This establishes the first claim in the next result.

**Lemma 3.7.1**  $\equiv_H$ *is an equivalence relation on* G. *Moreover, if* $[a]$ *is the class of* $a \in G$, *then*

$$[a] = \{ha \mid h \in H\}.$$

**Proof.**  It only remains to characterize $[a]$. We have

$$x \in [a] \iff a \equiv_H x$$

$$\iff x \equiv_H a$$

$$\iff xa^{-1} \in H$$

$$\iff xa^{-1} = h \qquad \text{for some } h \in H$$

$$\iff x = ha \qquad \text{for some } h \in H$$

so that

$$[a] = \{ha \mid h \in H\}. \qquad \square$$

Note that the relation $\equiv_H$ has a twin or *dual* relation, as we say, $_H\equiv$ that can be defined as follows:

$$a \,_H{\equiv}\, b \iff a^{-1}b \in H.$$

For every result about $\equiv_H$ there is a dual result for $_H\equiv$.

It seems fairly natural to define

$$Ha = \{ha \mid h \in H\}.$$

With this notation, we then have that

$$[a] = \text{the class of } a \text{ in } \equiv_H$$

$$= Ha.$$

Any set of the form $Ha$ is called a *right coset* (of $H$). Dually,

$$aH = \{ah \mid h \in H\}$$

is a *left coset* (of $H$). Of course, if $G$ is commutative, then the left and right cosets are the same: $aH = Ha$, for all $a \in G$.

Note that since the right cosets of $H$ are the classes of the equivalence relation $\equiv_H$, they form a partition of the group $G$. In particular, for any two right cosets $Ha$, $Hb$, we either have

$$Ha = Hb \quad \text{or} \quad Ha \cap Hb = \emptyset.$$

Let us see what these cosets look like in a specific example. Let $G = \mathbb{Z}_{12}$ and $H = 3\mathbb{Z}_{12}$. Then the cosets of $H$ are

$$H + 0 = H = \{0, 3, 6, 9\}$$
$$H + 1 = \{1, 4, 7, 10\}$$
$$H + 2 = \{2, 5, 8, 11\}.$$

Note that the left and right cosets are the same in this case since $G$ is abelian. Also,

$$H + 1 = H + 4 = H + 7 = H + 10.$$

In other words, it doesn't matter which element of the coset you use to describe it.

Consider another example. We have

$$S_3 = \{\epsilon, (12), (13), (23), (123), (132)\}.$$

Let

$$\alpha = (12) \quad \text{and} \quad H = \{\epsilon, \alpha\}.$$

Then

$$H\epsilon = H$$
$$H(13) = \{(13), (12)(13)\} = \{(13), (132)\}$$
$$H(23) = \{(23), (12)(23)\} = \{(23), (123)\}$$
$$H(123) = \{(123), (12)(123)\} = \{(123), (23)\} = H(23).$$

On the other (left) hand,

$$\epsilon H = H$$
$$(13)H = \{(13), (13)(12)\} = \{(13), (123)\}$$
$$(23)H = \{(23), (23)(12)\} = \{(23), (132)\}$$
$$(123)H = \{(123), (123)(12)\} = \{(123), (13)\} = (13)H.$$

Note that some of the left cosets are different from their right counterparts. Also note that for any subgroup $H$ of $G$, the only coset (left or right) of $H$ that is a subgroup of $G$ is $H$ itself. Here are some of the basic properties of cosets:

**Lemma 3.7.2** *Let H be a subgroup of G and $a \in G$.*

   (i) *$a \in Ha$.*
   (ii) *$Ha = H \iff a \in H$.*
  (iii) *For $a, b \in G$, either $Ha = Hb$ or $Ha \cap Hb = \emptyset$.*
  (iv) *$Ha = Hb \iff ab^{-1} \in H \iff ba^{-1} \in H$.*
   (v) *If G is finite, then $|H| = |Ha|$.*
  (vi) *$Ha = aH \iff H = a^{-1}Ha$.*

*(Note that $a^{-1}Ha = \{a^{-1}ha \mid h \in H\}$.)*

**Proof.** Parts (i) through (iv) and (vi) are left for you as exercises.
   (v) Define a mapping $\varphi : H \to Ha$ by

$$\varphi(h) = ha \qquad (h \in H).$$

Then, for $h, h' \in H$,

$$\varphi(h) = \varphi(h') \implies ha = h'a$$
$$\implies h = h' \qquad \text{by cancellation}$$

so that $\varphi$ is injective. Also, for any $y \in Ha$ we must have, for some $h \in H$,

$$y = ha = \varphi(h).$$

Thus, $\varphi$ is surjective and therefore a bijection. Note that we have not assumed that $G$ is finite. In other words, we always have a bijection from $H$ to $Ha$ even when $G$ is infinite. When $G$ is finite, so also are $H$ and $Ha$, and we can conclude that they have the same number of elements. $\quad\square$

The relationship between the size of a group and the sizes of its subgroups is extremely important.

**Theorem 3.7.3 (Lagrange's Theorem)** *Let H be a subgroup of the finite group G. Then $|H|$ divides $|G|$.*

**Proof.** Let $\equiv_H$ be the equivalence relation defined on $G$ prior to Lemma 3.7.1. Let $a \in G$. By Lemma 3.7.1,

$$[a] = Ha$$

so that, by Lemma 3.7.2 (v),

$$|[a]| = |Ha| = |H|.$$

Thus, every equivalence class of $\equiv_H$ has the same number of elements. Since these classes partition $G$, it follows that

$$|G| = \text{number of classes} \times |H|. \tag{3.4}$$

Therefore, $|H|$ divides $|G|$. $\quad\square$

Note that from equation (3.4) in the proof of Theorem 3.7.3 we can obtain a formula for the number of $\equiv_H$-classes. Indeed,

$$\text{number of right cosets} = \text{number of } \equiv_H \text{ -classes}$$
$$= |G|/|H|.$$

Dually, substituting left for right, we have

$$\text{number of left cosets} = \text{number of } {}_H\equiv \text{ -classes}$$
$$= |G|/|H|.$$

In particular, the numbers of left and right cosets of a given subgroup are the same. This is an important number in calculations involving groups. It is called the *index* of $H$ in $G$ and is denoted by $[G : H]$. We have

$$[G : H] = |G|/|H|.$$

However, it must be emphasized that although the number of left cosets is the same as the number of right cosets, the actual left and right cosets themselves can be different.

**Corollary 3.7.4** *Let $G$ be a finite group and $a \in G$. Then $|a|$ divides $|G|$.*

**Proof.** We have seen that $|a| = |\langle a \rangle|$. Since $\langle a \rangle$ is a subgroup of $G$, we know from Lagrange's Theorem that $|\langle a \rangle|$ divides $|G|$ and the result follows. $\quad\square$

**Corollary 3.7.5** *Let $G$ be a finite group and $a \in G$. Then*

$$a^{|G|} = e.$$

**Proof.** By Corollary 3.7.4, $|a|$ divides $|G|$. Therefore, by Lemma 3.1.7, $a^{|G|} = e$. $\quad\square$

Euler lived in an era preceding the introduction of groups, but it is interesting to see that Euler's Theorem can be obtained as a simple consequence of Corollary 3.7.5. The size of the group of invertible elements in $\mathbb{Z}_n$ is just $\varphi(n)$. Hence, by Corollary 3.7.5, for any invertible element $[a]$ in $\mathbb{Z}_n$, we must have

$[a]^{\varphi(n)} = [1]$. Since $[a]$ is invertible in $\mathbb{Z}_n$ if and only if $(a, n) = 1$, it follows that $a^{\varphi(n)} \equiv 1 \bmod n$ whenever $(a, n) = 1$, which is precisely Euler's Theorem.

**Corollary 3.7.6** *Groups of prime order are cyclic.*

**Proof.** Let $|G| = p$, where $p$ is a prime. Let $a \in G$, $a \neq e$. Then, $1 < |a|$. But by Corollary 3.7.4, $|a|$ divides $|G| = p$. Therefore, $|a| = p$. But then $|\langle a \rangle| = p$, so that $\langle a \rangle$ must contain all $p$ elements of $G$. Thus, $\langle a \rangle = G$ and $G$ is cyclic. $\square$

So far we have encountered the following groups of small order:

| Order | Group |
|-------|-------|
| 1 | $\{0\}$ |
| 2 | $\mathbb{Z}_2$ |
| 3 | $\mathbb{Z}_3$ |
| 4 | $\mathbb{Z}_4$, $\mathbb{Z}_2 \times \mathbb{Z}_2$ |
| 5 | $\mathbb{Z}_5$ |
| 6 | $\mathbb{Z}_6$, $S_3$ |
| 7 | $\mathbb{Z}_7$ |

We know that the two groups of order 4 are not isomorphic (Exercise 18 in Section 3.2) whereas the two groups of order 6 are not isomorphic since one is commutative and the other is not. By Corollary 3.7.6, we know that we have listed all the groups of orders 2, 3, 5, and 7 (up to isomorphism). By Exercise 8 in Section 3.1 we know that every group of order 4 is isomorphic to one of the two listed. In the exercises for this section we will see that $\mathbb{Z}_6$ and $S_3$ are also the only groups of order 6 (up to isomorphism).

We conclude this section with another interesting family of groups.

A pair of sets $(V, E)$ is called a *graph* if $V$ is a nonempty set and $E$ is a set with members that are pairs of vertices. The elements of $V$ are called *vertices* and the elements of $E$ are called *edges*. An *automorphism* $\alpha$ of a graph $(V, E)$ is a permutation of $V$ (so that $\alpha \in S_V$), satisfying the condition

$$(u, v) \in E \implies (\alpha(u), \alpha(v)) \in E.$$

It is easy to verify that the set of all automorphisms of a graph $(V, E)$ is a subgroup of $S_V$, called the *automorphism group* of the graph.

When a graph is small, it may be possible to represent it by a diagram. (For example, see Fig. 3.3)

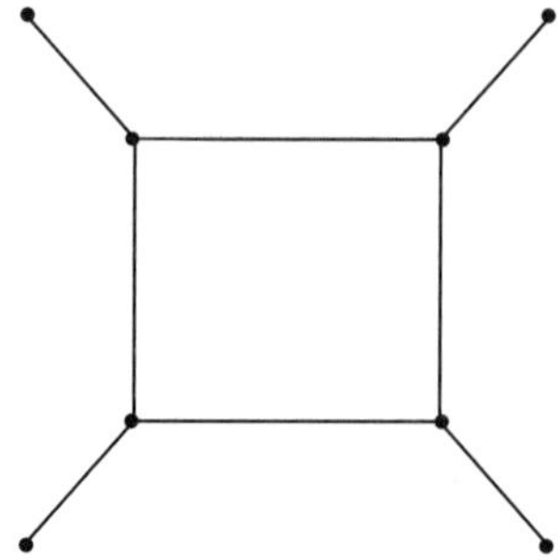

**Figure 3.3**

**Exercises 3.7**

1. (i) Find all cosets of the subgroup $H = \{0, 4, 8\}$ of $(\mathbb{Z}_{12}, +)$.
   (ii) Are the following pairs of elements related under $\equiv_H$?

   (a) $3, 7$.
   (b) $5, 11$.
   (c) $6, 9$.

2. Are the following pairs of elements related under $\equiv_H$ where $H = A_7$ in $S_7$?

   (i) $(12)(35)(27), (17)(26)(35)(47)$.
   (ii) $(23)(56), (1\,3\,5\,7\,4)$.
   (iii) $(2\,4\,6\,7), (5\,6\,4\,7\,1)$.

3. Let $H = \{\epsilon, (13)\}$ in $S_3$.

   (i) Find all the left cosets of $H$.
   (ii) Find all the right cosets of $H$.
   (iii) Describe $\equiv_H$ completely.
   (iv) Describe $_H\equiv$ completely.

4. Repeat Question 3 with $H = \{\epsilon, (123), (132)\}$.

5. Establish parts (i), (ii), (iii), (iv), and (vi) of Lemma 3.7.2.

6. Show that the converse of Corollary 3.7.4 is false, in the sense that the fact that $m$ divides $|G|$ does not imply that $G$ contains an element of order $m$. (Try $S_4$.)

7. (i) List all elements of $A_4$.
   (ii) List all subgroups of $A_4$. (There are 10 including $\{\epsilon\}$ and $A_4$.)
   (iii) Show that the converse of Lagrange's Theorem is false in $A_4$ in the sense that the fact that $m$ divides $|G|$ does not imply that $G$ contains a subgroup of order $m$.

8. Verify the conclusion of Corollary 3.7.4 for all elements of

    (i)  $(\mathbb{Z}_{12}, +)$
    (ii)  $(\mathbb{Z}_7{}^*, \cdot)$
    (iii)  $A_4$

9. Show that $D_5$ has subgroups of all possible orders.

10. Let $|G| = p$, where $p$ is a prime. Show that every nonidentity element of $G$ has order $p$.

11. Let $|G| = p^n$, where $p$ is a prime. Show that $G$ has an element of order $p$.

12. Let $|G| = p^2$. Show that either $G$ is cyclic or that $a^p = e$ for all $a \in G$.

13. Let $H, K$ be subgroups of a finite group $G$ such that $(|H|, |K|) = 1$. Show that $|H \cap K| = 1$.

14. Let $|G| = 35$.

    (i)  Show that $G$ has at most 8 subgroups of order 5.
    (ii)  Show that $G$ has at most 5 subgroups of order 7.
    (iii)  Deduce that $G$ has at least one element of order 5 and at least one element of order 7.

15. Use Corollary 3.7.5 to prove Euler's Theorem.

16. Let $H$ be a subgroup of a group $G$ with $|H| = \frac{1}{2}|G|$.

    (i)  Show that $a \notin H \Rightarrow G = H \cup Ha$.
    (ii)  Show that $a \notin H \Rightarrow H a^n \neq H a^{n+1}$.
    (iii)  Deduce that $H$ contains all elements of odd order.

17.  (i)  How many 3-cycles are there in $A_5$?
    (ii)  How many 5-cycles are there in $A_5$?
    (iii)  Use Exercise 16 to show that $A_5$ has no subgroup of order 30.

18. Repeat the argument of Exercise 17 to show that $A_4$ has no subgroup of size 6.

In the remaining exercises it is established that $\mathbb{Z}_6$ and $S_3$ are indeed the only groups of order 6 (to within isomorphism). First we consider abelian groups.

19. Let $G$ be an abelian group in which every nonidentity element is of order 2. Let $a, b, c \in G$ and $H = \langle a \rangle$. Show that if $H$, $Hb$, and $Hc$ are all distinct, then so also are $H$, $Hb$, $Hc$, and $Hbc$.

20. Let $G$ be an abelian group of order 6. Establish the following:

    (i)  Not every nonidentity element is of order 2. (Hint: Exercise 19 might help.)
    (ii)  Not every nonidentity element is of order 3.

21. Let $G$ be an abelian group of order 6. Establish the following:

   (i)  $G$ contains an element $a$ of order 2 and an element $b$ of order 3.
   (ii)  $ab$ has order 6.
   (iii)  $G$ is isomorphic to $\mathbb{Z}_6$.

*22. Let $G$ be a nonabelian group of order 6. Show that $G$ must contain an element $a$ of order 3 and an element $b$ of order 2. (Exercise 14 in section 3.1 may be helpful.)

*23. Let $a, b$ and $G$ be as in Exercise 22, and let $H = \langle a \rangle$. Establish the following.

   (i)  $G = H \cup Hb$.
   (ii)  $ba \notin H$.
   (iii)  $ba = a^2 b \quad (= a^{-1}b)$.

*24. Let $a, b,$ and $G$ be as in Exercise 22. Show that the mapping $\varphi : G \to S_3$ defined by

$$\varphi(a^i b^j) = (123)^i (12)^j \qquad (0 \le i \le 2, \ \ 0 \le j \le 1)$$

   is an isomorphism.


## 3.8 Orbits

In this section we will take a closer look at how groups of permutations act on various structures. We will assume throughout that $X$ is a nonempty set.

Let $G$ be a subgroup of $S_X$ and $x \in X$. Then the *orbit* of $x$ is

$$O(x) = \{\alpha(x) \mid \alpha \in G\}.$$

If $G = \{\epsilon, (12)\} \subseteq S_3$, then

$$O(1) = \{1, 2\}$$
$$O(3) = \{3\}.$$

Let $G$ be the automorphism group of the graph in Figure 3.4. Then it is clear that

$$O(a) = \{a, b, c, d\}$$
$$O(c_1) = \{c_1, c_2, c_3, c_4\}.$$

Consider $D_3$, the group of rotational symmetries of an equilateral triangle, as a group of permutations acting on all points within and on the boundary of the triangle (Fig. 3.5).

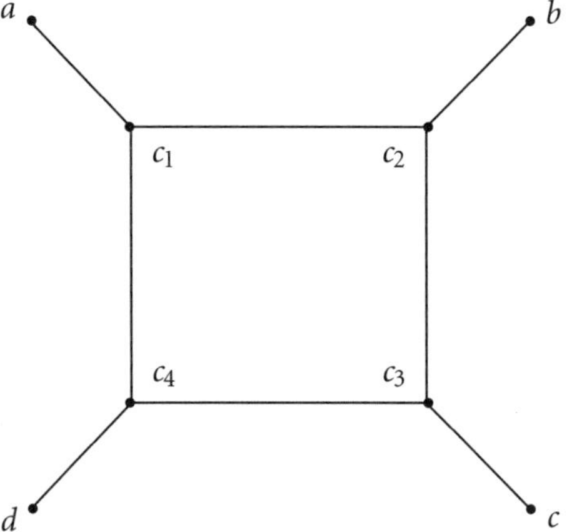



**Figure 3.4**

The point $O$ of intersection of the medians ($AM_1$, $BM_2$, and $CM_3$) constitutes a singleton orbit. Clearly, the three vertices constitute one orbit. The three midpoints of the sides constitute another orbit. Indeed, for every point $m$ on a median, other than $O$, $|O(m)| = 3$. Every other orbit intersects each of the smaller triangles in exactly one point and with a different point for each such triangle. Thus we have

one orbit of size one: $\{O\}$
infinitely many orbits of size three: $\{A, B, C\}$, $\{M_1, M_2, M_3\}$, and so forth
infinitely many orbits of size six.

**Lemma 3.8.1** *Let $G$ be a subgroup of $S_X$. Then the relation*

$$x \equiv y \iff y \in O(x)$$

*is an equivalence relation on $X$. Equivalently, $\{\mathcal{O}(x) \mid x \in X\}$ is a partition of $X$.*

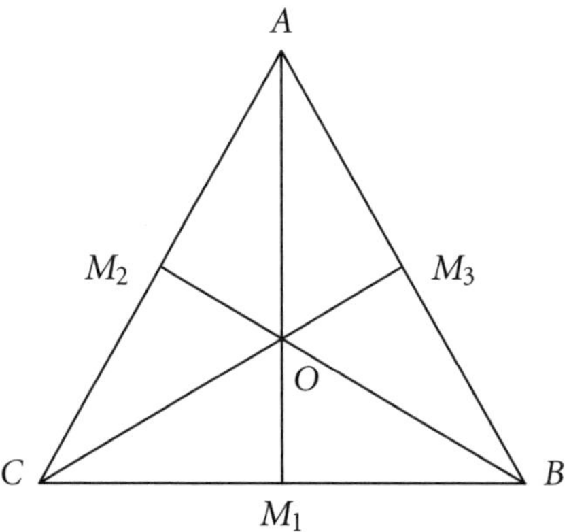



**Figure 3.5**

**Proof.** Let $x, y, z \in X$. Then

$$x = \epsilon(x) \implies x \equiv x$$

so that $\equiv$ is reflexive. Furthermore,

$$\begin{aligned}
x \equiv y &\implies y = \alpha(x), \text{ for some } \alpha \in G \\
&\implies \alpha^{-1}(y) = \alpha^{-1}\alpha(x) = \epsilon(x) = x \\
&\implies x \in O(y) \\
&\implies y \equiv x.
\end{aligned}$$

Therefore, $\equiv$ is symmetric. Last,

$$\begin{aligned}
x \equiv y, y \equiv z &\implies y = \alpha(x), z = \beta(y) \text{ for some } \alpha, \beta \in G \\
&\implies z = \beta(\alpha(x)) = \beta\alpha(x) \text{ where } \beta\alpha \in G \\
&\implies x \equiv z.
\end{aligned}$$

Therefore, $\equiv$ is transitive and so $\equiv$ is an equivalence relation. $\qquad\square$

If $G$ is a subgroup of $S_X$ such that there is only one orbit (that is, $O(x) = X$ for all $x \in X$), then $G$ is said to be a *transitive group* of permutations. For example, $S_X$ is a transitive group of permutations of the set $X$.

In many problems concerning the number of *essentially different* ways of doing something, we are really counting the number of different orbits of some group acting on some set. Elements in the same orbit are considered to be the same.

**Example 3.8.2** Suppose that we want to know in how many ways we can color the corners of a square with the colors red ($R$), white ($W$), blue ($B$), and green ($G$), where each color is used once. The answer depends on how we view the square.

**Case A.** Suppose that we consider the square as fixed in position like a picture on a wall. If we number the corners 1, 2, 3, and 4, then we will have

> 4 choices for corner 1
> 3 choices for corner 2
> 2 choices for corner 3
> 1 choice for corner 4

so that we have 4! choices in all. We might call these *oriented* colorings, since if we turn the square in some way, we consider that we have a new coloring.

**Case B.** Now suppose that the square (Fig. 3.6) can be picked up and turned over or around, so that we lose all record of particular corners. How many recognizably different colorings will there be? Suppose that we start with the coloring on the left as shown in Figure 3.6.

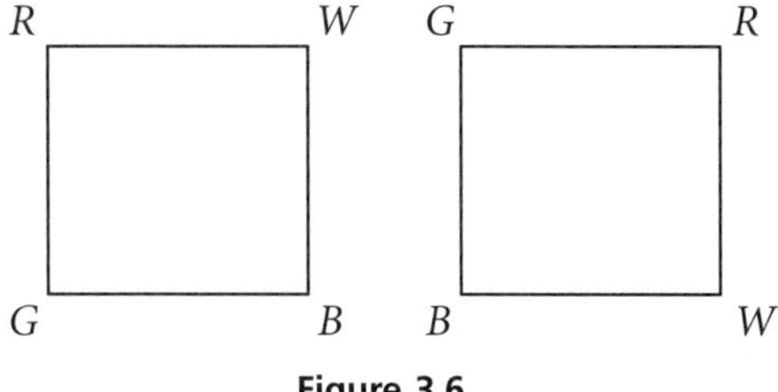

**Figure 3.6**

By rotating clockwise through 90°, we obtain the coloring on the right, which we consider to be essentially the same. Each rotation of the square provides a new coloring that we wish to identify with the first coloring. The same would apply to any other initial coloring.

So, what we have here is the group $D_4$ of rotations of the square acting as a group of permutations of the set of oriented colorings of the square—that is, on the $4! = 24$ colorings that we considered in Case A. Moreover, we want to identify any two colorings that are in the same orbit.

By Lemma 3.8.1, we know that the orbits partition the set of all colorings. Now starting with any particular oriented coloring, each application of $D_4$ gives us a fresh oriented coloring. Thus, each orbit contains $|D_4| = 8$ elements. Therefore,

$$24 = (\text{number of orbits}) \times 8$$

Hence, the number of orbits is 3. In other words, there are three essentially different ways of coloring the corners of a square, and it is not hard to identify them (Fig. 3.7).

**Example 3.8.3** As a second example, consider the number of (essentially different) ways of coloring the vertices of a hexagon such that one is white, one is red, and four are green.

We begin by calculating the number of oriented colorings with this combination of colors. To obtain such a coloring we must first choose two vertices

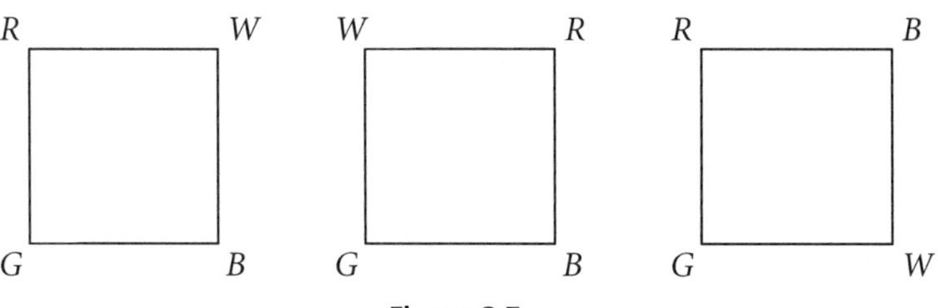

**Figure 3.7**

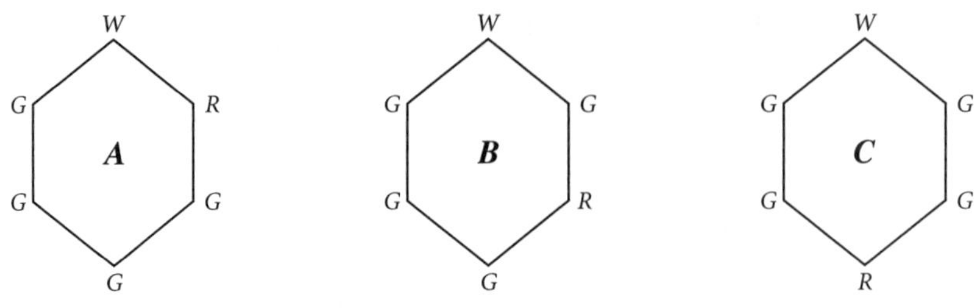

**Figure 3.8**

(to be colored white and red). There are $\binom{6}{2}$ ways of doing that. Then we have two ways of coloring one of them white and the other red. The remaining vertices are colored green. Thus, in total, we have

$$\binom{6}{2} \times 2 = 30$$

oriented colorings. Now we consider the number in each orbit under the action of $D_6$. Every element of $D_6$ gives a different oriented coloring when applied to **A**. Hence the orbit of **A** has 12 elements. Likewise, the orbit of **B** has 12 elements (Fig. 3.8).

On the other hand, the orbit of **C** has only 6 elements. This gives a total of 30. Hence, there are 3 distinct orbits and 3 essentially different colorings, as given by **A**, **B**, and **C**.

### Exercises 3.8

1. Let $G$ denote the group of invertible elements in $\mathbb{Z}_{12}$. For each $a \in G$, let $\rho_a$ denote the mapping of $\mathbb{Z}_{12}$ to $\mathbb{Z}_{12}$ defined by

$$\rho_a : x \to ax \qquad (x \in \mathbb{Z}_{12}).$$

   (i) Show that $H = \{\rho_a \mid a \in G\}$ is a subgroup of $S_{\mathbb{Z}_{12}}$.
   (ii) Describe the orbits in $\mathbb{Z}_{12}$ under $H$.

2. Let $D_4$, the group of rotations of a square act on the vertices, edges, and interior points of a square. Describe the orbits.

3. Let $G$ be the group of rotations of a rectangle (that is not a square), considered as acting on the vertices, edges, and interior points. Describe the orbits.

4. Let $G$ be the group of rotation of a box that has three unequal dimensions, considered as acting on the vertices, edges, sides, and interior points. Describe the orbits.

5. Let $\chi : (\mathbb{Z}_6, +) \to S_{\mathbb{Z}_6}$ be the Cayley representation of $\mathbb{Z}_6$. Describe the orbits of $\mathbb{Z}_6$ under

    (i) $\chi(\langle 2 \rangle)$
    (ii) $\chi(\langle 3 \rangle)$.

6. Let $\chi : S_3 \to S_{S_3}$ be the Cayley representation of $S_3$ and let $\alpha = (1\ 2)$. Describe the orbits of $S_3$ under $\chi(\langle \alpha \rangle)$.

7. Let $H$ be a subgroup of the group $G$ and let $\chi : G \to S_G$ be the Cayley representation of $G$. Let $x, y \in G$. Show that $x$ and $y$ lie in the same orbit with respect to $\chi(H)$ if and only if $Hx = Hy$.

8. Determine the number of ways of coloring the vertices of a square so that two are red and two are green.

9. Determine the number of different ways of arranging 6 keys on a key ring.

10. Determine the number of ways of coloring the vertices of a regular $n$-gon with $n$ different colors.

11. Determine the number of ways of seating $n$ dinner guests around a circular table.

12. Let $ABCD$ be a rectangle with $|AB| \neq |BC|$.

    (i)  How many essentially different ways are there to color the corners of the rectangle with 4 different colors?
    (ii) How many essentially different ways are there to color the corners of the rectangle if 2 are to be colored white, 1 red, and 1 green?

13. A stained glass ornament consists of two concentric circles with two perpendicular diagonals:

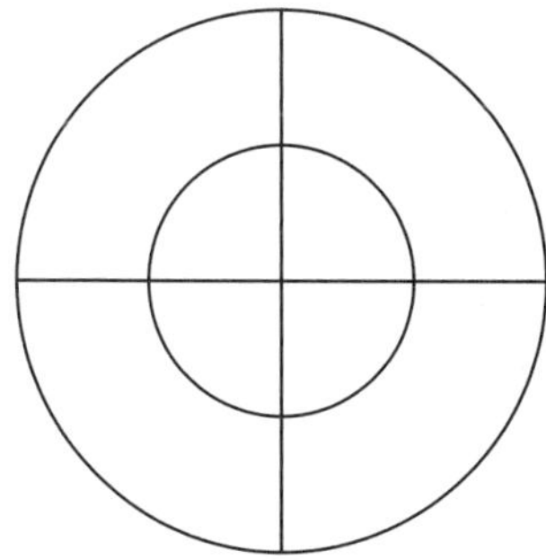

    It has 8 stained glass panels in total.

    (i) How many essentially different patterns are there if 8 different colors are used?

   (ii)  How many essentially different patterns are there if 1 panel is red and 7 are clear?

  (iii)  How many essentially different patterns are there if 2 panels are red and 6 are clear?

14. An equilateral triangle is divided into 6 inner triangles by the 3 medians. Determine the number of essentially different ways to color the 6 inner triangles using

   (i)  6 different colors

  (ii)  4 green and 2 red

 (iii)  3 green and 3 red.

## 3.9  Orbit/Stabilizer Theorem

In section 3.8 we saw how the action of a group $G$ of permutations on a set $X$ imposes some structure on that set by partitioning it into orbits. We begin this section by reversing the point of view to see that the elements of $X$ are reflected in the structure of $G$.

The *stabilizer* of $x$ (in $G$) is

$$S(x) = \{\alpha \in G \mid \alpha(x) = x\}.$$

For example, if $G = S_4$, then

$$S(2) = \{\epsilon, (13), (14), (34), (134), (143)\}$$

$$= S_{\{1,3,4\}}.$$

Let us consider again the rotations of an equilateral triangle as in Figure 3.5. We have a variety of stabilizer groups. Let $\rho_A$ (respectively, $\rho_B$, $\rho_C$) denote the rotation through $180°$ about the axis $AM_1$ (respectively, $BM_2$, $CM_3$). Then, where $P$ is any point that does not lie on a median,

$$S(A) = \{\epsilon, \rho_A\}, S(B) = \{\epsilon, \rho_B\}, S(C) = \{\epsilon, \rho_C\}$$

$$S(O) = D_3, \quad S(P) = \{\epsilon\}.$$

The fact that all these stabilizers turn out to be subgroups is no accident.

**Lemma 3.9.1** *Let $G$ be a subgroup of $S_X$ and $x \in X$. Then $S(x)$ is a subgroup of $G$.*

**Proof.**  Let $\alpha, \beta \in S(x)$. Then

$$
\begin{aligned}
(\alpha\beta)(x) &= \alpha(\beta(x)) \\
&= \alpha(x) \qquad \text{since } \beta \in S(x) \\
&= x \qquad\quad\ \text{since } \alpha \in S(x).
\end{aligned}
$$

Therefore, $\alpha\beta \in S(x)$. Also,

$$
\begin{aligned}
\alpha^{-1}(x) &= \alpha^{-1}(\alpha(x)) \quad \text{since } \alpha \in S(x) \\
&= (\alpha^{-1}\alpha)(x) \\
&= \epsilon(x) \\
&= x.
\end{aligned}
$$

Therefore, $\alpha^{-1} \in S(x)$ and $S(x)$ is a subgroup.  $\square$

The technique used in the previous section to count essentially different colorings requires us at each stage to find a new coloring that is not in any of the orbits previously listed. As the number of orbits increases, this becomes increasingly difficult. However, alternative approaches exist that shift the focus to aspects of the action of each group element. Since the group elements may be fewer in number than the elements of $X$, this has certain advantages.

**Theorem 3.9.2 (Orbit/Stabilizer Theorem)** *Let $X$ be a nonempty set, $G$ be a finite subgroup of $S_X$, and $x \in X$. Then*

$$
|G| = |O(x)||S(x)|.
$$

**Proof.**  Since $S(x)$ is a subgroup of $G$, we know from Lagrange's Theorem that

$$
|G|/|S(x)| = \text{the number of distinct left}
$$
$$
\text{cosets of } S(x) \text{ in } G.
$$

So it suffices to show that the number of left cosets equals the number of elements in $\mathcal{O}(x)$. To this end, define

$$
\varphi : \{\alpha S(x) \mid \alpha \in G\} \longrightarrow O(x)
$$

by

$$
\varphi(\alpha S(x)) = \alpha(x).
$$

Our goal is to show that $\varphi$ is a bijection.

(i) $\varphi$ *is well defined.* We have

$$\alpha S(x) = \beta S(x) \implies \quad \alpha = \beta\gamma \qquad \text{for some } \gamma \in S(x)$$
$$\implies \alpha(x) = \beta\gamma(x)$$
$$= \beta(x) \quad \text{since } \gamma \in S(x).$$

Therefore, $\varphi$ is well defined.

(ii) $\varphi$ *is injective.* For $\alpha, \beta \in G$, we have

$$\varphi(\alpha S(x)) = \varphi(\beta S(x)) \implies \alpha(x) = \beta(x)$$
$$\implies \beta^{-1}\alpha(x) = \beta^{-1}\beta(x) = x$$
$$\implies \beta^{-1}\alpha \in S(x)$$
$$\implies \alpha S(x) = \beta S(x)$$

which implies that $\varphi$ is injective.

(iii) $\varphi$ *is surjective.* Let $y \in O(x)$. Then for some $\alpha \in G$, we have $y = \alpha(x)$. Hence,

$$\varphi(\alpha S(x)) = \alpha(x) = y$$

and $\varphi$ is surjective.

From (i), (ii), and (iii), it follows that $\varphi$ is a bijection. Therefore,

$$|O(x)| = \text{number of left cosets of } S(x) \text{ in } G$$
$$= |G|/|S(x)|$$

which implies that

$$|G| = |O(x)|\,|S(x)|. \qquad \square$$

To illustrate the Orbit/Stabilizer Theorem, consider the group $D_3$ of rotations of an equilateral triangle (Fig. 3.5). We have the following:

| x | $\mid O(x) \mid$ | $\mid S(x) \mid$ | $\mid O(x) \mid\mid S(x) \mid$ |
|---|---|---|---|
| A | 3 | 2 | 6 |
| $M_1$ | 3 | 2 | 6 |
| O | 1 | 6 | 6 |
| p | 6 | 1 | 6 |

The Orbit/Stabilizer Theorem is a great help in determining the size of many groups.

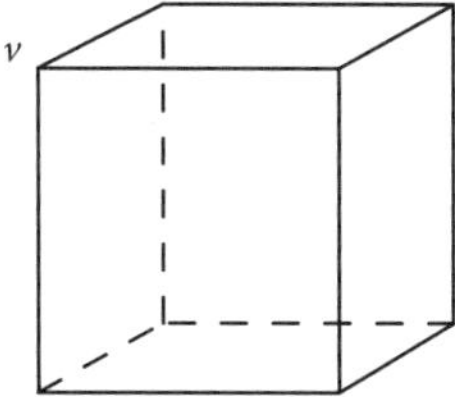

**Figure 3.9**

**Example 3.9.3** Let $G$ denote the group of rotations of a cube. Let $v$ denote any of the vertices (Fig. 3.9).

It is not hard to See that $O(v)$ consists of all vertices and that the number of rotations that fix the corner $v$ is 3. Thus,

$$|O(v)| = 8 \quad \text{and} \quad |S(v)| = 3.$$

Therefore, by the Orbit/Stabilizer Theorem,

$$|G| = |O(v)|\,|S(v)|$$
$$= 8 \times 3$$
$$= 24.$$

**Example 3.9.4** Balls used in various sports (for example, basketball, football, or soccer) often display interesting patterns that relate to various rotational groups. Suppose we wished to determine the size of the group $G$ of those rotations of a soccer ball (Fig. 3.10) that respect the pattern of panels. This pattern has 30 vertices, 12 pentagonal faces (shaded black), and 20 hexagonal faces (white).

There is more than one way to do this using the Orbit/Stabilizer Theorem.

Clearly, every rotation is completely determined by what it does to the vertices. So we can consider the group of rotations as a group of permutations of the vertices. It is not difficult to see that there is just one orbit that contains

**Figure 3.10**

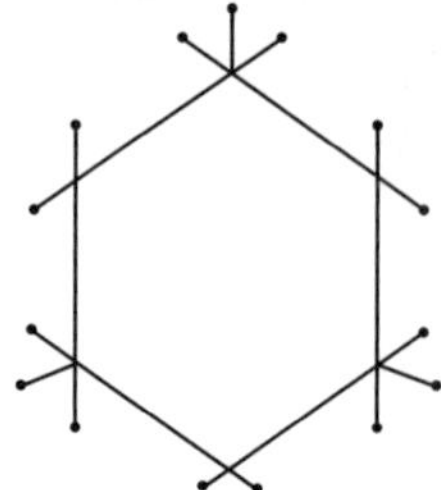

**Figure 3.11**

all the vertices. It is also easy to see that, for any vertex $v$, $S(v) = \{\epsilon\}$. Hence,

$$
\begin{aligned}
|G| &= |O(v)|\,|S(v)| \\
&= |O(v)| \\
&= \text{number of vertices} \\
&= 60.
\end{aligned}
$$

Alternatively, we could focus on the centers of the dark pentagonal faces. Again we have just 1 orbit. However, for any such face $f$, there will be 5 rotations that map that face to itself. Hence,

$$
\begin{aligned}
|G| &= |O(f)|\,|S(f)| \\
&= 12 \times 5 \\
&= 60.
\end{aligned}
$$

A similar calculation can be made relative to the hexagonal faces.

**Example 3.9.5** Let $G$ be the automorphism group of the graph in (Figure 3.11). Then it is not hard to see that, for any vertex $v$ on the inner hexagon,

$$
|O(v)| = 3, \quad |S(v)| = (6^3 \times 2^3) \times 2
$$

so that, by the Orbit/Stabilizer Theorem,

$$
\begin{aligned}
|G| &= |O(v)||S(v)| \\
&= 3 \times 6^3 \times 2^3 \times 2 \\
&= 3^4 \times 2^7.
\end{aligned}
$$

It is common to refer to the group of permutations of an object respecting its patterns or shape as the *group of symmetries* of that object. Thus, here we have been looking at the group of symmetries of a cube, a soccer ball, and a graph.

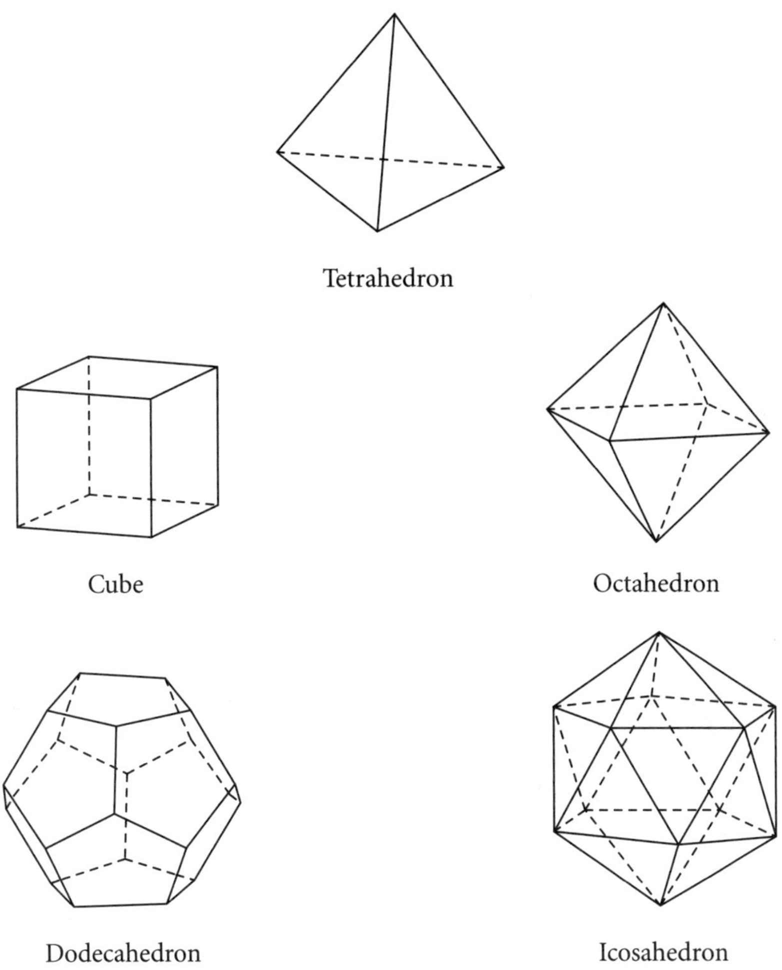

Tetrahedron

Cube

Octahedron

Dodecahedron

Icosahedron

**Figure 3.12**

The regular polyhedra displayed in Figure 3.12 display very interesting symmetries that will be considered in the exercises. Their arrangement in Figure 3.12 is deliberate to indicate the duality that exists among them. If you join the midpoint of each face of the cube, then these line segments form the edges for an octahedron. Dually, if you join the midpoint of each face of the octahedron, then these line segments form the edges of a cube. A similar duality exists between the dodecahedron and the icosahedron. The tetrahedron is self-dual. The 5 regular polyhedra are sometimes referred to as the *Platonic solids* in honor of the studies of them conducted by the Academy of Plato in Athens. Credit, however, for the proof of the fact that there are exactly five regular polyhedra is usually given to Theaetetus, born 414 B.C. The regular polyhedra fascinated the ancient Greek mathematicians. Book XIII of *Euclid's Elements*, the last book, is devoted to the study of the regular polyhedra.

**Exercises 3.9**

1.  (i)  Show that the mapping

$$\alpha : x \rightarrow 5x \qquad (x \in \mathbb{Z}_{12})$$

is a permutation of $\mathbb{Z}_{12}$.
    (ii)  Find the order of $G = \langle \alpha \rangle$.
    (iii)  Find the orbits and stabilizers under the action of $G$.

2. Repeat Exercise 1 with respect to the mapping

$$\alpha : x \rightarrow x + 8 \qquad (x \in \mathbb{Z}_{12}).$$

3. Let $R = \mathbb{Z}_3[y]/(y^2 - 1)$, $G = \{1, y\}$, and for each $a \in G$ define a mapping $\rho_a : R \rightarrow R$ by

$$\rho_a(f(y)) = af(y).$$

   (i)  Show that $G$ is a group with respect to multiplication (in $R$).
   (ii)  Show that $H = \{\rho_1, \rho_y\}$ is a subgroup of $S_R$.
   (iii)  Find the orbits of the elements of $R$ with respect to $H$.
   (iv)  Find the stabilizers in $H$ of the elements of $R$.

4. Let $R = \mathbb{Z}_2[y]/(y^4 - 1)$, $G = \{1, y, y^2, y^3\}$, and for each $a \in G$ define a mapping $\rho_a : R \rightarrow R$ by

$$\rho_a(f(y)) = af(y).$$

   (i)  Show that $G$ is a group with respect to multiplication (in $R$).
   (ii)  Show that $H = \{\rho_1, \rho_y, \rho_{y^2}, \rho_{y^3}\}$ is a subgroup of $S_R$.
   (iii)  Find the orbits of the elements of $R$ with respect to $H$.
   (iv)  Find the stabilizers in $H$ of the elements of $R$.
   (v)  Show that the sizes of the orbits and stabilizers conform to the Orbit/Stabilizer Theorem.

5. Describe the orbits and stabilizers for the following shapes under the action of the corresponding group of rotational symmetries:

   (i)  A circle
   (ii)  An isosceles triangle (that is not equilateral)
   (iii)  A square.

6. Determine the order of the group of rotational symmetries for each of the regular polyhedra (Fig. 3.12).

7. For each of the following objects, describe each element of the group of rotations as a single rotation:

   (i)  Tetrahedron.
   (ii)  Cube.
   (iii)  Octahedron.

8. Let $H$ be an abelian subgroup of $S_X$. Let $x, y \in X$ be such that $O(x) = O(y)$. Show that $S(x) = S(y)$.

9. Let $GF(9) = \mathbb{Z}_3[y]/(y^2 + 1)$. Define

$$\varphi : a \to a^3 \qquad (a \in GF(9))$$

(see Exercise 12 in section 3.3). Let $H = \langle \varphi \rangle$ be the subgroup of the group of permutations of $GF(9)$ generated by $\varphi$. Describe the orbits and stabilizers with respect to the action of $H$ on $GF(9)$.

## 3.10 The Cauchy-Frobenius Theorem

In the coloring problems that we considered in section 3.8, the variations were sufficiently limited that we were able to identify all the orbits by a process of exhaustion. This becomes much harder as the number of orbits increases. In this section we will consider a clever result that enables us to calculate the number of orbits as long as we have a good understanding of how each group element behaves.

Let $G$ be a subgroup of $S_X$ ($X \neq \varnothing$) and let $\alpha \in G$, $x \in X$. If $\alpha(x) = x$ then we say that $\alpha$ *fixes* $x$ or that $x$ is *fixed* by $\alpha$. The set

$$F(\alpha) = \{x \in X \mid \alpha(x) = x\}$$

is the *fixed set* of $\alpha$. Note that

$$x \in F(\alpha) \iff \alpha \in S(x).$$

Consider the group of rotations of the square $\{\epsilon, R = R, R^2, R^3, \rho, R\rho, R^2\rho, R^3\rho\}$ acting on the vertices, where $\rho$ is taken to be the rotation about the diagonal $BD$ (Fig. 3.13).

Then

$$F(\epsilon) = \{A, B, C, D\}$$

$$F(R) = F(R^2) = F(R^3) = \varnothing$$

$$F(\rho) = \{B, D\}.$$

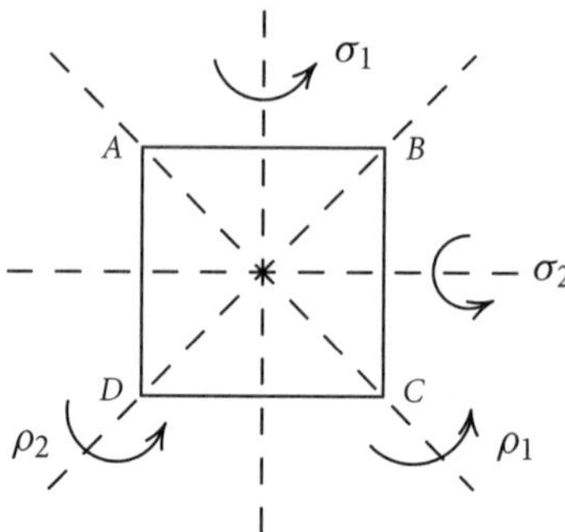


**Figure 3.13**

We already know from the Orbit/Stabilizer Theorem that there is a strong link between the sizes of the orbits and the stabilizers. The next result gives a connection between the stabilizers and the fixed sets.

**Lemma 3.10.1** *Let $X$ be a finite nonempty set and $G$ be a subgroup of $S_X$. Then*

$$\sum_{\alpha \in G} |F(\alpha)| \;=\; \sum_{x \in X} |S(x)|.$$

**Proof.** Consider the following array, where the top row lists the elements of $X$ and the first column lists the elements of $G$:

| $G \backslash X$ | $x_1$ | $\cdots$ | $x_j$ | $\cdots$ | $x_n$ | |
|---|---|---|---|---|---|---|
| $\alpha_1$ | | | | | | $F(\alpha_1)$ |
| $\vdots$ | | | | | | |
| $\alpha_i$ | | | $\checkmark$ | | | $F(\alpha_i)$ |
| $\vdots$ | | | | | | |
| $\alpha_m$ | | | | | | $F(\alpha_m)$ |
| | $S(x_1)$ | | $S(x_j)$ | | $S(x_n)$ | |

Now imagine that a check mark ($\checkmark$) is placed in the box in row $\alpha_i$ and column $x_j$ if and only if $\alpha_i(x_j) = x_j$. The number of check marks in the $\alpha_i$ row is just the number of $x \in X$ for which $\alpha_i(x) = x$—that is, $|F(\alpha_i)|$. Thus, the total number of check marks is

$$\sum_{\alpha \in G} |F(\alpha)|. \tag{3.5}$$

On the other hand, the number of check marks in the $x$ column is just the number of $\alpha \in G$ with $\alpha(x) = x$—that is, $|S(x)|$. Thus, the total number of check marks must be

$$\sum_{x \in X} |S(x)|. \tag{3.6}$$

Since both expressions (3.5) and (3.6) represent the total number of check marks in the array, they must be equal.  $\square$

The next theorem is often referred to as Burnside's Lemma. For an interesting article on the history of the naming and misnaming of this result, see [Neu].

**Theorem 3.10.2 (Cauchy-Frobenius Theorem)** *Let $X$ be a finite nonempty set and $G$ be a subgroup of $S_X$. Let $N$ denote the number of different orbits of $G$. Then*

$$N = \frac{1}{|G|} \sum_{\alpha \in G} |F(\alpha)|.$$

**Proof.**  Let the orbits be

$$\mathcal{O}_1 = \{a_1, \ldots, a_m\}$$
$$\mathcal{O}_2 = \{b_1, \ldots, b_n\}$$
$$\vdots$$
$$\mathcal{O}_N = \ldots.$$

Each element of $X$ appears in one and only one orbit. Then, by the Orbit/Stabilizer Theorem,

$$|S(a_1)| + \cdots + |S(a_m)| = \underbrace{\frac{|G|}{m} + \cdots + \frac{|G|}{m}}_{m} = |G|$$

$$|S(b_1)| + \cdots + |S(b_n)| = \underbrace{\frac{|G|}{n} + \cdots + \frac{|G|}{n}}_{n} = |G|$$

$$\cdots \qquad\qquad \cdots \qquad\qquad \cdots$$

summing all these equations, we obtain

$$\sum_{x \in X} |S(x)| = |G|N.$$

By Lemma 3.10.1, we have $|G| N = \sum_{\alpha \in G} |F(\alpha)|$. Hence, $N = \frac{1}{|G|} \sum_{\alpha \in G} |F(\alpha)|$, as required.  $\square$

To see how the concept of a fixed set works with colorings, consider the essentially different ways of coloring the vertices of a square with 1 white, 1 red, and 2 blue. To find the number of different oriented colorings, we see that we just have to pick one vertex to be white and then one to be red. Thus the total number of possibilities is

$$\binom{4}{1} \cdot \binom{3}{1} = 4 \times 3 = 12.$$

Again, we consider two coloring to be essentially the same if one is derived from the other by applying a rotation—that is, an element from

$$D_4 = \{\epsilon, R, R^2, R^3, \rho_1, \rho_2, \sigma_1, \sigma_2\}$$

where $R$ is a rotation through $90°$ about an axis perpendicular to the page, $\rho_1$ and $\rho_2$ are rotations about axes through $AC$ and $BD$ (in Fig. 3.13), respectively, and $\sigma_1$ and $\sigma_2$ are rotations about axes through the midpoints of the sides $AB$ and $BC$ (Fig. 3.13), respectively.

Any coloring in our set has only one vertex colored white. Applying any power of $R$ (other than $\epsilon$) to this coloring will move the white to a different vertex and so give a different oriented coloring. Thus, no oriented coloring is left fixed and $F(R^i) = \emptyset$ for $i = 1, 2, 3$. The same argument applies to both $\sigma_1$ and $\sigma_2$. However, the $\rho_i$'s are a bit different since $\rho_1$, for instance, will not move the colors assigned to the vertices $A$ and $C$. Hence, $\rho_1$ will leave the following colorings fixed:

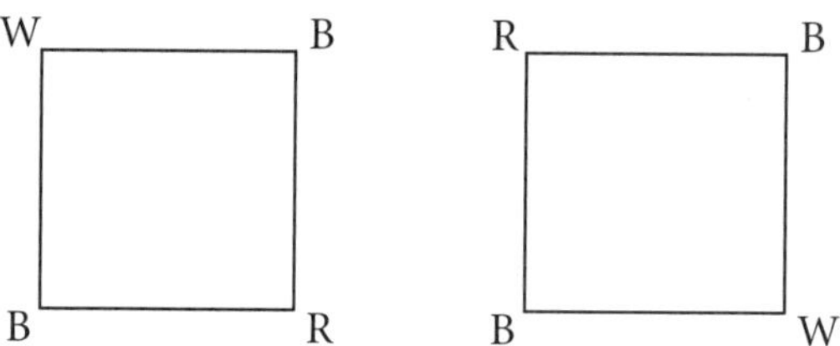


and it is not difficult to see that these are the only colorings left fixed by $\rho_1$. Thus, $|F(\rho_1)| = 2$. Similarly, $|F(\rho_2)| = 2$. Last, it is evident that the identity fixes all the colorings. Summarizing, we have

| $A$ | $|F(\alpha)|$ |
| --- | --- |
| $\epsilon$ | 12 |
| $R^i$ | 0 |
| $\rho_i$ | 2 |
| $\sigma_i$ | 0. |

Now to say that two oriented colorings are essentially the same is equivalent to saying that they lie in the same orbit. Thus, to find the number of essentially different colorings, what we want to do is find the number of distinct orbits. This is precisely what the Cauchy-Frobenius Theorem will do for us. Applying the formula from the Cauchy-Frobenius Theorem, we conclude that the number of orbits is

$$N = \frac{1}{8}(12 + 0 + 0 + 0 + 2 + 2 + 0 + 0)$$

$$= \frac{16}{8} = 2.$$

Thus there are only two essentially different colorings in this case and they are easily identified: Either the red and white vertices are adjacent or they are opposite to each other. Well, the answer is so simple in this case that we did not really need the Cauchy-Frobenius Theorem to solve it. However, it enabled us to illustrate the concepts before considering more challenging examples.

**Example 3.10.3** How many ways are there to color the 6 faces of a cube in such a way that 2 are red and 4 are white? The group of rotations of the cube has elements of the following types:

| | |
|---|---|
| 1 | identity $\epsilon$. |
| 6 | rotations $R$ through 90° or 270° about axes through the centers of the opposite faces. |
| 3 | rotations $S$ through 180° about axes through the centers of opposite faces. |
| 8 | rotations $\rho$ about axes through opposite corners. |
| 6 | rotations $\sigma$ about axes through midpoints of opposite edges. |

This gives us the required total of 24 rotations. The number of oriented colorings is just the number of ways of choosing 2 faces to color red—that is, $\binom{6}{2} = 15$. With a little thought we can calculate the sizes of the fixed sets as follows:

| $\alpha$ | $|F(\alpha)|$ |
|---|---|
| $\epsilon$ | 15 |
| $R$ | 1 |
| $S$ | 3 |
| $\rho$ | 0 |
| $\sigma$ | 3. |

By the Cauchy-Frobenius Theorem, the number of orbits is

$$N = \frac{1}{24}(15 + 6 \times 1 + 3 \times 3 + 8 \times 0 + 6 \times 3)$$

$$= \frac{1}{24} \times 48 = 2.$$

Thus, the number of essentially different colorings is 2.

**Example 3.10.4** Consider the number of ways of coloring the faces of a cube with 6 different colors. For this we think of the group of rotations of the cube as acting on the faces. The total number of oriented colorings is 6!. Then,

$$|F(\epsilon)| = 6!$$

whereas, for any $\alpha \in G, \alpha \neq \epsilon$,

$$|F(\alpha)| = \emptyset.$$

By the Cauchy-Frobenius Theorem, we have that

$$\text{the number of orbits} = \frac{1}{|G|} \sum_{\alpha \in G} |F(\alpha)|$$

$$= \frac{1}{24}(6! + 0 + 0 + \cdots + 0)$$

$$= \frac{6!}{24} = 30.$$

**Example 3.10.5** How many ways are there to color the triangular faces of an icosahedron using 20 distinct colors? An icosahedron has

20 triangular faces
30 edges
12 vertices.

By the Orbit/Stabilizer Theorem, we see that the size of the group $G$ of rotations is

$$20 \times 3 = 30 \times 2 = 12 \times 5 = 60.$$

The number of oriented colorings is 20!. We apply the Cauchy-Frobenius Theorem. We have

$$|F(\epsilon)| = 20!$$

whereas for any other $\alpha \in G$, $F(\alpha) = \emptyset$ so that $|F(\alpha)| = 0$. Hence the number of orbits is

$$N = \frac{1}{|G|} \sum_{\alpha \in G} F(\alpha)$$

$$= \frac{1}{60}(20! + 0 + 0 + \cdots + 0)$$

$$= \frac{20!}{60} \ .$$

Colorings are really just a device used by mathematicians to model some varying pattern of characteristics that might be present at different points of an "object" of some sort. The next example helps to illustrate this idea.

**Example 3.10.6** Benzene is a chemical compound, each molecule of which is made up of six carbon atoms and six hydrogen atoms. The carbon atoms are arranged in a hexagon and each carbon atom is bonded to a single hydrogen atom (Fig. 3.14). Related chemicals can be derived from benzene by replacing one or more of the hydrogen atoms by a cluster involving one new carbon atom and three new hydrogen atoms (Fig. 3.15).

Let us find the number of chemical derivatives that can be obtained from benzene by replacing three of the hydrogen atoms by $CH_3$ clusters. It is not difficult to see that this problem is really a coloring problem in a different guise—namely, how many essentially different ways are there to color the vertices of a hexagon if three of the vertices are to be colored blue ($\equiv$ a carbon atom is attached) and three of the vertices are to be colored green ($\equiv$ a hydrogen atom is to be attached)?

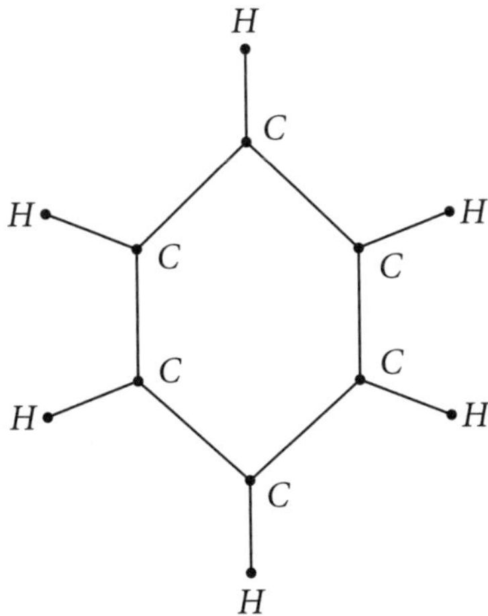

**Figure 3.14**

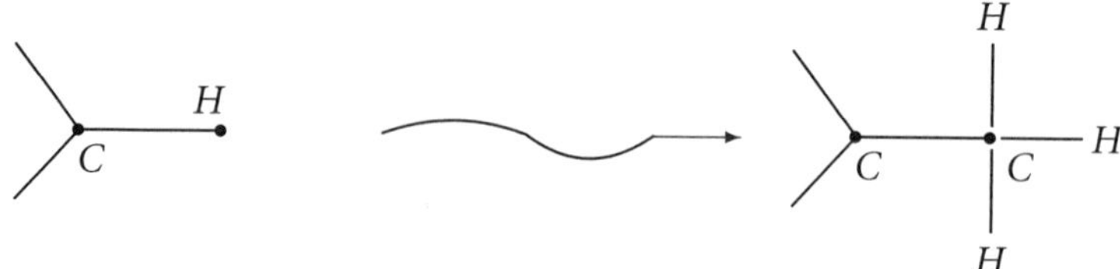

**Figure 3.15**

Taking into account orientation, the number of possibilities would just be the number of ways of choosing three hydrogen atoms for replacement from six possibilities—in other words, $\binom{6}{3} = 20$. However, those that are related by a rotational symmetry clearly correspond to the same derivative chemical. So, we wish to determine the group of rotational symmetries.

Since we are thinking of benzene and its derivatives as molecules in $\mathbb{R}^3$, the appropriate group of rotations is clearly $D_6$, which we may describe as

$$D_6 = \{\epsilon, R, R^2, R^3, R^4, R^5, \rho_1, \rho_2, \rho_3, \sigma_1, \sigma_2, \sigma_3\}$$

where $R$ denotes the rotation clockwise through $60°$ about an axis through the center and perpendicular to the page, each $\rho_i$ is a rotation about an axis through opposite vertices, and each $\sigma_i$ is a rotation about an axis through the midpoints of a pair of opposite sides.

For the current calculation, however, we are thinking of $D_6$ as acting on all the different oriented colorings of the vertices with three vertices colored blue and three vertices colored green. The number of oriented colorings is, as we saw earlier, just 20. It is then straightforward to calculate the sizes of the various fixed sets:

| $\alpha$ | $|F(\alpha)|$ |
|---|---|
| $\epsilon$ | 20 |
| $R, R^5$ | 0 |
| $R^2, R^4$ | 2 |
| $R^3$ | 0 |
| $\rho_i$ | 4 |
| $\sigma_i$ | 0. |

By the Cauchy-Frobenius Theorem, the number of orbits is (omitting zeros)

$$N = \frac{1}{12}\{20 + 2 + 2 + 4 + 4 + 4\}$$

$$= \frac{36}{12} = 3.$$

Thus, we might expect that there would be three different chemicals derivable from benzene in this way. This is indeed the case: they are 1,2,3-Trimethylbenzene, 1,2,4-Trimethylbenzene and 1,3,5-Trimethylbenzene (see [Var]).

**Exercises 3.10**

1. Determine the number of ways of coloring the vertices of a pentagon in each of the following ways:

    (i) With five distinct colors.
    (ii) Two are red and three are white.
    (iii) Two are red, two are white, and one is blue.

2. Find the number of ways of coloring the vertices of a septagon in each of the following combinations:

    (i) Three red and four white.
    (ii) One red, two white, and four blue.
    (iii) One red, three white, and three blue.

3. Find the number of ways of coloring the vertices of an octagon in each of the following combinations:

    (i) Four red and four white.
    (ii) Six red and two white.
    (iii) Two red, two white, and four blue.

4. A jeweller is making colored glass pendants by dividing an equilateral triangle into six small triangles using the medians. How many different designs can he make if

    (i) four panels are red and two are blue
    (ii) two panels are red, two are white, and two are blue?

5. Find the number of ways of coloring the vertices of a tetrahedron in the following combinations:

    (i) One red and three white.
    (ii) Two red and two white.

6. Find the number of ways of coloring the edges of a tetrahedron in the following combinations:

    (i) Three red and three white.
    (ii) Two red and four white.

7. Find the number of ways of arranging three large (identical) pearls and six small (identical) pearls loosely on a necklace.

8. Find the number of derivatives of benzene obtained by replacing two of the hydrogen atoms by $CH_3$.

9. How many different chemicals can be obtained by replacing two of the hydrogen atoms (H) in the following organic molecule (propane) by $CH_3$

$$
\begin{array}{ccc}
H & H & H \\
| & | & | \\
H-C-C-C-H \\
| & | & | \\
H & H & H
\end{array}
$$

where equivalence is taken with respect to the group that acts on the molecule as if it were a graph with vertices colored $C$ and $H$, with three of the $H$ vertices to be replaced by $CH_3$.

10. Repeat Exercise 9 but this time replace three of the hydrogen atoms.


## 3.11 K-Colorings

In the preceding sections we have been considering the number of different colorings where the number of times that each color is used is specified (two of this color, three of that color, ... ). In this and the next section, we consider a slightly different situation in which the *number* of available colors is specified, but there is no restriction on the number of times that each is used.

Let $X$ be a nonempty set where $|X| = n$. Let $K$ be a set of "colors", $|K| > 1$. Then, any mapping $\gamma : X \to K$ is called a *K-coloring*. We let $C$ denote the set of all possible $K$-colorings. Since we do not lose anything and since there is no loss in generality in doing so, we think of $K$ as just a set of integers $\{1, 2, \ldots, k\}$ where $k = |K|$. An important point to note is that a $K$-coloring may not actually use all $k$ colors of $K$ and some colors may be used many times. At the extreme, a coloring might only use one color.


**Lemma 3.11.1** $|C| = k^n$.


**Proof.** For each element of $X$ we have $k$ possible colors. Therefore, the total number of possibilities is

$$
\underbrace{k \times k \times \cdots \times k}_{n \text{ times}} = k^n. \qquad \square
$$


As before, if one coloring is obtained from another under the action of certain permutations (such as rotations), we may want to consider them to be equivalent. So our next step is to be a bit more precise regarding how permutations of the set $X$ determine permutations of the set $C$ of colorings.

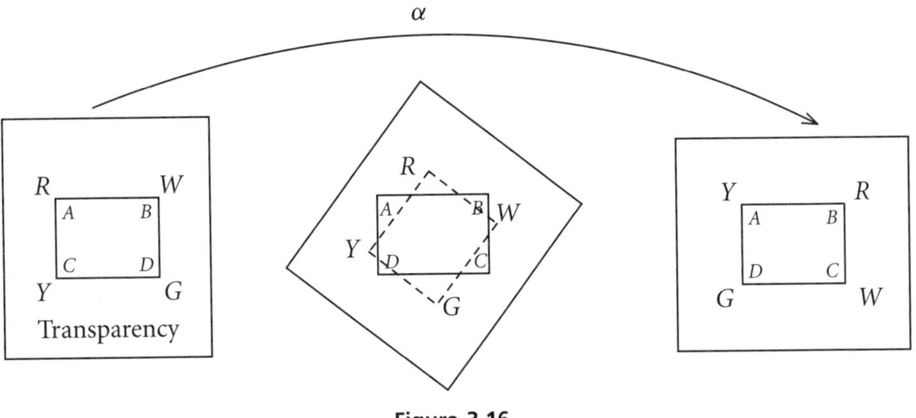


**Figure 3.16**

**Lemma 3.11.2** *Each element $\alpha \in S_X$ determines a permutation $\widehat{\alpha}$ of $C$ as follows: For every $\gamma \in C$,*

$$(\widehat{\alpha}(\gamma))(x) = \gamma(\alpha^{-1}(x)) \qquad (x \in X).$$

**Proof.** Clearly, $\widehat{\alpha}(\gamma) \in C$ so that $\widehat{\alpha}$ is a mapping of $C$ to itself. Also, for $\gamma, \gamma' \in C$.

$$\widehat{\alpha}(\gamma) = \widehat{\alpha}(\gamma') \implies \gamma(\alpha^{-1}(x)) = \gamma'(\alpha^{-1}(x)) \quad (x \in X)$$
$$\implies \gamma(y) = \gamma'(y) \quad (y \in X)$$
$$\implies \gamma = \gamma'$$

Hence, $\widehat{\alpha}$ is injective. Since $C$ is finite, it follows that $\widehat{\alpha}$ is bijective and therefore a permutation of $C$. $\square$

A useful way to visualize the action of $\widehat{\alpha}$ is as follows. Suppose that we are examining the colorings of the vertices of a square and suppose that $\alpha$ is the rotation through 90° in a clockwise direction. Let $\kappa$ be a coloring. Place a transparency over the drawing. Now color the vertices (according to $\kappa$), but on the transparency. We will now obtain the coloring $\alpha(\kappa)$ by rotating the *transparency* 90° in the clockwise direction.

**Theorem 3.11.3** *Let $X$ be a nonempty set with $|X| \geq 2$. The mapping*

$$\theta : \alpha \to \widehat{\alpha} \qquad (\alpha \in S_X)$$

*is an isomorphism of $S_X$ onto a subgroup of $S_C$.*

**Proof.** By Lemma 3.11.2, $\theta$ maps $S_X$ into $S_C$.

Let $\alpha, \beta \in S_X$. Then, for all $\gamma \in C$,

$$
\begin{aligned}
(\theta(\alpha\beta)(\gamma))(x) &= \gamma((\alpha\beta)^{-1}(x)) \qquad \text{for all } x \in X \\
&= \gamma(\beta^{-1}\alpha^{-1}(x)) \\
&= \gamma(\beta^{-1}(\alpha^{-1}(x)) \\
&= (\theta(\beta)(\gamma))(\alpha^{-1}(x)) \\
&= \theta(\alpha)(\theta(\beta)(\gamma))(x) \\
&= ((\theta(\alpha)\theta(\beta))(\gamma))(x).
\end{aligned}
$$

Therefore, $\theta(\alpha\beta)(\gamma) = (\theta(\alpha)\theta(\beta))(\gamma)$, for all $\gamma \in C$, which implies that $\theta(\alpha\beta) = \theta(\alpha)\theta(\beta)$. Hence, $\theta$ is a homomorphism.

Let $\widehat{S_X} = \theta(S_X)$. Then $\widehat{S_X}$ is a subgroup of $S_C$ (see Exercise 6 in this section 3.11) and $\theta$ is a surjective homomorphism of $S_X$ to $\widehat{S_X}$. It remains to show that $\theta$ is injective.

Let $\alpha, \beta \in S_X$ and $\alpha \neq \beta$. Then $\alpha^{-1} \neq \beta^{-1}$ so that there exists an element $x$ in $X$ with $\alpha^{-1}(x) \neq \beta^{-1}(x)$. Let $\gamma$ be any coloring with $\gamma(\alpha^{-1}(x)) \neq \gamma(\beta^{-1}(x))$. Then

$$
\begin{aligned}
(\theta(\alpha)(\gamma))(x) &= \gamma(\alpha^{-1}(x)) \neq \gamma(\beta^{-1}(x)) \\
&= (\theta(\beta)(\gamma))(x)
\end{aligned}
$$

so that $\theta(\alpha)(\gamma) \neq \theta(\beta)(\gamma)$ from which it follows that $\theta(\alpha) \neq \theta(\beta)$. Therefore, $\theta$ is injective and an isomorphism of $S_X$ onto $\widehat{S_X}$. $\square$

Now let $G$ be a subgroup of $S_X$. We will say that two colorings $\gamma_1, \gamma_2$ of $X$ are *G-equivalent* if $\widehat{\alpha}(\gamma_1) = \gamma_2$ for some $\alpha \in G$. For instance, if we were interested in colorings of the faces of a duodecahedron, it would be natural to consider two such colorings as equivalent if they are equivalent with respect to the group of rotations of the duodecahedron. When dealing with colorings associated with figures (vertices of a hexagon, faces of a tetrahedron, and so forth) we will assume that $G$ is the group of rotations.

Of course, what we will normally wish to find is the number of "inequivalent" colorings. This really means the number of orbits under the action of

$$
\widehat{G} = \{\widehat{\alpha} \mid \alpha \in G\}.
$$

When studying groups acting on sets, a concept that turns out to be quite helpful is that of faithfulness. Let $G$ be a subgroup of $S_X$ and $O$ be an orbit of

$X$ under the action of $G$. Then we say that $G$ *acts faithfully on* $O$ if it satisfies the following condition:

$$\forall \alpha, \beta \in G \; \exists x \in O \quad \text{such that} \quad \alpha(x) \neq \beta(x).$$

For example, if $G$ is the group of rotations of a cube, then it acts faithfully on the orbit consisting of the corners, the orbit consisting of the centers of the faces, and many other orbits. However, every element of $G$ fixes the center of the cube, and so $G$ does not act faithfully on the orbit consisting of the center.

### Exercises 3.11

1. Let $K = \{\,\text{red, white}\,\}$, $C$ be the set of $K$-colorings of the vertices of an isosceles triangle $T$, and $G$ be the group of rotations of $T$.

   (i)  List the elements of $C$.
   (ii) Determine the orbits of $C$ under $\widehat{G}$.

2. Let $K = \{\,\text{red, white}\,\}$, $C$ be the set of $K$-colorings of the vertices of an equilateral triangle $T$, and $G$ be the group of rotations of $T$.

   (i)   List the elements of $C$.
   (ii)  Determine the orbits of $C$ under $\widehat{G}$.
   (iii) Find an orbit on which $G$ acts faithfully.

3. Let $K = \{\,\text{red, white}\,\}$, $C$ be the set of $K$-colorings of a square $S$, and $G$ be the group of rotations of $S$.

   (i)   List the elements of $C$.
   (ii)  Determine the orbits of $C$ under $\widehat{G}$.
   (iii) Find an orbit on which $\widehat{G}$ acts faithfully.
   (iv)  Find a nontrivial orbit (in other words, with two or more elements) on which $\widehat{G}$ does not act faithfully.

*4. Let $|K| \geq 2$, $C$ denote the set of $K$-colorings of a set $X$, $G$ be a subgroup of $S_X$, and $X$ consist of just one orbit under the action of $G$. Show that there exists an orbit of $C$ under $\widehat{G}$ on which $\widehat{G}$ acts faithfully.

5. Let $G = D_9$, the group of rotations of a regular 9-gon $P$ with vertices $V(P) = \{1, 2, \ldots, 9\}$. Let $K = \{R, W, B\}$ and $\gamma$ be the coloring of $V(P)$ such that

$$\gamma(1) = \gamma(4) = \gamma(7) = R$$
$$\gamma(2) = \gamma(5) = \gamma(8) = W$$
$$\gamma(3) = \gamma(6) = \gamma(9) = B.$$

Establish the following:

(i) The orbit $O$ of $X$ under $\widehat{G}$ has three elements—say, $\gamma = \gamma_1, \gamma_2,$ and $\gamma_3$.

(ii) $\widehat{G}$ does not act faithfully on $O$.

(iii) The action of $\widehat{G}$ on $O$ is that of the full permutation group $S_3$.

6. Let $\theta$ be defined as in Theorem 3.11.3. Show that $\theta(S_X)$ is a subgroup of $S_C$.

## 3.12 Cycle Index and Enumeration

In this final section of this chapter we consider a way of enumerating the number of essentially different $K$-colorings of sets. Suppose that $G$ is a group of permutation of a set $X$ and that we wish to calculate the number of $K$-colorings. Let $g \in G$. Then a coloring is fixed by $g$ if and only if every element in every cycle of $g$ has the same color, though different cycles could have different colors. So, if $g$ has $c(g)$ cycles, then the number of $K$-colorings left fixed by $g$ is just $|K|^{c(g)}$. By the Cauchy-Frobenius Theorem, the number of different orbits in the set of all colorings must then be:

$$\frac{1}{|G|} \sum_{g \in G} |K|^{c(g)}$$

We will spell this argument in a little more detail by introducing an important concept concerning the pattern of cycles in permutations of finite sets.

Let $|X| = n$ and $\alpha \in S_X$ be written as a product of disjoint cycles:

$$\alpha = \alpha_1 \alpha_2 \ldots \alpha_k.$$

The *cycle pattern* of $\alpha$ is then

$$1^{r_1} 2^{r_2} \cdots n^{r_n}$$

where $\alpha$ has $r_1$ cycles of length 1, $r_2$ cycles of length 2, $\ldots$, $r_n$ cycles of length $n$. Note that

$$1 \cdot r_1 + 2 \cdot r_2 + \cdots + n \cdot r_n = n$$

since every element of $X$ appears in exactly one cycle of $\alpha$. For example, the cycle pattern of

$$\alpha = (1\,3\,5\,7)(2\,6\,8\,12)(4\,9\,10)(11\,13)(14\,15)(16)(17)(18)$$

as an element of $S_{18}$ is

$$1^3\ 2^2\ 3^1\ 4^2.$$

We now encode the cycle pattern of a permutation in a polynomial. For any $\alpha \in S_n$ with cycle pattern

$$1^{r_1}\ 2^{r_2}\ \ldots\ n^{r_n}$$

where

$$r_1 + 2r_2 + \cdots + nr_n = n$$

we define the *cycle index* of $\alpha$ to be

$$CI_\alpha(x_1, x_2, \ldots, x_n) = x_1{}^{r_1}\ x_2{}^{r_2}\ \ldots\ x_n{}^{r_n}.$$

For example, in $S_{10}$, we have the following:

| $\alpha$ | cycle pattern | $CI_\alpha$ |
|---|---|---|
| $\epsilon$ | $1^{10}$ | $x_1{}^{10}$ |
| $(1)(23)(45)(6789)(10)$ | $1^2 2^2 4$ | $x_1{}^2 x_2{}^2 x_4$ |
| $(123)(456)(789\,10)$ | $3^2 4$ | $x_3{}^2 x_4$ |

We now define the *cycle index* of $G$ to be the polynomial

$$CI_G(x_1, x_2, \ldots, x_n) = \frac{1}{|G|} \sum_{\alpha \in G} CI_\alpha(x_1, \ldots, x_n)$$

$$= \frac{1}{|G|} \sum c(r_1, r_2, \ldots, r_n) x^{r_1} x^{r_2} \cdots x^{r_n}$$

where $c(r_1, \ldots, r_n)$ is the number of permutations in $G$ with cycle pattern $1^{r_1} 2^{r_2} \cdots n^{r_n}$, and the summation runs over all vectors $(r_1, r_2, \ldots, r_n)$ such that $r_1 + 2r_2 + \cdots + nr_n = n$.

**Example 3.12.1** Let $\alpha = (1\,2\,3\,4) \in S_4$ and $G = \langle \alpha \rangle$. Then $G = \{\epsilon, \alpha, \alpha^2, \alpha^3\}$ and we have

| element | cycle pattern | $CI_\alpha$ |
|---|---|---|
| $\epsilon$ | $1^4$ | $x_1{}^4$ |
| $\alpha$ | $4^1$ | $x_4{}^1$ |
| $\alpha^2$ | $2^2$ | $x_2{}^2$ |
| $\alpha^3$ | $4^1$ | $x_4{}^1.$ |

Hence,

$$CI_G = \frac{1}{4}(x_1{}^4 + x_2{}^2 + 2x_4).$$

**Example 3.12.2** Consider the cycle index for $D_4$ acting on the vertices of a square. As before, we have $D_4 = \{\epsilon, R, R^2, R^3, \rho, R\rho, R^2\rho, R^3\rho\}$, where $\rho$ is a rotation about a diagonal.

| $\alpha$ | cyclic pattern | $CI_\alpha$ |
|---|---|---|
| $\epsilon$ | $1^4$ | $x_1{}^4$ |
| $R, R^3$ | $4^1$ | $x_4{}^1$ |
| $R^2$ | $2^2$ | $x_2{}^2$ |
| $\rho$ | $1^2\,2$ | $x_1{}^2\,x_2$ |
| $R\rho$ | $2^2$ | $x_2{}^2$ |
| $R^2\rho$ | $1^2\,2$ | $x_1{}^2\,x_2$ |
| $R^3\rho$ | $2^2$ | $x_2{}^2$ |

Therefore,

$$CI_G = \frac{1}{8}(x_1{}^4 + 2x_1{}^2x_2 + 3x_2{}^2 + 2x_4).$$

It is important to notice that the cycle index of a group depends very much on the set as well as the group. See Exercise 2 in this section.

There are some helpful observations regarding the cycle index of a group.

(1) If $G \subseteq S_n$, then there is a term $\frac{1}{|G|}x_1{}^n$.

(2) The sum of the coefficients must add up to 1.

(3) For each term $x_1{}^{r_1} x_2{}^{r_2} \ldots x_n{}^{r_n}$, we must have

$$r_1 + 2r_2 + 3r_3 + \cdots + nr_n = n.$$

(4) If $\alpha$ is written as a product of disjoint cycles and $CI_\alpha = x_1{}^{r_1} \ldots x_n^{r_n}$, then

$$r_1 + r_2 + \cdots + r_n = \text{ number of cycles in } \alpha.$$

(5) $\alpha$ and $\alpha^{-1}$ have the same cycle pattern.

(6) It is possible to calculate the cycle index of a group without having the full multiplication table written out.

In the next result we will see that there is an important connection between the cycle index of a group of permutations and certain kinds of counting problems. This is a very special case of a more general theory developed by G. Pólya.

**Theorem 3.12.3 (Pólya's Theorem)** *Let $G$ be a subgroup of $S_X$ and $K$ be a set of $k$ colors. Then the number of $G$-inequivalent $K$-colorings of $X$ is*

$$CI_G(k, k, \ldots, k).$$

**Proof.** As noted earlier, let $C$ denote the set of all oriented colorings using some or all of the $k$ colors. From the preceding section, we see that we wish to calculate the number $N$ of orbits under the action of $\widehat{G}$. By the Cauchy-Frobenius Theorem, we have

$$N = \frac{1}{|\widehat{G}|} \sum_{\widehat{\alpha} \in \widehat{G}} |F(\widehat{\alpha})|.$$

By Lemma 3.11.3, $|\widehat{G}| = |G|$ so that the challenge is to calculate $|F(\widehat{\alpha})|$ for each $\alpha \in G$. Let the cycle decomposition of $\alpha$ be

$$\alpha = (x_1 x_2 \ldots x_\ell)(y_1 y_2 \ldots y_m) \ldots$$

and let $\gamma \in F(\widehat{\alpha})$. Then

$$\gamma(x_1) = \gamma(\alpha^{-1}(x_2)) = (\widehat{\alpha}(\gamma))(x_2)$$
$$= \gamma(x_2).$$

Similarly,

$$\gamma(x_2) = \gamma(x_3) = \cdots = \gamma(x_\ell)$$

and

$$\gamma(y_1) = \gamma(y_2) = \cdots = \gamma(y_m)$$

and so on. Thus we see that

$\gamma \in F(\widehat{\alpha}) \iff \gamma$ assigns the same color to all the elements within each cycle of $\alpha$.

Therefore, the number of elements in $F(\widehat{\alpha})$ is equal to the number of ways of assigning colors to the different cycles of $\alpha$. If

$$CI_\alpha(x_1, \ldots, x_n) = x_1^{r_1} \ldots x_n^{r_n}$$

then this will be

$$k^{r_1} \ldots k^{r_n} = CI_\alpha(k, k, \ldots, k).$$

Thus,

$$|F(\widehat{\alpha})| = CI_\alpha(k, k, \ldots, k)$$

so the number of inequivalent colorings is

$$N = \frac{1}{|\widehat{G}|} \sum_{\widehat{\alpha} \in \widehat{G}} |F(\widehat{\alpha})| = \frac{1}{|G|} \sum_{\alpha \in G} CI_\alpha(k, k, \ldots, k) = CI_G(k, k, \ldots, k). \quad \square$$

**Example 3.12.4** Let us consider the number of ways of coloring the faces of a tetrahedron. By an application of the Orbit/Stabilizer Theorem, we see that the group $G$ of rotations of a tetrahedron has 12 elements. The nonidentity elements can be described as follows:

Two rotations of order 3 about each vertex
One rotation of order 2 about bisectors of opposite edges

| $\alpha$ | cycle pattern | $CI_\alpha$ |
|---|---|---|
| $\epsilon$ | $1^4$ | $x_1^4$ |
| rotation about vertex | $1 \cdot 3$ | $x_1 x_3$ |
| rotation about bisector | $2^2$ | $x_2^2$ |

Therefore,

$$CI_G(x_1, x_2, x_3) = \frac{1}{12}(x_1^4 + 3x_2^2 + 8x_1 x_3).$$

Hence, we have

| number of colors available | number of colorings $CI_G(k, k, k)$ |
|:---:|:---:|
| 2 | 5 |
| 3 | 15 |
| 4 | 36 |

**Example 3.12.5** Consider the number of inequivalent ways of coloring the six faces of a cube if six colors are available. For the group $G$ of rotations of a cube we have

| $\alpha$ | number of such elements | $CI_\alpha$ |
|:---|:---:|:---:|
| $\epsilon$ | 1 | $x_1^6$ |
| rotation about diagonal through opposite vertices | 8 | $x_3^2$ |
| rotation about axes through centers of opposite faces by 90° or 270° | 6 | $x_1^2 x_4$ |
| rotation about axes through centers of opposite faces by 180° | 3 | $x_1^2 x_2^2$ |
| rotation about axes through midpoints of opposite sides | 6 | $x_2^3$ |

Therefore,

$$CI_G = \frac{1}{24}(x_1^6 + 6x_2^3 + 8x_3^2 + 6x_1^2 x_4 + 3x_1^2 x_2^2)$$

so that the number of inequivalent colorings is

$$CI_G(6, 6, 6, 6, 6, 6) = 2226.$$

Of course, most of the hard work is in finding the cycle index for $G$. Once we have that, it is just a matter of substituting in the number of available colors. So if the number of available colors was five or seven instead of six, then the answers would be given by

$$CI_G(5, 5, 5, 5, 5) = 800$$

$$CI_G(7, 7, 7, 7, 7) = 5390$$

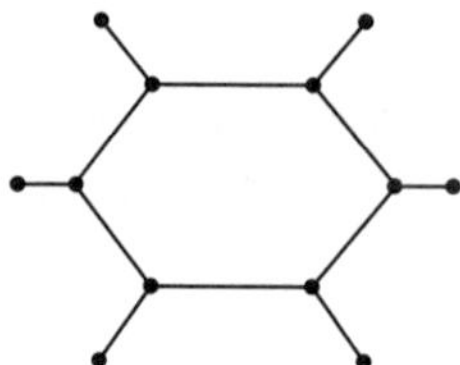

**Figure 3.17**

We conclude this section with an example that is a bit more complicated and that illustrates the application of both the Cauchy-Frobenius Theorem and the cycle index.

**Example 3.12.6** (1) How many essentially different ways are there to color the vertices in the graph in Figure 3.17 in such a way that six vertices are colored red and six vertices are colored white?

(2) How many essentially different ways are there to color the vertices of the graph in Figure 3.17 if two, three, or four colors are available?

The number of oriented colorings is $\binom{12}{6}$. Let $R$ denote the rotation through $60°$ in a clockwise direction, $\rho$ a rotation through $180°$ about a diagonal through opposite vertices, and $\sigma$ a rotation through $180°$ about a diagonal bisecting opposite sides. Now

$\epsilon$ has 12 cycles of length 1
$R, R^{-1}$ have 2 cycles of length 6
$R^2, R^4 = R^{-2}$, have 4 cycles of length 3
$R^3$ has 6 cycles of length 2
The 3 $\rho$'s have 4 cycles of length 2, 4 of length 1
The 3 $\sigma$'s have 6 cycles of length 2.

From the cycle pattern of each element, it is fairly straightforward to determine the size of the fixed set. For example, if $\gamma$ is a coloring that is fixed by $R$, then the elements in a given cycle must all have the same color. So it is really just a matter of deciding which cycle is red and there are two choices. Therefore, $|F(R)| = 2$. The trickiest case is for a $\rho$. Since within a cycle we must have the same color, the options are of three types

| | color 1 | color 2 |
|---|---|---|
| 1. | 3 cycles of length 2 | 1 cycle of length 2 |
| | | 4 cycles of length 1 |
| 2. | 2 cycles of length 2 | 2 cycles of length 2 |
| | 2 cycles of length 1 | 2 cycles of length 1 |
| 3. | Reverse of 1 | |

It suffices to count the number of ways of choosing which vertices to color red:

Option 1: There are $\binom{4}{3}$ ways of choosing 3 cycles of length 2.

Option 2: There are $\binom{4}{2} \times \binom{4}{2}$ ways of choosing 2 cycles of length 2 and 2 cycles of length 1.

Option 3: There are $\binom{4}{1}$ ways of choosing 1 cycle of length 2.

Hence,

$$|F(\rho)| = \binom{4}{3} + \binom{4}{2}^2 + \binom{4}{1}.$$

We can now tabulate the sizes of the fixed sets:

| $\alpha$ | $|F(\alpha)|$ | total |
|---|---|---|
| $\epsilon$ | $\binom{12}{6}$ | 924 |
| $R, R^{-1}$ | 2 | 4 |
| $R^2, R^4$ | $\binom{4}{2}$ | 12 |
| $R^3$ | $\binom{6}{3}$ | 20 |
| 3 $\rho$'s | $\binom{4}{1} + \binom{4}{3} + \binom{4}{2}^2$ | 132 |
| 3 $\sigma$'s | $\binom{6}{3}$ | 60 |

By the Cauchy-Frobenius Theorem, the number of orbits is

$$N = \frac{1}{12} \{924 + 4 + 12 + 20 + 132 + 60\}$$

$$= \frac{1}{12} \{1152\}$$

$$= 96.$$

Thus the answers to part (1) is 96.

We can also find the cycle index of $G$:

| $\alpha$ | cycle pattern | $CI_\alpha$ |
|---|---|---|
| $\epsilon$ | $1^{12}$ | $x_1^{12}$ |
| $R, R^{-1}$ | $6^2$ | $x_6^2$ |
| $R^2, R^4$ | $3^4$ | $x_3^4$ |
| $R^3$ | $2^6$ | $x_2^6$ |
| $3 \times \rho$ | $1^4 \cdot 2^4$ | $x_1^4 x_2^4$ |
| $3 \times \sigma$ | $2^6$ | $x_2^6$ |

Therefore,

$$CI_G = \frac{1}{12}\{x_1{}^{12} + 4x_2{}^6 + 2x_3{}^4 + 3x_1{}^4 x_2{}^4 + 2x_6{}^2\}.$$

Hence, the answers to part (2) are given in the following table:

| Number of colors available | Number of distinct colorings |
|---|---|
| 2 | 430 |
| 3 | 46,185 |
| 4 | 1,415,896 |

### Exercises 3.12

1. Let $G = \langle \alpha \rangle$ where $\alpha = (123)(45) \in S_5$. Find $CI_G$.

2. Let $G = \langle \alpha \rangle$ and $H = \langle \beta \rangle$ where

$$\alpha = (123) \in S_3 \quad \text{and} \quad \beta = (345)(678) \in S_8.$$

    (i)  Show that $G \cong H$.
    (ii)  Find $CI_G$ and $CI_H$.

3. Find the cycle index for each of the following groups:

    (i)  $S_2$.
    (ii)  $S_3$.
    (iii)  The group of rotations of a pentagon.
    (iv)  The group of rotations of a hexagon.

4. Find the cycle index for the groups of rotations of the following solids, acting on the vertices:

    (i)  A rectangular box with three unequal dimensions.
    (ii)  A rectangular box with two dimensions that are equal and distinct from the third.
    (iii)  Repeat (i) and (ii) with the groups acting on the faces.
    (iv)  Repeat (i) and (ii) with the groups acting on the edges.

5. Find the cycle index for the group of rotation of the tetrahedron, acting on

    (i)  the vertices
    (ii)  the faces
    (iii)  the edges.

6. Find the cycle index for $A_4$.

7. Find the cycle index for the Cayley representations of the following groups:

   (i) $(\mathbb{Z}_4, +)$.
   (ii) $(\mathbb{Z}_5, +)$.
   (iii) $(\mathbb{Z}_6, +)$.
   (iv) $S_3$.

8. Find the number of inequivalent colorings of the vertices of a pentagon if three colors are available.

9. If four colors are available, find the number of inequivalent colorings of

   (i) the vertices of a tetrahedron
   (ii) the edges of a tetrahedron.

10. Show that the group of rotations of a tetrahedron is $A_4$.

11. Show that the group of rotations of a cube is $S_4$. (Hint: Find some feature that a cube has four of.)

12. Show that the group of rotations of an octahedron is $S_4$. (Hint: Fit a tetrahedron inside a cube.)

# 4

## Groups: Homomorphisms and Subgroups

In this chapter we will probe a little deeper into certain groups that are closely related in some special way to a given group. In one direction, this leads to the concept of a homomorphism that enables us to consider groups that are related but not quite as strongly related as isomorphic groups. In another direction, we will begin the study of subgroups of a group. We will describe completely all finite abelian groups. We will see the wonderful theorems discovered by Sylow that provide us with a wealth of information regarding the presence of subgroups of prime power order. Combined with the direct product of groups, this enables us to describe completely all the groups of certain orders. We will conclude the chapter by considering two important classes of groups: solvable and nilpotent groups.

### 4.1 Homomorphisms

We have used the concept of an isomorphism to help us recognize groups that are "essentially" the same. Now we consider a slightly weaker concept that is important in the study of the relationships between groups that might not be isomorphic.

Let $G$ and $H$ be groups and $\varphi : G \to H$ be a mapping. Then, recalling the definition of a homomorphism from Section 3.2, $\varphi$ is a *homomorphism* if

$$\varphi(ab) = \varphi(a)\varphi(b) \quad \text{for all} \quad a, b \in G.$$

A surjective homomorphism is an *epimorphism.*
An injective homomorphism is a *monomorphism.*

An isomorphism is just a bijective homomorphism. We will occasionally use the notation $G \cong H$ to indicate that the groups $G$ and $H$ are isomorphic. If there exists a surjective homomorphism $\varphi : G \to H$, then $H$ is called a *homomorphic image* of G. Examples of homomorhisms and homorphic images abound.

**Example 4.1.1** Let $n \in N$ and define $\varphi : \mathbb{Z} \to \mathbb{Z}_n$ by

$$\varphi(a) = r_a$$

where $a = qn + r_a,\ 0 \le r_a < n$. Then $\varphi$ is an epimorphism.

**Example 4.1.2** Let $A$ be an $m \times n$ matrix of real numbers. Define a mapping $\varphi : \mathbb{R}^m \to \mathbb{R}^n$ by

$$\varphi(v) = vA.$$

Then $\varphi$ is a linear transformation. In particular, $\varphi$ is a homomorphism of $(\mathbb{R}^m, +)$ to $(\mathbb{R}^n, +)$.

**Example 4.1.3** Let $G$ denote the group of all invertible $n \times n$ matrices with real entries. Define

$$\varphi : G \to \mathbb{R}^*$$

by

$$\varphi(A) = \det A.$$

Then $\varphi$ is a homomorphism.

**Example 4.1.4** Let $G = D_4$, the group of rotations of a square. Every element $\alpha \in D_4$ determines a permutation of the set of diagonals $D = \{d_1, d_2\}$. Call this permutation $\alpha_d$. Then

$$\varphi : D_4 \to S_D$$

defined by

$$\varphi(\alpha) = \alpha_d \qquad (\alpha \in D_4)$$

is a homomorphism.

Homomorphisms are often used to relate *abstract* groups to *concrete* groups, where *concrete* might mean a group of matrices or permutations.

A homomorphism $\varphi : G \to GL_n(F)$ is a *representation by matrices.*
A homomorphism $\varphi : G \to S_X$ is a *representation by permutations.*

**Lemma 4.1.5** *Let G and H be groups and $\varphi : G \to H$ be a homomorphism.*
   (i)  $\varphi(e_G) = e_H.$
   (ii)  *For all $g \in G$, $\varphi(g^{-1}) = \varphi(g)^{-1}.$*

**Proof.** Exercise.  $\square$

Let $G$ and $H$ be groups, $h \in H$ and $A \subseteq H$, and $\varphi : G \to H$ be an epimorphism. Then we write

$$\varphi^{-1}(h) = \{g \in G \,|\, \varphi(g) = h\}$$

$$\varphi^{-1}(A) = \{g \in G \,|\, \varphi(g) \in A\}.$$

**Lemma 4.1.6** *Let G and H be groups, and $\varphi : G \to H$ be a homomorphism.*
   (i)  *For any subgroup A of G, $\varphi(A)$ is a subgroup of H.*
   (ii)  *For any subgroup B of H,*

$$\varphi^{-1}(B) = \{g \in G \,|\, \varphi(g) \in B\}$$

   *is a subgroup of G.*

**Proof.** (i) Since $e_H = \varphi(e_G) \in \varphi(A)$, $\varphi(A) \neq \emptyset$. We have

$$x, y \in \varphi(A) \implies \exists\, a, b \in A \quad \text{with} \quad x = \varphi(a),\ y = \varphi(b)$$
$$\implies xy^{-1} = \varphi(a)\varphi(b)^{-1} = \varphi(ab^{-1}) \quad \text{where}\ \ ab^{-1} \in A$$
$$\implies xy^{-1} \in \varphi(A).$$

Therefore, $\varphi(A)$ is a subgroup of $H$.

   (ii) Since $e_G \in \varphi^{-1}(e_H) \subseteq \varphi^{-1}(B)$, $\varphi^{-1}(B) \neq \emptyset$. Also we have

$$x, y \in \varphi^{-1}(B) \implies \varphi(x), \varphi(y) \in B$$
$$\implies \varphi(xy^{-1}) = \varphi(x)\varphi(y)^{-1} \in B$$
$$\implies xy^{-1} \in \varphi^{-1}(B).$$

Therefore, $\varphi^{-1}(B)$ is a subgroup of $G$.  $\square$

Let $\varphi : G \to H$ be a homomorphism between groups. Then $\{e_H\}$ is a subgroup of $H$ so that $\varphi^{-1}(\{e_H\})$ is a subgroup of $G$. We define the *kernel* of $\varphi$ to be

$$\ker(\varphi) = \varphi^{-1}(\{e_H\}).$$

This subgroup determines many of the properties of $\varphi$ and is a special kind of subgroup of $G$, in the sense that not every subgroup of $G$ can be the kernel of a homomorphism.

Let $\varphi : X \to Y$ be a mapping. Then $\varphi$ determines an equivalence relation $\sim$ on $X$ by

$$a \sim b \iff \varphi(a) = \varphi(b).$$

We denote this equivalence relation by

$$\overline{\varphi} \quad \text{or} \quad \varphi^{-1} \circ \varphi.$$

If $\varphi : G \to H$ is a homomorphism between groups, then $\overline{\varphi}$ is completely determined by $\ker(\varphi)$.

**Lemma 4.1.7** *Let $\varphi : G \to H$ be a homomorphism between groups with kernel $N$ and let $a, b \in G$. Then $\varphi(a) = \varphi(b) \iff Na = Nb$.*

**Proof.** We have

$$\varphi(a) = \varphi(b) \iff \varphi(a)\varphi(b)^{-1} = e_H$$
$$\iff \varphi(ab^{-1}) = e_H$$
$$\iff ab^{-1} \in N$$
$$\iff Na = Nb. \quad \square$$

**Corollary 4.1.8** *Let $\varphi : G \to H$ be a homomorphism between groups. Then $\varphi$ is a monomorphism if and only if $\ker(\varphi) = \{e_G\}$.*

**Proof.** Exercise. $\quad \square$

Recall that, for subsets $A, B \subseteq G$ and $g \in G$,

$$AB = \{ab \mid a \in A,\ b \in B\}$$
$$Ag = \{ag \mid a \in A\}$$
$$gAg^{-1} = \{gag^{-1} \mid a \in A\}.$$

Note that $Ag = A\{g\}$. In addition,

$$A(BC) = \{a(bc) \mid a \in A,\ b \in B,\ c \in C\}$$

$$(AB)C = \{(ab)c \mid a \in A,\ b \in B,\ c \in C\}.$$

Hence, since the multiplication in $G$ itself is associative,

$$A(BC) = (AB)C.$$

A subgroup $N$ of a group $G$ is *normal* if $gNg^{-1} = N$ for all $g \in G$, or, equivalently, if $gN = Ng$ for all $g \in G$. Note that

$$gNg^{-1} = N \ \ \forall g \in G \Longleftrightarrow g^{-1}Ng = N \ \ \forall g \in G$$

so that the positioning of the inverse on the left versus the right side of $N$ is not important.

Note also that *every subgroup of an abelian group is normal.* There are also nonabelian groups in which every subgroup is normal (see Exercise 8), but generally speaking, a nonabelian group will have subgroups that are not normal (see, for example, Exercise 5).

For $N$, a (normal) subgroup of $G$, let

$$G/N = \text{set of right cosets of } N \text{ in } G$$

$$= \{Ng \mid g \in G\}.$$

The next result provides some basic observations concerning normal subgroups.

**Theorem 4.1.9** *Let $N$ be a subgroup of a group $G$. Then the following conditions are equivalent:*

   (i)  *$N$ is normal.*
   (ii)  *$gNg^{-1} \subseteq N$,*       *for all $g \in G$.*
   (iii)  *$gN = Ng$,*         *for all $g \in G$.*
   (iv)  *$(Ng)(Nh) = Ngh$*   *for all $g, h \in G$.*
   (v)  *The rule*

$$(Ng) * (Nh) = Ngh$$

     *defines a binary operation on the set of right cosets $G/N$ of $N$.*
   (vi)  *There exists a homomorphism $\varphi : G \to H$ of $G$ onto a group $H$ such that $N = \ker(\varphi)$.*

**Proof.** (i) *implies* (ii). This follows immediately from the definition of normality.

(ii) *implies* (iii). Let $a \in N$. Then,

$$ga = (gag^{-1})g \in Ng \qquad \text{by (ii)}$$

so that $gN \subseteq Ng$. Similarly,

$$ag = g(g^{-1}ag) = g(g^{-1}a(g^{-1})^{-1}) \in gN \qquad \text{by (ii)}$$

so that $Ng \subseteq gN$ and equality follows.

(iii) *implies* (iv). We have

$$\begin{aligned}
(Ng)(Nh) &= N(gN)h \\
&= N(Ng)h \qquad \text{by (iii)} \\
&= Ngh.
\end{aligned}$$

(iv) *implies* (v). Let $Ng = Ng'$, $Nh = Nh'$. Then there exist $a, b \in N$ with $g' = ag$, $h' = bh$. Hence,

$$\begin{aligned}
Ng'h' &= Ng'\,Nh' \qquad \text{by (iv)} \\
&= (Nag)(Nbh) \\
&= Ng\,Nh \\
&= Ngh \qquad \text{by (iv).}
\end{aligned}$$

Therefore, $*$ is well-defined.

(v) *implies* (vi). By (v), $*$ is a well-defined binary operation on $G/N$.
*Associativity.* We have

$$\begin{aligned}
(Ng * Nh) * Nk &= (Ngh) * Nk = N(gh)k \\
Ng * (Nh * Nk) &= Ng * (Nhk) = Ng(hk) = N(gh)k
\end{aligned}$$

and the associativity of $*$ follows.

*Identity.* We have

$$Ng * Ne = Nge = Ng = Neg = Ne * Ng$$

so that $N = Ne$ is the identity with respect to the $*$ operation.

*Inverses.* For any $Ng \in G/N$,

$$(Ng) * (Ng^{-1}) = Ngg^{-1} = Ne = N = Ne = Ng^{-1}g = Ng^{-1} * Ng$$

so that

$$(Ng)^{-1} = Ng^{-1}.$$

Thus, $(G/N, *)$ is a group.

Now let $H = G/N$ and define $\varphi : G \to H$ by

$$\varphi(g) = Ng.$$

Then, it is clear that $\varphi$ is surjective and, furthermore, that

$$\begin{aligned}
\varphi(g) * \varphi(h) &= (Ng) * (Nh) \\
&= Ngh \\
&= \varphi(gh),
\end{aligned}$$

so that $\varphi$ is a homomorphism. Moreover,

$$\begin{aligned}
\ker(\varphi) &= \{g \in G \mid \varphi(g) = e_{G/N} = N\} \\
&= \{g \in G \mid Ng = N\} \\
&= \{g \in G \mid g \in N\} \\
&= N.
\end{aligned}$$

(vi) *implies* (i). By Lemma 4.1.6, $N = \varphi^{-1}(\{e_H\})$ is a subgroup of $G$. Now let $a \in N, \ g \in G$. Then,

$$\begin{aligned}
\varphi(gag^{-1}) = \varphi(g)\varphi(a)\varphi(g)^{-1} &= \varphi(g) \, e_H \, \varphi(g)^{-1} \\
&= e_H.
\end{aligned}$$

Thus, $gag^{-1} \in N$ so that $gNg^{-1} \subseteq N$ for all $g \in N$. Replacing $g$ with $g^{-1}$, we obtain $g^{-1}Ng \subseteq N$ and, multiplying on the left by $g$ and on the right by $g^{-1}$, we get $N \subseteq gNg^{-1}$. Therefore, $gNg^{-1} = N$ and $N$ is normal. $\quad\square$

The next result is contained in the proof of Theorem 4.1.9, but we want to make it explicit.

**Corollary 4.1.10** *Let $N$ be a normal subgroup of a group $G$.*

(i) *$(G/N, *)$ is a group and (if $G$ is finite) $G/N$ has order $|G| \, / \, |N|$.*
(ii) *The mapping $\pi : G \to G/N$ defined by*

$$\pi(g) = Ng \qquad (g \in G)$$

*is an epimorphism with $\ker(\pi) = N$.*

The homomorphism $\pi$ in Corollary 4.1.10 (ii) is called the *natural* homomorphism of $G$ onto $G/N$. Henceforth, we write simply $G/N$, for $(G/N, *)$, use juxtaposition for the operation in $G/N$, and refer to $G/N$ as "$G$ *mod(ulo)* $N$." Any group of the form $G/N$ is called a *quotient* group of $G$.

**Lemma 4.1.11** *Let $\varphi : G \to H$ be an epimorphism with $N = \ker(\varphi)$.*

(i) $A \to \varphi(A)$ *is a bijection of*

$$\{A \mid N \subseteq A \subseteq G, \ A \ \text{a subgroup of} \ G\}$$

*onto the set of subgroups of $H$.*

(ii) *For subgroups $A, B$ of $G$ with $N \subseteq A, B$,*

$$A \subseteq B \iff \varphi(A) \subseteq \varphi(B)$$

(iii) *For any subgroup $A$ of $G$ with $N \subseteq A$, $A$ is normal in $G$ if and only if $\varphi(A)$ is normal in $H$.*

**Proof.** Exercise. $\square$

We conclude this section with a very useful result concerning the number of elements in a product of subgroups.

**Proposition 4.1.12** *Let $A$ and $B$ be finite subgroups of a group $G$. Then*

$$|AB| = \frac{|A| \cdot |B|}{|A \cap B|}.$$

(*Note: There is no requirement here for the product $AB$ of the subgroups $A$ and $B$ to be itself a subgroup.*)

**Proof.** We will show that $\frac{|AB|}{|B|} = \frac{|A|}{|A \cap B|}$. We know that $A \cap B$ is a subgroup of $A$. So let

$$k = \frac{|A|}{|A \cap B|}$$

and $x_1, \ldots, x_k \in A$ be such that

$$x_1(A \cap B), \ldots, x_k(A \cap B)$$

are all the distinct left cosets of $A \cap B$ in $A$.

Now let $a \in A$, $b \in B$ so that $ab$ is an arbitrary element of $AB$. There must exist an integer $i$, $1 \le i \le k$, with $a \in x_i(A \cap B)$, so that $a = x_i g$ for some $g \in A \cap B$. Hence,

$$ab = x_i g b \in x_i B.$$

Therefore,

$$AB \subseteq \bigcup x_i B.$$

Clearly, $\bigcup x_i B \subseteq AB$ so that $AB = \bigcup x_i B$ and $AB$ is a union of left cosets of $B$. Furthermore,

$$
\begin{aligned}
x_i B = x_j B &\Longrightarrow x_i^{-1} x_j \in B \\
&\Longrightarrow x_i^{-1} x_j \in A \cap B \quad \text{since } x_i, x_j \in A \\
&\Longrightarrow x_i(A \cap B) = x_j(A \cap B) \\
&\Longrightarrow x_i = x_j.
\end{aligned}
$$

Therefore, $x_1 B, \ldots, x_k B$ are the distinct left cosets of $B$ contained in $AB$. Consequently,

$$k = \frac{|AB|}{B}$$

so that

$$\frac{|A|}{|A \cap B|} = k = \frac{|AB|}{|B|}. \qquad \square$$

**Exercises 4.1**

1. Let $H$ be a subgroup of the group $G$ and let $g \in G$.

   (i) Show that $gHg^{-1}$ is a subgroup of $G$.

   (ii) Show that the mapping $\varphi : H \to gHg^{-1}$ defined by

   $$\varphi(h) = ghg^{-1} \qquad (h \in H)$$

   is an isomorphism of $H$ to $gHg^{-1}$.

2. Let $(G, \cdot)$ be an abelian group. Show that the mapping $\varphi : G \to G$ defined by

   $$\varphi(g) = g^{-1} \qquad (g \in G)$$

   is an automorphism of $G$.

3. Let $(G, +)$ be an abelian group and $m \in \mathbb{Z}$.

   (i) Show that the mapping $\varphi : G \to G$ defined by

   $$\varphi(g) = mg \qquad (g \in G)$$

   is a homomorphism of $G$ to itself.
   (ii) Give an example where $\varphi$ is injective, but not surjective.
   (iii) Give an example where $G$ is finite and $\varphi$ is an automorphism.
   (iv) Give an example where $G$ is infinite and $\varphi$ is an automorphism.

4. Let $G, H$, and $K$ be groups and let $\theta : G \to H$, $\varphi : H \to K$ be homomorphisms. Establish the following:

   (i) $\varphi\theta$ is a homomorphism.
   (ii) $\varphi\theta$ a monomorphism $\Rightarrow$ $\theta$ a monomorphism.
   (iii) $\varphi\theta$ an epimorphism $\Rightarrow$ $\varphi$ an epimorphism.

5. Find all normal subgroups of

   (i) $S_3$
   (ii) $D_4$.

6. Show that $A_n$ is normal in $S_n$.

7. Let $H$ be a subgroup of the group $G$ with $[G : H] = 2$. Show that $H$ is a normal subgroup of $G$.

8. Show that every subgroup of the quaternion group is normal.

9. Find an example of a group $G$ with subgroups $A$ and $B$ such that $AB$ is not a subgroup.

10. Show that if $N_1, \ldots, N_k$ are normal subgroups of $G$, then $N_1 N_2 \cdots N_k$ is also a normal subgroup of $G$.

11. Let $X$ be a subset of a group $G$ and let $N$ denote the intersection of all the normal subgroups of $G$ containing $X$. Show that $N$ is a normal subgroup of $G$ and that $N$ is the smallest normal subgroup of $G$ containing $X$.

12. Let $A = \{\epsilon, (1\,2)(3\,4)\}$ and $V = \{\epsilon, (1\,2)(3\,4),\ (1\,3)(2\,4),\ (1\,4)(2\,3)\}$. Show that $A$ is a normal subgroup of $V$ and that $V$ is a normal subgroup of $S_4$ (and therefore $A_4$), but that $A$ is not a normal subgroup of $A_4$. Thus, a normal subgroup of a normal subgroup need not be normal.

13. Let $A, B$, and $C$ be normal subgroups of a group $G$ such that

    $$A \cap C = B \cap C, \quad AC = BC, \quad \text{and} \quad A \subseteq B.$$

    Show that $A = B$. (This is known as the *modular law*.)

14. Let $\alpha = (1\,2\,3)(4\,5)$, $\beta = (1\,7\,9\,2\,6\,8\,3\,5\,4) \in S_9$, $A = \langle\alpha\rangle$, and $B = \langle\beta\rangle$. Use Proposition 4.1.12 to calculate the number of elements in $AB$.

15. Show that $SL_n(F)$ is a normal subgroup of $GL_n(F)$.

16. Let $R$ denote the rotation of a regular $n$-gon about an axis perpendicular to the plane of the $n$-gon and through an angle of $\frac{360}{n}$ in a clockwise direction. Show that $\langle R\rangle$ is a normal subgroup of $D_n$.

17. Let $G_1 = Q_8$ (see Exercises 8 and 9 in section 3.2), the quaternion group, $G_2 = D_4$, $N_1 = \langle a^2\rangle$, and $N_2 = \langle R^2\rangle$. Establish the following:

  (i)  $N_1$ and $N_2$ are isomorphic.
  (ii)  $N_1$ and $N_2$ are normal subgroups of $G_1$ and $G_2$, respectively.
  (iii)  $|G_1/N_1| = |G_2/N_2| = 4$.
  (iv)  Show that $G_1/N_1$ and $G_2/N_2$ are isomorphic.

18.  (i)  Show that every element of $\mathbb{Q}/\mathbb{Z}$ has finite order.
  (ii)  Which elements of $\mathbb{R}/\mathbb{Z}$ have finite order and which have infinite order?

*19. Let $u = (2, 1)$, $v = (1, 2) \in \mathbb{R}^2$, $H = \langle u, v\rangle$ in $(\mathbb{R}^2, +)$, and

$$\mathrm{FR} = \{\alpha u + \beta v \mid 0 \le \alpha, \beta < 1\}.$$

(FR stands for *fundamental region*.)

  (i)  Plot the elements of $H$ in $\mathbb{R}^2$ and identify the set FR.
  (ii)  Show that each coset of $H$ contains a unique element of FR.

## 4.2 The Isomorphism Theorems

Homomorphisms have introduced us to new constructions and objects. These, in turn, lead to new relationships that we explore here.

**Theorem 4.2.1 (First Isomorphism Theorem)** *Let $\varphi : G \to H$ be an epimorphism with $\ker(\varphi) = N$. Let $\pi : G \to G/N$ be the natural homomorphism. Then there exists an isomorphism $\theta : G/N \to H$ such that $\theta \circ \pi = \varphi$.*

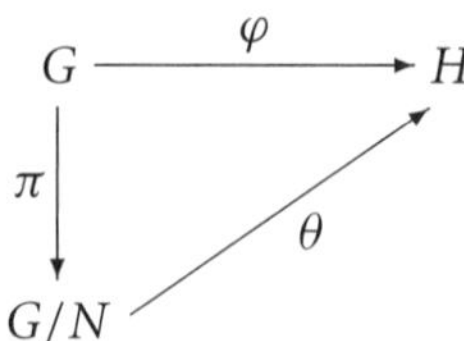

**Proof.** By Lemma 4.1.7, we know that

$$\varphi(a) = \varphi(b) \iff Na = Nb \qquad (a, b \in G). \qquad (4.1)$$

Therefore, we can define a mapping $\theta : G/N \to H$ by

$$\theta(Na) = \varphi(a).$$

Since $\varphi$ is surjective, so also is $\theta$. Moreover,

$$\theta(Na) = \theta(Nb) \implies \varphi(a) = \varphi(b)$$

$$\implies Na = Nb \qquad \text{by (4.1)}.$$

Thus, $\theta$ is injective and therefore $\theta$ is a bijection. Finally,

$$\begin{aligned}
\theta(Na)\,\theta(Nb) &= \varphi(a)\varphi(b) \\
&= \varphi(ab) && \text{since } \varphi \text{ is a homomorphism} \\
&= \theta(Nab) \\
&= \theta((Na)(Nb)) && \text{since } N \text{ is a normal subgroup.}
\end{aligned}$$

Thus, $\theta$ is a homomorphism and therefore an isomorphism. $\qquad \square$

This result can be viewed either way: (i) Given $N$, we can understand $G/N$ better if we can find a good model $H$; or (ii) given $\varphi$, $G/N$ gives us a description of $H$.

In the next result we collect several basic facts concerning combinations of subgroups and normal subgroups.

**Lemma 4.2.2** *Let $A, B$ be subgroups of $G$, and $N$ be a normal subgroup of $G$.*

(i) *$A \cap B$ is a subgroup of $G$.*
(ii) *$A \cap N$ is a normal subgroup of $A$.*
(iii) *$AN = NA$ is a subgroup of $G$.*
(iv) *$N$ is a normal subgroup of $AN$.*

**Proof.** (i), (ii) Exercises.

(iii) Let $a, b \in A$, $x, y \in N$. Then,

$$ax = (axa^{-1})a \in Na$$

so that $AN \subseteq NA$. Also,

$$xa = a(a^{-1}xa) \in aN.$$

Consequently, $NA \subseteq AN$, from which it follows that $AN = NA$. Furthermore,

$$(ax)(by) = ab(b^{-1}xb)y \in AN$$
$$(ax)^{-1} = x^{-1}a^{-1} = a^{-1}ax^{-1}a^{-1} \in AN$$

which implies that $AN$ is a subgroup of $G$.

    (iv) Exercise. $\quad\square$

It turns out that the quotient groups that we obtain from combinations of subgroups and normal subgroups are not always distinct.

**Theorem 4.2.3 (Second Isomorphism Theorem)** *Let A be a subgroup and N be a normal subgroup of a group G. Then,*

$$AN/N \cong A/(A \cap N).$$

**Proof.** Note that, by Lemma 4.2.2 (iii), $AN = NA$. Define $\varphi : A \to AN/N$ by

$$\varphi(a) = Na.$$

Since $AN = NA$, we have

$$AN/N = \{Na \mid a \in A\}.$$

Hence, $\varphi$ is surjective. For $a, b \in A$,

$$\varphi(a)\varphi(b) = Na \cdot Nb = Nab$$
$$= \varphi(ab).$$

Thus, $\varphi$ is a homomorphism. Furthermore,

$$a \in \ker(\varphi) \iff Na = e_{AN/N} = N$$
$$\iff a \in N.$$

Therefore, $\ker(\varphi) = A \cap N$. By Theorem 4.2.1,

$$AN/N \cong A/\ker(\varphi) = A/(A \cap N). \quad\square$$

In the Third Isomorphism Theorem we obtain a rather pleasing result, as it is reminiscent of the familiar cancellation that takes place in fractions of integers.

**Theorem 4.2.4 (Third Isomorphism Theorem)** *Let $M$ and $N$ be normal subgroups of a group $G$ with $M \subseteq N$. Define a mapping $\varphi : G/M \to G/N$ by*

$$\varphi : Ma \to Na \qquad (Ma \in G/M).$$

*Then $\varphi$ is an epimorphism of $G/M$ onto $G/N$ with kernel $N/M$. Consequently,*

$$(G/M)/(N/M) \cong G/N.$$

**Proof.** First let $\pi_M \colon G \to G/M$ be the natural homomorphism. Then, $\pi_M(N) = N/M$ and, since $N$ is normal in $G$, we see from Lemma 4.1.11 (iii) that $N/M$ is a normal subgroup of $G/M$.

Since $M \subseteq N$, it is easily verified that $\varphi$ is well defined and is surjective. Also,

$$\varphi((Ma)(Mb)) = \varphi(Mab) = Nab$$

$$= NaNb = \varphi(Ma)\varphi(Mb)$$

so that $\varphi$ is a homomorphism. Now,

$$Ma \in \ker(\varphi) \iff \varphi(Ma) = e_{G/N} = N$$

$$\iff Na = N$$

$$\iff a \in N.$$

Therefore,

$$\ker(\varphi) = \{Ma \mid a \in N\} = N/M.$$

The final claim now follows from the First Isomorphism Theorem. $\square$

**Exercises 4.2**

1. Use the First Isomorphism Theorem to identify the following quotient groups:

   (i)  $S_n/A_n$.
   (ii) $D_n/\langle R \rangle$.

2. By means of the First Isomorphism Theorem, or otherwise, identify the following quotient groups:

   (i)  $A_4/V$, where $V$ is defined as in Exercise 12, section 4.1.
   (ii) $GL_n(F)/SL_n(F)$.

*3. By means of the First Isomorphism Theorem, or otherwise, identify the group $S_4/V$.

4. Let $G$ denote the set of points on the unit circle in $\mathbb{R}^2$. Each point in $G$ can be represented in the form $(\cos\theta, \sin\theta)$. Define multiplication in $G$ by

$$(\cos\theta, \sin\theta) \cdot (\cos\varphi, \sin\varphi) = (\cos(\theta + \varphi), \sin(\theta + \varphi)).$$

(i) Show that $(G, \cdot)$ is a group.
(ii) Show that $G \cong (\mathbb{R}, +)/(\mathbb{Z}, +)$.

## 4.3 Direct Products

In this section we consider more fully a very simple way of combining two or more groups to form a new group that was introduced in chapter 3. This is not only important from the point of view of constructing new examples, it is also an extremely important tool for studying the structure of groups and for classifying groups.

**Theorem 4.3.1** *Let $A$ and $B$ be groups, and $G = A \times B$. Define a binary operation $*$ on $G$ by*

$$(a, b) * (a', b') = (aa', bb').$$

(i) *$(G, *)$ is a group with identity $(e_A, e_B)$ and with inverses given by $(a, b)^{-1} = (a^{-1}, b^{-1})$.*
(ii) *$A^* = A \times \{e_B\}$ is a normal subgroup of $G$ and the mapping $i_A : a \to (a, e_B)$ is an isomorphism of $A$ to $A^*$.*
(iii) *$B^* = \{e_A\} \times B$ is a normal subgroup of $G$ and the mapping $i_B : b \to (e_A, b)$ is an isomorphism of $B$ to $B^*$.*
(iv) *The mapping $\pi_A : (a, b) \to a$ of $G$ to $A$ is an epimorphism with kernel equal to $B^*$.*
(v) *The mapping $\pi_B : (a, b) \to b$ of $G$ to $B$ is an epimorphism with kernel equal to $A^*$.*
(vi) *$A^* \cap B^* = \{e_G\}$, $xy = yx$ for all $x \in A^*$, $y \in B^*$, and $A^* B^* = G$.*

**Proof.** There are many things to prove, but they are all quite straightforward and so we leave the verifications as exercises for you. $\quad\square$

Recall that the group $(G, *) = (A \times B, *)$ constructed in Theorem 4.3.1 is the (*external*) *direct product* of $A$ and $B$. Normally, we will just write $A \times B$ for the direct product of $A$ and $B$. Since the underlying set in the direct product is just the cartesian product, we can calculate the order of $A \times B$ from

$$|A \times B| = |A| \cdot |B|.$$

It is also easy to see that $A \times B \cong B \times A$.

Even in nonabelian groups there may be, as the next lemma suggests, many elements that commute with each other.

**Lemma 4.3.2** *Let* $M, N$ *be normal subgroups of a group* $G$ *such that* $M \cap N = \{e\}$. *Then* $mn = nm$ *for all* $m \in M$, $n \in N$.

**Proof.** We have

$$m^{-1}n^{-1}mn = (m^{-1}n^{-1}m)n \in NN = N$$

$$= m^{-1}(n^{-1}mn) \in MM = M$$

$$\in M \cap N = \{e\}.$$

Therefore,

$$m^{-1}n^{-1}mn = e$$

and

$$mn = nm. \quad \square$$

We are now in a position to obtain an "internal" characterization of a group that is isomorphic to a direct product.

**Theorem 4.3.3** *Let* $M, N$ *be normal subgroups of a group* $G$ *such that*

  (i) $M \cap N = \{e\}$
  (ii) $MN = G$.

*Then,* $G \cong M \times N$.

**Proof.** By (ii), every element $g \in G$ can be written as

$$g = mn \quad \text{where } m \in M, n \in N.$$

Also,

$$g = mn = m'n' \quad \text{where } m, m' \in M, n, n' \in N$$

$$\Rightarrow m^{-1}m' = n(n')^{-1} \in M \cap N = \{e\}$$

$$\Rightarrow m^{-1}m' = e = n(n')^{-1}$$
$$\Rightarrow m = m', \ n = n'.$$

Therefore, we can define a mapping

$$\varphi : G \to M \times N$$

as follows: For $g = mn \in G$,

$$\varphi(g) = (m, n).$$

For any $(m, n) \in M \times N$, we have $\varphi(mn) = (m, n)$ so that $\varphi$ is surjective. Also,

$$\varphi(mn) = \varphi(m'n') \implies (m, n) = (m', n')$$
$$\implies m = m', \ n = n'$$
$$\implies mn = m'n'.$$

Thus, $\varphi$ is injective and so $\varphi$ is bijective. Last, for $g = mn, \ h = m'n' \in G$,

$$gh = mn \, m'n' = (mm')(nn') \qquad \text{by Lemma 4.3.2}$$

from which it follows that

$$\varphi(gh) = \varphi((mm')(nn'))$$
$$= (mm', nn') = (m, n)(m', n')$$
$$= \varphi(g)\varphi(h).$$

Thus, $\varphi$ is an isomorphism. $\qquad\square$

When the conditions (i) and (ii) of Theorem 4.3.3 hold, $G$ is called the *internal direct product* of $M$ and $N$. When the context is clear, we will sometimes simply describe $G$ as the *direct product* of $M$ and $N$ without emphasizing the *internal* nature and will even write $G = M \times N$. As one important illustration of how Theorem 4.3.3 can be applied, we consider $\mathbb{Z}_n$, where $n$ is composite.

**Proposition 4.3.4** *Let $n, s, t \in \mathbb{N}$, $n = st$, and $(s, t) = 1$. Then*

$$\mathbb{Z}_n \cong \mathbb{Z}_t \times \mathbb{Z}_s.$$

**Proof.** Consider $s$ and $t$ as elements of $\mathbb{Z}_n$ and let

$$M = \langle s \rangle, \quad N = \langle t \rangle.$$

Since $\mathbb{Z}_n$ is abelian, $M$ and $N$ are both normal subgroups of $\mathbb{Z}_n$. Also, it is easily checked that

$$|M| = t, \quad |N| = s$$

so that $(|M|, |N|) = 1$. Hence, $M \cap N = \{0\}$ and, from Proposition 4.1.12,

$$|M + N| = |M| \cdot |N| = ts = n = |\mathbb{Z}_n|.$$

Therefore, $M + N = \mathbb{Z}_n$ and, by Theorem 4.3.3, $\mathbb{Z}_n \cong M \times N$. However, $M$ and $N$ are cyclic groups of orders $t$ and $s$, respectively, so that $M \cong \mathbb{Z}_t$, $N \cong \mathbb{Z}_s$. Consequently, $\mathbb{Z}_n \cong \mathbb{Z}_t \times \mathbb{Z}_s$. $\quad\square$

As we saw in chapter 3, the direct product construction can be generalized to an arbitrary number of factors. This has properties similar to those described earlier for the direct product of two groups, which will be explored in the exercises.

A group $G$ is said to be *decomposable* if there exist nontrivial groups $A$ and $B$ with $G \cong A \times B$. Equivalently, $G$ is decomposable if there exist proper nontrivial normal subgroups $M$ and $N$ with $M \cap N = \{e\}$ and $G = MN$. We say that a group is *indecomposable* if it is not decomposable. For example, since $\mathbb{Z}_6 \cong \mathbb{Z}_2 \times \mathbb{Z}_3$, $\mathbb{Z}_6$ is decomposable. On the other hand, it is not difficult to see that $S_3$ is indecomposable.

### Exercises 4.3

1. Let $A_1$ be a subgroup of $A$ and $B_1$ be a subgroup of $B$. Show that

    (i)  $A_1 \times B_1$ is a subgroup of $A \times B$
    (ii) if $A_1$ and $B_1$ are normal subgroups, then $A_1 \times B_1$ is a normal subgroup of $A \times B$
    (iii) $(A \times B)/(A_1 \times B_1)$ is isomorphic to $(A/A_1) \times (B/B_1)$.

2. Let $A_1, A_2, \ldots, A_n$ be groups.

    (i)  Show that $P = A_1 \times A_2 \times \cdots \times A_n$ is a group with respect to the operation

$$(a_1, \ldots, a_n)(b_1, \ldots, b_n) = (a_1 b_1, \ldots, a_n b_n).$$

(ii)  Show that, for $1 \leq i \leq n$, $A_i^* = \{(e_1, \ldots, a_i, \ldots, e_n) \mid a_i \in A_i\}$ is a normal subgroup of $P$.

(iii)  Show that

$$A_i^* \cap A_1^* \cdots A_{i-1}^* A_{i+1}^* \cdots A_n^* = \{e_P\}.$$

(iv)  Show that $xy = yx$ for all $x \in A_i^*$, $y \in A_j^*$ where $i \neq j$.

$P$ is called the *(external) direct product* of the groups $A_1, \ldots, A_n$.

*3.  Let $G$ be a group with normal subgroups $N_1, \ldots, N_n$ such that

(i)  $N_i \cap N_1 \cdots N_{i-1} N_{i+1} \cdots N_n = \{e_G\}$, for all $i$ with $1 \leq i \leq n$
(ii)  $G = N_1 N_2 \cdots N_n$.

$G$ is then called the *internal direct product* of $N_1, \ldots, N_n$. Show that $G \cong N_1 \times N_2 \times \cdots \times N_n$.

4.  Let $G$ be the internal direct product of the subgroups $A_1$ and $B$, and let $B$ be the internal direct product of the subgroups $A_2, \ldots, A_m$. Show that $G$ is the internal direct product of $A_1, \ldots, A_m$.

5.  Let $G$ be a group and $P = G \times \cdots \times G$, the direct product of $n$ copies of $G$. Show that

$$D = \{(g, g, \ldots, g) \mid g \in G\}$$

is a subgroup of $P$.

6.  Let $G, H_i$ $(1 \leq i \leq n)$ be groups and for each $i$, let $\varphi_i : G \to H_i$ be a homomorphism. Define $\varphi : G \to H_1 \times H_2 \times \cdots \times H_n$ by

$$\varphi(g) = (\varphi_1(g), \varphi_2(g), \ldots, \varphi_n(g)).$$

Establish the following:

(i)  $\varphi$ is a homomorphism.
(ii)  $\ker(\varphi) = \bigcap_{1 \leq i \leq n} \ker(\varphi_i)$.
(iii)  $\varphi$ is a monomorphism if and only if $\bigcap_{1 \leq i \leq n} \ker(\varphi_i) = \{e\}$.
(iv)  The assumption that each $\varphi_i$ is surjective is not sufficient to conclude that $\varphi$ is surjective.

7.  Let $B = \langle (1\,2) \rangle$ in $S_3$. Establish the following:

(i)  $A_3 \cap B = \{\varepsilon\}$.
(ii)  $S_3 = A_3 B$.
(iii)  $S_3$ is not the internal direct product of $A_3$ and $B$.
(iv)  $S_3$ is not isomorphic to $A_3 \times B$.

8. Show that the following groups are indecomposable:

    (i) $\mathbb{Z}_{p^n}$ ($p$ a prime, $n \in \mathbb{N}$).
    (ii) $Q_8$.

9. Show that for any group $G$, the set of automorphisms of $G$, $\mathrm{Aut}(G)$, is a subgroup of $S_G$.

10. Let $N$ and $G$ be groups and $\varphi : G \to \mathrm{Aut}(N)$ be a homomorphism. Define a binary operation on $N \times G$ by

$$(n, g)(n', g') = (n\varphi(g)(n'), gg').$$

    (i) Show that $N \times G$ is a group with respect to this operation. It is called a *semidirect product* of $N$ and $G$ and is denoted by $N \rtimes_{\varphi} G$.
    (ii) Show that $N^* = N \times \{e_G\}$ is a normal subgroup of $N \rtimes_{\varphi} G$.
    (iii) Show that $G^* = \{e_N\} \times G$ is a subgroup of $N \rtimes_{\varphi} G$.
    (iv) Show that $N \rtimes_{\varphi} G = N^* G^*$.

## 4.4 Finite Abelian Groups

There are many proofs of the Fundamental Theorem for Finite Abelian Groups. The one that we follow here is based on the nice proof by Gabrial Navarro in [Nav]. For any finite abelian group $G$, and any prime $p$, let

$$G_p = \left\{ g \in G \mid g^{p^k} = e, \text{ for some } k \in \mathbb{N} \right\}$$

$$G'_p = \{ g \in G \mid (|g|, p) = 1 \}.$$

**Lemma 4.4.1** *Let $G$ be a finite abelian group. Then $G_p$ and $G'_p$ are both subgroups of $G$, and $G$ is the direct product of $G_p$ and $G'_p$.*

**Proof.** The verification of the facts that $G_p$ and $G'_p$ are subgroups of $G$ is left to you as an exercise. To see that $G$ is the direct product of $G_p$ and $G'_p$, first note that $G_p \cap G'_p = \{e\}$ since elements in $G_p$ and $G'_p$ have relatively prime orders. So, it remains to show that $G = G_p G'_p$. Let $g \in G$ and let $|g| = p^k m$, where $(p, m) = 1$. Then $\langle g \rangle$ is a cyclic group of order $p^k m$. By Proposition 4.3.4, $\langle g \rangle$ is the direct product of a cyclic group $A$ of order $p^k$ and a cyclic group $B$ of order $m$. Hence, there exist elements $a \in A$, $b \in B$ such that $g = ab$. But, clearly, $a \in G_p$ and $b \in G'_p$. Therefore, $g \in G_p G'_p$, $G \subseteq G_p G'_p$ so that $G = G_p G'_p$. We have now shown that $G$ is the direct product of $G_p$ and $G'_p$. $\square$

A more general version (known as Cauchy's Theorem) of the next lemma will appear later.

**Lemma 4.4.2** *Let G be a finite abelian group, p be a prime number, and p divide $|G|$. Then G contains an element of order p.*

**Proof.** We argue by induction on the order of $G$. Let $H$ be a maximal proper subgroup of $G$. Then there is no subgroup $K$ of $G$ such that $H \subset K \subset G$. Hence, by Lemma 4.1.11(i), there is no proper nontrivial subgroup of $G/H$, which means that any nonidentity element of $G/H$ is a generator. In other words, $G/H$ must be a cyclic group of prime order. If $p$ divides $|H|$, then we are done by induction. Otherwise, $p$ must divide $G/H$ and, since $G/H$ is cyclic of prime order, we must have $|G/H| = p$. Let $y \in G \backslash H$. Then $Hy \neq H$, so that we must have $G/H = \langle Hy \rangle$ and $|\langle Hy \rangle| = p$. Let $\pi : G \longrightarrow G/H$ be the natural homomorphism. Then $\pi(y) = Hy$ so that $\pi(\langle y \rangle) = \langle Hy \rangle$. By the First Isomorphism Theorem, we then have that $\langle Hy \rangle \cong \langle y \rangle / N$ for some (normal) subgroup $N$ of $\langle y \rangle$. Hence,

$$p = |\langle Hy \rangle| = |\langle y \rangle| / |N|.$$

Therefore, $p$ must divide $|\langle y \rangle| = |y|$. Hence, some suitable power of $y$ is of order $p$. $\square$

Let $|G| = p^{\alpha}$ where $p$ is a prime number and $\alpha \in \mathbb{N}$. Then $G$ is a called a *p-group*. Since the order of any element in a finite group divides the order of the group, it is clear that the order of any element in a $p$-group must be a power of $p$. From Lemma 4.4.2, we know, conversely, that if every element of a finite abelian group $G$ has order equal to a power of some fixed prime $p$, then $G$ is a $p$-group. In Corollary 4.6.4, we will see that this is also true in arbitrary finite groups. When infinite groups are included in the discussion, it is customary to define a $p$-group to be a group in which every element has order equal to a power of $p$.

**Lemma 4.4.3** *Let G be a cyclic group of order $p^n$, for some prime p. Then G has a unique subgroup K of order p. Moreover, K is contained in every nontrivial subgroup of G. Conversely, if G is a finite abelian p-group that is not cyclic, then G contains two distinct subgroups of order p.*

**Proof.** Let $G = \langle a \rangle$. Since $(a^{p^{n-1}})^p = a^{p^n} = e$, it follows that $K = \langle a^{p^{n-1}} \rangle$ is a subgroup of order $p$. It will suffice to show that $K$ is contained in every other nontrivial subgroup of $G$. Let $H$ be a nontrivial subgroup. Then $H$ contains a nontrivial element $b$, say. Let $b = a^m$ where $m = sp^k, (s, p) = 1, k < n$. Then there exists $t \in \mathbb{N}$ such that $st \equiv 1 \bmod p^n$ so that

$$b^t = a^{mt} = a^{(st)p^k} = (a^{st})^{p^k} = a^{p^k}.$$

Therefore,

$$a^{p^{n-1}} = (a^{p^k})^{p^{n-1-k}} = (b^t)^{p^{n-1-k}} \in \langle b \rangle$$

and

$$K = \langle a^{p^{n-1}} \rangle \subseteq \langle b \rangle \subseteq H.$$

Consequently, $K$ is contained in every nontrivial subgroup and, in particular, in every subgroup of order $p$. But $K$ itself has order $p$ and so it must be the only subgroup of order $p$.

Conversely, suppose that $G$ is a finite abelian $p$-group that is not cyclic. Let $A = \langle a \rangle$ be a cyclic subgroup of $G$ with the maximum order among all cyclic subgroups of $G$. Let $|A| = p^n$. Since $G$ is not cyclic, $G \neq A$ and there must exist an element $b \in G \backslash A$. Let $b$ be one such element of the smallest possible order. Then we must have $b^p \in A$ (since the order of $b^p$ is $|b|/p$). Let $b^p = a^k$, for some integer $k$. First suppose that $p$ does not divide $k$. Then, since $p$ is prime, $(p, k) = 1$ and there exist integers $x, y$ with $xp + yk = 1$ so that

$$a = a^{xp+yk} = a^{xp}(a^k)^y = a^{xp}b^{py} = (a^x b^y)^p.$$

Hence,

$$p^n = |a| = |(a^x b^y)^p| = |a^x b^y|/p.$$

But then $|a^x b^y| = p^{n+1} > |a|$, contradicting the choice of a. Hence $p$ must divide $k$. Let $k = pm$. Now we have

$$(b(a^m)^{-1})^p = b^p(a^{pm})^{-1} = b^p(a^k)^{-1} = 1$$

which implies that $b(a^m)^{-1}$ has order dividing $p$. But $b \notin A$, which implies that $b(a^m)^{-1} \notin A$ and, in particular, $b(a^m)^{-1} \neq 1$. Consequently, we must have $|b(a^m)^{-1}| = p$. Thus $\langle b(a^m)^{-1} \rangle$ is a cyclic group of order $p$ so that it is generated by any non-identity member. Since $b(a^m)^{-1} \notin A$, it follows that the only element in common between $A$ and $\langle b(a^m)^{-1} \rangle$ is 1. Therefore, $\langle a^{p^{n-1}} \rangle$ and $\langle b(a^m)^{-1} \rangle$ are two distinct subgroups of $G$ of order $p$. $\quad\square$

**Lemma 4.4.4** *Let $G$ be a finite abelian p-group. Let $A$ be a cyclic subgroup of $G$ of maximal order for a cyclic subgroup. Then there is a subgroup $B$ of $G$ such that $G$ the direct product of $A$ and $B$.*

**Proof.** We proceed by induction. If $G$ is cyclic, then $G = A$ and we are done. So we assume that $G$ is not cyclic. By Lemma 4.4.3, $G$ must have at least two subgroups of order $p$. However, since $A$ is cyclic, it also follows from

Lemma 4.4.3 that $A$ has only one subgroup of order $p$. Let $P$ denote a subgroup of $G$ of order $p$ that is not contained in $A$. Then $A \cap P = \{e\}$. Now $P \subseteq AP$, so that we can form the quotient group $AP/P$. By the Second Isomorphism Theorem, Theorem 4.2.3, we have

$$AP/P \cong A/(A \cap P) = A/\{e\} \cong A.$$

Thus, $AP/P$ is cyclic and the order of $AP/P$ ($\cong A$) must be a maximum among the orders of cyclic subgroups of $G/P$. By the induction hypothesis,

$$G/P = (AP/P) \times C$$

for some subgroup $C$ of $G/P$. By Lemma 4.1.11, there exists a subgroup $B$ of $G$ with $P \subseteq B$ and $C = B/P$. Then,

$$G/P = (AP)/P \times B/P \implies G/P = ((AP)/P)(B/P) = (APB)/P$$
$$\implies G = APB = AB$$

where the final assertion follows from Lemma 4.1.11. However,

$$A \cap B = A \cap AP \cap B \subseteq AP \cap B$$

where

$$x \in AP \cap B \implies Px \in ((AP)/P) \cap (B/P) = \{P\}$$
$$\implies x \in P.$$

Therefore, $A \cap B \subseteq P$ and

$$A \cap B = A \cap (A \cap B) \subseteq A \cap P = \{e\}$$

so that $G$ is the direct product of $A$ and $B$. $\quad\square$

**Theorem 4.4.5 (Fundamental Theorem for Finite Abelian Groups)** *Let $G$ be a finite abelian group. Then $G$ is isomorphic to a direct product of cyclic groups.*

**Proof.** We again proceed by induction on the order of $G$. The claim is clearly true if the order of $G$ is a prime, since then $G$ is cyclic. Let $p$ be any prime dividing the order of $G$. With the notation introduced at the beginning of the section we have, by Lemma 4.4.1, that $G$ is the direct product of $G_p$ and $G_p'$. By Lemma 4.4.2, $G_p$ is nontrivial. If $G_p \neq G$, then $G_p'$ must be nontrivial and a proper subgroup of $G$. By the induction hypothesis applied to $G_p$ and $G_p'$, we see that $G_p$ and $G_p'$ are both direct products of cyclic groups. Hence, $G$ must also be a direct product of cyclic groups. Therefore, it only remains to consider the case $G = G_p$—that is, when $G$ is a $p$-group. If $G$ is cyclic, then we are finished. So suppose that $G$ is not cyclic. Let $A$ be a cyclic subgroup of $G$ of maximum possible order. By Lemma 4.4.4, $G$ is the direct product of $A$ and another proper subgroup $B$. By the induction hypothesis, $B$ is also a direct product of cyclic groups and so the proof is complete. $\square$

In the next two results, we just give the fundamental theorem a more exact formulation, taking advantage of the fact that we know that every finite cyclic group is isomorphic to one of the groups $(\mathbb{Z}_n, +)$. As a bonus, we obtain some uniqueness in the representation.

**Proposition 4.4.6** *Let $p$ be a prime number, $n \in \mathbb{N}$, and $G$ be an abelian group of order $p^n$. Then there exist $a_1, \ldots, a_m \in G$ of orders $p^{\alpha_1}, \ldots, p^{\alpha_m}$ $(\alpha_i \in \mathbb{N})$ such that $\alpha_1 \geq \alpha_2 \geq \cdots \geq \alpha_m$, $n = \alpha_1 + \alpha_2 + \cdots + \alpha_m$, and $G$ is the internal direct product of the subgroups $\langle a_i \rangle$, $1 \leq i \leq m$. The sequence of integers $\alpha_1, \alpha_2, \ldots, \alpha_m$ is unique.*

**Proof.** By the Fundamental Theorem of Finite Abelian Groups, $G$ is a direct product of cyclic groups (which can be assumed to be subgroups) each of which must have an order dividing the order of $G$ (that is, $p^n$) and therefore of the form $p^\alpha$ for some integer $\alpha$ with $1 < \alpha \leq n$. The first part of Proposition 4.4.6 then just consists of listing the exponents $\alpha$ in descending order.

Concerning the uniqueness of the integers $\alpha_1, \alpha_2, \ldots, \alpha_m$, we will not present a complete proof, but we will indicate how the number of elements of each order determines the values of $\alpha_1, \alpha_2, \ldots, \alpha_m$. First note that we have shown that

$$G \cong \langle a_1 \rangle \times \langle a_2 \rangle \times \cdots \times \langle a_m \rangle$$
$$\cong \mathbb{Z}_{p^{\alpha_1}} \times \mathbb{Z}_{p^{\alpha_2}} \times \cdots \times \mathbb{Z}_{p^{\alpha_m}}. \tag{4.2}$$

So it suffices to show how the number of elements of different orders determines the number of factors of each order in the product in (4.2).

First consider $\mathbb{Z}_{p^\alpha}$. Then

$$a \in \mathbb{Z}_{p^\alpha}, \ a \neq 0, \ |a| = p \Longleftrightarrow pa = 0$$
$$\Longleftrightarrow p^\alpha \mid pa$$
$$\Longleftrightarrow p^{\alpha-1} \mid a$$
$$\Longleftrightarrow a = kp^{\alpha-1}, \ k = 0, 1, \ldots, p - 1.$$

Thus, the number of elements of order $p$ is $p - 1$ and the number of elements of order $\leq p$ is $p$. Now consider (4.2). There,

$$|(a_1, \ldots, a_m)| \leq p \iff |a_i| \leq p, \qquad 1 \leq i \leq m.$$

Therefore, the number of elements of order $\leq p$ is $p^m$. Thus, the number of elements of order, at most, $p$ in $G$, which is quite independent of any way that we might like to represent $G$, completely determines the number of factors in any representation of the form (4.2). Similarly, the number of elements of order $\leq 2$ determines the number of factors with $\alpha_i \geq p^2$, the number of elements of order $\leq p^3$ determines the number of factors with $\alpha_i \geq 3$, and so forth. The details will be given in the exercises. Thus, the representation in (4.2) is unique. $\quad\square$

This leads to the following common formulation of the Fundamental Theorem of Finite Abelian Groups.

**Theorem 4.4.7** *Let $G$ be a finite abelian group.*

  (i) *$G$ is isomorphic to a direct product of the form*

$$\mathbb{Z}_{p_1^{\alpha_{11}}} \times \mathbb{Z}_{p_1^{\alpha_{12}}} \times \cdots \times \mathbb{Z}_{p_2^{\alpha_{21}}} \times \cdots \times \mathbb{Z}_{p_m^{\alpha_{m1}}}$$

  *where the $p_i$ are prime numbers. The factors are unique to within the order in which they are written.*

  (ii) *$G$ is isomorphic to a direct product of the form*

$$\mathbb{Z}_{n_1} \times \mathbb{Z}_{n_2} \times \cdots \times \mathbb{Z}_{n_r}$$

  *where $n_{i+1}$ divides $n_i$. The integers $n_1, n_2, \ldots, n_r$ are unique.*

**Proof.** By the fundamental theorem, we know that $G$ is a direct product of finite cyclic groups. By virtue of the fact that if $(m, n) = 1$ then $\mathbb{Z}_{mn} \cong \mathbb{Z}_m \times \mathbb{Z}_n$, we can write any finite cyclic group as a direct product of cyclic groups of prime power order—that is, in the form given in (i)—and therefore we can also write $G$ in this form. Uniqueness follows from the uniqueness in Proposition 4.4.6.

To obtain the description in (ii), we take advantage of the fact that if $(m, n) = 1$, then $\mathbb{Z}_m \times \mathbb{Z}_n \cong \mathbb{Z}_{mn}$. We then combine the factor in (i) involving the largest power of $p_1$ with the factor involving the largest power of $p_2$, and so on. Then we combine all the factors with the second largest powers of each prime, and so on. This clearly leads to a description of $G$ of the form given in (ii). Uniqueness follows from the uniqueness in (i). $\quad\square$

**Example 4.4.8** The claim in Theorem 4.4.7 (ii) is best understood by considering an example. Let

$$G = \mathbb{Z}_{2^4} \times \mathbb{Z}_{2^2} \times \mathbb{Z}_{3^3} \times \mathbb{Z}_3 \times \mathbb{Z}_3 \times \mathbb{Z}_5.$$

Then $G$ is written in the form of Theorem 4.4.7 (i). Since $2^4, 3^3$, and $5$ are relatively prime, we have

$$\mathbb{Z}_{2^4} \times \mathbb{Z}_{3^3} \times \mathbb{Z}_5 \cong \mathbb{Z}_{2^4 \cdot 3^3} \times \mathbb{Z}_5$$
$$\cong \mathbb{Z}_{2^4 \cdot 3^3 \cdot 5}.$$

Similarly,

$$\mathbb{Z}_{2^2} \times \mathbb{Z}_3 \cong \mathbb{Z}_{2^2 \cdot 3}.$$

Hence

$$G \cong \mathbb{Z}_{2^4 \cdot 3^3 \cdot 5} \times \mathbb{Z}_{2^2 \cdot 3} \times \mathbb{Z}_3$$

where $3$ divides $2^2 \cdot 3$ and $2^2 \cdot 3$ divides $2^4 \cdot 3^3 \cdot 5$. Thus we now have $G$ written in the form of Theorem 4.4.7 (ii).

The description of finite abelian groups in the Fundamental Theorem is satisfyingly complete and it would be unrealistic to expect to find similarly detailed and complete descriptions of larger classes of groups. However, we do find parallels of certain aspects of the theory of finite abelian groups in other classes. One feature that is fruitful in the study of nonabelian groups is the following corollary, which we will refer to in later sections.

**Corollary 4.4.9** *Let $G$ be a finite abelian group. Then there exists a sequence of subgroups*

$$\{e\} = G_{r+1} \subseteq G_r \subseteq G_{r-1} \subseteq \cdots \subseteq G_0 = G$$

*such that $G_i \,/\, G_{i+1}$ is cyclic for all $i$, $0 \leq i \leq r$.*

**Proof.** By the fundamental theorem, we may write $G$ as

$$G = \mathbb{Z}_{n_1} \times \mathbb{Z}_{n_2} \times \cdots \times \mathbb{Z}_{n_r}.$$

So let

$$
\begin{aligned}
G_r &= \{e\} \\
G_{r-1} &= \mathbb{Z}_{n_1} \times \{0\} \times \cdots \times \{0\} \\
G_{r-2} &= \mathbb{Z}_{n_1} \times \mathbb{Z}_{n_2} \times \{0\} \times \cdots \times \{0\} \\
&\ \ \vdots \\
G_1 &= \mathbb{Z}_{n_1} \times \cdots \times \mathbb{Z}_{n_{r-1}} \times \{0\} \\
G_0 &= G.
\end{aligned}
$$

Then

$$
\{e\} = G_r \subseteq G_{r-1} \subseteq \cdots \subseteq G_1 \subseteq G_0 = G
$$

and $G_i \,/\, G_{i+1} \cong \mathbb{Z}_{n_{r-i}}$ is cyclic for all $i$ with $0 \le i \le r-1$.  $\square$

## Exercises 4.4

1. List the elements of different orders in $\mathbb{Z}_2 \times \mathbb{Z}_2 \times \mathbb{Z}_4$.

2. List the elements of different orders in $\mathbb{Z}_2 \times \mathbb{Z}_2 \times \mathbb{Z}_3$.

3. List the elements of different orders in $\mathbb{Z}_3 \times \mathbb{Z}_4$.

4. List the elements of different orders in $\mathbb{Z}_2 \times \mathbb{Z}_3 \times \mathbb{Z}_4$.

5. How many elements are there of the different possible orders in $G = \mathbb{Z}_2 \times \mathbb{Z}_2 \times \mathbb{Z}_4$?

6. How many elements are there of the different possible orders in $G = \mathbb{Z}_4 \times \mathbb{Z}_4$?

7. How many elements are there of the different possible orders in $G = \mathbb{Z}_2 \times \mathbb{Z}_2 \times \mathbb{Z}_4$?

8. Let $\alpha, n \in \mathbb{N}$, $\alpha \le n$, and $p$ be a prime. Show that the number of elements of order $p^\alpha$ in $\mathbb{Z}_{p^n}$ is $p^\alpha - p^{\alpha-1}$.

9. Let $p$ be a prime and

$$
G = \mathbb{Z}_{p^{\alpha_1}} \times \mathbb{Z}_{p^{\alpha_2}} \times \cdots \times \mathbb{Z}_{p^{\alpha_n}}.
$$

   Show that the number of elements in $G$ of order $p$ is $p^n - 1$.

10. Let $p$ be a prime and

$$
G = \mathbb{Z}_{p^{\alpha_1}} \times \mathbb{Z}_{p^{\alpha_2}} \times \cdots \times \mathbb{Z}_{p^{\alpha_n}}.
$$

Assume that $\alpha_i \geq \alpha$ for $1 \leq i \leq m$ and that $\alpha_i < \alpha$ for $m+1 \leq i \leq n$. Show that the number $0_\alpha$ of elements in $G$ of order $p^\alpha$ is

$$0_\alpha = p^{m(\alpha-1)+\alpha_{m+1}+\cdots+\alpha_n}(p^m - 1).$$

11. In Exercise 10, assume that $n$ is fixed. Now show that the value of $m$ in Exercise 10 is uniquely determined by the value of $0_\alpha$.

12. Let $p$ be a prime number. How many elements are there of the different possible orders in $G = \mathbb{Z}_p \times \mathbb{Z}_{p^2} \times \mathbb{Z}_{p^3}$?

13. Find all the subgroups of $\mathbb{Z}_2 \times \mathbb{Z}_2 \times \mathbb{Z}_3$.

14. Find all the subgroups of $\mathbb{Z}_2 \times \mathbb{Z}_3 \times \mathbb{Z}_4$.

15. Write the group $G = \mathbb{Z}_2 \times \mathbb{Z}_2 \times \mathbb{Z}_3 \times \mathbb{Z}_3 \times \mathbb{Z}_4$ in the form $\mathbb{Z}_{m_1} \times \mathbb{Z}_{m_2} \times \cdots \times \mathbb{Z}_{m_k}$, where $m_{i+1}$ divides $m_i$.

16. Write the group $G = \mathbb{Z}_2 \times \mathbb{Z}_2 \times \mathbb{Z}_3 \times \mathbb{Z}_3 \times \mathbb{Z}_4 \times \mathbb{Z}_5 \times \mathbb{Z}_6$ in the form $\mathbb{Z}_{m_1} \times \mathbb{Z}_{m_2} \times \cdots \times \mathbb{Z}_{m_k}$, where $m_{i+1}$ divides $m_i$.

17. How many nonisomorphic abelian groups are there of order 24?

18. How many nonisomorphic abelian groups are there of order 30?

19. How many nonisomorphic abelian groups are there of order 72?

20. Let $G$ be an abelian group of order $2^\alpha$ with a unique element of order 2. Describe $G$ as a product of cyclic groups.

21. Let $G$ be an abelian group of order $p^\alpha$, where $p$ is a prime, with $p-1$ elements of order $p$. Describe $G$ as a product of cyclic groups.

22. Let $G = \mathbb{Z}_4 \times \mathbb{Z}_8$, $a = (1, 1)$, and $A = \langle a \rangle$. Describe $G/A$ as a direct product of cyclic groups.

23. Let $G = \mathbb{Z}_2 \times \mathbb{Z}_4 \times \mathbb{Z}_8$, $a = (1, 1, 1)$, and $A = \langle a \rangle$. Describe $G/A$ as a direct product of cyclic groups.

24. Let $G$ denote the group (with respect to matrix multiplication) of all invertible $m \times m$ diagonal matrices over $GF(p^n)$. Describe $G$ as an internal and external direct product of cyclic groups.

25. Let $G$ denote the group (with respect to matrix multiplication) of all $m \times m$ diagonal matrices over $GF(p^n)$ with determinant equal to one. Describe $G$ as an internal and external direct product of cyclic groups.

26. Describe $(GF(p^n), +)$ as a direct product of cyclic groups.

27. Let $G$ be a finite abelian $p$-group. Show that every nonidentity element in $G/pG$ has order $p$.

28. Let $G = \mathbb{Z}_2 \times \mathbb{Z}_2 \times \mathbb{Z}_3$. Find $2G$ and $G_2$.

29. Let $G = \mathbb{Z}_2 \times \mathbb{Z}_3 \times \mathbb{Z}_4$. Find $2G$ and $G_2$.

30. Let $G$ be a finite abelian group and $m$ be a positive integer that divides $|G|$. Prove that $G$ has a subgroup of order $m$.

31. Let $G$ be a finite abelian group of odd order. Show that the product of all the elements is equal to the identity.

32. Let $G$ be an abelian group and $T$ denote the set of elements in $G$ of finite order.

   (i) Show that $T$ is a subgroup of $G$.
   (ii) Show that the identity is the only element in $G/T$ of finite order.

   $T$ is called the *torsion subgroup* of $G$ and is denoted by Tor $(G)$.

33. Let $p$ be a prime, $m \in \mathbb{N}$, $(p, m) = 1$, $n = pm$, and

$$G(p, m) = \{[mx]_n \mid x \in \mathbb{Z}, \ (x, p) = 1\}.$$

   (i) What is $|G(p, m)|$?
   (ii) Show that $(G(p, m), \cdot)$ is an abelian group.
   (iii) Find $G(5, 8)$, its identity, and determine the inverse of each element.

34. Let $p, q, r$ be distinct primes and $n = pqr$. Let

$$H(p, q, r) = \{[a]_n \mid p \ \text{divides} \ a, \ (a, q) = 1, \ (a, r) = 1\}.$$

   (i) Show that $(H(p, q, r), \cdot)$ is an abelian group.
   (ii) Determine the group $H(2; 3, 5)$.

## 4.5 Conjugacy and the Class Equation

Elements $a, b$ of a group $G$ are said to be *conjugate elements* if there exists an element $g \in G$ with $b = gag^{-1}$. Likewise, subgroups $A$ and $B$ of a group $G$ are *conjugate subgroups* if there exists an element $g \in G$ with $B = gAg^{-1}$.

**Lemma 4.5.1** *Let $G$ be a group. The relation $\sim$ defined on $G$ by*

$$a \sim b \quad \text{if} \quad a \text{ and } b \text{ are conjugate}$$

*is an equivalence relation on $G$.*

**Proof.** Exercise. $\square$

   Let $C[a]$ denote the class of $a$ with respect to $\sim$. Then $C[a]$ is the *conjugacy class* of $a$ and the elements of $C[a]$ are the *conjugates* of $a$. Similar terms and notation are applied to subgroups. In particular, for any subgroup $A$, we denote by $C[A]$ the conjugacy class of $A$, and the elements of $C[A]$ are the conjugates of $A$.

**Theorem 4.5.2** *Let $\alpha, \beta \in S_n$. Then $\alpha$ and $\beta$ are conjugate if and only if they have the same cycle pattern.*

**Proof.** First suppose that there exists $\sigma \in S_n$ with $\beta = \sigma\alpha\sigma^{-1}$. Consider any cycle

$$\gamma = (a_1 a_2 \cdots a_m)$$

and its conjugate $\sigma\gamma\sigma^{-1}$. For $1 \leq i < m$,

$$(\sigma\gamma\sigma^{-1})\,\sigma(a_i) = \sigma\gamma(a_i) = \sigma(a_{i+1})$$

while

$$(\sigma\gamma\sigma^{-1})\,\sigma(a_m) = \sigma\gamma(a_m) = \sigma(a_1).$$

For $1 \leq a \leq n$, and $a \neq a_i$,

$$(\sigma\gamma\sigma^{-1})\,\sigma(a) = \sigma\gamma(a) = \sigma(a).$$

Since $\sigma$ is a permutation, for all $b$ with $1 \leq b \leq n$, there exists an element $a$, $1 \leq a \leq n$, with $\sigma(a) = b$. Hence, we have described the action of $\sigma\gamma\sigma^{-1}$ completely:

$$\sigma\gamma\sigma^{-1} = (\sigma(a_1)\,\sigma(a_2)\cdots\sigma(a_m))$$

so that $\sigma\gamma\sigma^{-1}$ is another cycle of length $m$. Now let

$$\alpha = \alpha_1\alpha_2\cdots\alpha_k$$

as a product of disjoint cycles. Then

$$\sigma\alpha\sigma^{-1} = (\sigma\alpha_1\sigma^{-1})(\sigma\alpha_2\sigma^{-1})\cdots(\sigma\alpha_k\sigma^{-1})$$

where, by the preceding argument, $\sigma\alpha_i\sigma^{-1}$ is a cycle of length equal to the length of $\alpha_i$. Also, for $i \neq j, \sigma\alpha_i\sigma^{-1}$ and $\sigma\alpha_j\sigma^{-1}$ are disjoint permutations. Hence, $\alpha$ and $\sigma\alpha\sigma^{-1}$ have the same cycle pattern.

For the converse, now suppose that $\alpha$ and $\beta$ have the same cycle pattern. Then we can write $\alpha$ and $\beta$ as products of disjoint cycles

$$\alpha = \alpha_1\cdots\alpha_t$$
$$\beta = \beta_1\cdots\beta_t$$

where $\alpha_i$ and $\beta_i$ are cycles of the same length for each $i$. Let

$$\alpha = (a_{11}a_{12}\cdots a_{1m_1})(a_{21}a_{22}\cdots a_{2m_2})\cdots(a_{t1}a_{t2}\cdots a_{tm_t})$$

$$\beta = (b_{11}b_{12}\cdots b_{1m_1})(b_{21}b_{22}\cdots b_{2m_2})\cdots(b_{t1}b_{t2}\cdots b_{tm_t}).$$

Then, $a_{11}a_{12}\cdots a_{1m_1}a_{21}\cdots a_{2m_2}\cdots a_{t1}\cdots a_{tm_t}$ is just a rearrangement of $1, 2, 3, \ldots, n$, as is $b_{11}\cdots b_{tm_t}$. So we can define a permutation $\sigma \in S_n$ by

$$\begin{pmatrix} a_{11}a_{12}\cdots a_{1m_1}\,a_{21}\cdots a_{2m_2}\cdots a_{t1}\cdots a_{tm_t} \\ b_{11}b_{12}\cdots b_{1m_1}\,b_{21}\cdots b_{2m_2}\cdots b_{t1}\cdots b_{tm_t} \end{pmatrix}.$$

Then,

$$\sigma\alpha\sigma^{-1} = (\sigma\alpha_1\sigma^{-1})(\sigma\alpha_2\sigma^{-1})\cdots(\sigma\alpha_t\sigma^{-1})$$

$$= \beta_1\beta_2\cdots\beta_t$$

$$= \beta.$$

Thus, $\alpha$ and $\beta$ are conjugate in $S_n$. $\square$

C. Deavours, in his review of [Rej] accompanying the article, describes Theorem 4.5.2 as "the theorem that won World War II" because of its importance in the cryptoanalytic work of the Polish mathematician M. Rejewski (see [Rej]) in breaking the Enigma code being used in the years prior to the outbreak of World War II. We will return to this topic in Section 4.10.

Recall that

$$Z(G) = \{a \in G \mid ax = xa, \ \forall x \in G\}$$

is called the *center* of $G$. Note that

$$|C[a]| = 1 \Longleftrightarrow C[a] = \{a\}$$

$$\Longleftrightarrow gag^{-1} = a, \quad \text{for all } g \in G$$

$$\Longleftrightarrow ga = ag, \qquad \text{for all } g \in G$$

$$\Longleftrightarrow a \in Z(G).$$

Thus, we may also write

$$Z(G) = \left\{a \in G \,\middle|\, |C[a]| = 1\right\}.$$

For $a \in G$,

$$Z_G(a) = \{x \in G \mid ax = xa\}$$

is the *centralizer* of $a$ in $G$. Note that

$$Z_G(a) = G \iff a \in Z(G).$$

**Lemma 4.5.3** *Let $G$ be a group and $a \in G$.*

   (i) *$Z(G)$ is a normal subgroup of $G$.*
   (ii) *$Z_G(a)$ is a subgroup of $G$.*

**Proof.** Exercise. $\square$

For any subgroup $H$ of $G$,

$$N_G(H) = \{x \in G \mid xHx^{-1} = H\}$$

is called the *normalizer in $G$ of $H$*. Clearly,

$$N_G(H) = G \iff H \text{ is normal in } G.$$

**Lemma 4.5.4** *Let $H$ be a subgroup of $G$.*

   (i) *$N_G(H)$ is a subgroup of $G$.*
   (ii) *$H$ is a normal subgroup of $N_G(H)$.*
   (iii) *$N_G(H)$ is the largest subgroup of $G$ in which $H$ is normal.*

**Proof.** Exercise. $\square$

What can we say about the sizes of the nontrivial conjugacy classes? It turns out that these numbers must divide the order of the group.

**Lemma 4.5.5** *Let $G$ be a finite group $a \in G$ and $H$ be a subgroup of $G$.*

   (i) *$|C[a]| = [G : Z_G(a)]$.*
   (ii) *$|C[H]| = [G : N_G(H)]$.*

**Proof.** (i) For any $x, y \in G$, we have

$$xax^{-1} = yay^{-1} \iff ax^{-1}y = x^{-1}ya$$

$$\iff x^{-1}y \in Z_G(a)$$

$$\iff xZ_G(a) = yZ_G(a).$$

Therefore,

$$|C[a]| = \text{number of left cosets of } Z_G(a)$$
$$= |G| \,/\, |Z_G(a)|$$
$$= [G : Z_G(a)].$$

The proof of part (ii) is similar.  $\square$

Let $G$ be a finite group and select one element from each conjugacy class to get a set of representatives $\{a_1, \ldots, a_n\}$. For any $a \in Z(G)$, we have $C[a] = \{a\}$ so that $a$ must be the representative of $C[a]$. Since each element of $Z(G)$ is the representative of its class, let us list these first as $a_1, a_2, \ldots, a_m$. Then,

$$|G| = \sum_{i=1}^{n} |C[a_i]|$$

$$= \sum_{i=1}^{m} |C[a_i]| + \sum_{i=m+1}^{n} |C[a_i]|$$

$$|G| = |Z(G)| + \sum_{i=m+1}^{n} |G| \,/\, |Z_G(a_i)|.$$

This equation is the *class equation for G*. A very important application of the class equation is to $p$-groups.

**Theorem 4.5.6** *Every nontrivial $p$-group has a nontrivial center.*

**Proof.** Let $G$ be a nontrivial $p$-group—say, $|G| = p^\alpha$—where $p$ is a prime and $\alpha \in \mathbb{N}$. From the class equation for $G$, we have

$$p^\alpha = |G| = |Z(G)| + \sum_{i=m+1}^{n} |G| \,/\, |Z_G(a_i)|. \tag{4.3}$$

For $m + 1 \leq i \leq n$, since $a_i \notin Z(G)$, we have $|Z_G(a_i)| < |G|$ so that

$$|G| \,/\, |Z_G(a_i)| > 1.$$

At the same time,

$$|G| \,/\, |Z_G(a_i)| \quad \text{divides} \quad |G| = p^\alpha.$$

Hence,

$$|G| \, / \, |Z_G(a_i)| = p^{\alpha_i} \quad \text{for some } 1 \le \alpha_i < \alpha$$

from which it follows that $p$ divides

$$\sum_{i=m+1}^{n} |G| \, / \, |Z_G(a_i)|.$$

It now follows from equation (4.3) that $p \mid |Z(G)|$.  $\square$

**Example 4.5.7**  Let us consider the class equation for $S_4$.
The conjugacy classes and their sizes are as follows:

$$\epsilon$$

| | | | |
|---|---|---|---|
| 2 cycles | $(ab)$ | $\dbinom{4}{2} = \dfrac{4 \cdot 3}{1 \cdot 2}$ | $= 6$ |
| 3 cycles | $(abc)$ | $\dfrac{4 \cdot 3 \cdot 2}{3}$ | $= 8$ |
| 4 cycles | $(abcd)$ | $\dfrac{4 \cdot 3 \cdot 2 \cdot 1}{4}$ | $= 6$ |
| $2 \times 2$ cycles | $(ab)(cd)$ | $6/2$ | $= 3.$ |

The centralizers for each of the different cycle patterns are as follows:

$$
\begin{aligned}
Z_{S_4}(\epsilon) \quad &= S_4 \\
Z_{S_4}(ab) \quad &= \{\epsilon, (ab), (cd), (ab)(cd)\} \\
Z_{S_4}(abc) \quad &= \{\epsilon, (abc), (acb)\} \\
Z_{S_4}(abcd) \quad &= \langle (abcd) \rangle \\
Z_{S_4}((ab)(cd)) \quad &= \{\epsilon, (ab), (cd), (ab)(cd)\} \cup \\
&\quad\ \{\epsilon, (ab), (cd), (ab)(cd)\} \, (ac)(bd).
\end{aligned}
$$

Thus the class equation becomes

$$24 = |G| = |Z(G)| + \sum \frac{|G|}{|Z_G(a_i)|}$$

$$= 1 + \frac{24}{4} + \frac{24}{3} + \frac{24}{4} + \frac{24}{8}$$

$$= 1 + 6 + 8 + 6 + 3.$$

## Exercises 4.5

1. Determine the conjugacy classes in each of the following groups:

    (i) $S_3$.
    (ii) $D_4$.
    (iii) $Q_8$.

2. Find the center of each of the following groups:

    (i) $S_3$.
    (ii) $D_4$.
    (iii) $Q_8$.
    (iv) $A_4$.
    (v) $S_4$.

3. How many elements are there in the conjugacy class of the element

$$\alpha = (2\,3)(4\,5\,6)(7\,8\,9\,10) \in S_{10}?$$

*4. How many elements are there in the conjugacy class of $\alpha \in S_n$ if $\alpha$ has cycle pattern

$$1^{r_1}\,2^{r_2}\cdots n^{r_n}?$$

5. Let $G$ be a group such that $G/Z(G)$ is cyclic. Prove that $G$ is abelian.

6. Let $G$ be a group such that $G/Z(G)$ is abelian. Show, in contrast to the preceding exercise, that $G$ need not be abelian. (Consider $Q_8$ or $D_4$.)

7. Let $p$ be a prime. Show that any group of order $p^2$ is abelian.

*8. Let $|G| = p^n$ where $p$ is a prime. Let $0 \le k \le n$. Prove that $G$ has a normal subgroup of order $p^k$.

*9. Let $|G| = p^n$ and $H$ be a nontrivial normal subgroup of $G$. Prove that $H \cap Z(G) \ne \{e\}$.

*10. Let $Y$ and $Z$ be nonempty subsets of the finite set $X$ such that $|Y| = |Z|$. Let

$$H_Y = \{\alpha \in S_X \mid \alpha(x) = x, \ \forall\, x \in X \backslash Y\}.$$

Let $H_Z$ be defined similarly. Prove the following:

    (i) $H_Y$ is a subgroup of $S_X$.
    (ii) $H_Y \cong S_Y$.
    (iii) $H_Y$ and $H_Z$ are conjugate subgroups of $S_X$.

*11. Let $G$ be a finite group and $a, b \in G$ be such that $|a| = |b|$. Show that there exists a group $H$ and a monomorphism $\varphi : G \to H$ such that $\varphi(a)$ and $\varphi(b)$ are conjugate in $H$.

*12. Let $G$ and $H$ be finite groups and $a \in G$, $b \in H$ be such that $|a| = |b|$. Show that there exists a group $K$ and monomorphisms $\theta : G \to K$, $\varphi : H \to K$ such that $\theta(a)$ and $\varphi(b)$ are conjugate.

## 4.6 The Sylow Theorems 1 and 2

We know from Lagrange's Theorem that the order of a subgroup of a finite group $G$ must divide the order of $G$. A natural related question is the following: If $m \in \mathbb{N}$ and $m$ divides $|G|$, does there exist a subgroup of $G$ of order $m$? The answer, in general, is no. For example, $A_4$ (of order 12) has no subgroup of order 6. However, there are circumstances in which the answer will be yes. We consider one such situation here and consider three fundamental theorems of the Norwegian mathematician Ludwig Sylow (1832–1918).

Before we consider the first theorem, we require a little combinatorial observation. First recall that we can write the binomial coefficient $\binom{n}{r}$ as follows:

$$\binom{n}{r} = \frac{n!}{(n-r)!\, r!} = \frac{n(n-1)\ldots(n-r+1)}{1 \cdot 2 \ldots r}$$

$$= \frac{n(n-1)\ldots(n-(r-1))}{r \cdot 1 \cdot 2 \ldots (r-1)}$$

**Lemma 4.6.1** *Let $p$ be a prime, $r, \alpha \in \mathbb{N}$, and $(p, r) = 1$. Then $p$ does not divide*

$$n = \binom{p^\alpha r}{p^\alpha}.$$

**Proof.** We have

$$n = \binom{p^\alpha r}{p^\alpha} = \frac{p^\alpha r(p^\alpha r - 1)(p^\alpha r - 2)\ldots(p^\alpha r - (p^\alpha - 1))}{p^\alpha \cdot 1 \cdot 2 \ldots (p^\alpha - 1)}$$

$$= r \prod_{k=1}^{p^\alpha - 1} \frac{p^\alpha r - k}{k}.$$

We want to see that $(n, p) = 1$. Consider the factors $\frac{p^\alpha r - k}{k}$. There are two cases:

(a)  $(p, k) = 1$. Then $p$ does not divide $p^\alpha r - k$.

(b)  $(p, k) > 1$, say $k = p^\beta s$ where $1 \le \beta < \alpha$ and $(p, s) = 1$. Then,

$$\frac{p^\alpha r - k}{k} = \frac{p^\beta(p^{\alpha-\beta} r - s)}{p^\beta s} = \frac{p^{\alpha-\beta} r - s}{s}$$

and $p$ does not divide $p^{\alpha-\beta} r - s$. Hence, after cancellation, $\frac{p^{\alpha} r - k}{k}$ is a rational fraction with no factor of $p$ in the numerator. Thus, $p$ cannot divide $n$. $\quad\square$

**Theorem 4.6.2 (Sylow's First Theorem)** *Let $G$ be a group of order $p^{\alpha} r$, where $p$ is a prime, $\alpha \geq 1$ and $(p, r) = 1$. Then $G$ contains a subgroup of order $p^{\alpha}$.*

**Proof.** Let $\mathbf{S}$ denote the set of all subsets of $G$ consisting of exactly $p^{\alpha}$ elements, say

$$\mathbf{S} = \{S_1, S_2, \ldots, S_n\}$$

where

$$n = \binom{p^{\alpha} r}{p^{\alpha}}.$$

By Lemma 4.6.1, $p$ does not divide $n$. Now, for all $x \in G$, we have $|xS_i| = p^{\alpha}$. Hence, for all $i = 1, 2, \ldots, n$ and all $x \in G$, there must exist an integer $j$ with $xS_i = S_j$. We clearly have that

$$xS_i = xS_j \Rightarrow S_i = S_j$$

In addition, for all $j$ with $1 \leq j \leq n$ and for all $x \in G$, we have $x(x^{-1}S_j) = S_j$. Therefore, each element of $G$ defines a permutation of $\mathbf{S}$. Also, for distinct $x, y \in G$, let $S_i$ be such that it contains the identity of $G$, but not the element $xy^{-1}$. Clearly such a subset must exist, since $r > 1$. Then $xy^{-1}S_i \neq S_i$ so that $xy^{-1}$ does not define the identity permutation and, equivalently, $x$ and $y$ define distinct permutations. In this way, we can consider $G$ as a group of permutations of $\mathbf{S}$.

Let the orbits in $\mathbf{S}$ under the action of $G$ be $O_1, O_2, \ldots, O_k$ and $M = |\mathbf{S}|$. Then

$$M = \sum_{i=1}^{k} |O_i|.$$

Since $p$ does not divide $|\mathbf{S}|$, there must exist $k$ such that $p$ does not divide $|O_k|$. Let $S_t$ be any element of $O_k$. By the Orbit/Stabilizer Theorem (Theorem 3.9.2), we have

$$|G| = |O_k| \cdot |S(S_t)|$$

But $p$ does not divide $|O_k|$, so we must have $p^{\alpha}$ dividing $|S(S_t)|$. In particular, that means that $p^{\alpha} \leq |S(S_t)|$. On the one hand, for any element $z$ of $S_t$, viewing $S(S_t)z$ as a right coset of $S(St)$, we get $|S(S_t)| = |S(S_t)z|$. On the other hand, viewing $S(S_t)$ as the stabilizer of $S_t$, we must have $S(S_t)z \subseteq S_t$ and therefore $|S(S_t)z| \leq |S_t| = p^\alpha$. Consequently, $|S(S_t)| = |S(S_t)z| \leq p^\alpha$ and we have equality $|S(S_t)| = p^\alpha$. Thus $S(S_t)$ is a subgroup of order $p^\alpha$ and the proof is complete. $\square$

If $|G| = p^\alpha r$ where $p$ is a prime and $(p, r) = 1$, then any subgroup $H$ of $G$ of order $p^\alpha$ is called a *Sylow p-subgroup* of $G$. By Theorem 4.6.2, we know that such subgroups always exist.

**Example 4.6.3** $|S_3| = 6$. By Sylow's First Theorem, $S_3$ has subgroups of orders 2 and 3. These are easy to recognize. For example, $A = \{\epsilon, (1\,2\,3), (1\,3\,2)\}$ is a subgroup of order 3 and

$$B_1 = \{\epsilon, (1\,2)\}$$
$$B_2 = \{\epsilon, (1\,3)\}$$
$$B_3 = \{\epsilon, (2\,3)\}$$

are all subgroups of order 2. Now $A$ is the only subgroup of $S_3$ of order 3, so that we see immediately that a finite group may have a unique Sylow $p$-subgroup for some prime $p$ and several Sylow $p$-subgroups for other primes.

We are now ready to give the generalization of Lemma 4.4.2 that was promised.

**Corollary 4.6.4 (Cauchy's Theorem)** *Let $G$ be a finite group and $p$ be a prime dividing $|G|$. Then $G$ contains an element of order $p$.*

**Proof.** Let $H$ be a Sylow $p$-subgroup of $G$ and $a \in H$, $a \neq e$. Then the order of $a$ must divide the order of $H$. Since $|H| = p^\alpha$, for some $\alpha \in \mathbb{N}$, we must have $|a| = p^\beta$ for some $\beta \in \mathbb{N}$, $\alpha \leq \beta$. Then, $a^{p^{\beta-1}}$ has order $p$. $\square$

Now suppose that $H$ is a Sylow $p$-subgroup of a finite group $G$ and that $a \in G$. Then

$$|aHa^{-1}| = |H|$$

so that $aHa^{-1}$ must also be a Sylow $p$-subgroup of $G$. In other words, any conjugate of a Sylow $p$-subgroup is also a Sylow $p$-subgroup. As we shall soon see, this is the only way to obtain "new" Sylow $p$-subgroups. However, once again we require a preliminary counting lemma.

**Lemma 4.6.5** *Let $H$ and $K$ be subgroups of a group $G$ and define a relation $\sim$ on $G$ by*

$$a \sim b \iff \exists\, h \in H,\ k \in K \quad \text{with} \quad b = hak.$$

*Then, $\sim$ is an equivalence relation. For any $a \in G$, the $\sim$ class of $a$ is $HaK$.*

**Proof.** Exercise $\quad\square$

The subsets of the form $HaK$ that show up as the equivalence classes under $\sim$ are sometimes referred to as *double cosets*. Since the double cosets with respect to two subgroups are classes of an equivalence relation, it follows that any two are either identically equal or disjoint. However, double cosets are not quite as "regular" as the usual cosets, in the sense that different double cosets may have different sizes.

**Example 4.6.6** Let $H = \{\epsilon, (1\,2)\}$, $K = \{\epsilon, (1\,3)\}$ in $S_3$, and let $a = (2\,3)$. Then

$$
\begin{aligned}
HaK &= \{\epsilon, (1\,2)\}\,(2\,3)\,\{\epsilon, (1\,3)\} \\
&= \{(2\,3), (2\,3)(1\,3), (1\,2)(2\,3), (1\,2)(2\,3)(1\,3)\} \\
&= \{(2\,3), (1\,2\,3), (1\,2\,3), (2\,3)\} \\
&= \{(2\,3), (1\,2\,3)\}.
\end{aligned}
$$

However, since $(1\,2) \in H$,

$$
\begin{aligned}
H(1\,2)K &= HK \\
&= \{\epsilon, (1\,2)\}\,\{\epsilon, (1\,3)\} \\
&= \{\epsilon, (1\,3), (1\,2), (1\,2)(1\,3)\} \\
&= \{\epsilon, (1\,2), (1\,3), (1\,3\,2)\}.
\end{aligned}
$$

Thus,

$$
S_3 = H(2\,3)K \cup H(1\,2)K
$$

where

$$
|H(2\,3)K| = 2, \quad |H(1\,2)K| = 4.
$$

**Theorem 4.6.7 (Sylow's Second Theorem)** *Let $G$ be a finite group and $p$ be a prime with $p \mid |G|$. Let $H$ and $K$ be Sylow $p$-subgroups of $G$. Then $H$ and $K$ are conjugate in $G$.*

**Proof.** Let $|G| = p^\alpha r$ where $\alpha \in \mathbb{N}$ and $(p, r) = 1$. Then $|H| = |K| = p^\alpha$. Let the distinct double cosets of $H$ and $K$ in $G$ be

$$
Hx_1 K, \, Hx_2 K, \ldots, Hx_n K
$$

where $x_i \in G$. Now every double coset $Hx_iK$ is a union of left cosets of $K$. So the number of left cosets of $K$ in $Hx_iK$ is, by Proposition 4.1.12, just

$$\frac{|Hx_iK|}{p^\alpha} = \frac{|H| \cdot |x_iKx_i^{-1}|}{p^\alpha d_i} = \frac{p^\alpha}{d_i} \tag{4.4}$$

where $d_i = |H \cap x_iKx_i^{-1}|$. The total number of left cosets of $K$ in $G$ is $p^\alpha r/p^\alpha = r$. Hence, from (4.4), we have

$$r = \sum_{i=1}^{n} \frac{p^\alpha}{d_i}.$$

But, $(p, r) = 1$, so there must exist some $i$ with $d_i = p^\alpha$. Without loss of generality, we can take $i = 1$. Then

$$H \cap x_1Kx_1^{-1} \subseteq H$$

and

$$|H \cap x_1Kx_1^{-1}| = d_1 = p^\alpha = |H|.$$

Thus,

$$H \cap x_1Kx_1^{-1} = H.$$

But we also have $|xKx_1^{-1}| = p^\alpha = |H|$. Therefore,

$$H = x_1Kx_1^{-1}$$

and $H$ and $K$ are conjugate subgroups.  $\square$

The situation in which a group has unique Sylow $p$-subgroups deserves special mention.

**Corollary 4.6.8** *Let $p$ be a prime number that divides the order of the finite group $G$. Then $G$ has just one Sylow $p$-subgroup if and only if it has a normal Sylow $p$-subgroup.*

**Proof.** First suppose that $P$ is a unique Sylow $p$-subgroup of $G$. Since any conjugate of $P$ would also be a Sylow $p$-subgroup, it follows that $P = xPx^{-1}$, for all $x \in G$, and therefore $P$ is normal.

Conversely, suppose that $P$ is a normal Sylow $p$-subgroup of $G$. Then the only conjugate of $P$ is $P$ itself and, therefore, by Sylow's Second Theorem, $P$ is the unique Sylow $p$-subgroup. $\square$

For example, $S_3$ has a unique Sylow 3-subgroup $H = \{\epsilon, (1\,2\,3), (1\,3\,2)\}$, which is normal in $S_3$ but has three Sylow 2-subgroups listed in Example 4.6.3. By Theorem 4.6.7, these Sylow 2-subgroups must be conjugate. In Section 4.9, we will take a closer look at finite groups in which every Sylow subgroup is normal.

## Exercises 4.6

1. Find the unique Sylow 2-subgroup of $A_4$.

2. Find all Sylow 3-subgroups of $A_4$. Show that they are conjugate.

3. Let $N$ be a normal subgroup of a group $G$ and $a \in G$ be an element of order 2. Show that $N \cup Na$ is a subgroup of $G$.

4. Find all the Sylow 2-subgroups of $S_4$ and show that they are conjugate. (Hint: Use Exercise 3 in this section and the fact that the Sylow 2-subgroup of $A_4$ found in Exercise 1 in this section is normal in $S_4$.)

5. Let $G$ be a group of order $p^\alpha r$ where $p$ is a prime, $r \in \mathbb{N}$, and $(p, r) = 1$. Let $H$ be a subgroup of $G$ of order $p^\beta$ where $\beta \leq \alpha$. Show that $H \subseteq P$, for some Sylow $p$-subgroup $P$. (Hint: Adapt the proof of Sylow's Second Theorem.)

6. Let $H, K$ be subgroups of a group $G$ and define the relation $\sim$ on $G$ by

$$a \sim b \iff \exists\, h \in H,\ k \in K \quad \text{with} \quad b = hak.$$

   Show that $\sim$ is an equivalence relation on $G$ and that the $\sim$ class containing $a$ is $HaK$.

7. Let $H = \langle (1\,2\,3) \rangle$ and $K = \langle (1\,2)(3\,4) \rangle$ in $A_4$. Determine the classes of the relation $\sim$ as defined in Exercise 6.

8. Let $H$ be a subgroup of a group $G$ and define a relation $\sim$ on the set $\mathcal{S}$ of all subgroups of $G$ by

$$A \sim B \iff \exists\, h \in H \quad \text{with} \quad B = hAh^{-1}.$$

   Show that $\sim$ is an equivalence relation on $\mathcal{S}$.

9. Let $p$ be a prime and $G$ be a finite group such that the order of every element in $G$ is a power of $p$. Show that $G$ is a $p$-group.

## 4.7 Sylow's Third Theorem

Sylow's Third Theorem gives us information regarding the number of Sylow subgroups that may occur in a finite group. This is extremely valuable information in the search for features of a group, such as normal subgroups, that might be used in the analysis of a group's structure.

**Theorem 4.7.1 (Sylow's Third Theorem)** *Let $G$ be a group of order $p^{\alpha} r$ where $p$ is a prime, $\alpha, r \in \mathbb{N}$, and $(p, r) = 1$. Then the number of Sylow $p$-subgroups is of the form $1 + kp$, for some integer $k$, and divides $r$.*

**Proof.** If there is only one Sylow $p$-subgroup, then we can take $k = 0$ and the claim holds trivially.

So suppose that there is more than one and let $P$ be one of them. By Theorem 4.6.7 (Sylow's Second Theorem), the Sylow $p$-subgroups of $G$ are then just the conjugates of $P$. By Lemma 4.5.5 (ii), the number of distinct conjugates of $P$ is

$$[G : N_G(P)]$$

which is a number that divides $|G|$.

To see that the number of Sylow $p$-subgroups is of the form $1 + kp$, let the Sylow $p$-subgroups *other* than $P$ be $P_1, P_2, \ldots, P_m$. We will show that $p$ divides $m$. Define a relation $\sim$ on $\mathcal{P} = \{P_1, \ldots, P_m\}$ by

$$P_i \sim P_j \iff \exists x \in P \quad \text{with} \quad P_j = xP_ix^{-1}.$$

Note that only elements from $P$ are used here to determine the relation $\sim$. It is again straightforward to verify that $\sim$ is an equivalence relation on $\mathcal{P}$.

Let the equivalence classes be $O_1, O_2, \ldots, O_K$. Then

$$m = \sum_{i=1}^{K} |O_i|.$$

Let $P_i \in O_i$ and

$$S(P_i) = \left\{ x \in P | xP_ix^{-1} = P_i \right\}.$$

Clearly $S(P_i)$ is a subgroup of $P$ and, just as in the Orbit/Stabilizer Theorem (Theorem 3.9.2), we have

$$|P| = |O_i| \cdot |S(P_i)|.$$

Thus $|O_i|$ divides $p^\alpha$. Suppose that there exists an orbit $O_i$ with just a single element, $O_i = \{P_i\}$. Then $P_i = xP_ix^{-1}$, for all $x \in P$. Therefore, $P \subseteq N_G(P_i)$.

Then, $P$ and $P_i$ are both subgroups of $N_G(P_i)$. Hence, $p^\alpha = |P| = |P_i|$ divides $|N_G(P_i)|$. However, $p^\alpha$ is the largest power of $p$ dividing $|G|$, so it must also be the largest power of $p$ dividing $|N_G(P_i)|$. Therefore, $P$ and $P_i$ are Sylow $p$-subgroups of $N_G(P_i)$. Invoking Sylow's Second Theorem, we may assert the existence of $h \in N_G(P_i)$ with $P = hP_ih^{-1}$. But, $h \in N_G(P_i)$. Consequently, $hP_ih^{-1} = P_i$ and therefore $P = P_i$. This contradicts the assumption that $P$ and $P_i$ are distinct. It follows that $|O_i| > 1$, for all $i$ and therefore that $|O_i| = p^{\alpha_i}$ for some integer $\alpha_i > 0$. Hence we have

$$m = \sum_{i=1}^{K} p^{\alpha_i} = kp$$

for some integer $k$. Therefore, when we add 1 to include $P$ we find that the total number of Sylow $p$-subgroups is of the form $1 + kp$.

From the beginning of the proof, we know that $1 + kp$ must divide $|G| = p^\alpha r$. However, $1 + kp$ and $p$ are relatively prime. Therefore $1 + kp$ must divide $r$. $\quad\square$

If $p$ is a prime dividing the order of a group $G$, then we denote by $s_p$ the number of Sylow $p$-subgroups in $G$. The Sylow theorems enable us to gather a wealth of information regarding groups of small order.

**Example 4.7.2** Let $|G| = 48 = 2^4 \cdot 3$. Then, $s_2 = 1 + k2$ and divides 3. Therefore, $s_2 = 1$ or 3. Also, $s_3 = 1 + k3$ and divides 16. Therefore, $s_3 = 1, 4,$ or 16.

**Example 4.7.3** Let $|G| = 1225 = 5^2 \cdot 7^2$. Then, $s_5 = 1 + k5$ and divides $7^2$, so that $s_5 = 1$. Similarly, $s_7 = 1 + k7$ and divides $5^2$ so that $s_7 = 1$ as well. Thus, $G$ has a unique normal Sylow 5-subgroup $P_5$, say, and a unique normal Sylow 7-subgroup, $P_7$, say, where $|P_5| = 5^2$ and $|P_7| = 7^2$. Hence, $P_5 \cap P_7 = \{e\}$ and, by Proposition 4.1.12,

$$|P_5 \, P_7| = \frac{|P_5| \cdot |P_7|}{|P_5 \cap P_7|} = \frac{5^2 \cdot 7^2}{1} = |G|.$$

Thus, $G = P_5 \, P_7$ and $G$ is the internal direct product of $P_5$ and $P_7$.

By Theorem 4.3.3, $G \cong P_5 \times P_7$. Now $|P_5| = 5^2$ so that, by Exercise 7 in section 4.5, $P_5$ is abelian. Likewise $P_7$ is abelian, and therefore $G$ itself is abelian.

**Example 4.7.4** Let $|G| = 15$. In this example we can go even further than in the previous one. We have $s_3 = 1 + k3$ and must divide 5, so that $s_3 = 1$.

Likewise $s_5 = 1 + k5$ and divides 3, so that $s_5 = 1$. Thus, $G$ has a unique normal Sylow 3-subgroup $P_3$, say, and a unique normal Sylow 5-subgroup $P_5$, say. As in the previous example, $P_3 \cap P_5 = \{e\}$ so that $G$ is the internal direct product of $P_3$ and $P_5$. Hence, $G \cong P_3 \times P_5$. Now, $|P_3| = 3$ and $|P_5| = 5$ so that both $P_3$ and $P_5$ are cyclic; $P_3 \cong \mathbb{Z}_3$ and $P_5 \cong \mathbb{Z}_5$. Thus, $G \cong \mathbb{Z}_3 \times \mathbb{Z}_5 \cong \mathbb{Z}_{15}$.

The observations in these examples concerning direct products of Sylow $p$-subgroups generalize nicely to a powerful theorem.

**Theorem 4.7.5** *Let $G$ be a finite group with $|G| = p_1^{\alpha_1} \cdots p_m^{\alpha_m}$ where, for $1 \leq i \leq m$, the $p_i$ are distinct prime numbers and $\alpha_i \in \mathbb{N}$. Suppose that for each $i$, there exists a unique (normal) Sylow $p_i$-subgroup $P_i$. Then $G$ is isomorphic to*

$$P_1 \times P_2 \times \cdots \times P_m.$$

**Proof.** Let $t$ be such that $1 \leq t \leq m$. Since each $P_i$ is normal in $G$, it follows that $P_1 P_2 \cdots P_t$ is also a normal subgroup of $G$ (Exercise 10 in section 4.1). We will show by induction on $t$ that

$$|P_1 P_2 \ldots P_t| = |P_1| \cdot |P_2| \ldots |P_t|.$$

The claim is trivially true for $t = 1$. So suppose that it holds for $t - 1$. Then

$$|P_1 \ldots P_{t-1}| = p_1^{\alpha_1} \cdots p_{t-1}^{\alpha_{t-1}}, \qquad |P_t| = p_t^{\alpha_t}$$

from which it follows that

$$P_1 \ldots P_{t-1} \cap P_t = \{e\}. \tag{4.5}$$

Hence, by Proposition 4.1.12 and the induction hypothesis,

$$\begin{aligned}
|P_1 P_2 \ldots P_t| &= |(P_1 P_2 \ldots P_{t-1})P_t| \\
&= \frac{|P_1 P_2 \ldots P_{t-1}| \cdot |P_t|}{|P_1 P_2 \ldots P_{t-1} \cap P_t|} \\
&= |P_1||P_2| \ldots |P_{t-1}| \cdot |P_t|.
\end{aligned}$$

By induction, we have

$$|P_1 P_2 \ldots P_m| = |P_1| \ldots |P_m| = |G|.$$

Therefore, $P_1 P_2 \ldots P_m = G$. In addition, the argument that led to (4.5) will also show that

$$P_i \cap P_1 \ldots P_{i-1} P_{i+1} \ldots P_m = \{e\}$$

for all $1 \leq i \leq m$. Hence, $G$ is the internal direct product of the $P_i$ and therefore, by Exercise 3 in section 4.3, the proof is complete. $\quad\square$

**Example 4.7.6** Let $|G| = 5^2 \cdot 7 \cdot 19 = 3325$. Then $G$ has a unique (normal) Sylow 5-subgroup $P_5$, Sylow 7-subgroup $P_7$, and Sylow 19-subgroup $P_{19}$. Therefore,

$$G \cong P_5 \times P_7 \times P_{19}.$$

Since $|P_5| = 5^2$, $|P_7| = 7$, and $|P_{19}| = 19$, these are all abelian groups and therefore $G$ is also abelian.

## Exercises 4.7

1. Let $G$ be a group of order $2^4 7^2$. How many Sylow 2-subgroups and Sylow 7-subgroups might it have?

2. Let $G$ be a group of order 20. Show that $G$ has a proper nontrivial normal subgroup.

3. Let $G$ be a group of order 225. Show that $G$ has a proper nontrivial normal subgroup.

4. Let $G$ be a group of order $pq$, where $p$ and $q$ are distinct primes. Let $G$ have a unique Sylow $p$-subgroup $P$ and a unique Sylow $q$-subgroup $Q$. Show that $G$ is the internal direct product of $P$ and $Q$. Deduce that $G$ is cyclic.

5. Let $G$ be a group of order 35. Show that $G$ is cyclic.

6. Let $G$ be a group of order $p^\alpha r$, where $p$ is a prime, $r \in \mathbb{N}$, and $r < p$. Show that $G$ has a proper nontrivial normal subgroup.

7. Let $p$ be a prime dividing the order of a group $G$. Show that the intersection of all the Sylow $p$-subgroups is a normal subgroup of $G$.

8. Let $G$ be a group of order $p^\alpha r$ with $X = \{P_1, P_2, \ldots, P_k\}$ as its set of Sylow $p$-subgroups. For each element $g \in G$, define a mapping $\alpha_g : X \to X$, by

$$\alpha_g(P_i) = gP_ig^{-1}.$$

   (i) Show that $\alpha_g$ is a permutation of $X$.
   (ii) Show that the mapping $\alpha : g \to \alpha_g$ is a homomorphism of $G$ into $S_X$.
   (iii) Show that $\alpha(G)$ acts transitively on $X$.

## 4.8 Solvable Groups

The *commutator* of the elements $x, y$ in a group is the element

$$[x, y] = x^{-1}y^{-1}xy.$$

Note that

$$xy = yx[x, y]$$

so that, in some sense, the commutator $[x, y]$ reflects or measures the extent to which the elements $x$ and $y$ do not commute. The subgroup generated by all the commutators of a group $G$ is the *commutator subgroup* or the *derived subgroup*. It is denoted variously by $[G, G]$ or $G'$ or $G^{(1)}$.

We begin with some simple observations concerning commutators. First, the inverse of a commutator is a commutator:

$$[x, y]^{-1} = (x^{-1}y^{-1}xy)^{-1} = y^{-1}x^{-1}yx = [y, x].$$

Therefore, the inverse of a product of commutators is still just a product of commutators. Hence the commutator subgroup is generated simply by forming all products of commutators without regard to inverses:

$$[G, G] = \{c_1 c_2 \cdots c_m \mid c_i \text{ is a commutator}\}.$$

Also, any conjugate of a commutator is a commutator:

$$a^{-1}[x, y]a = a^{-1}x^{-1}y^{-1}xya$$

$$= a^{-1}x^{-1}aa^{-1}y^{-1}aa^{-1}xaa^{-1}ya$$

$$= [a^{-1}xa, a^{-1}ya].$$

Consequently, for any commutators $c_1, \ldots, c_m$,

$$a^{-1}c_1 \cdots c_m a = a^{-1}c_1 aa^{-1}c_2 a \cdots a^{-1}c_m a$$

$$= c_1' c_2' \cdots c_m'$$

where each $c_i' = a^{-1}c_i a$ is a commutator. Hence,

$$a^{-1}[G, G]a \subseteq [G, G] \qquad \text{for all } a \in G$$

which establishes the following lemma.

**Lemma 4.8.1** $[G, G]$ *is a normal subgroup of* $G$.

It is on account of the next result that the commutator subgroup is so important.

**Theorem 4.8.2** *Let N be a normal subgroup of a group G. Then the following statements are equivalent:*

   (i)  *$G/N$ is abelian.*
   (ii) *$[G, G] \subseteq N$.*

*In particular, G is abelian if and only if $[G, G] = \{e\}$.*

**Proof.** We have

$$
\begin{aligned}
G/N \text{ is abelian} &\Longleftrightarrow NxNy = NyNx && \forall\, x, y \in G \\
&\Longleftrightarrow Nxy = Nyx && \forall\, x, y \in G \\
&\Longleftrightarrow (Nyx)^{-1}Nxy = N && \forall\, x, y \in G \\
&\Longleftrightarrow Nx^{-1}y^{-1}xy = N && \forall\, x, y \in G \\
&\Longleftrightarrow [x, y] = x^{-1}y^{-1}xy \in N && \forall\, x, y \in G \\
&\Longleftrightarrow [G, G] \subseteq N. && \square
\end{aligned}
$$

Theorem 4.8.2 tells us that $[G, G]$ is the smallest normal subgroup of $G$ such that the quotient is abelian. So it should come as no surprise that the quotient $G/[G, G]$ should, in a certain sense, be the largest abelian homomorphic image of $G$. We make this precise, as follows.

**Lemma 4.8.3** *Let $\varphi : G \to H$ be an epimorphism from a group G to an abelian group H. Let $\pi : G \to G/[G, G]$ be the natural homomorphism. Then there exists an epimorphism $\theta : G/[G, G] \to H$ such that $\theta\pi = \varphi$.*

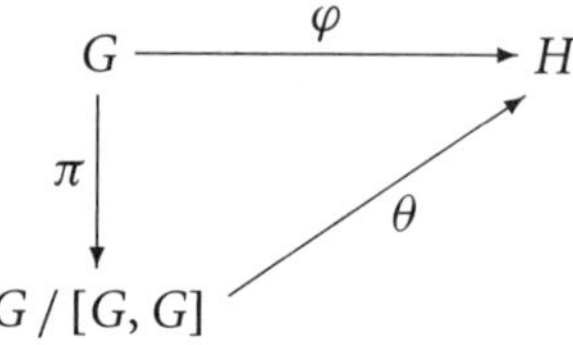

**Proof.** By the First Isomorphism Theorem, we know that $H$ is isomorphic to $G/\ker(\varphi)$. Since $H$ is abelian, so also is $G/\ker(\varphi)$. By Theorem 4.8.2, this implies that $[G, G] \subseteq \ker(\varphi)$. Now define a mapping $\theta : G/[G, G] \to H$ by

$$
\theta\left([G, G]g\right) = \varphi(g).
$$

We have

$$
\begin{aligned}
[G, G]g = [G, G]h &\Longrightarrow gh^{-1} \in [G, G] \subseteq \ker(\varphi) \\
&\Longrightarrow \varphi(gh^{-1}) = e_H \\
&\Longrightarrow \varphi(g) = \varphi(h)
\end{aligned}
$$

from which it follows that $\theta$ is indeed well defined. For any elements $x = [G, G]g$, $y = [G, G]g' \in G/[G, G]$,

$$\begin{aligned}
\theta(xy) &= \theta\left([G, G]gg'\right) \\
&= \varphi(gg') \\
&= \varphi(g)\,\varphi(g') \\
&= \theta(x)\,\theta(y)
\end{aligned}$$

so that $\theta$ is a homomorphism. Finally, for any $h \in H$, there exists a $g \in G$ with $\varphi(g) = h$, since $\varphi$ is surjective. Then

$$\theta\left([G, G]g\right) = \varphi(g) = h.$$

Thus, $\theta$ is surjective and, therefore, an epimorphism. Finally, for any $g \in G$, we have

$$\theta\pi(g) = \theta([G, G]g) = \varphi(g).$$

Thus, $\theta\pi = \varphi$. $\quad\square$

The process of forming the derived subgroup can be repeated to yield the derived subgroup of the derived subgroup, and so on. Let us write

$$\begin{aligned}
G^{(0)} &= G \\
G^{(1)} &= [G, G] = [G^{(0)}, G^{(0)}] \\
G^{(2)} &= \left[[G, G], [G, G]\right] = [G^{(1)}, G^{(1)}]
\end{aligned}$$

$$\vdots$$

$$G^{(n+1)} = [G^{(n)}, G^{(n)}].$$

In this way we obtain a descending sequence of subgroups

$$G = G^{(0)} \supseteq G^{(1)} \supseteq G^{(2)} \cdots.$$

In a finite group, this sequence must obviously cease to change from some point onward. However, in some infinite groups it will continue indefinitely.

**Lemma 4.8.4** *Let $G$ be a group. Then, for all $n \geq 0$, $G^{(n)}$ is a normal subgroup of $G$.*

**Proof.** By Lemma 4.8.1, we know that each $G^{(n)}$ is a subgroup of $G$. We will show by induction on $n$ that each $G^{(n)}$ is actually a normal subgroup in $G$. Since $G^{(0)} = G$, the claim is certainly true for $n = 0$. Now assume that the claim is true for $n$. Let $a, b \in G^{(n)}$, $x \in G$. Then, by the induction hypothesis,

$$xax^{-1} = a', \ xbx^{-1} = b' \in G^{(n)}$$

so that

$$x[a, b]x^{-1} = [xax^{-1}, xbx^{-1}]$$

$$= [a', b']$$

$$\in G^{(n+1)}.$$

However, $G^{(n+1)}$ is the subgroup of $G$ generated by elements of the form $[a, b]$ where $a, b \in G^{(n)}$. Hence,

$$xG^{(n+1)}x^{-1} \subseteq G^{(n+1)} \qquad \text{for all } x \in G.$$

Thus, $G^{(n+1)}$ is normal in $G$. $\quad \square$

A group $G$ is said to be *solvable* if there exist normal subgroups $G_i$, $\ 1 \leq i \leq r$, such that

$$\{e\} \subseteq G_r \subseteq G_{r-1} \subseteq \cdots \subseteq G_1 \subseteq G$$

and $G_i / G_{i+1}$ is abelian for all $i$. The class of solvable groups is not only important in the development of group theory, it also plays an important role in the study of the solution of equations by radicals—a central topic in Galois Theory.

**Theorem 4.8.5** *Let $G$ be a group. If $G$ is a finite group then the following conditions are equivalent. If $G$ is an infinite group, then conditions* (i), (iii), *and* (iv) *are equivalent:*

(i)  *$G$ is solvable.*
(ii)  *$G$ has a sequence of subgroups*

$$\{e\} = G_{s+1} \subseteq G_s \subseteq \cdots \subseteq G_1 \subseteq G_0 = G$$

*such that $G_{i+1}$ is a normal subgroup of $G_i$ for $i = 0, 1, \ldots, s$ and $G_i / G_{i+1}$ is cyclic.*

(iii)  *G has a sequence of subgroups*

$$\{e\} = G_{s+1} \subseteq G_s \subseteq \cdots \subseteq G_1 \subseteq G_0 = G$$

*such that $G_{i+1}$ is a normal subgroup of $G_i$ and $G_i \, / \, G_{i+1}$ is abelian.*
(iv)  *There exists an integer $n$ such that $G^{(n)} = \{e\}$.*

**Proof.**  It is clear that, for an arbitrary group, (i) implies (iii) and (ii) implies (iii). Now let $G$ be finite and we will show that (i) implies (ii).

By (i), there exists a sequence of subgroups

$$\{e\} = G_r \subseteq G_{r-1} \subseteq \cdots \subseteq G_1 \subseteq G_0 = G$$

where $G_i$ is normal in $G$ and $G_i \, / \, G_{i+1}$ is abelian, $0 \le i \le r - 1$. Since $G$ is finite, $G_i \, / \, G_{i+1}$ is a finite abelian group and so, by Corollary 4.4.9, there exists a chain of subgroups

$$\{G_{i+1}\} = H_t \subseteq \cdots \subseteq H_1 \subseteq H_0 = G_i \, / \, G_{i+1}$$

such that $H_j \, / \, H_{j+1}$ is cyclic. Let

$$G_{i+1,j} = \{g \in G \mid G_{i+1} \, g \in H_j\}, \qquad 0 \le j \le t.$$

In other words, $G_{i+1,j}$ is the preimage of $H_j$ with respect to the natural homomorphism

$$\pi_i : G_i \to G_i \, / \, G_{i+1}$$

so that $G_{i+1,0} = G_i$, $G_{i+1,t} = G_{i+1}$. Then $G_{i+1,j}$ is a subgroup of $G$ and

$$G_{i+1,j} \, / \, G_{i+1} = H_j.$$

Since $G_i \, / \, G_{i+1}$ is abelian, $H_j$ is a normal subgroup of $G_i \, / \, G_{i+1}$. Hence, as the preimage of $H_j$ under $\pi_i$, $G_{i+1,j}$ is a normal subgroup of $G_i$, and therefore a normal subgroup of $G_{i+1,j-1}$ where, by the Third Isomorphism Theorem,

$$G_{i+1,j-1} \, / \, G_{i+1,j} \cong \frac{G_{i+1,j-1} \, / \, G_{i+1}}{G_{i+1,j} \, / \, G_{i+1}}$$

$$\cong H_{j-1} \, / \, H_j$$

which is cyclic. Repeating the argument for each $i, 1 \leq i \leq r - 1$, we obtain the sequence of subgroups

$$\{e\} \subseteq \cdots \subseteq G_{r-1,1} \subseteq G_{r-1,0} \subseteq \cdots \subseteq G_{i+1} \subseteq \cdots G_{i+1,j} \subseteq \cdots G_{i+1,0}$$

$$= G_i \cdots \subseteq G_{0,1} \subseteq G_{0,0} = G$$

with the required properties. Thus (i) implies (ii).

(iii) *implies* (iv). We will show, by induction on $i$ that $G^{(i)} \subseteq G_i$, for all $i$.

Since $G/G_1$ is abelian, it follows from Theorem 4.8.2 that $G^{(1)} \subseteq G_1$ so that the claim is true for $i = 1$. Now suppose that the claim is true for $i$—that is, $G^{(i)} \subseteq G_i$. Then, $G_i / G_{i+1}$ is abelian so that, again by Theorem 4.8.2, $G_i^{(1)} \subseteq G_{i+1}$. Hence,

$$\begin{aligned} G^{(i+1)} &= [G^{(i)}, G^{(i)}] \\ &\subseteq [G_i, G_i] \qquad \text{(by the induction hypothesis)} \\ &= G_i^{(1)} \\ &\subseteq G_{i+1}. \end{aligned}$$

Therefore, by induction,

$$G^{(s+1)} \subseteq G_{s+1} = \{e\}.$$

(iv) *implies* (i). The chain

$$\{e\} = G^{(n)} \subseteq G^{(n-1)} \subseteq \cdots \subseteq G^{(1)} \subseteq G$$

has the required properties. $\square$

We conclude the section with one last result that requires volumes to justify and so is far beyond the scope of this book.

**Theorem 4.8.6 (Feit/Thompson Theorem, 1963)** *Every group of odd order is solvable.*

This chapter has been devoted, in large measure, to the study of normal subgroups. Of course, every nontrivial group $G$ has at least two normal subgroups: namely, $\{e\}$ and $G$. For some groups there are no others. If $G$ is a nontrivial group and the only normal subgroups of $G$ are $\{e\}$ and $G$, then $G$ is said to be *simple*. By the Feit/Thompson Theorem, the order of any finite simple nonabelian group must be even. The search for a characterization of all finite simple groups was, for many years, the Holy Grail of finite group theory. It was finally considered to be completed in 1983. The proof involved an estimated 15,000 pages in 500 journal articles involving approximately 100 authors. Not surprisingly, it has sometimes been referred to as the *enormous theorem.*

**Exercises 4.8**

1. Find the derived subgroup of each of the following groups:

   (i) Any abelian group $G$.
   (ii) $S_3$.
   (iii) $A_4$.
   (iv) $S_4$.
   (v) $D_4$.
   (vi) $Q_8$.
   (vii) $D_n$.

2. Show that all groups listed in Exercise 1 are solvable.

3. Let $H$ be a subgroup of a solvable group $G$. Show that $H$ is solvable.

4. Let $G$ be a solvable group and $\varphi : G \to H$ be an epimorphism. Show that $H$ is also solvable.

5. Let $G$ and $H$ be solvable groups. Show that $G \times H$ is solvable.

6. Let $G$ be a nontrivial, simple abelian group. Show that there exists a prime $p$ with $G \cong \mathbb{Z}_p$.

## 4.9 Nilpotent Groups

We have seen in sections 4.6 and 4.7 that the ability to show that a finite group is the direct product of its Sylow subgroups can be a major step in describing the structure of the group. The class of finite groups with precisely that property will be the focus of this section.

By a *central series* in a group $G$ we mean a series of subgroups

$$\{1\} = G_n \subseteq G_{n-1} \subseteq G_{n-2} \subseteq \cdots \subseteq G_0 = G \tag{4.6}$$

such that

   (i) $G_i$ is normal in $G$ for all $i = 0, \ldots, n$
   (ii) $G_i / G_{i+1} \subseteq Z(G/G_{i+1})$ for $i = 0, 1, \ldots, n-1$.

The *length* of the series in (4.6) is $n$. If the group $G$ has a central series as in (4.6), then it is called a *nilpotent group* and is said to be of *class n* if $n$ is the length of the shortest such series.

Note that $G$ is nilpotent of class 1 if and only if it is abelian and that, from their respective definitions, it is evident that a nilpotent group must be solvable. Thus,

$$\text{abelian} \implies \text{nilpotent} \implies \text{solvable}.$$

Before coming to the main results concerning nilpotent groups, we need to do some preparatory work with "commutator" subgroups.

For any subsets $X, Y$ of a group $G$, let

$$[X, Y] = \langle [x, y] \mid x \in X, \ y \in Y \rangle.$$

Thus, $[X, Y]$ is the subgroup generated by all the commutators that can be formed using an element from $X$ and an element from $Y$. In particular, we should note that

$$X \subseteq Z(G) \implies [X, Y] = \{1\}$$

since $[x, y] = 1$ for all $x \in Z(G)$.

**Lemma 4.9.1** *Let $N$ and $H$ be subgroups of $G$ with $N \subseteq H \subseteq G$ and $N$ normal in $G$.*

(i) $H/N \subseteq Z(G/N) \iff [H, G] \subseteq N.$
(ii) *If $H$ is also normal in $G$, then $[H, G] \subseteq H$ and $[H, G]$ is normal in $G$.*

**Proof.** (i) We have that

$$\begin{aligned}
H/N \subseteq Z(G/N) &\iff aN \cdot hN = hN \cdot aN \\
&\qquad \text{for all } h \in H, \ a \in G \\
&\iff ahN = haN \\
&\qquad \text{for all } h \in H, \ a \in G \\
&\iff N = h^{-1}a^{-1}haN \\
&\qquad \text{for all } h \in H, \ a \in G \\
&\iff h^{-1}a^{-1}ha \in N \\
&\qquad \text{for all } h \in H, \ a \in G \\
&\iff [H, G] \subseteq N.
\end{aligned}$$

(ii) Now suppose that $H$ is a normal subgroup of $G$. Let $h \in H$ and $g, x \in G$. Then, by the normality of $H$,

$$[h, g] = h^{-1}g^{-1}hg = h^{-1}(g^{-1}hg) \in H.$$

Therefore, $[H, G] \subseteq H$. Also,

$$
\begin{aligned}
x[h, g]x^{-1} &= x(h^{-1}g^{-1}hg)x^{-1} \\
&= (xhx^{-1})^{-1}(xgx^{-1})^{-1}(xhx^{-1})(xgx^{-1}) \\
&= a^{-1}b^{-1}ab \\
&= [a, b]
\end{aligned}
$$

where

$$
a = xhx^{-1} \in H \quad \text{and} \quad b = xgx^{-1} \in G.
$$

Thus,

$$
x[h, g]x^{-1} \in [H, G]
$$

for all $h \in H$, $g \in G$. Therefore $[H, G]$ is normal in $G$. $\quad\square$

A minor difficulty with the definition of a nilpotent group is that it depends on the existence of a series of subgroups with a certain property with no indication as to how one might find such a series. The following will help to resolve that difficulty.

For any group $G$, the following series of subgroups is known as the *lower central series*:

$$
Z^1(G) = G, \ Z^2(G) = [G, G], \ Z^3(G) = \Big[[G, G], G\Big] =
$$

$$
[Z^2(G), G], \ldots, Z^{n+1}(G) = [Z^n(G), G].
$$

Clearly,

$$
Z^1(G) \supseteq Z^2(G) \supseteq Z^3(G) \supseteq \cdots .
$$

In general, this series can be infinite, it can stop at some nontrivial subgroup, or it can stop when it reaches the trivial subgroup.

**Lemma 4.9.2** *Let $N$ be a normal subgroup of a group $G$ and $n \in \mathbb{N}$.*

   (i) *$Z^n(G)$ is a normal subgroup of $G$.*
   (ii) *$Z^n(G) / Z^{n+1}(G) \subseteq Z(G/Z^{n+1}(G))$.*
   (iii) *$Z^n(G)N/N \subseteq Z^n(G/N)$.*

**Proof.** (i) Since $Z^1(G) = G$, the claim is true trivially for $n = 1$. Arguing by induction, let us assume that the claim is true for $n$—that is, that $Z^n(G)$ is normal in $G$. Then

$$Z^{n+1}(G) = [Z^n(G), G]$$

and, by Lemma 4.9.1 (ii), $Z^{n+1}(G)$ is a normal subgroup of $G$. Part (i) now follows by induction.

(ii) By definition, $[Z^n(G), G] \subseteq Z^{n+1}(G)$ and so, by Lemma 4.9.1 (i), we have

$$Z^n(G) \,/\, Z^{n+1}(G) \subseteq Z(G/Z^{n+1}(G)).$$

(iii) The assertion is trivially true for $n = 1$ and it is possible to proceed by induction. We leave the details for you as an exercise. $\square$

We can now describe one possible approach to testing whether a group is nilpotent.

**Theorem 4.9.3** *Let $G$ be a group. Then $G$ is nilpotent if and only if there exists an integer $n$ such that $Z^n(G) = \{1\}$.*

**Proof.** First suppose that $n$ is an integer such that $Z^n(G) = \{1\}$. Then, by Lemma 4.9.2,

$$\{1\} = Z^n(G) \subseteq Z^{n-1}(G) \subseteq \cdots \subseteq Z^1(G) = G$$

is a central series. Therefore, $G$ is nilpotent.

Conversely, suppose that $G$ is nilpotent. Then there exists a central series

$$\{1\} = G_n \subseteq G_{n-1} \subseteq \cdots \subseteq G_0 = G$$

as in (4.6). We claim that, for each $i \geq 1$, $Z^i(G) \subseteq G_{i-1}$. For $i = 1$, we have

$$Z^1(G) = G = G_0 = G_{1-1}.$$

Thus, the claim is true for $i = 1$. So suppose that the claim is true for $i$—that is, that $Z^i(G) \subseteq G_{i-1}$. Then, since $\{G_i\}$ is a central series,

$$G_{i-1} \,/\, G_i \subseteq Z(G/G_i).$$

By Lemma 4.9.1 (i), this implies that

$$[G_{i-1}, G] \subseteq G_i. \tag{4.7}$$

Therefore,

$$Z^{i+1}(G) = [Z^i(G), G] \qquad \text{(definition)}$$
$$\subseteq [G_{i-1}, G] \qquad \text{(induction hypothesis)}$$
$$\subseteq G_i \qquad \text{(by (4.7))}.$$

Thus, the claim is true for $i + 1$. Hence, by induction, the claim holds. In particular, with $i = n + 1$, we obtain

$$Z^{n+1}(G) \subseteq G_n = \{1\}$$

so that $Z^{n+1}(G) = \{1\}$. $\qquad \square$

The next result provides some indication of the complexity of the class of finite nilpotent groups.

**Lemma 4.9.4** *Every finite p-group is nilpotent.*

**Proof.** Let $G$ be a finite $p$-group, say $|G| = p^\alpha$ where $p$ is a prime and $\alpha \in \mathbb{N}$. We argue by induction on $\alpha$. If $\alpha = 1$, then $|G| = p$ so that $G$ is cyclic and abelian and, therefore, nilpotent. So suppose that the claim is true for any group $H$ with $|H| = p^\beta$ where $\beta < \alpha$.

By the class equation for $G$, we know that $G$ has a nontrivial center $Z(G)$. If $Z(G) = G$, then $G$ is abelian and therefore nilpotent. So suppose that $\{1\} \subsetneq Z(G) \subsetneq G$. Then $|G/Z(G)| = p^\beta$ where $\beta < \alpha$. Therefore, by the induction hypothesis, $G/Z(G)$ is nilpotent. Hence, by Theorem 4.9.3, there exists an integer $n$ such that $Z^n(G/Z(G)) = \{1\}$. Lemma 4.9.2 (iii) tells us that

$$Z^n(G) Z(G)/Z(G) \subseteq Z^n(G/Z(G)) = \{1\}$$

which implies that

$$Z^n(G)Z(G) \subseteq Z(G)$$

and therefore that $Z^n(G) \subseteq Z(G)$. Hence,

$$Z^{n+1}(G) = [Z^n(G), G]$$
$$\subseteq [Z(G), G]$$
$$= \{1\}.$$

Therefore, $G$ is nilpotent. $\qquad \square$

One consequence of Lemma 4.9.4 is that every Sylow $p$-subgroup of a finite group is nilpotent, so we now take a closer look at Sylow $p$-subgroups.

**Lemma 4.9.5** *Let $P$ be a Sylow $p$-subgroup of a finite group $G$. Then $N_G(P)$ is its own normalizer—that is, $N_G(N_G(P)) = N_G(P)$.*

**Proof.** Let $g \in G$. Then,

$$g \in N_G(N_G(P)) \implies g N_G(P) g^{-1} = N_G(P)$$

$$\implies g P g^{-1} \subseteq g N_G(P) g^{-1} \subseteq N_G(P)$$

$$\implies g P g^{-1} \text{ is a Sylow } p\text{-subgroup of } N_G(P)$$

$$\implies g P g^{-1} \text{ is conjugate to } P \text{ in } N_G(P) \text{ (by Theorem 4.6.7)}$$

$$\implies g P g^{-1} = a P a^{-1} \text{ for some } a \in N_G(P)$$

$$\implies g P g^{-1} = P \text{ since } a \in N_G(P)$$

$$\implies g \in N_G(P).$$

Therefore, $N_G(N_G(P)) \subseteq N_G(P)$ whereas the reverse containment is obvious. $\square$

It is possible for a proper subgroup to be equal to its own normalizer (see the exercises). However, this will never happen in a nilpotent group.

**Lemma 4.9.6** *Let $H$ be a proper subgroup of the nilpotent group $G$. Then $H$ is also a proper subgroup of $N_G(H)$.*

**Proof.** Since $G$ is nilpotent, there exists an integer $n \in \mathbb{N}$ with $Z^n(G) = \{1\}$. Since $Z^1(G) = G$, it follows that there must exist an integer $i$ with

$$Z^{i+1}(G) \subseteq H \quad \text{but} \quad Z^i(G) \not\subseteq H.$$

Then,

$$[Z^i(G), H] \subseteq [Z^i(G), G] = Z^{i+1}(G) \subseteq H.$$

Therefore, for any $g \in Z^i(G)$, $h \in H$ we have

$$g^{-1} h g = g^{-1} h g h^{-1} h$$

$$= [g, h^{-1}] h$$

$$\in [Z^i(G), H]h$$

$$\subseteq Hh$$

$$= H.$$

Hence, $Z^i(G) \subseteq N_G(H)$ so that $H$ is a proper subgroup of $N_G(H)$. $\quad\square$

We are now ready for the characterization of finite nilpotent groups in terms of their Sylow subgroups.

**Theorem 4.9.7** *Let $G$ be a finite group with $|G| = p_1^{\alpha_1} \cdots p_k^{\alpha_k}$ where the $p_i$ are distinct primes. Then the following statements are equivalent:*

   (i)  *$G$ is nilpotent.*
   (ii)  *For each $i$, $G$ has a unique Sylow $p_i$-subgroup, $P_i$ say.*
   (iii)  *Every Sylow subgroup of $G$ is normal in $G$.*
   (iv)  *$G \cong P_1 \times P_2 \times \cdots \times P_k$ where, for each $i$, $P_i$ is the unique Sylow $p_i$-subgroup of $G$.*

**Proof.** (i) *implies* (ii). Let $P$ be any Sylow subgroup of $G$. By Lemma 4.9.5, we have

$$N_G(N_G(P)) = N_G(P).$$

But, by Lemma 4.9.6, we know that $N_G(P)$ is a proper subgroup of $N_G(N_G(P))$ unless $N_G(P) = G$. Hence, to avoid a contradiction, we must have

$$N_G(P) = G.$$

Thus every Sylow subgroup of $G$ is normal and therefore unique.
   (ii) and (iii) are equivalent. This follows immediately from Corollary 4.6.8.
   (ii) *implies* (iv). This follows immediately from Theorem 4.7.5.
   (iv) *implies* (i). This is left for you as an exercise. $\quad\square$

**Exercises 4.9**

1. Which of the following groups are nilpotent?

   (i)  $\mathbb{Z}_n$.
   (ii)  $S_3$.
   (iii)  $Q_8$.
   (iv)  $D_4$.
   (v)  $D_5$.

2. Find the normalizer of each subgroup in $S_3$.

3. Show that every subgroup of a nilpotent group is nilpotent.

4. Show that any homomorphic image of a nilpotent group is nilpotent.

5. Let $P_1, \ldots, P_n$ be nilpotent groups. Show that $P_1 \times P_2 \times \cdots \times P_n$ is nilpotent.

## 4.10 The Enigma Encryption Machine

Recall that a substitution cipher is one in which the encoding process is determined by a permutation of the letters of the alphabet. Such a scheme, based on a single permutation, applied to a complete message, would be considered extremely insecure these days. However, there was one fairly recent and historically very important substitution cipher that was based on refinements of this idea. It was called the *Enigma*. The scheme was designed by German engineer Arthur Scherbius at the end of World War I, around 1918. The basic model was used commercially and by the German armed forces between the two world wars. Various cryptographic improvements were introduced by the German armed forces prior to and during the World War II.

The cipher was implemented by means of a very ingenious machine (the *Enigma Machine*) that consisted of a mechanical keyboard, a plugboard, an entry wheel, three (sometimes four) rotors, a reflection board, and a lamp board together with a battery and some wired circuits that connected the components. Figure 4.1 contains a photograph of an Enigma Machine with three rotors. The entry wheel, rotors, and reflection board were on the one shaft or axle and touching each other via pins and contact plates. The 26 letters

**Figure 4.1** Enigma Machine This is a standard German military (3-wheel, with plugboard) Enigma machine, from the Collection in the Museum of the Government Communications Headquarters (GCHQ), the United Kingdom's Signals Intelligence Agency and is used by the permission of the Director, GCHQ; UK Crown Copyright Reserved.

of the alphabet were equally spaced round the rim of each rotor, and there was a small window for each rotor through which one letter of the alphabet was visible. The entry and reflection boards were fixed, but the rotors could rotate through the 26 positions corresponding to the letters of the alphabet, and whichever letter was visible on a rotor then identified the position of that rotor. The order in which the three rotors were positioned on the shaft was variable, with six possible arrangements in all.

Each rotor had 26 small contact plates on one side and 26 spring loaded brass pins on the other, with the pins of one rotor contacting the plates of the neighboring rotor. Each plate and pin on a rotor corresponded to a letter of the alphabet and internal wiring in each rotor connected the plates to the pins in a scrambled manner that was different for each rotor. In that way, an electrical current entering the $a$ plate, for instance, might flow to the $k$ pin, so that each rotor implemented a substitution cipher.

The plugboard implemented a simple substitution cipher consisting of a series of as many as 13, but usually 6–10, pairings of letters. If $f$ was paired with $q$, then the plugboard would replace $f$ by $q$ and $q$ by $f$. Some letters would remain unpaired and therefore be transmitted to the next stage unchanged. Thus viewed as a permutation $P$, the plugboard consisted of a product of disjoint 2-cycles. The actual pairings in the plugboard were not fixed, but were set by the operator each day simply by plugging in (hence the name "plugboard") connecting cables between the sockets corresponding to each set of paired letters, rather like an old-fashioned telephone switchboard. The introduction of the plugboard complicated the task of the cryptanalyst as it increased the total number of available encryption codes by an enormous factor.

The contribution of the reflection board was twofold. First it applied a substitution cipher consisting, like the plugboard, of a series of pairings of letters, but here the pairings were fixed and did not vary from day to day. Also the pairings included the whole alphabet. So the reflection board could be described as a permutation $M$ consisting of a product of 13 disjoint 2-cycles. The second function of the reflection board was to act a bit like a mirror and send the resulting letter back through the rotors and plug board in reverse order. In the early commercial versions, the entry wheel also involved an encryption based on the arrangement of keys on a keyboard: $a$ to $q$, $b$ to $w$, $c$ to $e$, and so on. So it was natural to assume that the military version would also involve an encrypting entry wheel. This caused quite some delay in the analysis of the Enigma encryption until it was realized that the entry wheel in the military version involved no encryption.

Since a composition of substitution ciphers is still a substitution cipher, the entry wheel, three rotors, and reflection plate all constituted one complex substitution cipher. What moved this beyond being just a simple substitution cipher was the fact that the overall configuration was constantly being changed. The starting positions of the plug board and rotation wheels were determined each day by a code book that was distributed monthly. Then the position of the rotors would change with *each letter* so that successive letters were encoded according to different encryption schemes. This constant change was achieved in a manner similar to that in which the digits change in an odometer in a car. The first rotor rotated one twenty-sixth of a revolution with every letter typed, the second rotor would then turn one twenty-sixth after each full rotation of the first rotor and the third rotor would turn one twenty-sixth after each full rotation of the second rotor. However, this pattern was not followed exactly, since there was also a feature that resulted in the second and third rotors occasionally moving through two twentysixths, instead of just one twentysixth of a revolution. We will ignore this added twist in our discussion.

In addition, in later versions, the three or four rotors would be chosen each day from a set of six, seven or eight available rotors. Since the three chosen rotors could be placed in any order, even once the daily choice of rotors and the configuration of plugboard were known, the rotors themselves provided $6 \times 26 \times 26 \times 26 = 105,456$ different encryption schemes so that a message would have to contain over 105,456 letters before the same scheme would be used twice.

So, the parts that contributed to the encryption were:

the plugboard: $P$
the rotors: $R_1, R_2, R_3$
the reflection board:$M$.

Any letter entered on the keyboard would be encrypted successively by the plug board, the rotors in the order $R_1, R_2, R_3$, followed by the reflection board which would send the result back through the rotors in reverse order, $R_3, R_2, R_1$, then to the plugboard once more, at the end of which a bulb would light up on the lampboard (there being one bulb for each letter of the alphabet) indicating the final encrypted output.

Thus the basic encryption, omitting for the moment the rotating feature of the rotors, was obtained as the composition of nine individual encryptions and can be represented as the product of the corresponding permutations as

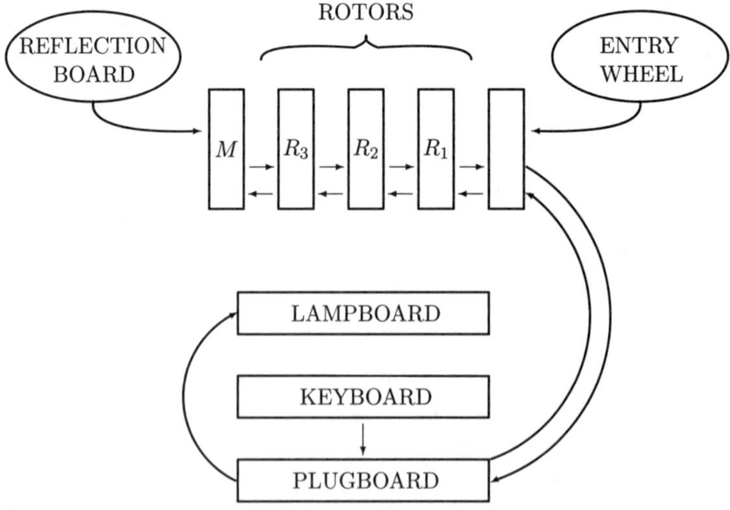


**Figure 4.2** Enigma encryption scheme

(where, since we write our functions on the left, the product should be read from right to left):

$$E = P^{-1}R_1^{-1}R_2^{-1}R_3^{-1}MR_3R_2R_1P.$$

The action of the reflecting board gave the Enigma scheme three interesting properties. First, the decryption process was exactly the same as the encryption process. In other words, for any letter $x$, $EE(x) = x$, that is, as a permutation, $E$ is of order two. That meant that the operator could use the same settings for both sending and receiving messages. It also meant that, if the Enigma machine encrypted $a$ as $t$, then it encrypted $t$ as $a$ and so on. In addition, no letter would ever be left fixed by the encryption process: $E(x) \neq x$, for all letters $x$. In other words, E itself, as a permutation, had to be a product of 13 disjoint 2-cycles. These latter two features proved to be a serious weakness in the security of the Enigma encryption as they imposed some structure on the encryption process that aided cryptanalysis and eliminated many possible configurations from the calculations of the Polish and British cryptanalysts.

Polish mathematicians, led by M. Rejewski, were able to determine the wiring of the first three rotors prior to the commencement of World War II. We will now consider some of the important mathematical features that made this possible.

In order to see the effect of rotating the first rotor through 1/26th of a revolution, consider first the effect on the encryption of the letter $a$ in the initial position. Suppose that $R_1(a) = p$, $R_1(z) = t$. Then the action of the first rotor is illustrated in Figure 4.3.

After $R_1$ is rotated through 1/26th of a rotation clockwise, the situation changes to that illustrated in Figure 4.4.

Let $S = (abc\ldots yz)$, the permutation consisting of one cycle of length 26. Then

$$a \to u = S(t) = SR_1(z) = SR_1S^{-1}(a).$$

Thus the first rotor now encrypts $a$ as $SR_1S^{-1}(a)$. Clearly the same reasoning applies to any letter, so that after a rotation through 1/26th of a revolution

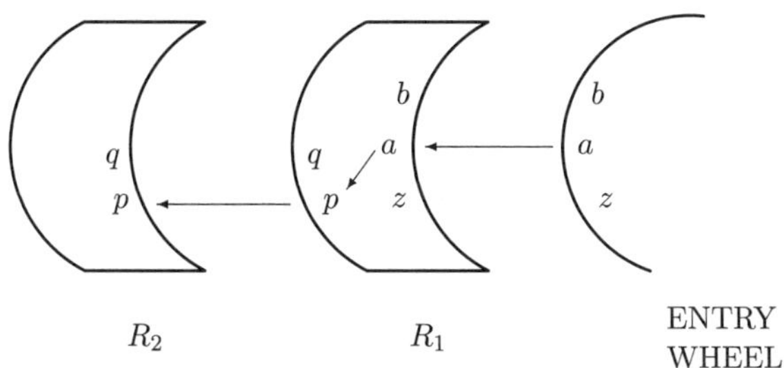


**Figure 4.3** Encryption by $R_1$ before rotation

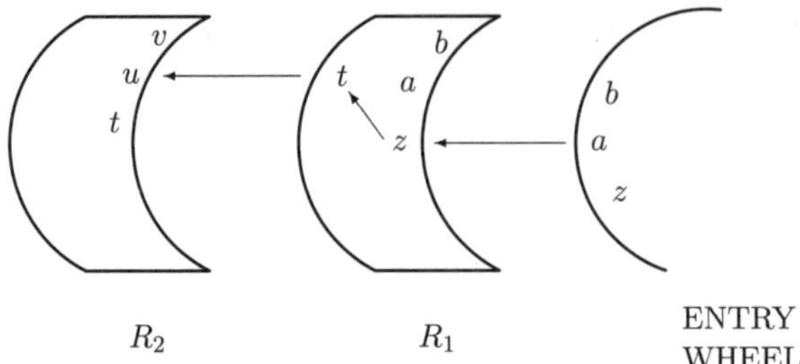


**Figure 4.4** Encryption by $R_1$ after rotation

the first rotor now encrypts the whole alphabet via $SR_1S^{-1}$. Repeating the argument, we see that if the first rotor has been rotated through 1/26ths of a revolution, then the resulting encryption from the first rotor will be $S^iR_1S^{-i}$.

Extending the above argument to all the rotors, assuming that at some point in a transmission the rotors had been moved respectively $i, j$ and $k$ times, then the resulting encryption for that letter would be given by:

$$E_{ijk} = P^{-1}S^iR_1^{-1}S^{-i}S^jR_2^{-1}S^{-j}S^kR_3^{-1}S^{-k}MS^kR_3S^{-k}S^jR_2S^{-j}S^iR_1S^{-i}P.$$

A protocol introduced by the German armed forces turned out to be the System's Achilles heel. By this protocol, the Enigma operators would choose their own starting position for the rotors. The problem was that they had to communicate to their recipients what that starting position would be. That was done by sending the three letters representing the starting position of the three rotors *twice*. So if the starting position was going to be *abc*, then the operator would send *abcabc* (which we will refer to as a *prefix*) encoded using the prescribed starting position for that day, according to the code book. The message that followed would then be sent using *abc* as the starting positions for the three rotors. The objective in sending the prefix with the duplication of the three letters was to ensure that they were properly received. The unintended consequence was the introduction of a fatal weakness in the procedure.

All the transmitting stations would be using the same standard position for the transmission of these first six letters. By piecing together the transmissions from many different stations on a given day, Rejewski could see the resulting encryptions of many different letters being repeated. Let us denote the resulting encryptions that were being applied to these first six letters by $A$, $B, C, D, E,$ and $F$.

As before, each of these encryptions is of order two. If the operator was encrypting *abcabc*, then the transmission would consist of $A(a)$, $B(b)$, $C(c)$, $D(a)$, $E(b)$, $F(c)$. Suppose now that there was another prefix from another station that *began* with $D(a)$. Then that must come from a prefix of the form *pqrpqr* where $A(p) = D(a) = DA(A(a))$. Now add a third prefix *uvwuvw* starting with $A(u) = D(p)$. We now have $A(u) = D(p) = DA(A(p)) =$

$DA(DA(A(a)))$. Thus we are building up the cycle in the permuation $DA$ starting with $A(a) : (A(a)\,DA(A(a))\,(DA)^2(A(a))\ldots)$. With enough messages to work with on a given day, Rejewski was able to determine, from the encoded prefixes, all the cycles of the permutations $DA$, $EB$, and $FC$.

The next step was to extract the individual permutations $A$, $B$, $C$ and so on from these products. As it turned out, finding just one correct encryption $A(x)$ of any letter $x$ could provide a great deal of information about $A, B, C, \ldots$ Recall that the permutations $A, B, C, \ldots$ are all products of 13 disjoint transpositions. So, if $A(a_1) = b_1$, for instance, then that means that $(a_1\,b_1)$ is one of the transpositions in $A$. It also means that $a_1$ and $b_1$ belong to different cycles in $DA$ (see exercises). Let the disjoint cycles in $DA$ containing $a_1$ and $b_1$ be

$$\alpha = (a_1 a_2 \ldots a_m), \quad \beta = (b_1 b_2 \ldots b_n)$$

respectively. Then, for any $x = a_i$ or $b_j$, $DA(x) = \beta\alpha(x) = \alpha\beta(x)$. Hence

$$D(a_1) = DA(b_1) = \alpha\beta(b_1) = b_2.$$

Since $D$ is a product of disjoint 2-cycles, $(a_1\,b_2)$ must be one of those 2-cycles. Furthermore,

$$DA(a_m) = \alpha\beta(a_m) = a_1 = D(b_2) = DA(A(b_2))$$

which, since $DA$ is a permutation, implies that $A(b_2) = a_m$, so that $(a_m b_2)$ is a cycle of $A$. Continuing in this way,

$$D(a_m) = D(A(b_2)) = \alpha\beta(b_2) = b_3$$

and

$$DA(A(b_3)) = D(b_3) = a_m = \alpha\beta(a_{m-1}) = DA(a_{m-1})$$

so that $(a_m b_3)$ is a 2-cycle in $D$ and $(a_{m-1} b_3)$ is a 2-cycle in $A$. Thus, the pattern emerges and we see that if we just line up the cycle $\alpha$ above the inverse of the cycle $\beta$ and line up $a_1$ with $b_1$ for $A$ and $a_1$ with $b_2$ for $D$ then we can read off all the other cycles in $A$ and $D$. In particular, we see that the cycles $\alpha$ and $\beta$ must have the same length, $m = n$. Thus writing

$$(a_1 a_2 \ldots a_m) \quad (a_1 a_2 \ldots a_m)$$
$$(b_1 b_m \ldots b_2) \quad (b_2 b_1 \ldots b_3)$$

we can read off

$$A = (a_1 b_1)(a_2 b_m)\ldots(a_m b_2)$$
$$B = (a_1 b_2)(a_2 b_1)\ldots(a_m b_3)$$

That means that the cycles in $DA$ must pair off into cycles of equal length and to find the individual permutations $A$ and $D$, it suffices to find one single value of $A$ in each pair of cycles of $DA$. The same approach applies to the permutations $B$, $C$, $D$, $E$, and $F$. The key to this was Rejewski's intuition that certain prefixes, such as *aaaaaa*, *abcabc*, possible names of girl friends and so on would be especially popular. A combination of trial and error and intelligent guessing yielded up the values of $A$, $B$, $C$, $D$, $E$, and $F$ and therefore the settings for the day.

Satisfying as this must have been, it was only the first hurdle. Knowing only the permutations $A, \ldots, F$ would not enable the decryption of whole messages. For that it would still be necessary to find the values of $P$, $R_1$, $R_2$, $R_3$, and $M$. Rejewski knew the value of

$$A = P^{-1}SR_1^{-1}S^{-1}R_2^{-1}R_3^{-1}MR_3R_2SR_1S^{-1}P.$$

French espionage had managed to obtain the standard daily settings (including the value of $P$) for a period of two months in 1932 and had given them to the Polish secret service. Hence the value

$$U = R_1^{-1}S^{-1}R_2^{-1}R_3^{-1}MR_3R_2SR_1 = S^{-1}PAP^{-1}S$$

was also now known. Similarly, the following permutations could be calculated from B, C and D, respectively:

$$V = R_1^{-1}S^{-2}R_2^{-1}R_3^{-1}MR_3R_2S^2R_1$$
$$W = R_1^{-1}S^{-3}R_2^{-1}R_3^{-1}MR_3R_2S^3R_1$$
$$X = R_1^{-1}S^{-4}R_2^{-1}R_3^{-1}MR_3R_2S^4R_1.$$

Now let

$$T = R_1^{-1}SR_1 \tag{4.8}$$

Then it was quite straightforward to calculate that

$$WV = T^{-1}(VU)T \quad and \quad XW = T^{-1}(WV)T. \tag{4.9}$$

From the knowledge of $U$, $V$, $W$ and $X$ it was possible to calculate the permutations $VU$, $WV$ and $XW$. It was at this point that Theorem 4.5.2 played a critical role. Using the technique in Theorem 4.5.2, it was possible to extract the value of $R_1$ from these equations. In order to solve the above equations, it is sufficient to be able to find, for any two conjugate permutations $\rho, \sigma$ in $S_n$ a permutation $\gamma$ such that $\rho = \gamma^{-1}\sigma\gamma$. The proof of Theorem 4.5.2 tells us exactly how to do that. Since the permutations $\rho$ and $\sigma$ are conjugate, they have the same cycle pattern and so we may write them one above the other in such a way that cycles that are above each other are of equal length. We can then read off a conjugating element. However, since a cycle can be written in many different ways and there may be more than one cycle of the same length, there was still the question of finding the correct conjugating element. At the first step, solving (4.9), there are two equations and so the solution had to work for both equations and, it turned out, that that sufficed to guarantee a unique solution. Then in order to complete the final step of solving (4.8) for $R_1$, there are only 26 possible ways to position $S$ over $T$, since $S$ is a permutation with just one cycle of length 26. So trial and error would suffice, although Rejewski claimed to have some additional tricks that would reduce the work involved at this stage.

By a stroke of good luck, the daily starting configurations provided by the French covered a period where the positioning of the three rotors was changed. Consequently the above methods could be employed again to provide the wiring/encoding of a second rotor and from the knowledge of two of the rotors it was possible to deduce the wiring of the third and the reflection board.

The Polish mathematicians passed this information on to the British on the eve of the war. From their center at Bletchley Park, the British then mounted a continual and growing attack on German communications throughout the war. It is claimed that the success of these efforts shortened the war by as many as two years. This work was led by Alan Turing, who is renowned for his fundamental contributions to the theory of computing and is known to many as the father of computing. Despite suspicions of security leaks, the Germans never doubted the security of the Enigma. Their confidence appears to have been based on the enormous number (approximately $3 \times 10^{114}$, see [Mil]) of possible configurations of the various settings. For details on the contributions of the various components in the Enigma machine to the total number of possible configurations, see [Mil].

For background on the vital Polish contribution to the cracking of the code and the critical role played by the theory of permutations, see [Rej]. For more information on the history of the Enigma machine, the workings of Bletchley Park and Turing's life, see [Hod].

The author gratefully acknowledges having consulted with the GCHQ Historian regarding the history of the Enigma. This and the provision of the photograph in Figure 4.1 should not be taken as confirming or refuting any aspect, nor the completeness, of the analysis of the mathematics of the Enigma machine provided here.

## Exercises 4.10

1. Prove that, in the Enigma encryption scheme, the decryption function is the same as the encryption function.

2. Prove that the Enigma encryption scheme encrypts every letter to a different letter.

3. How many possible configurations are there of the plugboard in an enigma machine assuming that there are 7 pairings?

4. Show that the Enigma encryption E can be represented as a product of disjoint 2-cycles.

5. Let $\lambda, \mu, \nu, \rho, \sigma \in S_6$, be such that

$$\lambda = (13)(25)(46)$$
$$\mu = (16)(24)(35)$$
$$\nu = (15)(26)(34)$$

$\sigma = (1\,2\,3\,4\,5\,6)$ and $\rho(1) = 3$. Now solve the equations:

$$\mu = \tau^{-1}\lambda\tau, \quad \nu = \tau^{-1}\mu\tau, \quad \tau = \rho^{-1}\sigma\rho$$

for $\tau$ and $\rho$.

*6. Let $\alpha, \beta \in S_{2m}$, where both $\alpha$ and $\beta$ are products of $m$ disjoint transpositions:

$$\alpha = (a_1 a_2)(a_3 a_4) \cdots (a_{m-1} a_m)$$
$$\beta = (b_1 b_2)(b_3 b_4) \cdots (b_{m-1} b_m).$$

Let $\gamma = \beta\alpha$. Show that, when $\gamma$ is expressed as a product of disjoint cycles, $a_1$ and $a_2$ lie in different cycles.

# 5

# Rings and Polynomials

In chapter 2 we encountered some of the basic properties of rings and fields. In particular, we considered the ring of polynomials in a single variable and saw how essential that theory is to the study of finite fields. In this chapter we return to the study of rings and polynomials, but this time we will be interested in some geometric aspects. After the introduction of the necessary ring theoretical concepts (such as an ideal) we turn our attention to polynomials in several variables and to various properties of the curves and surfaces that they define.

## 5.1 Homomorphisms and Ideals

Let $(R, +, \cdot)$ and $(S, +, \cdot)$ be rings. A mapping $\varphi : R \to S$ is a (*ring*) *homomorphism* if, for all $a, b \in R$,

$$\varphi(a + b) = \varphi(a) + \varphi(b)$$

$$\varphi(a \cdot b) = \varphi(a) \cdot \varphi(b)$$

that is, $\varphi$ must respect both operations $+$ and $\cdot$. If $\varphi$ is surjective, injective, or bijective then it is an *epimorphism, monomorphism,* or *isomorphism,* respectively. If $R = S$, then $\varphi$ is an *endomorphism.* If $R = S$ and $\varphi$ is bijective, then $\varphi$ is an *automorphism.*

Let $\varphi : (R, +, \cdot) \to (S, +, \cdot)$ be a ring homomorphism. Then, necessarily, $\varphi : (R, +) \to (S, +)$ is a group homomorphism so that we should expect that an important feature of $\varphi$ will be its *kernel*:

$$\ker(\varphi) = \{a \in R \mid \varphi(a) = 0\}.$$

Obviously, the fact that a homomorphism respects the multiplication must be reflected in its kernel in some way. This leads to the following concept.

A subset $I$ of a ring $R$ is an *ideal* if it satisfies the following conditions:

(i)  $(I, +)$ is a subgroup of $(R, +)$.
(ii)  For all $r \in R$, $a \in I$, we have $ra, ar \in I$.

Clearly, $\{0\}$ and $R$ are ideals in any ring $R$. In addition, for any $m \in \mathbb{Z}$, $m\mathbb{Z}$ is an ideal in the ring $\mathbb{Z}$. Note that condition (ii) implies that an ideal $I$ will be closed under multiplication. Hence, an ideal is always a subring. The converse of this is false; a subring need not be an ideal. For example, $\mathbb{Z}$ is a subring of $\mathbb{Q}$ and $\mathbb{Q}$ is a subring of $\mathbb{R}$ but $\mathbb{Z}$ is not an ideal in $\mathbb{Q}$ and $\mathbb{Q}$ is not an ideal in $\mathbb{R}$. However, since any ideal must be a subring, the following condition can be substituted for condition (i):

(i)  $I$ is a subring of $R$.

Substituting the usual test for a subgroup in the definition of an ideal, we arrive at the following test for an ideal.

**Lemma 5.1.1** *Let $R$ be a ring and $I \subseteq R$. Then $I$ is an ideal if and only if it satisfies the following conditions:*

(i)  $I \neq \emptyset$.
(ii)  $a, b \in I \implies a - b \in I$.
(iii)  $a \in I, r \in R \implies ra, ar \in I$.

**Proof.** Exercise. $\square$

It is possible to combine ideals in simple ways to obtain new ideals. For any ideals $I, J$ in a ring $R$, let

$$I + J = \{a + b \mid a \in I, \ b \in J\}.$$

**Lemma 5.1.2** *Let $I, J$ be ideals in a ring $R$. Then $I \cap J$ and $I + J$ are both ideals in $R$.*

**Proof.** Exercise. $\square$

We can also generate ideals by picking elements from $R$ and then taking all possible combinations of those elements. For any nonempty subset $A$ of $R$, let $\langle A \rangle$ denote the set of all elements of the form

$$a_1 + a_2 + \cdots + a_k +$$
$$r_{k+1}a_{k+1} + r_{k+2}a_{k+2} + \cdots + r_l a_l +$$
$$a_{l+1}r_{l+1} + \cdots + a_m r_m +$$
$$r_{m+1}a_{m+1}s_{m+1} + \cdots + r_n a_n s_n$$

where $a_i \in A$, $r_i, s_i \in R$, for all $i$. If $A$ is finite, say, $A = \{a_1, \ldots, a_m\}$, then we also write $\langle a_1, \ldots, a_m \rangle$ for $\langle A \rangle$. It is also convenient to extend the definition of $\langle A \rangle$ to include $A = \emptyset$ by defining $\langle \emptyset \rangle = \{0\}$.

**Lemma 5.1.3** *Let $R$ be a ring and $A \subseteq R$. Then $\langle A \rangle$ is an ideal in R and is the smallest ideal of R containing A.*

**Proof.** Exercise. $\square$

We refer to $\langle A \rangle$ as the *ideal generated by* $A$ (or the *ideal generated by* $a_1, \ldots, a_m$ if $A = \{a_1, \ldots, a_m\}$). Moreover, if $I$ is an ideal and there exist $a_1, \ldots, a_m \in I$ such that

$$I = \langle a_1, \ldots, a_m \rangle$$

then we say that $I$ is a *finitely generated ideal.*

If $R$ is a commutative ring, then the description of $\langle A \rangle$ simplifies considerably, as it will consist of all elements of the form

$$a_1 + \cdots + a_k + r_{k+1}a_{k+1} + \cdots + r_l a_l \qquad (a_i \in A, \ r_i \in R).$$

If $R$ is commutative and has an identity, then the description of $\langle A \rangle$ becomes even simpler as $\langle A \rangle$ will consist of all elements of the form

$$r_1 a_1 + \cdots + r_k a_k \qquad (a_i \in A, \ r_i \in R).$$

Finally, if $R$ is a commutative ring with identity and $a \in R$, then

$$\langle a \rangle = \{ra \mid r \in R\}.$$

An ideal of the form $\langle a \rangle$ is called a *principal ideal*. If $R$ is an integral domain and every ideal in $R$ is principal, then $R$ is a *principal ideal domain*. The most familiar example of a principal ideal domain is $\mathbb{Z}$. Another example is $F[x]$ for any field $F$. Any field $F$ is a principal ideal domain in a trivial sort of way since the only ideals in $F$ are $F = \langle 1 \rangle$ and $\{0\} = \langle 0 \rangle$.

Since an ideal $I$ of a ring $R$ is also a subgroup of the abelian group $(R, +)$, we can form the quotient group $(R/I, +)$ which consists of the cosets of $(I, +)$ in $(R, +)$. The fact that $I$ is an ideal makes it possible to define a further operation on $R/I$ as follows: For any $a, b \in R$, we define

$$(I + a) \cdot (I + b) = I + ab. \tag{5.1}$$

Suppose that $I + a = I + a'$ and $I + b = I + b'$. Then there must exist elements $x, y \in I$ with

$$a' = x + a, \qquad b' = y + b$$

so that

$$\begin{aligned} a'b' &= (x + a)(y + b) \\ &= xy + xb + ay + ab \\ &\in I + ab \end{aligned}$$

since $I$ is an ideal. Thus, $I + a'b' = I + ab$, and the operation in (5.1) is well defined.

**Lemma 5.1.4** *Let $I$ be an ideal in a ring $R$. Then $R/I$ is also a ring and the $0$ element in $R/I$ is $I$. If $R$ is commutative, then so also is $R/I$. If $R$ has an identity $1$, then so also does $R/I$—namely, $I + 1$.*

**Proof.** Exercise. $\square$

So we see more evidence that ideals play the role with respect to rings that normal subgroups play with respect to groups. Now $n\mathbb{Z}$ is an ideal in $\mathbb{Z}$ and $\mathbb{Z}/n\mathbb{Z}$ consists of the set of cosets of $n\mathbb{Z}$—that is, the sets of the form

$$\begin{aligned} n\mathbb{Z} + a &= \{a + nk \mid k \in \mathbb{Z}\} \\ &= [a]_n. \end{aligned}$$

Thus, $(\mathbb{Z}_n, +, \cdot) = (\mathbb{Z}/n\mathbb{Z}, +, \cdot)$. In a similar fashion, it follows that, for any $f(x) \in F[x]$,

$$(F[x]/f(x), +, \cdot) = (F[x]/\langle f(x)\rangle, +, \cdot).$$

where $F[x]/f(x)$ is as defined in chapter 2. The use of the ideal notation, $\langle f(x)\rangle$, on the right hand side of the above equation is, in fact, the more traditional and conventional notation.

**Lemma 5.1.5**  *Let R be a ring.*

(i)  *If $\varphi : R \to S$ is a ring homomorphism, then $\ker(\varphi)$ is an ideal in R.*

(ii)  *If I is an ideal in R, then the mapping $\pi : R \to R/I$ defined by*

$$\pi(r) = I + r \qquad (r \in R)$$

*is an epimorphism with kernel equal to I.*

**Proof.**  (i) Since $\varphi : (R, +) \to (S, +)$ is a group homomorphism, we know that $\ker(\varphi)$ is a subgroup of $(R, +)$. Now let $r \in R,\ a \in I$. Then,

$$\begin{aligned}
\varphi(ra) &= \varphi(r)\varphi(a) \\
&= \varphi(r) \cdot 0 \qquad \text{since } a \in I \\
&= 0.
\end{aligned}$$

Thus, $ra \in I$. Similarly, $ar \in I$ so that $I$ is an ideal.

(ii) Since $(R/I, +)$ is a quotient group of $(R, +)$, we know that $\pi$ is surjective and respects addition. For any $r, s \in R$,

$$\begin{aligned}
\pi(rs) = I + rs &= (I + r)(I + s) \\
&= \pi(r)\pi(s)
\end{aligned}$$

so that $\pi$ also respects multiplication and is a ring homomorphism. Moreover,

$$\begin{aligned}
r \in \ker(\varphi) &\Leftrightarrow \varphi(r) = 0 \\
&\Leftrightarrow I + r = I \\
&\Leftrightarrow r \in I
\end{aligned}$$

that is, $\ker(\varphi) = I$.  $\square$

The homomorphism $\pi : R \to R/I$ in Lemma 5.1.5 (ii) is known as the *natural homomorphism.*

**Theorem 5.1.6 (First Homomorphism Theorem for Rings)**  *Let R, S be rings and $\varphi : R \to S$ be a surjective ring homomorphism. Let $I = \ker(\varphi)$ and*

$\pi : R \to R/I$ *be the natural homomorphism. Then there exists an isomorphism* $\theta : R/I \to S$ *such that* $\theta\pi = \varphi$:

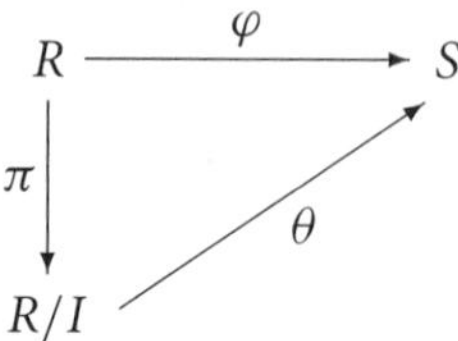

**Proof.** For any $r \in R$, define

$$\theta(I + r) = \varphi(r).$$

To see that this defines a mapping $\theta : R/I \to S$, let $s \in R$ be such that $I + r = I + s$. Then,

$$I + r = I + s \Rightarrow r - s \in I$$
$$\Rightarrow \varphi(r - s) = 0$$
$$\Rightarrow \varphi(r) - \varphi(s) = 0$$
$$\Rightarrow \varphi(r) = \varphi(s).$$

Thus, $\theta$ is well defined.

For any $r, s \in R$, we have

$$\theta((I + r) + (I + s)) = \theta(I + (r + s))$$
$$= \varphi(r + s)$$
$$= \varphi(r) + \varphi(s)$$
$$= \theta(I + r) + \theta(I + s)$$

so that $\theta$ respects addition. A similar argument shows that $\theta$ respects multiplication:

$$\theta((I + r)(I + s)) = \theta(I + rs)$$
$$= \varphi(rs)$$
$$= \varphi(r)\varphi(s)$$
$$= \theta(I + r) \cdot \theta(I + s).$$

Therefore, $\varphi$ is a homomorphism. For any $s \in S$, there exists an $r \in R$ with $\varphi(r) = s$, since $\varphi$ is surjective by the hypothesis. Hence,

$$s = \varphi(r) = \theta(I + r)$$

and $\theta$ is also surjective. Furthermore,

$$\theta(I + r) = \theta(I + s) \Rightarrow \varphi(r) = \varphi(s)$$
$$\Rightarrow \varphi(r - s) = 0$$
$$\Rightarrow r - s \in I$$
$$\Rightarrow I + r = I + s$$

so that $\varphi$ is also injective and therefore an isomorphism. Finally, for any $r \in R$, we have

$$\theta\pi(r) = \theta(I + r) = \varphi(r)$$

so that $\theta\pi = \varphi$. $\square$

We conclude this section with the observation that at least in some situations (namely, $\mathbb{Z}$ and $F[x]$) we know all the ideals.

**Theorem 5.1.7** (i) *Every ideal in $\mathbb{Z}$ is of the form $\langle n \rangle$, for some $n \in \mathbb{Z}$.*
  (ii) *Every ideal in $F[x]$ is of the form $\langle f(x) \rangle$, for some $f(x) \in F[x]$.*

**Proof.** This is a simple application of the division algorithms for $\mathbb{Z}$ and $F[x]$, and is left for you as an exercise. $\square$

**Exercises 5.1**

1. Show that the set of all noninvertible $2 \times 2$ matrices over a field $F$ is not a subring of $M_2(F)$.

2. Let $D$ denote the set of all diagonal $n \times n$ matrices with entries from $\mathbb{Q}$. Show that $D$ is a subring, but not an ideal, of $M_n(\mathbb{Q})$, the ring of all $n \times n$ matrices over $\mathbb{Q}$.

3. Let $F$ be a field. Prove that $F$ contains only two ideals: namely, $F$ itself and $\{0\}$.

4. Characterize all ideals in $\mathbb{Z}$.

5. Characterize all ideals in $\mathbb{Z}_n$, $n > 1$.

6. Let $F$ be a field. Show that every ideal in $F[x]$ is principal.

7. Let $F$ be a field and $f_i(x) \in F[x]$, $1 \le i \le n$. Let $I = \langle f_1(x), f_2(x), \ldots, f_n(x) \rangle$. Describe an algorithm to find a single generator for $I$.

8. Let $\{I_\alpha \mid \alpha \in A\}$ be a nonempty family of ideals in a ring $R$. Show that $\bigcap_{\alpha \in A} I_\alpha$ is an ideal in $R$.

9. Let $I, J$ be ideals in a ring $R$. Prove that $I + J$ is an ideal in $R$.

10. Let $I$ and $J$ be ideals in a ring $R$ such that $I \cap J = \{0\}$. Show that $IJ = \{0\}$.

11. Let $I$ and $J$ be ideals in a ring $R$ such that

   (i)  $I \cap J = \{0\}$
   (ii) $I + J = R$.

   Show that $R$ is isomorphic to the *direct product $I \times J$* of $I$ and $J$ where the operations in $I \times J$ are defined componentwise:

$$(i, j) + (i', j') = (i + i', j + j'),$$

$$(i, j)(i', j') = (ii', jj').$$

12. Let $\varphi : R \to S$ be a ring epimorphism.

   (i)  Show that for any ideal $I$ in $R$, $\varphi(I)$ is an ideal in $S$.
   (ii) Show that for any ideal $J$ in $S$, $\varphi^{-1}(J)$ is an ideal in $R$.

*13. Show that, for any field $F$, $M_n(\mathbb{F})$ has no proper nontrivial ideals.

A subset $I$ of a ring $R$ is said to be a *left ideal* if it is a subring and satisfies the condition $r \in R$, $a \in I \Rightarrow ra \in I$. A *right ideal* is defined similarly.

*14. Find left and right ideals in $M_n(\mathbb{Q})$ that are not (two-sided) ideals.

*15. Let $R$ consist of all $2 \times 2$ matrices $(a_{ij})$ with $a_{11} \in \mathbb{Z}$, $a_{12}, a_{22} \in \mathbb{Q}$ and $a_{21} = 0$.

   (i)   Show that $R$ is a subring of $M_n(\mathbb{Q})$.
   (ii)  Find an infinite chain of ideals: $I_1 \supseteq I_2 \ldots \supseteq I_n \ldots$ in $R$.
   (iii) Find an infinite chain of left ideals: $J_1 \subseteq J_2 \ldots \subseteq J_n \ldots$ in $R$.

16. Let $R$ be a ring, $a \in R$, $L_a = \{x \in R \mid xa = 0\}$, and $R_a = \{x \in R \mid ax = 0\}$.

   (i)   Show that $L_a$ is a left ideal of $R$.
   (ii)  Show that $R_a$ is a right ideal of $R$.
   (iii) Use $M_2(\mathbb{Q})$ to show that $L_a$ and $R_a$ can be distinct.

17. Let $R$ be a commutative ring and $N = \{a \in R \mid \exists\, n \in \mathbb{N} \text{ with } a^n = 0\}$. (Note that the value of $n$ may vary depending on the particular element $a$. Such an element $a$ is called a *nilpotent element*.) Show that $N$ is an ideal in $R$.

18. For $R = \mathbb{Z}_{36}$, find $N$ as described in Exercise 17.

19. Characterize the nilpotent elements in $\mathbb{Z}_n$.

20. Let $R$ be a commutative ring and $I$ be an ideal of $R$. Let $\sqrt{I} = \{a \in R \mid \exists\, n \in \mathbb{N} \text{ with } a^n \in I\}$. (Again, the particular value of $n$ may depend on the particular $a$.) Show that $\sqrt{I}$ is an ideal of $R$.

21. Let $I = 360\,\mathbb{Z}$. Find $\sqrt{I}$.

22. Let $R$ be an integral domain and $M$ be a *maximal* ideal in $R$ (that is, if $A$ is an ideal in $R$ with $M \subseteq A \subseteq R$, then either $A = M$ or $A = R$). Show that $R/M$ is a field.

## 5.2 Polynomial Rings

Let $R$ be a commutative ring with identity and $\{x_1, \ldots, x_n\}$ be a set of variables. Then we denote by $R[x_1, \ldots, x_n]$ the set of all *polynomials* in $x_1, \ldots, x_n$ with coefficients in $R$—that is, the set of all formal expressions of the form

$$\sum_{\alpha = (\alpha_1, \ldots, \alpha_n)} a_\alpha x_1^{\alpha_1} x_2^{\alpha_2} \cdots x_n^{\alpha_n}$$

where $a_\alpha \in R$, the summation runs over

$$\alpha = (\alpha_1, \ldots, \alpha_n) \in (\mathbb{N} \cup \{0\})^n$$

and, at most, a finite number of the $a_\alpha$ are nonzero. For simplicity, we often omit expressions of the form $x_i^0$. For example,

$$x_1 + 3x_2 x_3 + x_4^2, \qquad x_1^2 + x_2^2 + x_3^2 + x_4^2$$

are elements of $\mathbb{Z}[x_1, x_2, x_3, x_4]$. It is usually convenient to adopt a simpler notation for an arbitrary polynomial, such as $f(x_1, \ldots, x_n)$, $g(x_1, \ldots, x_n)$, and so forth. By this we mean that

$$f(x_1, \ldots, x_n) = \sum_\alpha a_\alpha x_1^{\alpha_1} \cdots x_n^{\alpha_n} \tag{5.2}$$

for suitable $a_\alpha \in R$, and so on. Indeed, we may go one step further and write simply $f$ for $f(x_1, \ldots, x_n)$. We refer to the elements of $R[x_1, \ldots, x_n]$ as *polynomials in $x_1, \ldots, x_n$ over $R$*. When the number of variables involved is small, we may use the symbols $x, y, z, \ldots$ in place of $x_1, x_2, x_3, \ldots$.

Extending the terminology of chapter 2, we refer to the $a_\alpha$ as the *coefficients* of $f$ and say that $f$ is a *constant polynomial* if $a_\alpha = 0$ for all $\alpha \neq (0, 0, \ldots, 0)$.

Each expression of the form

$$x_1^{\alpha_1} x_2^{\alpha_2} \cdots x_n^{\alpha_n}$$

is a *monomial* and each expression of the form

$$a_\alpha x_1^{\alpha_1} x_2^{\alpha_2} \cdots x_n^{\alpha_n}$$

is a (*monomial*) *term.* We denote by $\mathcal{M}$ the set of all monomials in $R[x_1, \ldots, x_n]$. The *degree* of the previous term is

$$|\alpha| = \alpha_1 + \alpha_2 + \cdots + \alpha_n.$$

The *degree*, $\deg(f(x_1, \ldots, x_n))$, of $f(x_1, \ldots, x_n)$ is

$$\max\{|\alpha| \mid a_\alpha \neq 0\}.$$

It is clear how we can define the sum of two such polynomials in $n$ variables over $R$:

$$\sum_\alpha a_\alpha x_1^{\alpha_1} \cdots x_n^{\alpha_n} + \sum_\alpha b_\alpha x_1^{\alpha_1} \cdots x_n^{\alpha_n} = \sum_\alpha (a_\alpha + b_\alpha) x_1^{\alpha_1} \cdots x_n^{\alpha_n}.$$

A formal definition of the product of two polynomials in $n$ variables is a bit more complicated:

$$\left( \sum_\alpha a_\alpha x_1^{\alpha_1} \cdots x_n^{\alpha_n} \right) \left( \sum_\alpha b_\alpha x_1^{\alpha_1} \cdots x_n^{\alpha_n} \right) = \sum_\alpha c_\alpha x_1^{\alpha_1} \cdots x_n^{\alpha_n}$$

where

$$c_\alpha = \sum_{\alpha = \gamma + \delta} a_\gamma b_\delta$$

where $\gamma + \delta$ is the sum of vectors performed in $R^n$. When applied to a specific example, this looks exactly as one would expect.

Note that it is implicit in the definition of multiplication of polynomials that the variables commute with each other and with the elements of $R$:

$$x_i x_j = x_j x_i, \quad a x_i = x_i a \quad (a \in R).$$

**Lemma 5.2.1** *Let $F$ be a field and $f(x_1, \ldots, x_n)$, $g(x_1, \ldots, x_n) \in F[x_1, \ldots, x_n]$. Then*

$$\deg(f(x_1, \ldots, x_n) g(x_1, \ldots, x_n)) = \deg(f(x_1, \ldots, x_n)) + \deg(g(x_1, \ldots, x_n)).$$

**Proof.** Exercise. $\square$

**Theorem 5.2.2** (i) *For any commutative ring $R$ (with identity), $R[x_1, x_2, \ldots, x_n]$ is a commutative ring (with identity).*
(ii) *For any field $F$, $F[x_1, \ldots, x_n]$ is an integral domain.*

**Proof.** Exercise. $\square$

Henceforth, let us focus on polynomials over a field $F$.

At this point, it is helpful to take a look at rational expressions in polynomials. They turn out to be quite reasonable to work with. In fact, the main observation at this point applies to any integral domain.

**Theorem 5.2.3** *Let $R$ be an integral domain and*

$$Q = \{(a,\, b) \in R \times R \,|\, b \neq 0\}.$$

*Define a relation $\sim$ on $Q$ by*

$$(a,\, b) \sim (c,\, d) \Leftrightarrow ad = bc.$$

*Then $\sim$ is an equivalence relation on $Q$. Let $[a, b]$ denote the $\sim$ class of $(a, b)$ and $Q/\!\sim$ denote the set of all $\sim$-classes. Define addition and multiplication in $Q/\!\sim$ by*

$$[a,\, b] + [c,\, d] = [ad + bc,\, bd]$$
$$[a,\, b] \cdot [c,\, d] = [ac,\, bd].$$

*Then $(Q/\!\sim, +, )$ is a field with identity $[1, 1]$ and zero $[0, 1]$. For any nonzero element $[a, b]$, $[a, b]^{-1} = [b, a]$.*

**Proof.** Exercise. $\square$

The field $Q/\!\sim$ is called the *field of quotients* of $R$.

If you apply Theorem 5.2.3 to the ring $\mathbb{Z}$, then, of course, we obtain a field $Q/\!\sim$ isomorphic to $\mathbb{Q}$, where $[a, b]$ corresponds to the rational number $\frac{a}{b}$. In like manner, if we take $R = F[x_1, \ldots, x_n]$, where $F$ is a field, then by Theorem 5.2.2, $R$ is an integral domain and we may apply Theorem 5.2.3 to obtain its field of quotients. It is customary to denote this field by

$$F(x_1, \ldots, x_n).$$

As with the rational numbers, we adopt the familiar notation for fractions when convenient and write

$$\frac{f(x_1, \ldots, x_n)}{g(x_1, \ldots, x_n)}$$

in place of

$$[f(x_1, \ldots, x_n), g(x_1, \ldots, x_n)].$$

A nonconstant polynomial in $F[x_1, \ldots, x_n]$ is said to be *reducible* if it can be written as the product of two nonconstant polynomials; otherwise, it is said to be *irreducible*.

Let $f, g \in F[x_1, \ldots, x_n]$. We say that $g$ *divides* $f$ or that $g$ is a *divisor* of $f$ if there exists a polynomial $h \in F[x_1, \ldots, x_n]$ with $f = gh$. When $g$ divides $f$, we write $g \mid f$.

**Lemma 5.2.4** *Let $f, g, h \in F[x_1, \ldots, x_n]$, $f$ be irreducible, and $f \mid gh$. Then $f$ divides $g$ or $h$.*

**Proof.** We argue by induction on the number of variables $n$. By Exercise 2.6.15, the claim is true for $n = 1$. So let us assume that the claim holds for $n - 1$ variables.

First assume that $f$ does not involve one of the variables, so that it is a polynomial in $n - 1$ variables. Without loss of generality, we may assume that it is $x_1$ that does not appear in $f$. Let us write $g$, $h$ and $gh$ as polynomials in $x_1$ with coefficients in $F[x_2, \ldots, x_n]$ as follows:

$$g = \sum_{i=1}^{k} a_i x_1^i, \quad h = \sum_{i=1}^{l} b_i x_1^i, \quad gh = \sum_{i=1}^{l+k} c_i x_1^i.$$

Note that since $f$ does not involve the variable $x_1$, the only way that $f$ can divide $gh$ is if $f$ divides every coefficient $c_i$. If $f$ is a factor of every coefficient $a_i$, then clearly $f$ divides $g$ and if $f$ is a factor of every coefficient $b_i$, then clearly $f$ divides $h$. Either way, we have the desired result. We will now show that the alternative leads to a contradiction. So suppose that there exists a coefficient $a_r$ and a coefficient $b_s$ such that $f$ divides neither $a_r$ nor $b_s$. Again, without loss of generality, we can assume that $r$ and $s$ are the smallest such integers. In other words, $f$ divides $a_i$ and $b_j$ for any $i < r$ and any $j < s$. We now employ essentially the same technique that we used in the proof of Eisenstein's criterion. Consider the coefficient $c_{r+s}$ (in $gh$). We have

$$c_{r+s} = \sum_{i+j=r+s} a_i b_j = (a_0 b_{r+s} + \cdots + a_{r-1} b_{s+1}) + a_r b_s$$

$$+ (a_{r+1} b_{s-1} + \cdots + a_{r+s} b_0).$$

Now $f$ divides at least one factor in each of the terms in $(a_0 b_{r+s} + \cdots + a_{r-1} b_{s+1})$ and at least one factor in each of the terms in $(a_{r+1} b_{s-1} + \cdots + a_{r+s} b_0)$. Therefore $f$ must divide $a_r b_s$ and, by the induction hypothesis, $f$ must divide either $a_r$ or $b_s$. But this contradicts the choice of $a_r$ and $b_s$.

Consequently, $f$ must involve $x_1$, indeed, all the variables. Suppose for the moment that $f$ is not just irreducible in $F[x_1, \ldots, x_n]$ but even in $F(x_2, \ldots, x_n)[x_1]$. From the single variable case (Exercise 2.6.15), we know that in $F(x_2, \ldots, x_n)[x_1]$ either $f$ divides $g$, or $f$ divides $h$. Without loss of generality, we can assume that $f$ divides $g$. Then there exists a polynomial $r$ in $F(x_2, \ldots, x_n)[x_1]$ such that $rf = g$. Let $d$ be the product of all the denominators of the coefficients in $r$, where $r$ is considered as a polynomial in $x_1$ with coefficients in the field $F(x_2, \ldots, x_n)$. Then $d$, $dr$ are elements of $F[x_2, \ldots, x_n]$ and

$$(dr)f = dg.$$

Now let $p$ be any irreducible factor of $d$ in $F[x_2, \ldots, x_n]$. Then $p$ divides $(dr)f$. If $p$ divided $f$ then, since $f$ is irreducible in $F[x_1, x_2, \ldots, x_n]$, we would have that $f = ap$, for some element $a$ in $F$. But $f$ involves the variable $x_1$ while $p$ does not. So this is impossible and we must have that $p$ divides $dr$. We can now write

$$\frac{dr}{p} f = \frac{d}{p} g$$

Continuing in this way with all the irreducible factors of $d$, we finally express $g$ as a multiple of $f$ in $F[x_1, \ldots, x_n]$. Thus $f$ divides $g$. It only remains to show that whenever $f(x_1, \ldots, x_n)$ is irreducible in $F[x_1, \ldots, x_n]$, it is also irreducible in $F(x_2, \ldots, x_n)[x_1]$. But this argument follows the argument of Proposition 2.6.7 almost word for word and so we leave the details to the reader.   $\square$

With the help of Lemma 5.2.4, we can establish unique factorization rather easily.

**Theorem 5.2.5** *Let* $f(x_1, \ldots, x_n) \in F[x_1, \ldots, x_n]$ *be a nonconstant polynomial. Then* $f(x_1, \ldots, x_n)$ *can be written as a product of irreducible polynomials in* $F[x_1, \ldots, x_n]$. *Moreover, any such expression is unique to within the order in which the factors are written and scalar multiples.*

**Proof.** We argue by induction on the degree of $f = f(x_1, \ldots, x_n)$. If $\deg(f) = 1$, then $f$ is itself irreducible. So suppose that $\deg(f) = m$ and that the claim is true for all polynomials in $F[x_1, \ldots, x_n]$ of degree less than $m$.

If $f$ is irreducible, then we have the desired conclusion. If $f$ is reducible, then there exist nonconstant polynomials $g(x_1, \ldots, x_n)$, $h(x_1, \ldots, x_n) \in F[x_1, \ldots, x_n]$ with

$$f(x_1, \ldots, x_n) = g(x_1, \ldots, x_n)\, h(x_1, \ldots, x_n). \tag{5.3}$$

Now, $\deg(g) < m$ and $\deg(h) < m$ so that, by the induction hypothesis, $g$ and $h$ can be written as products of irreducible polynomials in $F[x_1, \ldots, x_n]$. It then follows from (5.3) that $f$ can be written as a product of irreducible polynomials. The uniqueness of the product now follows easily from Lemma 5.2.4. $\quad\square$

There is one more familiar idea from the theory of polynomials in one variable that extends to several variables. Let $f, g, d \in F[x_1, \ldots, x_n]$. Then $d$ is a *greatest common divisor* of $f$ and $g$ if

(i) $d$ divides $f$ and $g$
(ii) if $h \in F[x_1, \ldots, x_n]$ divides $f$ and $g$ then $h$ divides $d$.

**Corollary 5.2.6** *Any two polynomials $f, g \in F[x_1, \ldots, x_n]$ have a greatest common divisor.*

**Proof.** This is a simple exercise based on Theorem 5.2.5. $\quad\square$

The parallel with the theory of polynomials in a single variable breaks down at this point as the greatest common divisor may not be a linear combination of the two polynomials. For example, the greatest common divisor of $xy$ and $yz$ is clearly $y$. However, $y$ is not a combination of $xy$ and $yz$.

When working with polynomials in one variable of degree 2 or 3 over a field $F$, there is a very simple test for irreducibility. If $f(x) \in F[x]$ and $\deg(f(x)) = 2, 3$, then $f(x)$ is irreducible if and only if $f(x)$ has no roots in $F$. Life is not quite that simple for polynomials in several variables. There are some important differences between the theory of irreducible polynomials in one variable and the theory of irreducible polynomials in several variables. Let $f(x) \in F[x]$. Then,

$$f(x) \text{ irreducible, } \deg(f(x)) \geq 2 \implies f(x) \text{ has no roots in } F.$$

However, the fact that $f \in F[x_1, \ldots, x_n]$ is irreducible does not necessarily imply that $f$ has no zeros. For example let $f(x, y) = x^2 + y \in F[x, y]$. Then, $f(x, y)$ has many zeros. For example, $(a, -a^2)$, for any $a \in F$. However, $f(x, y)$ is irreducible. To see this, suppose that $f(x, y)$ were reducible, then it would have to be expressible as a product of two polynomials of degree 1, say

$$x^2 + y = (ax + by + c)(px + qy + r).$$

Equating certain of the coefficients, we find

$$\begin{aligned}
\text{constant:} \quad & 0 = cr \\
\text{coefficient of } x: \quad & 0 = ar + cp.
\end{aligned}$$

If $c = r = 0$, then the right side becomes $(ax + by)(px + qy)$, in which every nonzero term will have degree 2. Hence, one of $c, r$ must be zero and one nonzero. We can assume that $c \neq 0$, $r = 0$. Then, from the coefficient of $x$, we have $cr = 0$ and, therefore, $r = 0$. The right-hand side now becomes $(ax + by + c) \cdot qy$, which has no term $x^2$. Thus, we have a contradiction and $f(x, y)$ is irreducible. Notice that the argument is quite independent of the field $F$. Disappointing as this observation might be, there are some simple and helpful observations. The first is that the method that we used here can be applied to many simple situations. We assume that a polynomial is factorizable in some way and then equate the coefficients of different terms to determine whether it is possible to find a solution or to show that none exists. Another useful observation is the following.

Let $f(x_1, \ldots, x_n) \in F[x_1, \ldots, x_n]$ have degree 2 or 3 and suppose that $f$ is reducible as $f = gh$ where $g, h \in F[x_1, \ldots, x_n]$ are nonconstant polynomials. Then,

$$\deg(f) = \deg(g) + \deg(h)$$

so that either $\deg(g) = 1$ or $\deg(h) = 1$. Without loss of generality, let us suppose that $\deg(g) = 1$. Then there exist $c_i$, $c \in F$ with

$$g = c_1 x_1 + c_2 x_2 + \cdots + c_n x_n + c$$

where at least one $c_i$ is nonzero. The equation $g = 0$ is then the equation of a line (if $n = 2$), plane (if $n = 3$), or, in general, a hyperplane. In particular, we must have $f(x_1, \ldots, x_n) = 0$ wherever $g(x_1, \ldots, x_n) = 0$ so that the solutions to $f(x_1, \ldots, x_n) = 0$ must include the solutions to $g(x_1, \ldots, x_n) = 0$. If it is evident that the solutions to $f(x_1, \ldots, x_n)$ do not include a line, plane, and so forth, then $f$ must indeed be irreducible.

In this way, it is easy to see that the following polynomials are irreducible:

$$x^2 + y^2 - r^2 \in \mathbb{R}[x]$$

$$y - ax^2 + k \in \mathbb{R}[x]$$

$$\frac{x^2}{a^2} - \frac{y^2}{b^2} + c \in \mathbb{R}[x].$$

There is one rather special situation that will, nevertheless, be important to us later, where a polynomial turns out to be highly reducible for reasons that are not immediately transparent. A *Vandermonde matrix* is an $n \times n$ matrix of the form

$$A = \begin{bmatrix} x_1^{n-1} & x_1^{n-2} & \cdots & x_1 & 1 \\ x_2^{n-1} & x_2^{n-2} & \cdots & x_2 & 1 \\ \vdots & & & & \\ x_n^{n-1} & x_n^{n-2} & \cdots & x_n & 1 \end{bmatrix} \tag{5.4}$$

where the $x_i$ are variables. Then det $(A)$ will be a polynomial in $x_1, x_2, \ldots, x_n$ over $F$—that is, det $(A) \in F[x_1, \ldots, x_n]$. Subtracting the second row from the first, we obtain a matrix with first row

$$[x_1^{n-1} - x_2^{n-1} \quad x_1^{n-2} - x_2^{n-2} \quad \cdots \quad x_1 - x_2 \quad 0]$$

$$= (x_1 - x_2)[x_1^{n-2} + \cdots + x_2^{n-2} \quad x_1^{n-3} + \cdots + x_2^{n-3} \quad \cdots \quad 1 \quad 0].$$

It follows that $x_1 - x_2$ must be a factor of det $(A)$.

In the same way, by subtracting the $j$th row from the $i$th row, we see that $x_i - x_j$ must divide det $(A)$, for all $1 \le i < j \le n$. Hence,

$$\det (A) = \left( \prod_{1 \le i < j \le n} (x_i - x_j) \right) q(x_1, \ldots, x_n) \tag{5.5}$$

for some $q(x_1, \ldots, x_n) \in F[x_1, \ldots, x_n]$.

Now, since each element in the expansion of det $(A)$ is a product of terms with exactly one element from each row and column, it follows that

$$\deg (\det (A)) = (n - 1) + (n - 2) + \cdots + 2 + 1$$

$$= \frac{(n - 1)n}{2}.$$

However,

$$\deg \left( \prod_{1 \le i < j \le n} (x_i - x_j) \right) = \frac{(n - 1)n}{2}.$$

Consequently, $q(x_1, \ldots, x_n)$ is a constant $c$. Now one term in det $(A)$ is

$$x_1^{n-1} x_2^{n-2} \cdots x_{n-1}$$

so this must equal the corresponding term from (5.5):

$$c x_1^{n-1} x_2^{n-2} \cdots x_{n-1}.$$

Hence, $c = 1$ and we have proved the following lemma.

**Lemma 5.2.7** *Let $A$ be a Vandermonde matrix as given in (5.4). Then*

$$\det (A) = \prod_{1 \le i < j \le n} (x_i - x_j).$$

Now the equality in Lemma 5.2.7 is an equality of two polynomials. Consequently, they will yield the same result no matter what values in $F$ we assign to the variables $x_1, \ldots, x_n$. In other words, the result also holds when we consider the elements $x_i$ to be elements of any field. Of course, we are then just evaluating the two polynomials for certain values of the variables.

Associated in a natural way with each polynomial $f(x_1, \ldots, x_n) \in R[x_1, \ldots, x_n]$, we have an evaluation function $f : R^n \to R$ where

$$f : (a_1, \ldots, a_n) \to f(a_1, \ldots, a_n)$$

and by $f(a_1, \ldots, a_n)$ we mean the element of $R$ obtained by replacing each occurrence of $x_1$ in $f(x_1, \ldots, x_n)$ by $a_1$, each occurrence of $x_2$ in $f(x_1, \ldots, x_n)$ by $a_2$, and so on.

Now, in some situations, distinct polynomials will yield the same evaluation function. For example, the polynomials $x + y$, $x^p + y$, and $x + y^p$ in $\mathbb{Z}_p[x, y]$ yield the same function from $\mathbb{Z}_p^2$ to $\mathbb{Z}_p$ since $a^p = a$ for all $a \in \mathbb{Z}_p$. We saw this sort of thing in section 2.5. However, there is one important situation regarding the underlying field that will guarantee that distinct polynomials determine distinct functions. (See Exercise 10 in section 2.5 for the corresponding result in $F[x]$.)

**Theorem 5.2.8** *Let $F$ be an infinite field and $f, g \in F[x_1, \ldots, x_n]$. Then $f = g$ if and only if the associated functions $F^n \to F$ are also equal.*

**Proof.** Assume that $f$ and $g$ induce the same functions $F^n \to F$ and let $h = f - g$. Then

$$h(a_1, \ldots, a_n) = f(a_1, \ldots, a_n) - g(a_1, \ldots, a_n)$$

$$= 0 \tag{5.6}$$

for all $a_1, \ldots, a_n \in F$.

*Claim*: $h$ is the zero polynomial. We will argue by induction on the degree of $h$. If $\deg(h) = 0$—that is, $h$ is a constant—then the claim is obvious.

Also, if $n = 1$ then we have a polynomial in the single variable $x_1$. However, any nonzero polynomial in a single variable can only have finitely many zeros. So, once again, we must have $h = 0$. So let us assume that $n \geq 2$ and that for any $k \in F[x_1, \ldots, x_t]$, where $t < n$, if $k$ induces the zero function $F^t \to F$, then $k$ is the zero polynomial.

Now let us consider $h$ as a polynomial in $x_n$ with coefficients in $F[x_1, \ldots, x_{n-1}]$:

$$h(x_1, \ldots, x_n) = p_0 + p_1 x_n + \cdots + p_m x_n^m$$

where $p_i = p_i(x_1, \ldots, x_{n-1}) \in F[x_1, \ldots, x_{n-1}]$. Let $a_1, \ldots, a_{n-1} \in F$. Substituting these values into the polynomials $p_i$, we obtain a polynomial of degree $m$ in $F[x_n]$:

$$h(a_1, a_2, \ldots, a_{n-1}, x_n) = c_0 + c_1 x_n + \cdots + c_m x_n^m \qquad (5.7)$$

where $c_i = p_i(a_1, a_2, \ldots, a_{n-1})$. By (5.6) we have that

$$h(a_1, a_2, \ldots, a_{n-1}, a_n) = 0$$

for all $a_n \in F$. Thus, the polynomial in (5.7) has infinitely many zeros. Since this is a polynomial in just one variable, it must be the zero polynomial:

$$h(a_1, a_2, \ldots, a_{n-1}, x_n) = 0.$$

This means that every coefficient is zero—that is,

$$p_i(a_1, a_2, \ldots, a_{n-1}) = 0$$

for all $i$. But, the $a_1, \ldots, a_{n-1}$ were chosen arbitrarily and $p_i(x_1, \ldots, x_{n-1}) \in F[x_1, \ldots, x_{n-1}]$ for all $i$. Hence, by the induction hypothesis, we must have

$$p_i(x_1, \ldots, x_{n-1}) = 0$$

for all $i = 1, 2, \ldots, m$. Thus,

$$h(x_1, \ldots, x_n) = p_0 + p_1 x_n + \cdots + p_m x_n^m$$
$$= 0$$

the zero polynomial. Consequently,

$$f = h + g = 0 + g = g$$

as required. $\qquad \square$

**Exercises 5.2**

1. Let char $(F) \neq 3$. Show that

$$\langle 3x^3 y^5, 2y^2 x^2 + 3 \rangle = F[x, y].$$

2. Show that

$$\langle x^2 y - 1, y^2 x - x \rangle = \langle x^2 - y, y^2 - 1 \rangle$$

in $F[x, y]$.

3. Show that

$$\langle x^2 + y, y^2 + z, z^2 + x \rangle = \langle x + z^2, y + z^4, z + z^8 \rangle$$

in $F[x, y, z]$.

4. Let $F$ be a field with characteristic not equal to 2 and let $a \in F^*$ be such that $F$ contains a solution to $x^2 = a$. Show that $F$ must contain two distinct roots of $x^2 = a$.

5. Let $F$ be a field with characteristic different from 3 and let $b \in F^*$. Show that $x^3 = b$ cannot have two repeated roots in $F$.

6. Show that $x^2 + y^2 - 1 \in \mathbb{C}[x, y]$ is irreducible.

7. Let $f(x, y) = y^2 - g(x)$, where $g(x) \in F[x]$, be a polynomial in $F[x, y]$. Show that $f(x, y)$ is reducible if and only if there exists $h(x) \in F[x]$ with $g(x) = (h(x))^2$.

8. Show that $x^n + y \in F[x, y]$ is irreducible for all $n \in \mathbb{N}$.

9. Let $g(x_1, \ldots, x_{n-1}) \in F[x_1, \ldots, x_{n-1}]$ be nonconstant and $f(x_1, \ldots, x_n) = x_n^{2k} - g(x_1, \ldots, x_{n-1})$. Show that $f$ is reducible if and only if there exists $h(x_1, \ldots, x_{n-1}) \in F[x_1, \ldots, x_{n-1}]$ such that $g = h^2$.

10. Show that $x^{2k} + y^{2k} \in F[x, y]$ is reducible if $F$ contains a root of the polynomial $x^2 + 1$.

11. Show that $y^2 - x^3 - ax - b \in F[x, y]$ is irreducible for all $a, b \in F$.

12. Show that $x^2 + xy + y^2 \in F[x, y]$ is reducible if and only if the polynomial $x^2 - x + 1$ has a root in $F$.

## 5.3 Division Algorithm in $F[x_1, x_2, \ldots, x_n]$ : Single Divisor

In chapter 2 we saw many applications of the division algorithm for polynomials in one variable. To those applications, we can add one more. Let $I$ be an ideal in $F[x]$ and suppose that we would like an algorithm that would decide, given any $f(x) \in F[x]$, whether $f(x) \in I$. To achieve this we must, of course, have a reasonable description of $I$. We know (see Exercise 6 in section 5.1) that $I$ is a principal ideal so that there exists $g(x) \in F[x]$ such that $I = g(x)F[x]$. It is then a simple matter to determine whether $f(x) \in I$. We simply apply the division algorithm to obtain

$$f(x) = q(x) g(x) + r(x)$$

where $r(x) = 0$ or $\deg(r(x)) < \deg(g(x))$. Then $f(x) \in I$ if and only if $r(x) = 0$.

Thus, we can use the division algorithm to solve the *membership problem*—that is, to determine whether a polynomial $f(x)$ chosen at random from $F[x]$ is a member of $I$.

This suggests the possibility that a division algorithm in $F[x_1, \ldots, x_n]$ might also have many applications. However, we have to proceed with some caution because there are pitfalls. The theory of division of polynomials in several variables is a good deal more subtle than that for polynomials in a single variable.

The problem is well illustrated by the following example. Suppose that we want to divide $x^3 + y^2$ by $x + y$. Using the method of long division, we might calculate

$$
\begin{array}{r}
x^2 - xy \ + y^2 \phantom{+ y^2} \\
x + y \overline{)x^3 \phantom{+ x^2y} + y^2} \\
\underline{x^3 + x^2y} \phantom{+ y^2} \\
- x^2y \phantom{+ y^2} \\
\underline{- x^2y - xy^2} \phantom{+} \\
xy^2 + y^2 \\
\underline{xy^2 + y^3} \\
y^2 - y^3
\end{array}
$$

which yields

$$x^3 + y^2 = (x^2 - xy + y^2)(x + y) + y^2 - y^3. \tag{5.8}$$

On the other hand, since $x + y = y + x$, we might just as easily calculate

$$
\begin{array}{r}
y \ - x \phantom{+ x^3} \\
y + x \overline{)y^2 \phantom{+ xy} + x^3} \\
\underline{y^2 + xy} \phantom{+ x^3} \\
- xy \phantom{+ x^3} \\
\underline{- xy - x^2} \\
x^2 + x^3
\end{array}
$$

which yields

$$y^2 + x^3 = (y - x)(y + x) + x^2 + x^3. \tag{5.9}$$

Thus, there is no well-defined notion of *quotient* and *remainder* without the introduction of some new features or constraints. The concepts required to bring some order to this situation appeared relatively recently (1960s), compared with the centuries that mathematicians have been studying polynomials. Yet, the key idea can be found if we take a look at how the division algorithm works in $F[x]$. Let us revisit Example 2.4.5, where $f(x) = x^5 + 4x^3 + 4x^2 + 3$

and $g(x) = x^2 + 3x + 2$ are polynomials in $\mathbb{Z}_5[x]$. Dividing $f(x)$ by $g(x)$ we obtain

$$
\begin{array}{r}
x^3 \quad\;\; +2x^2 \;\; +x+2 \\[2pt]
\hline
x^2+3x+2\,)\,\overline{\;x^5 \;+0\cdot x^4 \;+4\cdot x^3 \;+4\cdot x^2 \;+0\cdot x+3\;} \\[2pt]
\underline{x^5 \;+3x^4 \quad +2x^3} \\[2pt]
2x^4 \;\;+2x^3 \;\;+4x^2 \\[2pt]
\underline{2x^4 \;\;+x^3 \quad\;\;+4x^2} \\[2pt]
x^3 \;\;+0\cdot x^2 +0\cdot x \\[2pt]
\underline{x^3 \quad +3x^2 \quad +2x} \\[2pt]
2x^2 \;\;+3x+3 \\[2pt]
\underline{2x^2 \;\;+x+4} \\[2pt]
2x+4.
\end{array}
$$

Working through the calculation, we see that all the action is dictated by the $x^2$ term in the divisor. It is the leading term in deciding what happens at each stage. The other terms have a secondary role only. This suggests that if we want to design a division algorithm for polynomials in several variables, then we should fix on a leading term in the divisor and use that to determine what happens at each stage in the division. This leads us to the question of how to pick a leading term in a consistent way. For example, suppose that we wish to divide by

$$
g(x, y, z) = x^7 y^2 z^3 + x^4 y^8 z^4 + y^6 z^7,
$$

What term should we choose? We need an orderly approach.

One approach (and we emphasize that there are others) is to introduce an ordering of the variables. For example, if we are dealing with polynomials in $F[x_1, x_2, \ldots, x_n]$, then we might choose the ordering

$$
x_1 > x_2 > x_3 > \cdots > x_{n-1} > x_n,
$$

although any ordering would do. We extend this to an ordering of the set $\mathcal{M}$ of monomials. Consider two monomials

$$
m_1 = x_{i_1}^{\alpha_1} \cdots x_{i_k}^{\alpha_k}, \qquad m_2 = x_{j_1}^{\beta_1} \cdots x_{j_\ell}^{\beta_\ell}.
$$

If any variable, $x_i$ say, does not appear in $m_1$ or $m_2$, then we add a factor $x_i^0$ and write the variables in the order that we have chosen for the variables. In our case,

$$
m_1 = x_1^{\alpha_1} x_2^{\alpha_2} \cdots x_n^{\alpha_n}, \qquad m_2 = x_1^{\beta_1} x_2^{\beta_2} \cdots x_n^{\beta_n}.
$$

Note that every variable now appears in its proper place in $m_1$ and $m_2$. Now define

$$m_1 > m_2 \iff \alpha_i > \beta_i \qquad \text{where } i \text{ is the least integer with } \alpha_i \neq \beta_i.$$

This ordering of the set $\mathcal{M}$ of monomials is known as the *lexicographic ordering*. For example, if we use $x, y, z$ to denote our variables and order them as

$$x > y > z$$

then in the lexicographic ordering,

$$x^8 y^3 z^2 > x^7 > x^6 y^5 z^5 > x^6 y^4 z^8.$$

Of course, if we choose a different ordering for the variables, then we will obtain a different ordering for the monomials. For instance, if we have

$$y > z > x$$

then

$$y^5 z^5 x^6 > y^4 z^8 x^6 > y^3 z^2 x^8 > x^7.$$

The lexicographic ordering of monomials has three important properties:

M1: The order relation $<$ is a *total order*—that is, for any monomials $m_1$ and $m_2$, exactly one of the following must hold:

$$m_1 < m_2 \quad \text{or} \quad m_1 = m_2 \quad \text{or} \quad m_1 > m_2.$$

M2: The order relation $<$ is a *compatible order* relation—that is, if $m, m_1$, and $m_2$ are monomials, then

$$m_1 > m_2 \implies mm_1 > mm_2.$$

M3: The set of monomials is *well ordered* with respect to the relation $<$—that is, every nonempty subset of monomials has a smallest member.

Any order relation on the set $\mathcal{M}$ of monomials satisfying M1, M2, and M3 is called a *monomial order*.

Properties M1 and M2 are pretty straightforward. With regard to property M3, it is easy to make the mistake of thinking that it is obvious because there are only finitely many monomials smaller than any chosen monomial. After all, this is basically the reason that $\mathbb{N}$ is well ordered. However, in this case, most monomials are bigger than infinitely many monomials. For example, for monomials in $x, y, z$ with $x > y > z$, we have

$$x > y^m z^n \qquad \text{(for all } m, n \in \mathbb{N}\text{)}.$$

So the well-ordering property is not quite so obvious. However, consider any nonempty set $M$ of monomials $x^\alpha y^\beta z^\gamma$. Let

$$E_x = \{\alpha \in \mathbb{N} \cup \{0\} \mid \exists\, x^\alpha y^\beta z^\gamma \in M\}.$$

Since $\mathbb{N} \cup \{0\}$ is well ordered and $E_x$ is a nonempty subset of $\mathbb{N} \cup \{0\}$, $E_x$ must have a smallest member $\alpha_0$, say. Now let

$$E_y = \{\beta \in \mathbb{N} \cup \{0\} \mid \exists\, x^{\alpha_0} y^\beta z^\gamma \in M\}.$$

Since $\mathbb{N} \cup \{0\}$ is well ordered, this set must also have a smallest element $\beta_0$, say. Finally, let

$$E_z = \{\gamma \in \mathbb{N} \cup \{0\} \mid \exists\, x^{\alpha_0} y^{\beta_0} z^\gamma \in M\}.$$

Again, this set must have a smallest element $\gamma_0$, say, and then it is evident that

$$m = x^{\alpha_0} y^{\beta_0} z^{\gamma_0}$$

is the smallest element in $M$. Thus, the set of all monomials in $x, y,$ and $z$ is well ordered. It is clear that the argument will extend to any number of variables.

We can now introduce some terminology for polynomials in several variables corresponding to the terminology that we had for polynomials in a single variable. Let

$$f(x_1, \ldots, x_n) = \sum_\alpha c_\alpha m_\alpha \in F[x_1, \ldots, x_n]$$

where each $m_\alpha$ is a monomial. Assume that we have adopted a monomial order. Then we define

$$\begin{aligned}
\textit{leading monomial} &= \mathrm{lm}\,(f) = m_\alpha \;\; \text{where} \;\; c_\alpha \\
&\neq 0 \;\; \text{and} \;\; m_\alpha > m_\beta \;\; \forall \beta \neq \alpha \;\; \text{with} \;\; c_\beta \neq 0 \\
\textit{leading coefficient} &= \mathrm{lc}\,(f) = c_\alpha \;\; \text{where} \;\; \mathrm{lm}\,(f) = m_\alpha \\
\textit{leading term} &\quad = \mathrm{lt}\,(f) = c_\alpha m_\alpha \;\; \text{where} \;\; \mathrm{lm}\,(f) = m_\alpha.
\end{aligned}$$

For example, if

$$f(x, y, z) = 10x^8 y^3 z^2 + 11x^7 + 12x^6 y^5 z^5 + 13x^6 y^4 z^8$$

and we adopt the lexicographic ordering based on

$$y > z > x$$

then

$$\mathrm{lm}\,(f) = y^5 z^5 x^6$$
$$\mathrm{lc}\,(f) = 12$$
$$\mathrm{lt}\,(f) = 12 y^5 z^5 x^6.$$

Whenever there is a reference to the leading term, monomial, or coefficient of a polynomial, it is to be understood that it is with reference to some monomial order.

In the next lemma, we list some simple observations concerning these concepts.

**Lemma 5.3.1** *Let $f, g \in F[x_1, \ldots, x_n]$.*

(i) $\mathrm{lt}\,(f) = \mathrm{lc}\,(f) \cdot \mathrm{lm}\,(f)$.
(ii) $\mathrm{lt}\,(f \cdot g) = \mathrm{lt}\,(f) \cdot \mathrm{lt}\,(g)$.

**Proof.** (i) This follows immediately from the definitions.

(ii) This follows from property M2 of the ordering of monomials. $\square$

**Theorem 5.3.2 (Division Algorithm in $n$ variables for a single divisor)** *Let $f, g \in F[x_1, \ldots, x_n]$. Then there exist $q, r \in F[x_1, \ldots, x_n]$ such that*

$$f = qg + r$$

*where, if $r = \sum_{\alpha} c_\alpha r_\alpha$ $(c_\alpha \in F,\ r_\alpha \in \mathcal{M})$, then $\mathrm{lt}\,(g)$ does not divide $r_\alpha$ for any $\alpha$ (with $c_\alpha \neq 0$). Moreover, $q$ and $r$ are unique.*

**Proof.** First compare the leading terms of $f$ and $g$. If $\mathrm{lt}\,(g)$ divides $\mathrm{lt}\,(f)$, then let

$$f_1 = f - \frac{\mathrm{lt}\,(f)}{\mathrm{lt}\,(g)} \cdot g, \qquad q_1 = \frac{\mathrm{lt}\,(f)}{\mathrm{lt}\,(g)}, \qquad r_1 = 0.$$

If $\mathrm{lt}\,(g)$ does not divide $\mathrm{lt}\,(f)$, then let

$$f_1 = f - \mathrm{lt}\,(f), \qquad q_1 = 0, \qquad r_1 = \mathrm{lt}\,(f).$$

Note that in both cases,

$$f = f_1 + q_1 g + r_1 \tag{5.10}$$

where $\mathrm{lt}\,(g)$ does not divide any term in $r_1$. Also, in both cases, we must have

$$\mathrm{lt}\,(f_1) < \mathrm{lt}\,(f).$$

Now repeat this procedure with $f_1$ to obtain

$$f_1 = f_2 + q_2 g + r_2$$

where $\mathrm{lt}\,(g)$ does not divide any term in $r_2$ and

$$\mathrm{lt}\,(f_2) < \mathrm{lt}\,(f_1).$$

We now have from equation (5.10) that

$$\begin{aligned}
f &= f_1 + q_1 g + r_1 \\
&= f_2 + (q_1 + q_2)\,g + r_1 + r_2
\end{aligned}$$

where

$$\mathrm{lt}\,(f_2) < \mathrm{lt}\,(f_1) < \mathrm{lt}\,(f)$$

and $\mathrm{lt}\,(g)$ does not divide any term in $r_1 + r_2$. Now repeat the process with $f_2$, and so on. The critical question now is whether the process will stop. This is not immediately obvious for the following reason. Each time we perform the step

$$f_{i+1} = f_i - \frac{\mathrm{lt}\,(f_i)}{\mathrm{lt}\,(g)}\,g$$

although we eliminate the leading term of $f_i$, we introduce a whole collection of terms corresponding to the other nonleading terms in $g$ so that we may end up with more terms in $f_{i+1}$ than we had in $f_i$. The critical observation here is that the leading terms in $f_i$ and $\frac{\mathrm{lt}\,(f_i)}{\mathrm{lt}\,(g)} \cdot g$ are the same. Therefore, the leading term in $f_{i+1} = f_i - \frac{\mathrm{lt}\,(f_i)}{\mathrm{lt}\,(g)} \cdot g$ must be smaller. The same applies when the leading term in $f_i$ is transferred to the remainder. So suppose that we let

$$A = \{\mathrm{lm}\,(f_i) \mid i = 1, 2, \dots\}.$$

This is a subset of monomials and so, by property M3, must contain a smallest member. Hence, the process must stop. Let it stop at $i = m$.

Then we must have $f_m = 0$; otherwise, we could proceed one step further, so that we find

$$f = (q_1 + q_2 + \cdots + q_m)\,g + (r_1 + r_2 + \cdots + r_m)$$

where the leading term of $g$ does not divide any of the terms in $r_1 + r_2 + \cdots + r_m$. This establishes the main claim.

Now let $f = qg + r = q_1 g + r_1$. Then

$$r_1 - r = (q - q_1)\, g.$$

Suppose that $q - q_1 \neq 0$ and let $\mathrm{lt}\,(q - q_1) = cm \neq 0$ where $c \in F$, $m \in M$. Then, by Lemma 5.3.1 (ii),

$$\mathrm{lt}\,(r - r_1) = -\mathrm{lt}\,((q - q_1)\, g)$$
$$= -\mathrm{lt}\,(q - q_1) \cdot \mathrm{lt}\,(g)$$
$$= -cm \cdot \mathrm{lt}\,(g).$$

But any term in $r - r_1$ is just the difference of two terms involving the same monomials from $r$ and $r_1$, so that, in particular,

$$\mathrm{lt}\,(r - r_1) = at - a_1 t = (a - a_1)\, t$$

where, for some $a, a_1 \in F$ and $t \in M$, $at$ is a term in $r$ and $a_1 t$ is a term in $r_1$. Hence, $\mathrm{lt}\,(g)$ divides $(a - a_1)\, t$ and, therefore, $t$. This is a contradiction. Consequently, $q - q_1 = 0$, which leads to $q = q_1$ and $r = r_1$. $\quad\square$

For any polynomial $f \in F[x_1, \ldots, x_n]$, we define the $x_i$ *degree* of $f$, which we denote by $\deg_{x_i}(f)$, to be the degree of $f$ when considered as a polynomial in $x_i$ with coefficients in $F[x_1, \ldots, x_{i-1}, x_{i+1}, \ldots, x_n]$. For the sake of simplicity, the next two results are set in $F[x, y, z]$, but clearly extend to an arbitrary number of variables. We will use the techniques of this section to obtain the results, though other methods would also work.

**Corollary 5.3.3** *Let $f, g \in F[x, y, z]$ and let $g = ax^n + p\,(x, y, z)$ where $a \in F^*$, $n \geq 1$, and $\deg_x(p\,(x, y, z)) < n$. Then there exist $q, r \in F[x, y, z]$ such that*

$$f = qg + r \quad and \quad \deg_x(r) < n.$$

*Moreover, $q$ and $r$ are unique.*

**Proof.** We adopt the lexicographic order for monomials based on $x > y > z$. Then,

$$\mathrm{lt}\,(g) = ax^n.$$

By the division algorithm (Theorem 5.3.2), there exist unique $q, r \in F[x, y, z]$ with

$$f = qg + r$$

where $\mathrm{lt}\,(g)$ does not divide any term in $r$. Hence, $\deg_x(r) < n$. $\quad\square$

The next result should look a little bit familiar.

**Corollary 5.3.4** *Let $f(x, y, z) \in F[x, y, z]$, $a \in F$, and $f(a, y, z) = 0$. Then there exists $q(x, y, z) \in F[x, y, z]$ with*

$$f(x, y, z) = (x - a)\, q(x, y, z).$$

**Proof.** We adopt the lexicographic order for monomials based on $x > y > z$. Let $g(x, y, z) = x - a$. By the division algorithm, there exist $q, r \in F[x, y, z]$ such that

$$f(x, y, z) = (x - a)\, q + r$$

where $x = \operatorname{lt}(g)$ does not divide any term in $r$. Hence, $r = r(y, z) \in F[y, z]$. Consequently,

$$0 = f(a, y, z) = (a - a)\, q + r = r.$$

Therefore,

$$f(x, y, z) = (x - a)\, q$$

as required.    $\square$

In the previous corollaries, we already see some applications of our division algorithm for polynomials in several variables. However, we have further extensions of the division algorithm to consider in the next section.

### Exercises 5.3

1. Show that it would not be sufficient in Theorem 5.2.8 to assume only that the functions associated with $f$ and $g$ agree on an infinite subset of $F^n$ (as opposed to the whole of $F^n$).

2. Let $f(x, y)$, $g(x, y) \in F[x, y]$ and let $g(x, y) = ax + p(y)$ where $a \neq 0$, $p(y) \in F[y]$. Show that there exists $q(x, y) \in F[x, y]$, $r(y) \in F[y]$ such that

$$f(x, y) = g(x, y)\, q(x, y) + r(y).$$

3. Let $f(x, y) \in \mathbb{R}[x, y]$ and let the line $L$, given by $ax + by + c = 0$, $a \neq 0$, be such that for infinitely many points $(x_1, y_1)$ on $L$, we have $f(x_1, y_1) = 0$. Show that $ax + by + c$ divides $f(x, y)$.

## 5.4 Multiple Divisors: Groebner Bases

When working with polynomials in a single variable (that is, in $F[x]$), we have seen in Theorem 5.1.7 that every ideal is generated by a single polynomial. This does not carry over to ideals in several variables. For example, consider

$$I = \{f(x, y) \in F[x, y] \mid f(0, 0) = 0\}.$$

This just consists of all polynomials in $x, y$ for which the constant term is zero and is clearly an ideal in $F[x, y]$. It is evident that $I = \langle x, y \rangle$ and that no single polynomial will generate I. However, we will see in this section that at least a finite number of polynomials will do the job for us.

Let $I$ be an ideal in $F[x_1, \ldots, x_n]$. As we shall see, there exist $g_1, g_2, \ldots, g_m \in F[x_1, \ldots, x_n]$ with

$$I = \langle g_1, g_2, \ldots, g_m \rangle.$$

Now consider the membership problem for $I$. In other words, given $f \in F[x_1, \ldots, x_n]$, how can we determine whether $f \in I$? Following the pattern for one variable, we would like to write

$$f = \sum_{i=1}^{m} q_i g_i + r$$

in such a way that $f \in I$ if and only if $r = 0$. This suggests that we need an algorithm that "divides" $f$ by several polynomials instead of the usual single divisor.

**Lemma 5.4.1** *Let* $f, g_1, g_2, \ldots, g_m \in F[x_1, x_2, \ldots, x_n]$. *Then there exist* $q_1, q_2, \ldots, q_m, r \in F[x_1, x_2, \ldots, x_n]$ *such that*

$$f = q_1 g_1 + q_2 g_2 + \cdots + q_m g_m + r$$

*where either* $r = 0$ *or* $r$ *is a sum of monomial terms, none of which is divisible by any of* $\mathrm{lt}(g_1), \ldots, \mathrm{lt}(g_m)$.

**Proof.** The proof is very similar to that of Theorem 5.3.2. The only difference is that before assigning a term to the remainder $r$, we try to divide it by all the leading terms $\mathrm{lt}(g_i)$, $1 \leq i \leq m$, in turn according to their order under the monomial ordering. $\square$

There is one desirable feature that is noticeably absent in Lemma 5.4.1— namely, uniqueness. Whenever $m \geq 2$, uniqueness of the $q_i$ becomes a lost cause because we can always replace

$$q_1 g_1 + q_2 g_2$$

with

$$(q_1 + g_2)\, g_1 + (q_2 - g_1)\, g_2.$$

The possibility that $r$ might be unique remains, and this is important. However, the uniqueness of $r$ will not come without some constraint on the $g_i$.

**Example 5.4.2**

$$g_1 = x^3 y^2 + 1, \qquad g_2 = x^2 y^3 + 1 \in F[x, y].$$

Then

$$
\begin{aligned}
0 &= 0 \cdot g_1 + 0 \cdot g_2 + 0 \\
&= y \cdot g_1 - x \cdot g_2 + x - y
\end{aligned}
$$

which gives us two distinct ways to write zero in the form $q_1 g_1 + q_2 g_2 + r$ where no term of $r$ is divisible by any of the leading terms of $g_1$ and $g_2$. $\qquad\square$

The critical idea comes from comparing the leading monomials in $I = \langle g_1, \ldots, g_m \rangle$ with $\langle \mathrm{lm}\,(g_1), \ldots, \mathrm{lm}\,(g_m) \rangle$. Note that in the previous example that

$$y - x = y \cdot g_1 - x \cdot g_2 \in \langle g_1, g_2 \rangle.$$

However,

$$\mathrm{lt}\,(g_1) = x^3 y^2, \qquad \mathrm{lt}\,(g_2) = x^2 y^3$$

so that neither term in $x - y$ is divisible by $\mathrm{lt}\,(g_1)$ or $\mathrm{lt}\,(g_2)$. In some sense this has allowed $y - x$ to slip through the net of the division algorithm. What is critical here is that $y - x$ is still an element of $\langle g_1, g_2 \rangle$. Before we get to the solution of this conundrum, we need the following observation.

**Lemma 5.4.3** *Any nonzero ideal $I \subseteq F[x_1, \ldots, x_n]$ that is generated by monomials is generated by a finite set of monomials.*

**Proof.** The general argument proceeds by induction on $n$. To keep it reasonably simple, we will just consider the cases $n = 1$ and $2$. These cases will illustrate the elements of the general proof quite well. So first let $n = 1$ and $I = \langle x^i \mid i \in A \rangle$ where $A \subseteq \mathbb{N} \cup \{0\}$. Let

$$m = \min\{i \mid i \in A\}.$$

Then $x^m \in I$ and we have

$$\langle x^m \rangle \subseteq I = \langle x^i \mid i \in A \rangle \subseteq \langle x^m \rangle$$

so that $I = \langle x^m \rangle$, and we have the desired result.

Now let $I \subseteq F[x, y]$ and $A \subseteq (\mathbb{N} \cup \{0\}) \times (\mathbb{N} \cup \{0\})$ be such that

$$I = \langle x^i y^j \mid (i, j) \in A \rangle.$$

Let

$$J = \langle x^i \mid \exists\, j \text{ with } x^i y^j \in I \rangle.$$

If $J = \{0\}$, then $I \subseteq F[y]$ and we are back in the single-variable case. So we will assume that $J \neq \{0\}$. Clearly, $J$ is an ideal in $F[x]$ that is generated by monomials and so, by the preceding, $J = \langle x^m \rangle$ for some integer $m$. Now let $k$ be such that $x^m y^k \in I$. Define

$$I_0 = \langle x^i \mid x^i \in I \rangle$$
$$I_1 = \langle x^i \mid x^i y \in I \rangle$$
$$I_2 = \langle x^i \mid x^i y^2 \in I \rangle$$
$$\vdots$$
$$I_{k-1} = \langle x^i \mid x^i y^{k-1} \in I \rangle.$$

Each of these ideals is generated by monomials in $F[x]$ and so each is generated by a single monomial, say

$$I_j = \langle x^{m_j} \rangle.$$

Now set

$$K = \langle x^m y^k, x^{m_0}, x^{m_1} y, \ldots, x^{m_{k-1}} y^{k-1} \rangle.$$

Since each generator for $K$ is a monomial in $I$, it follows immediately that $K \subseteq I$. Conversely, let $x^s y^t$ be any monomial in $I$. We have

$$t \geq k \Longrightarrow x^s \in J = \langle x^m \rangle$$
$$\Longrightarrow m \leq s$$
$$\Longrightarrow x^s y^t = x^{s-m} y^{t-k} (x^m y^k) \in K$$

whereas

$$t < k \implies x^s \in I_t = \langle x^{m_t} \rangle$$
$$\implies m_t \leq s$$
$$\implies x^s y^t = x^{s-m_t}(x^{m_t} y^t) \in K.$$

Thus, in all cases, $x^s y^t \in K$. Since $I$ is generated by monomials, it follows that $I \subseteq K$ and therefore that $I = K$, yielding the desired conclusion. $\quad\square$

**Definition 5.4.4** *Let $I$ be an ideal in $F[x_1, \ldots, x_n]$, $g_1, \ldots, g_k \in F[x_1, \ldots, x_n]$. Then $\{g_1, \ldots, g_k\}$ is a* Groebner basis *for $I$ (with respect to a monomial order) if*

(i) $I = \langle g_1, \ldots, g_k \rangle$
(ii) $g \in I, g \neq 0 \Rightarrow \exists\, i$ *such that* lt $(g_i)$ *divides* lt $(g)$.

Note that, by definition, a Groebner basis is necessarily finite.

**Theorem 5.4.5 (Buchberger)** *Every nonzero ideal $I$ in $F[x_1, \ldots, x_n]$ has a Groebner basis.*

**Proof.** Let

$$M = \langle m \mid m = \text{lm}\,(f), \quad \text{for some } f \in I \rangle.$$

By Lemma 5.4.3, there exists a finite set $\{m_1, m_2, \ldots, m_k\}$ of monomials that generates $M$:

$$M = \langle m_1, m_2, \ldots, m_k \rangle.$$

It is easy to see that each of these monomials is the leading monomial of a polynomial in $I$, say

$$m_i = \text{lm}\,(g_i) \qquad i = 1, 2, \ldots, k$$

where each $g_i \in I$. The polynomials $g_i$ are going to be our generators for $I$.
Let $f \in I$. Dividing $f$ by $g_1, \ldots, g_k$ as in Lemma 5.4.1, we obtain

$$f = a_1 g_1 + a_2 g_2 + \cdots + a_k g_k + r$$

where no term in $r$ is divisible by any of the leading terms of $g_1, \ldots g_k$. However,

$$r = f - a_1 g_1 - a_2 g_2 - \cdots - a_k g_k \in I$$

which implies that either $r = 0$ or

$$\operatorname{lm}(r) \in M = \langle m_1, \ldots, m_k \rangle.$$

However, the latter would imply that $m_i \mid \operatorname{lm}(r)$, for some $i$, and therefore that $\operatorname{lt}(g_i) \mid \operatorname{lt}(r)$, which would be a contradiction. So we must have $r = 0$ and

$$f = a_1 g_1 + \cdots + a_k g_k \in \langle g_1, \ldots, g_k \rangle.$$

Therefore, $I \subseteq \langle g_1, \ldots, g_k \rangle$ and the reverse containment is trivial. Thus

$$I = \langle g_1, \ldots, g_k \rangle.$$

Finally,

$$
\begin{aligned}
f \in I &\Longrightarrow \operatorname{lm}(f) \in M \\
&\Longrightarrow m_i \mid \operatorname{lm}(f), \quad \text{for some } i \\
&\Longrightarrow \operatorname{lt}(g_i) \mid \operatorname{lt}(f), \quad \text{for some } i.
\end{aligned}
$$

Therefore, $\{g_1, \ldots, g_k\}$ is a Groebner basis for $I$. $\quad\square$

One important consequence of Theorem 5.4.5, is a famous result known as Hilbert's Basis Theorem which states that every nonzero ideal in $F[x_1, \ldots, x_n]$ has a finite basis.

From this we can attain our goal of a unique remainder under division by multiple divisors, when properly executed.

**Theorem 5.4.6 (Division Algorithm with Multiple Divisors in $F[x_1, \ldots, x_n]$)**
*Let $\{g_1, \ldots, g_k\}$ be a Groebner basis for the ideal $I$ in $F[x_1, \ldots, x_n]$. Let $f \in F[x_1, \ldots, x_n]$. Then there exist $q_1, \ldots, q_k, r \in F[x_1, \ldots, x_n]$ such that*

$$f = q_1 g_1 + \cdots + q_k g_k + r$$

*and no monomial term in $r$ is divisible by any of $\operatorname{lt}(g_1), \ldots, \operatorname{lt}(g_k)$. Moreover, $r$ is unique.*

**Proof.** In light of Lemma 5.4.1, it only remains to verify the uniqueness of $r$. Let

$$
\begin{aligned}
f &= q_1 g_1 + \cdots + q_k g_k + r \\
&= q_1^* g_1 + \cdots + q_k^* g_k + r^*
\end{aligned}
$$

where no term in $r$ or $r^*$ is divisible by any of lt $(g_1), \ldots,$ lt $(g_k)$. Then,

$$r^* - r = (q_1 - q_1^*)g_1 + \cdots + (q_k - q_k^*)g_k \in I.$$

If $r^* - r = 0$, then we are done. So suppose that $r^* - r \neq 0$. Since $\{g_1, \ldots, g_k\}$ is a Groebner basis for $I$, there must exist $g_i$ such that lt $(g_i)$ divides lt $(r^* - r)$. Now, lt $(r^* - r)$ is just the difference between two monomial terms $c^*m$, $cm$, say, in $r^*$ and $r$, respectively. Thus we can write

$$\text{lt } (r^* - r) = (c^* - c)m.$$

But then the lt $(g_i)$ must divide both $c^*m$ and $cm$, contradicting the assumption that no term in $r$ or $r^*$ is divisible by any of lt $(g_1), \ldots,$ lt $(g_k)$. Hence, $r^* - r = 0$ and $r^* = r$, as required.  $\square$

Theorem 5.4.6 provides the solution to the membership problem for ideals in $F[x_1, \ldots, x_n]$ that we were looking for.

**Corollary 5.4.7** *Let* $\{g_1, \ldots, g_k\}$ *be a Groebner basis for the ideal $I$ in* $F[x_1, \ldots, x_n]$ *and* $f \in F[x_1, \ldots, x_n]$. *Let*

$$f = \sum q_\alpha g_\alpha + r$$

*as in Theorem 5.4.6. Then* $f \in I$ *if and only if* $r = 0$.

Theorem 5.4.5 may be very satisfying, but it asserts the *existence* of a Groebner basis leaving one question unanswered. Given an ideal $I = \langle g_1, \ldots, g_k \rangle$, how do we find a Groebner basis for $I$? From Example 5.4.2, we can see that this is not a simple question. However, the very calculation that we performed in Example 5.4.2 is how we find the solution.

For any monomial terms

$$m_1 = c_1 x_1^{\alpha_1} \cdots x_n^{\alpha_n}, \qquad m_2 = c_2 x_1^{\beta_1} \cdots x_n^{\beta_n} \qquad (c_1, c_2 \neq 0)$$

we define the *least common monomial multiple* of $m_1$ and $m_2$ to be

$$\text{lcmm } (m_1, m_2) = x_1^{\max(\alpha_1, \beta_1)} \cdots x_n^{\max(\alpha_n, \beta_n)}.$$

For any $g_1, g_2 \in F[x_1, \ldots, x_n]$, we define

$$g_1 * g_2 = \frac{\text{lcmm}\,(\text{lt}\,(g_1), \text{lt}\,(g_2))}{\text{lt}\,(g_1)}\, g_1 - \frac{\text{lcmm}\,(\text{lt}\,(g_1), \text{lt}\,(g_2))}{\text{lt}\,(g_2)}\, g_2.$$

The idea is that we take the smallest multiples of $g_1$ and $g_2$ that will enable us to cancel the leading terms through subtraction. For example, with $g_1$ and $g_2$ as in Example 5.4.2,

$$\text{lt}\,(g_1) = x^3 y^2, \qquad \text{lt}\,(g_2) = x^2 y^3$$

$$\text{lcmm}\,(\text{lt}\,(g_1),\, \text{lt}\,(g_2)) = x^3 y^3$$

and

$$g_1 * g_2 = y g_1 - x g_2 = y - x.$$

This is the basis for the following result, which we present without proof (see [CLO]).

**Theorem 5.4.8 (Buchberger's Algorithm)** *Let $I = \langle g_1, \ldots, g_k \rangle \subseteq F[x_1, \ldots, x_n]$. Let $B_1 = \{g_1, \ldots, g_k\}$ and for $i = 1, 2, \ldots$ define $R_{i+1}, B_{i+1}$ as follows:*

$$R_{i+1} = \text{set of nonzero remainders obtained from all the polynomials}$$
$$g * h, \ g, h \in B_i \text{ on division by } B_i$$

$$B_{i+1} = B_i \cup R_{i+1}.$$

*Then there exists an integer $m$ such that $B_m = B_{m+1}$, and $B_m$ is a Groebner basis for $I$.* $\square$

The Groebner basis produced by Buchberger's algorithm may contain some obvious redundancies that can be deleted. There are various refinements that can be introduced and that lead to a unique form of the basis.

The Groebner basis generated by Buchberger's algorithm can look quite different from the original basis. In Example 5.4.2 we have the basis

$$B_1 = \{a, b\} \quad \text{for} \quad I = \langle a, b \rangle$$

where

$$a = a(x, y) = x^3 y^2 + 1, \qquad b = b(x, y) = x^2 y^3 + 1.$$

With the lexicographic order based on $x > y$, we compute

$$a * b = y(x^3 y^2 + 1) - x(x^2 y^3 + 1)$$

$$= y - x.$$

Division of $c = -x + y$ by $B_1$ yields $c$ as the remainder. Hence,

$$B_2 = \{a, b, c\}.$$

We have already computed $a * b$, so there is no need to recompute it. At this step we just consider $a * c$ and $b * c$. We have

$$a * c = 1 \cdot (x^3 y^2 + 1) + x^2 y^2 \cdot (-x + y)$$

$$= x^2 y^3 + 1$$

$$= b$$

so that the remainder will be zero. Next

$$b * c = 1 \cdot (x^2 y^3 + 1) + xy^3(-x + y)$$

$$= 1 + xy^4$$

where

$$1 + xy^4 = -y^4(-x + y) + 1 + y^5.$$

Thus, the remainder in this case is

$$d = 1 + y^5$$

and

$$B_3 = \{a, b, c, d\}.$$

Continuing,

$$a * d = y^3(x^3 y^2 + 1) - x^3(y^5 + 1)$$

$$= y^3 - x^3 = (y - x)(y^2 + xy + x^2),$$

$$b * d = y^2(x^2 y^3 + 1) - x^2(y^5 + 1)$$

$$= y^2 - x^2 = (y - x)(y + x),$$

$$c * d = -y^5(-x + y) - x(y^5 + 1)$$
$$= -(y^6 + x)$$
$$= -y(y^5 + 1) + (-x + y).$$

Thus, the remainders from $a * d$, $b * d$, and $c * d$ are all zero. Thus,

$$B_3 = \{x^3 y^2 + 1, \; x^2 y^3 + 1, \; -x + y, \; 1 + y^5\}$$

is a Groebner basis for

$$I = \langle x^3 y^2 + 1, \; x^2 y^3 + 1 \rangle.$$

Now any monomial that is divisible by $x^3 y^2$ or $x^2 y^3$ is also divisible by the leading term $-x$ of $-x + y$. Thus, $a$ and $b$ are not needed for any contribution from their leading terms. Moreover,

$$x^3 y^2 + 1 = (-x^2 y^2 - xy^3 - y^4)(-x + y) + y^5 + 1$$
$$x^2 y^3 + 1 = (- xy^3 - y^4)(-x + y) + y^5 + 1$$

so that

$$x^3 y^2 + 1, \; x^2 y^3 + 1 \in \langle -x + y, \; 1 + y^5 \rangle.$$

Therefore,

$$I = \langle -x + y, \; 1 + y^5 \rangle$$

and

$$B_4 = \{-x + y, \; 1 + y^5\}$$

is a simpler Groebner basis for $I$. If we had chosen to order the variables with $y > x$, we would have obtained the Groebner basis

$$B_5 = \{x - y, \; 1 + x^5\}.$$

**Exercises 5.4**

1. Divide the polynomial $x^2 z - y^3$ by the polynomials $x^2 + y$, $y^2 + z$, $x + z^2$ and the polynomials $x + z^2$, $y^2 + z$, $x^2 + z$ (as in Lemma 5.4.1) to show that the remainder depends on the order in which the divisors are applied.

2. Show that $\{x + z^2, y + z^4, z + z^8\}$ is a Groebner basis for

$$I = \langle x^2 + y, \ y^2 + z, \ z^2 + x \rangle \subseteq F[x, y, z]$$

   using Buchberger's algorithm with the lexicographic order based on $x > y > z$.

3. Let $I$ be an ideal in $F[x_1, \ldots, x_n]$ and $g_1, \ldots, g_k \in I$ be such that

$$g \in I \implies \exists\, i \ \text{ such that lt}\,(g_i) \text{ divides lt}\,(g).$$

   Show that $\{g_1, \ldots, g_k\}$ is a Groebner basis for $I$.

4. Show that the following three conditions on a ring $R$ are equivalent:

   (i)  For every ascending chain of ideals $I_1 \subseteq I_2 \subseteq \ldots$, there exists an integer $N$ such that $I_N = I_{N+1} = \ldots$.
   (ii)  Every nonempty set $\mathcal{C}$ of ideals contains a maximal ideal.
   (iii)  Every ideal in $R$ is finitely generated.


## 5.5 Ideals and Affine Varieties

One of the important reasons for studying ideals in $F[x_1, \ldots, x_n]$ is their connection with the zeros of polynomials. We begin to explore this relationship in this section. We remind you that, if $f = f(x_1, \ldots, x_n), g = g(x_1, \ldots, x_n) \in [x_1, \ldots, x_n]$, then $fg$ denotes the product of polynomials, not the composition of polynomials.


**Lemma 5.5.1** *Let $V \subseteq F^n$. Then*

$$I(V) = \{f \in F(x_1, \ldots, x_n) \mid f(a_1, \ldots, a_n) = 0$$
$$\forall\, (a_1, \ldots, a_n) \in V\}$$

*is an ideal of $F(x_1, \ldots, x_n)$.*


**Proof.** Clearly, $I(V) \neq \emptyset$ because it contains the zero polynomial. Let $f, g \in I(V)$ and $(a_1, \ldots, a_n) \in V$. Then,

$$(f - g)(a_1, \ldots, a_n) = f(a_1, \ldots, a_n) - g(a_1, \ldots, a_n)$$
$$= 0 - 0$$
$$= 0$$

and, for any $h \in F[x, \ldots, x_n]$, we have

$$
\begin{aligned}
(hf)(a_1, \ldots, a_n) &= h(a_1, \ldots, a_n) f(a_1, \ldots, a_n) \\
&= h(a_1, \ldots, a_n) \cdot 0 \\
&= 0.
\end{aligned}
$$

Thus, $f - g$, $hf \in I(V)$. Since $F[x_1, \ldots, x_n]$ is a commutative ring, it follows that $I(V)$ is an ideal. $\quad\square$

We refer to $I(V)$ as the *ideal of V* or the *ideal defined by V*. Whenever we come across a new operator like $I(-)$, it is always helpful to pinpoint one or two values in simple situations. The two extreme values for $V$ would be $V = \emptyset$ and $V = F^n$. Let $f \in F[x_1, \ldots, x_n]$. Then the following implication holds:

$$
(a_1, \ldots, a_n) \in \emptyset \implies f(a_1, \ldots, a_n) = 0
$$

since there are no points in $\emptyset$ to contradict the conclusion. Hence, $f \in V(\emptyset)$ so that

$$
I(\emptyset) = F[x_1, \ldots, x_n].
$$

At the other extreme, when $V = F^n$, the outcome depends on the field $F$. If $F$ is finite, say $F = \mathrm{GF}(p^m)$, then $I(F) \neq \{0\}$ since we know from Lemma 2.9.1 that

$$
x^{p^m} - x \in I(F).
$$

It is a simple exercise to show that

$$
I(F) = \langle x^{p^m} - x \rangle.
$$

On the other hand, suppose that $F$ is infinite. Then,

$$
f \in I(F^n) \implies f(a_1, \ldots, a_n) = 0 \quad \forall\, (a_1, \ldots, a_n) \in F^n
$$

$$
\implies f = 0 \quad \text{by Theorem 5.2.8.}
$$

Thus, in this case,

$$
I(F^n) = \{0\}.
$$

For another example, suppose that we let $P = (a, b) \in \mathbb{R}^2$ and set $V = \{P\}$. What is $I(V)$? Clearly, $I(V)$ does not contain any nonzero constants. Looking at polynomials of degree 1 we see that

$$
x - a, \ y - b \in I(V)
$$

so that $\langle x - a, y - b \rangle \subseteq I(V)$.

Now consider any $f(x) \in I(V)$. By Lemma 5.4.1, there exist $q(x,y)$, $r(x,y) \in F[x,y]$ and $c \in F$ such that

$$f(x,y) = (x - a)\,q(x,y) + (y - b)r(x,y) + c$$

where no term in $c$ is divisible by either $\mathrm{lt}(x - a) = x$ or $\mathrm{lt}(y - b) = y$. Therefore $c$ must be a constant.

Then $0 = f(a,b) = (a - a)\,q(x,y) + (b - b)r(x,y) + c = c$. Thus, $f(x,y) \in \langle x - a, y - b \rangle$. Therefore, $I(V) \subseteq \langle x - a, y - b \rangle$, from which it follows that $I(V) = \langle x - a, y - b \rangle$. This argument is easily extended to yield the following useful observation concerning $F^n$. Let $P = (a_1, \ldots, a_n)$ be a point in $F_n$ and set $V = P$. Then

$$I(V) = \langle x_1 - a_1, x_2 - a_2, \ldots, x_n - a_n \rangle$$

From Lemma 5.5.1, we see that subsets of $F^n$ define ideals in $F[x_1, \ldots, x_n]$. Conversely, subsets of $F[x_1, \ldots, x_n]$ can be used to identify special subsets of $F^n$. For any subset $A \subseteq F[x_1, \ldots, x_n]$, we define

$$V(A) = \{(a_1, \ldots, a_n) \in F^n \mid f(a_1, \ldots, a_n) = 0 \quad \forall f \in A\}.$$

Another way of viewing $V(A)$ is as the solution set for the set of polynomial equations

$$f(x_1, \ldots, x_n) = 0 \qquad (f \in A).$$

Any subset of $F^n$ of the form $V(A)$, where $A \subseteq F[x_1, \ldots, x_n]$ is called an *affine variety*. If $A = \{f\}$ has only one element, we may also write $V(A) = V(f)$. If $n = 2$, that is if $f \in F[x_1, x_2]$ and is nonconstant, we will refer to $V(f)$, or simply to $f$ or $f = 0$, as (defining) a *curve*. If the degree of $f$ is one, then the curve is a line and if the degree of $f$ is two, then it is a conic.

**Example 5.5.2** Any linear equation $ax + by + c = 0$ defines a line in $\mathbb{R}^2$ and a linear equation $a_1 x_1 + a_2 x_2 + \cdots + a_n x_n = 0$ defines a hyperplane in $\mathbb{R}^n$.

**Example 5.5.3** The familiar conics provide examples of curves in $\mathbb{R}^2$ defined by quadratic polynomials.

(1) The empty set: $x^2 + y^2 + 1 = 0$.

(2) A point: $(x - a)^2 + (y - b)^2 = 0$.

(3) Two intersecting lines: $(a_1 x + b_1 y + c_1)(a_2 x + b_2 y + c_2) = 0$, where $a_1 b_2 \neq a_2 b_1$.

(4) A circle: $(x - a)^2 + (y - b)^2 = r^2$.

(5) A parabola: $(y - b)^2 = c(x - a)$, $c \neq 0$.

(6) An ellipse: $\dfrac{(x - x_1)^2}{a^2} + \dfrac{(y - y_1)^2}{b^2} = 1.$

(7) A hyperbola: $\dfrac{(x - x_1)^2}{a^2} - \dfrac{(y - y_1)^2}{b^2} = 1.$

We are a good deal more familiar with affine varieties, as the previous examples illustrate, than with the corresponding ideals. However, it is still useful to identify the extreme values. In this case, it is not difficult to see that

$$V(0) = F^n, \qquad V(F[x_1, \ldots, x_n]) = \emptyset.$$

We now have mappings from subsets of $F^n$ to subsets of $F[x_1, \ldots, x_n]$ and back again. What happens when we compose them?

**Proposition 5.5.4** *Let $U, V \subseteq F^n$ and $A, B \subseteq F[x_1, \ldots, x_n]$. Then*

(i) $U \subseteq V \implies I(V) \subseteq I(U),$

$\quad\ A \subseteq B \implies V(B) \subseteq V(A).$

(ii) $V \subseteq V(I(V)), \quad A \subseteq I(V(A)).$

**Proof.** (i) Let $f \in I(V)$ and $(a_1, \ldots, a_n) \in U$. Then $(a_1, \ldots, a_n) \in V$ so that

$$f(a_1, \ldots, a_n) = 0.$$

Hence, $f \in I(U)$ and so $I(V) \subseteq I(U)$.

Now let $(a_1, \ldots, a_n) \in V(B)$ and $f \in A$. Then, $f \in B$ and so

$$f(a_1, \ldots, a_n) = 0.$$

Hence, $(a_1, \ldots, a_n) \in V(A)$ and so $V(B) \subseteq V(A)$.

(ii) Let $(a_1, \ldots, a_n) \in V$. Then, for any $f \in I(V)$ we have $f(a_1, \ldots, a_n) = 0$. Hence, $(a_1, \ldots, a_n) \in V(I(V))$ so that $V \subseteq V(I(V))$.

Similarly, let $f \in A$. Then, for any $(a_1, \ldots, a_n) \in V(A)$ we have $f(a_1, \ldots, a_n) = 0$. Hence, $f \in I(V(A))$ so that $A \subseteq I(V(A))$. $\square$

As it turns out, we do not lose any affine varieties if we restrict our attention to those of the form $V(I)$ for some ideal $I$.

**Corollary 5.5.5** *Let $A \subseteq F[x_1, \ldots, x_n]$. Then*

$$V(A) = V(\langle A \rangle).$$

**Proof.** Since $A \subseteq \langle A \rangle$, it follows from Proposition 5.5.4 (i) that $V(\langle A \rangle) \subseteq V(A)$. So let $(a_1, \ldots, a_n) \in V(A)$. Consider any $f \in \langle A \rangle$. Then there exist

$f_1, \ldots, f_m \in A$ and $g_1, \ldots, g_m \in F[x_1, \ldots, x_n]$ with

$$f = g_1 f_1 + \cdots + g_m f_m.$$

Hence, since $f_i \in A$ for all $i$,

$$\begin{aligned}
f(a_1, \ldots, a_n) &= g_1(a_1, \ldots, a_n) f_1(a_1, \ldots, a_n) + \cdots \\
&\quad + g_m(a_1, \ldots, a_n) f_m(a_1, \ldots, a_n) \\
&= g_1(a_1, \ldots, a_n) \cdot 0 + \cdots + g_m(a_1, \ldots, a_n) \cdot 0 \\
&= 0.
\end{aligned}$$

Since this holds for all $f \in \langle A \rangle$, it follows that $(a_1, \ldots, a_n) \in V(\langle A \rangle)$. Therefore, $V(A) \subseteq V(\langle A \rangle)$ and equality prevails. $\quad\square$

**Corollary 5.5.6** *The mapping I from the set of affine varieties in $F^n$ to the set of ideals in $F[x_1, \ldots, x_n]$ is injective.*

**Proof.** Let $V_1$ and $V_2$ be distinct affine varieties in $F^n$, say $V_1 = V(A_1)$, $V_2 = V(A_2)$ where $A_1, A_2 \subseteq F[x_1, \ldots, x_n]$. Since $V_1 \neq V_2$, there is no loss in assuming that there exists $(a_1, \ldots, a_n) \in V_1 \backslash V_2$. By the definition of $V_1$ and $V_2$ it follows that there must be an element $f \in A_2$ such that $f(a_1, \ldots, a_n) \neq 0$. Consequently $f \notin I(V_1)$. But, by Proposition 5.5.4 (ii),

$$f \in A_2 \subseteq I(V(A_2)) = I(V_2).$$

Therefore, $I(V_1) \neq I(V_2)$. $\quad\square$

In contrast to Corollary 5.5.6, it is easy to see that the mapping $V$ is not injective on ideals. For example,

$$V(\langle x \rangle) = V(\langle x^2 \rangle) = \cdots = V(\langle x^n \rangle).$$

Certain simple combinations of affine varieties are also affine varieties.

**Lemma 5.5.7** *Let $U, V \subseteq F^n$ be affine varieties. Then $U \cup V$ and $U \cap V$ are also affine varieties.*

**Proof.** Let $A, B \subseteq F[x_1, \ldots, x_n]$ be such that

$$U = V(A), \qquad V = V(B).$$

Let

$$AB = \{fg \mid f \in A,\ g \in B\}.$$

We will show that

(i) $U \cap V = V(A \cup B)$

(ii) $U \cup V = V(AB)$.

(i) Let $x = (a_1, \ldots, a_n) \in U \cap V = V(A) \cap V(B)$. Then for any $f \in A \cup B$, we have either $f \in A$ and $x \in V(A)$ or $f \in B$ and $x \in V(B)$. Either way, $f(a_1, a_2, \ldots, a_n) = 0$. Thus, $x \in V(A \cup B)$ and $U \cap V \subseteq V(A \cup B)$.

Conversely, $x \in V(A \cup B)$ implies that $x \in V(A) = U$ and $x \in V(B) = V$ so that $x \in U \cap V$. Thus, $V(A \cup B) \subseteq U \cap V$ and equality prevails.

(ii) Let $x = (a_1, \ldots, a_n) \in U \cup V$. Then for any $f \in A$, $g \in B$ we have that either $x \in U$ so that

$$fg(x) = f(x)\, g(x) = 0 \cdot g(x) = 0$$

or $x \in V$, yielding

$$fg(x) = f(x)\, g(x) = f(x) \cdot 0 = 0.$$

Either way, $fg(x) = 0$ so that $x \in V(AB)$. Thus, $U \cup V \subseteq V(AB)$.

Conversely, let $x = (a_1, \ldots, a_n) \in V(AB)$. Suppose that $x \notin U = V(A)$. Then there must exist $f \subset A$ such that $f(x) \neq 0$. However, $x \in V(AB)$. Hence, for all $g \in B$,

$$f(x)\, g(x) = fg(x) = 0.$$

Since $F$ is a field and $f(x) \neq 0$, it follows that $g(x) = 0$. Hence, $x \in V(B) = V$. Therefore, $V(AB) \subseteq U \cup V$ and equality prevails. $\square$

We say that an affine variety $U$ is *reducible* if there exist affine varieties $V$ and $W$ such that

$$U = V \cup W, \quad V \neq U, \; W \neq U.$$

Otherwise, $U$ is said to be *irreducible*. Thus, $U$ is irreducible if, whenever $U = V \cup W$, for affine varieties $V$ and $W$ then either $U = V$ or $U = W$. We will now see that every affine variety is just a union of irreducible affine varieties. The proof involves a nice application of Theorem 5.4.5.

**Theorem 5.5.8** *Let $V \subseteq F^n$ be an affine variety. Then $V$ is the union of a finite number of irreducible affine varieties.*

**Proof.** We argue by contradiction. Suppose that $V$ is not the union of a finite number of irreducible affine varieties. Then $V$ itself must be reducible, say

$$V = V_1 \cup W_1$$

where $V_1$ and $W_1$ are both affine varieties properly contained in $V$. They cannot both be a union of a finite number of irreducible varieties (otherwise, $V$ will be the union of a finite number of irreducible varieties). Without loss of generality, we can assume that $V_1$ is not a union of a finite number of irreducible varieties. Then we can write $V_1 = V_2 \cup W_2$ where $V_2$ and $W_2$ are affine varieties properly contained in $V_1$ and at least one of $V_2$, $W_2$ is not a union of a finite number of irreducible varieties. Without loss of generality, we may assume that it is $V_2$. Now,

$$V = V_1 \cup V_2 \cup W_2.$$

Continuing in this way, we obtain a strictly decreasing sequence of affine varieties:

$$V_0 = V \supset V_1 \supset V_2 \supset \cdots .$$

By Proposition 5.5.4 and Corollary 5.5.6, this yields the strictly increasing sequence

$$I(V_0) \subset I(V_1) \subset I(V_2) \subset \cdots$$

of ideals. Let

$$I = \bigcup_{i=0}^{\infty} I(V_i).$$

It is quite straightforward to verify that $I$ is an ideal in $F[x_1, \ldots, x_n]$. By Theorem 5.4.5, this means that there exists a finite set $\{f_1, \ldots, f_m\}$ of polynomials such that

$$I = \langle f_1, \ldots, f_m \rangle.$$

But then, for each $i$, there exists $\alpha_i$ with $f_i \in I(V_{\alpha_i})$. Consequently, with $\alpha = \max\{\alpha_1, \ldots, \alpha_m\}$ we have

$$f_1, \ldots, f_m \in I(V_{\alpha_m}).$$

Therefore,

$$I \subseteq I(V_{\alpha_m})$$

so that

$$I(V_{\alpha_m}) = I(V_{\alpha_{m+1}}) = \cdots = I$$

which is a contradiction. Therefore the theorem holds. $\quad\square$

**Exercises 5.5**

1. Sketch the curves in $\mathbb{R}^2$ defined by the following polynomials.

   (i) $f_1 = x - 1$.
   (ii) $f_2 = y + 2$.
   (iii) $f_3 = 3x - 2y + 4$.
   (iv) $f_4 = x^2 + y^2 + 2x - 4y + 5$.
   (v) $f_5 = y^2 - 16x + 4$.
   (vi) $f_6 = y^2 - x^2$.

2. With the notation as in the preceding exercise, sketch the following affine varieties in $\mathbb{R}^2$:

   (i) $V(f_1 f_3)$.
   (ii) $V(f_2 f_4)$.
   (iii) $V(f_3 f_5)$.
   (iv) $V(f_1, f_4)$.
   (v) $V(f_2, f_6)$.

3. Find a polynomial $f(x, y) \in \mathbb{R}[x, y]$ such that the points $(1, 1)$, $(2, -1)$, $(3, 1)$ belong to $V(f)$.

4. Find a polynomial $f(x, y) \in \mathbb{R}[x, y]$ such that the points $(1, 1)$, $(2, -1)$, $(3, 1)$, $(3, 2)$ belong to $V(f)$ but $(1, 2)$ does not belong to $V(f)$.

5. Let $P_i$, $1 \le i \le n$ be distinct points in $\mathbb{R}^2$. Show that there exists a polynomial $f(x, y) \in \mathbb{R}[x, y]$ such that $V(f) = \{P_1, \ldots, P_n\}$.

6. Find all points in $\mathbb{Z}_7^2$ on the curve $C : y^2 - x(x - 1)(x - 2) = 0$.

7. Let $n \in \mathbb{N}$. Show that $I_n = \{f(x, y) \in F[x, y] \mid \text{degree } f(x, y) \ge n\}$ is not an ideal in $F[x, y]$.

8. Let $n \in \mathbb{N}$. Show that $J_n = \{f(x, y) \in F[x, y] \mid \text{the degree of every monomial in } f(x, y) \ge n\}$ is an ideal in $F[x, y]$.

9. Let $I = I(V)$ where $V \subseteq F^n$. Let $f \in F[x_1, \ldots, x_n]$ be such that $f(x_1, \ldots, x_n)^m \in I$ for some $m \in \mathbb{N}$. Show that $f \in I$.

10. Show that $\langle x^2 + 2xy + y^2 \rangle \ne I(V)$ for any $V \subseteq \mathbb{R}^2$.

11. Show that $J_1$, as defined in Exercise 8, is of the form $I(V)$ for some $V \subseteq F^2$.

12. Show that, for $m \ge 2$, $J_m$, as defined in Exercise 8, is not of the form $I(V)$ for any $V \subseteq F^2$.

13.   (i) Sketch the curve in $\mathbb{R}^2$ defined by the polynomial
   $$y - x(x^2 - 3x + 2).$$
   (ii) Sketch the curve in $\mathbb{R}^2$ defined by the polynomial
   $$y^2 - x(x^2 - 3x + 2).$$

    (iii) Determine the slope of the tangents to the curve in (ii) at each point where the curve crosses the $x$-axis.

14.   (i) Sketch the curve in $\mathbb{R}^2$ defined by the polynomial
$$y - x(x^2 - 2x + 1).$$
   (ii) Sketch the curve in $\mathbb{R}^2$ defined by the polynomial
$$y^2 - x(x^2 - 2x + 1).$$
  (iii) Determine the slope of the tangents to the curve in (ii) at each point where the curve crosses the $x$-axis.

15.   (i) Sketch the curve in $\mathbb{R}^2$ defined by the polynomial $y - (x - 1)^3$.
   (ii) Sketch the curve in $\mathbb{R}^2$ defined by the polynomial $y^2 - (x - 1)^3$.
  (iii) Determine the slope of the tangents to the curve in (ii) at each point where the curve crosses the $x$-axis.

16. Show that any line segment in $\mathbb{R}^2$ of finite nonzero length is not an affine variety.

17. Show that any arc of finite nonzero length on a parabola in $\mathbb{R}^2$ (with the $x$-axis as its principal axis) is not an affine variety.

18. Let $f(x, y) \in \mathbb{R}[x, y]$ and let the parabola $P$, defined by the polynomial $g(x, y)$, be such that $P \cap V(f)$ is infinite. Show that $g(x, y)$ divides $f(x, y)$.

*19. Let $V, W \subseteq F^n$. Establish the following:

    (i) $I(V \cup W) = I(V) \cup I(W)$.
   (ii) $I(V \cap W) \supseteq I(V) \cup I(W)$.
  (iii) Find an example of affine varieties $V, W \subseteq \mathbb{R}^2$ such that
$$I(V) \cup I(W) \neq I(V \cap W).$$

*20. Let $A, B$ be ideals in $F[x_1, \ldots, x_n]$. Establish the following:

    (i) $V(A \cup B) = V(A) \cap V(B)$.
   (ii) $V(A \cap B) = V(A) \cup V(B)$.

21. Let $\{V_i \mid i \in I\}$ be a (possibly) infinite family of affine varieties in $F^n$.

    (i) Show that $\bigcap_{i \in I} V_i$ is an affine variety.
   (ii) Show that for every subset $P$ of $F^n$ there is a smallest affine variety $V$ that contains $P$.

22. Find an infinite family $\{V_i \mid i \in I\}$ of affine varieties such that $\bigcup_{i \in I} V_i$ is not an affine variety.

23. Let $P = (a_1, \ldots, a_n)$ be a point in $F^n$. Show that $I(P) = \langle x_1 - a_1, x_2 - a_2, \ldots, x_n - a_n \rangle$ is a maximal ideal in $F[x_1, \ldots, x_n]$.

## 5.6 Decomposition of Affine Varieties

We have seen earlier that every polynomial $f$ in $F[x_1, \ldots, x_n]$ is a product of irreducible polynomials and we now see that every affine variety, such as $V(f)$, is a union of irreducible affine varieties. Is there any connection? First a few general trivialities.

**Lemma 5.6.1** *Let $f, g$, and $f_i$, $1 \le i \le m$, be polynomials in $F[x_1, \ldots, x_n]$.*

(i) $V(fg) = V(f) \cup V(g)$.
(ii) $I = \langle f_1, \ldots, f_m \rangle \implies V(I) = V(f_1) \cap V(f_2) \cap \cdots \cap V(f_m)$.

**Proof.** Exercise (see Lemma 5.5.7). $\square$

At first glance it would appear that Lemma 5.6.1 (i) tells us that

$$f \text{ reducible} \implies V(f) \text{ reducible.}$$

Unfortunately, it is not quite that simple in general. Consider

$$f = (x - y)(x^2 + y^2 + 2) \in \mathbb{R}[x, y].$$

Then $f$ is certainly reducible and

$$V(f) = V(x - y) \cup V(x^2 + y^2 + 2).$$

However, $V(x^2 + y^2 + 2) = \emptyset$ in $\mathbb{R}^2$. The outcome is different if we consider $f \in \mathbb{C}[x, y]$. Then $V(x^2 + y^2 + 2) \neq \emptyset$ and $V(f)$ is indeed reducible. Similarly, it is possible to have $f$ irreducible but $V(f)$ reducible. Let $f(x, y) = y^2 + x^2(x^2 - 1)^2 \in \mathbb{R}[x, y]$. Then $f(x, y)$ is irreducible but $V(f)$ is reducible. We can write $f(x, y) = y^2 - [-x^2(x^2 - 1)^2]$ and it then follows from Exercise 7 in section 5.2 that $f(x, y)$ is irreducible. However,

$$V(f) = \{(0, 0), (-1, 0), (1, 0)\}$$
$$= \{(0, 0)\} \cup \{(-1, 0)\} \cup \{(1, 0)\}$$

and so is reducible.

Thus, the picture is a little bit confused when we work over an arbitrary field, but it will be a little bit simpler if we work over an algebraically closed field. One important consequence of working over an algebraically closed field is that every nonconstant polynomial has a zero.

**Lemma 5.6.2** *Let $F$ be an algebraically closed field and $f$ be a nonconstant polynomial in $F[x_1, \ldots, x_n]$. Then $V(f) \neq \emptyset$.*

**Proof.** We argue by induction on $n$. The claim is true when $n = 1$, since that is just the definition of an algebraically closed field. Now suppose that it is true for polynomials in $x_1, \ldots, x_{n-1}$ and write $f$ as a polynomial in $x_n$

$$f = a_m x_n^m + a_{m-1} x_n^{m-1} + \cdots + a_1 x_n + a_0$$

where $a_0, \ldots, a_m \in F[x_1, \ldots, x_{n-1}]$, and $a_m \neq 0$. If $a_m \in F$, then assign any values $x_i \to \alpha_i \in F$ to $x_i$ $(1 \leq i \leq n - 1)$. Then we obtain a polynomial

$$f_1 = a_m x_n^m + a_{m-1}(\alpha_1, \ldots, \alpha_{n-1}) x_n^{m-1} + \cdots + a_0(\alpha_1, \ldots, \alpha_{n-1}) \in F[x_n].$$

Since $F$ is algebraically closed, $f_1$ has a root $\alpha_n$ so that $(\alpha_1, \ldots, \alpha_n) \in V(f)$.

Now suppose that $a_m$ is nonconstant. Then, so is $a_m(x_1, \ldots, x_{n-1}) - 1 \in F[x_1, \ldots, x_{n-1}]$ and, by the induction hypothesis, there exists $(\alpha_1, \ldots, \alpha_{n-1}) \in V(a_m - 1)$. Consequently

$$a_m(\alpha_1, \ldots, \alpha_{n-1}) = 1$$

so that

$$f_2(x_n) = f(\alpha_1, \ldots, \alpha_{n-1}, x_n)$$

$$= x_n^m + a_{m-1}(\alpha_1, \ldots, \alpha_{n-1}) x_n^{m-1} + \cdots + a_0(\alpha_1, \ldots, \alpha_{n-1}) \in F[x_n].$$

Since $F$ is algebraically closed, there exists $\alpha_n \in V(f_2)$. Then $(\alpha_1, \ldots, \alpha_n) \in V(f)$. $\square$

For any $f \in F[x_1, \ldots, x_n]$, we know that $f \in I(V(f))$ so that $\langle f \rangle \subseteq I(V(f))$. In general, this inclusion will be proper (consider $f(x) = x^2$). In one important circumstance, however, we do have equality. The next result is a special case of a theorem known as *Hilbert's Nullstellensatz* (see Theorem 5.6.7 below). It will help us to shed some light on the relationship between the factorization of a polynomial $f$ and the decomposition of $V(f)$.

**Theorem 5.6.3** *Let $F$ be an algebraically closed field and $f \in F[x_1, \ldots, x_n]$ be an irreducible polynomial. Then*

$$I\,V(f) = \langle f \rangle.$$

Even Theorem 5.6.3 does not hold over an arbitrary field. Consider $f(x, y) = x^2 + y^4$, $g(x, y) = x^2 + y^2 \in \mathbb{R}[x, y]$. Both of these polynomials are irreducible while

$$V(f) = \{(0, 0)\} = V(g).$$

Thus, $g \in I(V(f))$, but $g \notin \langle f \rangle$.

The next observation will be extremely useful, especially in chapter 6.

**Corollary 5.6.4** *Let $F$ be an algebraically closed field and $f, g \in F[x_1, \ldots, x_n]$.*

   (i)  *If $f$ is irreducible and $V(f) \subseteq V(g)$, then $f$ divides $g$.*
  (ii)  *If $f$ is irreducible then $V(f)$ is also irreducible.*

**Proof.** (i) We have

$$
\begin{aligned}
g \in I(V(g)) \qquad & \text{by Proposition 5.5.4 (ii)} \\
\subseteq I(V(f)) \qquad & \text{by Proposition 5.5.4 (i)} \\
= \langle f \rangle \qquad & \text{by Theorem 5.6.3}
\end{aligned}
$$

so that $f$ divides $g$.

(ii) By way of contradiction, suppose that $V(f) = V(A) \cup V(B)$, for some affine varieties $V(A)$, $V(B)$ where $V(A)$ and $V(B)$ are both proper subsets of $V(f)$. By Corollary 5.5.5, we may assume that $A$ and $B$ are ideals in $F[x_1, \ldots, x_n]$. By Theorem 5.4.5, there exist polynomials $f_i, g_j$ with

$$
A = \langle f_1, \ldots, f_k \rangle, \qquad B = \langle g_1, \ldots, g_\ell \rangle.
$$

If $f$ divides $f_i$ for all $i$, then $V(f) \subseteq V(f_i)$ for all $i$, so that $V(f) \subseteq V(A)$, which is a contradiction. Without loss of generality, assume that $f$ does not divide $f_1$.

For any $g_j$, we have

$$
\begin{aligned}
V(f) &= V(f_1, \ldots, f_k) \cup V(g_1, \ldots, g_\ell) \\
&= \left( \bigcap_{i=1}^{k} V(f_i) \right) \cup \left( \bigcap_{j=1}^{\ell} V(g_j) \right) \\
&\subseteq V(f_1) \cup V(g_j) \\
&= V(f_1 g_j) \qquad \text{by Lemma 5.6.1.}
\end{aligned}
$$

By part (i), this yields that $f$ divides $f_1 g_j$, but not $f_1$. By Lemma 5.2.4, $f \mid g_j$. Since $g_j$ was chosen arbitrarily, we have that $f \mid g_j$ for all $j = 1, \ldots, \ell$. Hence,

$$
V(f) \subseteq \bigcap V(g_j) = V(g_1, \ldots, g_\ell) = V(B)
$$

which is a contradiction. Therefore the claim holds. $\square$

Note that the converse of Corollary 5.6.4 (ii) does not hold since, if $V(f)$ is irreducible, then so is $V(f^2) = V(f^3) = \cdots = V(f)$, but $f^k$ is reducible for $k > 1$. However, if we eliminate powers, then we can do much better.

**Theorem 5.6.5** *Let $F$ be an algebraically closed field and $f \in F[x_1, \ldots, x_n]$. Let*

$$f = \prod_{i=1}^{m} f_i^{\alpha_i}$$

*where the $f_i$ are distinct (in the sense that no two are scalar multiples of each other) irreducible polynomials in $F[x_1, \ldots, x_n]$.*

(i) $V(f) = V(f_1) \cup V(f_2) \cup \cdots \cup V(f_m)$.
(ii) $I\,V(f) = \langle f_1 \cdots f_m \rangle$.
(iii) *If $V(f) = W_1 \cup W_2 \cup \cdots \cup W_\ell$ where each $W_j$ $(1 \le j \le \ell)$ is an irreducible affine variety, then $\ell = m$ and the $W_j$ can be reordered so that $W_i = V(f_i)$, $i = 1, \ldots, m$.*

**Proof.** (i) By Lemma 5.6.1,

$$V(f) = \bigcup_{i=1}^{m} V(f_i^{\alpha_i}) = \bigcup_{i=1}^{m} V(f_i).$$

(ii) From (i),

$$V(f) = \bigcup_{i=1}^{m} V(f_i) = V(f_1 \cdots f_m)$$

$$\implies f_1 \cdots f_m \in I\,V(f_1 \cdots f_m) = I\,V(f)$$

$$\implies \langle f_1 \cdots f_m \rangle \subseteq I\,V(f).$$

Conversely,

$$g \in I\,V(f) \implies V(f) \subseteq V(g)$$

$$\implies V(f_i) \subseteq V(f) \subseteq V(g)$$

$$\implies f_i \mid g \qquad \text{by Corollary 5.6.4 (i)}$$

$$\implies f_1 \cdots f_m \mid g$$

$$\implies g \in \langle f_1 \cdots f_m \rangle.$$

Therefore equality prevails.
(iii) See the exercises. $\square$

We are now in a position to round out the result in Corollary 5.6.4 (ii).

**Corollary 5.6.6** *Let $F$ be an algebraically closed field and $f \in F[x_1, \ldots, x_n]$. Then $V(f)$ is irreducible if and only if $f = g^\alpha$ for some irreducible polynomial $g \in F[x_1, \ldots, x_n]$ and some $\alpha \in \mathbb{N}$.*

**Proof.** If $f = g^\alpha$ where $g$ is irreducible, then $V(f) = V(g^\alpha) = V(g)$ is irreducible by Corollary 5.6.4 (ii). Conversely, assume that $V(f)$ is irreducible and let $f = f_1^{\alpha_1} \cdots f_m^{\alpha_m}$ where the $f_i$ are distinct irreducible polynomials in $F[x_1, \ldots, x_n]$. Then, by Theorem 5.6.5,

$$V(f) = V(f_1) \cup \cdots \cup V(f_m)$$

where the $V(f_i)$ are the irreducible components of $V(f)$. Since $V(f)$ is irreducible, we must have $m = 1$ and $f = f_1^{\alpha_1}$, where $f_1$ is irreducible, as required. $\quad\square$

One concluding remark is appropriate here. We have seen that for every ideal $J$ in $F[x_1, \ldots, x_n]$, we have $J \subseteq I(V(J))$ and that the containment may be proper. For instance, for any polynomial $f$ in $F[x_1, \ldots, x_n]$ and for every positive integer $m$, it is clear that $f \in I(V(\langle f^m \rangle))$. The following remarkable result shows that the difference between $J$ and $I(V(J))$, when working over an algebraically closed field, does not get much more complicated than that.

**Theorem 5.6.7** *Hilbert's Nullstellensatz. Let F be an algebraically closed field and let J be an ideal in $F[x_1, \ldots x_n]$. Then*

$$I(V(J)) = \left\{ f \in F[x_1, \ldots, x_n] | \exists m, f^m \in I \right\}.$$

**Proof.** See ([Ful], page 21) or ([Hu], Chapter 1).

**Exercises 5.6** In the exercises to follow we will show that the representation in Theorem 5.6.5 of $V(f)$ as a union of irreducible varieties is essentially unique. We continue with the notation of Theorem 5.6.5.

1. Let $g, g_i, h_i \in F[x_1, \ldots, x_n]$, where $F$ is algebraically closed, be such that $g_i = gh_i$ for $i = 1, \ldots, k$ and $\deg(g) \geq 1$. Establish the following:

   (i) $V(g_1, \ldots, g_k) = V(g) \cup V(h_1, \ldots, h_k)$.
   (ii) If $V(g_1, \ldots, g_k)$ is irreducible, then

   $$V(g_1, \ldots, g_k) = V(g).$$

2. Let $f$ be as in Theorem 5.6.5 and

   $$V(f) = W_1 \cup W_2 \cup \cdots \cup W_\ell$$

   where each $W_j$ is an irreducible affine variety and no $W_j$ is superfluous. Establish the following:

   (i) For each $i$ $(1 \leq i \leq m)$ there exists a $\beta_j$ $(1 \leq \beta_j \leq \ell)$ such that $V(f_i) \subseteq W_{\beta_i}$.

(ii) If $W_{\beta_i} = V(g_1, \ldots, g_k)$, then $f_i \mid g_j$ for all $j$.

(iii) $V(f_i) = W_{\beta_i}$.

(iv) $\ell = m$ and the decomposition of $V(f)$ in Theorem 5.6.5 is unique.

3. Let $F$ be an algebraically closed field and $f, g \in F[x_1, \ldots, x_n]$. Show that

$$V(f) = V(g) \iff f \text{ and } g \text{ have the same irreducible factors.}$$

## 5.7 Cubic Equations in One Variable

By way of background to the study of certain cubic equations in two variables that will be the focus of our attention later, we consider here the solution of the general cubic equation

$$a_3 x^3 + a_2 x^2 + a_1 x + a_0 = 0 \qquad (a_3 \neq 0) \qquad (5.11)$$

in one variable over a field of characteristic different from 2 and 3 and in which it is possible to construct square and cube roots. As the argument is the same in all these cases and because the complex case is of greatest interest to us, we assume the base field to be $\mathbb{C}$.

The first reduction is to a monic polynomial equation of the form

$$x^3 + b_2 x^2 + b_1 x + b_0 = 0 \qquad (5.12)$$

by the simple expedient of multiplying both sides of (5.11) by $a_3^{-1}$. Now, substituting $x = y - b_2/3$ in (5.12) our polynomial becomes

$$(y - \frac{1}{3} b_2)^3 + b_2 (y - \frac{1}{3} b_2)^2 + b_1 (y - \frac{1}{3} b_2) + b_0$$
$$= y^3 + 0 \cdot y^2 + py + q$$

for suitable $p$ and $q$ since the terms in $y^2$ cancel. Thus, our problem reduces to solving the equation

$$y^3 + py + q = 0. \qquad (5.13)$$

Now, substituting

$$y = z - \frac{p}{3z} \qquad (5.14)$$

we have to solve

$$0 = y^3 + py + q = \left(z - \frac{p}{3z}\right)^3 + p\left(z - \frac{p}{3z}\right) + q$$

$$= z^3 - pz + \frac{p^2}{3z} - \frac{p^3}{27z^3} + pz - \frac{p^2}{3z} + q$$

$$= z^3 - \frac{p^3}{27z^3} + q.$$

Multiplying through by $z^3$ we obtain

$$(z^3)^2 + q(z^3) - \frac{p^3}{27} = 0$$

a quadratic in $z^3$. Applying the standard solution for a quadratic equation we obtain

$$z^3 = \frac{-q + \alpha}{2} \tag{5.15}$$

where $\alpha$ is cither of the roots of the equation

$$w^2 = q^2 + \frac{4p^3}{27}. \tag{5.16}$$

Taking cube roots in (5.15) yields six possible values for $z$. We could write these as

$$z = \left(-\frac{q}{2} + \sqrt{\frac{q^2}{4} + \frac{p^3}{27}}\right)^{1/3} \tag{5.17}$$

where the $\sqrt{\phantom{x}}$ sign stands for either root of (5.16).

Thus we have shown that we can solve equation (5.13) by proceeding as follows:

(1) Solve (5.16).

(2) Substitute the solutions from (5.16) in (5.17) and solve (5.17).

(3) Substitute the solutions from (5.17) in (5.14).

(4) Substitute the solutions from (5.14) in the equation $x = y - b_2/3$.

Of course we know from the general theory that there can be, at most, three distinct solutions to equation (5.13). Thus, our six possible solutions to equation (5.17) must collapse to, at most, three distinct solutions when we make the substitution in (5.14). Note that we can rewrite equation (5.14) as

$$z^2 - 3yz - p = 0 \tag{5.18}$$

which is a quadratic in $z$. Thus, for each of our solutions for $y$ there will, in general, be two values of $z$ that provide solutions to (5.18) and, therefore, the same value of $y$ when substituted into (5.14).

There is one observation that can be made when we are working over the field $\mathbb{C}$.

**Proposition 5.7.1** *If* $4p^3 + 27q^2 \neq 0$, *then the equation* $y^3 + py + q = 0$ *has three distinct solutions in* $\mathbb{C}$.

**Proof.** If $4p^3 + 27q^2 \neq 0$, then $q^2 + \frac{4p^3}{27} \neq 0$ and the equation (5.16) has two distinct solutions. Hence, equation (5.17) must have six distinct solutions leading, in (5.14), to three distinct solutions to (5.13). $\square$

The expression $4p^3 + 27q^2$ will figure prominently in our discussions of elliptic curves later.

**Exercises 5.7**

1. (i) Let $p \in \mathbb{C}$. Show that the mapping

$$\varphi : z \to z - \frac{p}{3z} \qquad (z \in \mathbb{C})$$

   is surjective (from $\mathbb{C}$ to $\mathbb{C}$) for all $p \in \mathbb{C}$.
   (ii) Show that, for $p \neq 0$, $\varphi$ is not injective.

2. Let $\alpha, \beta \in \mathbb{C}$ and $\alpha \neq \beta$. Show that

$$V(x^3 - \alpha) \cap V(x^3 - \beta) = \emptyset.$$

## 5.8 Parameters

Given a set of polynomial equations $f_i(x_1, \ldots, x_n) = 0$, natural questions to ask are: In what form would we hope or like to describe the solution set? What is it that we would like to achieve? After all, the equations as they stand, assuming that there are only finitely many of them, are well suited to testing whether a given point $x = (a_1, \ldots, a_n)$ is in the solution set or not. All that we have to do is substitute the values of the $a_i$'s for the $x_i$'s in each $f_i$ and see if the result reduces to zero. However, this does not help us to find or produce solutions. What we want is a way to list, or at least generate, solutions. A good way to do this is with parameters. We say that we have a *parametric solution* to the system of equations $f_i(x_1, \ldots, x_n) = 0$ $(i = 1, \ldots, m)$ if the solutions are given by the equations

$$x_i = p_i(t_1, \ldots, t_k) \qquad (1 \leq i \leq n)$$

where the *parameters* $t_j$ range freely over sets of values $A_j$   $(1 \leq j \leq k)$ for suitable $A_j$.

A parametric solution makes it easy to generate many solutions; we just feed in as many different values of the parameters $t_j$ as we please. In particular, this makes it easy to plot solutions if we are working with polynomials in two or three variables over $\mathbb{R}$ or $\mathbb{Q}$.

This concept is almost certainly not new to you.

A good illustration is the way that we solve systems of linear equations. Consider the system of equations in $F[w, x, y, z]$:

$$
\begin{aligned}
w + \ x + y + \ z &= 0 \\
w + 2x \quad\ - \ z &= 0 \\
2w + 3x + y \quad\quad &= 0 \\
3w + 7x - y - 5z &= 0.
\end{aligned}
$$

Gaussian elimination yields the solution

$$
\begin{aligned}
w &= -\ 2y - 3z \\
x &= \quad y + 2z.
\end{aligned}
$$

We can write this in parametric form as

$$
\begin{aligned}
w &= -2s - 3t \\
x &= \quad s + 2t \\
y &= \quad s \\
z &= \quad t.
\end{aligned}
\tag{5.19}
$$

Then every choice of the parameters $s, t \in \mathbb{R}$ yields a solution, and every solution can be obtained in this way.

A very important special case of this leads to a parametric description of a line. For any field $F$ we define a *line* in $F^2$ to be the set of solutions to an equation of the form

$$
L : \ lx + my + n = 0 \qquad (l, m, n \in F)
$$

where at least one of $l, m$ is nonzero. If $l \neq 0$, then we can solve this equation as

$$
x = -l^{-1}my - l^{-1}n.
$$

Since the value of $y$ is unconstrained, we can easily write this solution in parametric form as

$$
\begin{aligned}
x &= -l^{-1}mt - l^{-1}n \\
y &= t.
\end{aligned}
$$

In a similar manner, if $m \neq 0$, then we can obtain a parametric solution of the form

$$x = t$$

$$y = -m^{-1}lt - m^{-1}n.$$

Either way, we obtain a solution in the form

$$\left.\begin{array}{l} x = a + ct \\ y = b + dt \end{array}\right\} \qquad (5.20)$$

where $a, b, c, d \in F$, $(a, b)$ is a point on $L$, and $lc + md = 0$. Conversely, given any point $(a, b)$ on $L$ and $c, d \in F$ with $lc + md = 0$, any point with coordinates given by (5.20) will lie on $L$ and every point on $L$ can be so described.

Henceforth, when we speak of the *parametric equations* of a line or a *line in parametric form*, we will mean a system of parametric equations as in (5.20).

Note that when a line $L$ is given in parametric form, as in (5.20), then

(1) the point $(a, b)$ lies on $L$

(2) if the base field is $\mathbb{R}$ and $c \neq 0$, the slope of $L$ is given by $c^{-1}d$.

If $c = 0$, then the line is parallel to the $y$-axis and is given by the equation $x = a$.

Note that distinct values of $c$ and $d$ do not necessarily define different lines in (5.20). If $c' = \lambda c$, $d' = \lambda d$ ($\lambda \neq 0$), then the line

$$x = a + c't = a + c(\lambda t)$$

$$y = b + d't = b + d(\lambda t)$$

is clearly the same line as that defined in (5.20). Conversely, suppose that $L$ is also defined by

$$x = a + c't$$

$$y = b + d't.$$

With $t = 1$, we see that $(a + c', b + d') \in L$. Hence there must exist a value $t = \lambda$ with

$$(a + c', b + d') = (a + c\lambda, b + d\lambda)$$

so that we have

$$c' = \lambda c, \qquad d' = \lambda d$$

where, since one of $c'$, $d'$ is nonzero, $\lambda \neq 0$. Thus, the line $L$ is really defined by the pair $(c, d)$ to within nonzero scalar multiples—an idea to which we will return shortly.

For any two points $P_1 = (x_1, y_1, z_1)$, $P_2 = (x_2, y_2, z_2)$, in $\mathbb{R}^3$, the standard form for the line through $P_1$ and $P_2$ is

$$\frac{x - x_1}{x_2 - x_1} = \frac{y - y_1}{y_2 - y_1} = \frac{z - z_1}{z_2 - z_1}.$$

If $x_2 = x_1$, then we drop the first expression and add the equation $x = x_1$, with similar treatment if $y_2 = y_1$ or $z_2 = z_1$. Assuming that $x_2 - x_1 = a \neq 0$, $y_2 - y_1 = b \neq 0$, and $z_2 - z_1 = c \neq 0$, we obtain the familiar form

$$\frac{x - x_1}{a} = \frac{y - y_1}{b} = \frac{z - z_1}{c}.$$

If we set this quantity to equal $t$, then we immediately obtain the parametric equations for a line in $\mathbb{R}^3$:

$$x = x_1 + at$$
$$y = y_1 + bt$$
$$z = z_1 + ct.$$

The nondegenerate conics (circles, ellipses, parabolas, hyperbolas) in $\mathbb{R}^2$ can all be nicely parameterized. The secret is to choose a suitable point and line to achieve the parameterization, by mapping the line to the conic. Intuitively, we are simply wrapping a line around the conics.

Consider the ellipse

$$E: \frac{x^2}{9} + \frac{y^2}{4} = 1 \tag{5.21}$$

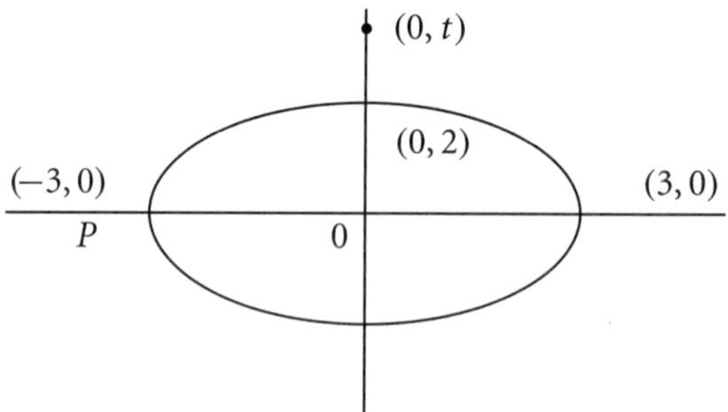

Let $L_t$ denote the line through the points $(0, t)$, $(-3, 0)$. This line has slope $t/3$ and $y$-intercept $t$ so that

$$L_t : y = \frac{t}{3}x + t. \qquad (5.22)$$

To see where this intersects the ellipse, we substitute for $y$ from (5.22) into (5.21):

$$\frac{x^2}{9} + \frac{t^2(\frac{1}{3}x + 1)^2}{4} = 1$$

or

$$4x^2 + t^2(x + 3)^2 = 36$$

or

$$(4 + t^2)x^2 + 6t^2x + (9t^2 - 36) = 0.$$

However, since $L_t$ passes through the point $(-3, 0)$, we know that $(x + 3)$ must be a factor of the left-hand side. Thus we obtain

$$(x + 3)((4 + t^2)x + (3t^2 - 12)) = 0$$

so that $L_t$ intersects $E$ in a second point where

$$(4 + t^2)x + 3t^2 - 12 = 0$$

or

$$x = \frac{12 - 3t^2}{4 + t^2} \qquad (5.23)$$

and from (5.22)

$$y = \frac{t}{3} \cdot \frac{12 - 3t^2}{4 + t^2} + t$$

$$= \frac{4t - t^3 + 4t + t^3}{4 + t^2}$$

$$= \frac{8t}{4 + t^2}. \qquad (5.24)$$

Thus we obtain the parameterization

$$x = \frac{12 - 3t^2}{4 + t^2}, \; y = \frac{8t}{4 + t^2}.$$

To check our work, we can substitute the parameterization from (5.23) and (5.24) back in (5.21) and confirm that the points with coordinates given by (5.23) and (5.24) do indeed lie on $E$. Thus, in (5.23) and (5.24) we, have a parameterization of $E$. Note that the parameterization is not a perfect match as one point (namely, $P$) is not captured. For this we need to be able to substitute the value $t = \infty$. In the world of projective spaces (see subsequent sections), this will make perfectly good sense.

To obtain the previous parameterization of the ellipse, we basically, wrapped a line ($x = 0$ in this case) around the ellipse. There are other ways of obtaining parameterizations. For example, consider the circle $x^2 + y^2 = 1$. Here we can resort to the familiar polar coordinates to characterize the points:

$$x = \cos\theta, \quad y = \sin\theta \qquad (0 \le \theta < 2\pi).$$

This parameterization of the circle is fine for the purposes of calculus. However, it uses trigonometric functions when we would prefer to stick to polynomials. We can't quite do this with this example, but we can get a parametric solution in terms of rational expressions in polynomials.

Let $t = \tan\theta/2$. Then standard trigonometric formulas yield

$$\cos\theta = \frac{1 - t^2}{1 + t^2}, \quad \sin\theta = \frac{2t}{1 + t^2}.$$

Thus we obtain the parametric solution

$$x = \frac{1 - t^2}{1 + t^2}, \quad y = \frac{2t}{1 + t^2}$$

for $t \in \mathbb{R}$. Unfortunately, this solution is not quite perfect either. To have $y = 0$, we must substitute $t = 0$. When we do this we also must have $x = 1$. Thus, there is no way to obtain the solution $(-1, 0)$, which is definitely one of the points on the circle. However, it turns out that the fact that we miss one point or even several points is not always important.

Let $V$ be an affine variety in $F^n$. Let $m \in \mathbb{N}$, and

$$p_i(t_1, \ldots, t_m), \, q_i(t_1, \ldots, t_m) \in F[t_1, \ldots, t_m] \qquad (1 \le i \le n)$$

be such that

$$V^* = \left\{ \left( \frac{p_1(t_1, \ldots, t_m)}{q_1(t_1, \ldots, t_m)}, \ldots, \frac{p_n(t_1, \ldots, t_m)}{q_n(t_1, \ldots, t_m)} \right) \,\middle|\, t_i \in F, \, q_j(t_1, \ldots, t_m) \ne 0 \right\} \subseteq V.$$

If $V$ is the smallest affine variety containing $V^*$ then we say that $V$ is *parameterized* by the equations

$$x_i = \frac{p_i(t_1, \ldots, t_m)}{q_i(t_1, \ldots, t_m)} \qquad (1 \leq i \leq n) \qquad (5.25)$$

or that (5.25) is a *parameterization* of $V$. Parameterizable affine varieties are rather special, but are nice to work with. In (5.23) and (5.24), we found a parameterization of the ellipse. If $q_i(t_1, \ldots, t_m) = 1$ for $1 \leq i \leq n$, then we refer to (5.25) as a *polynomial parametrization*. It is the following result that makes $V^*$ so useful.

**Lemma 5.8.1** *With $V$ and $V^*$ as noted earlier, $I(V) = I(V^*)$.*

**Proof.** Since $V^* \subseteq V$, we immediately have $I(V) \subseteq I(V^*)$. However,

$$\begin{aligned}
f \in I(V^*) &\implies V^* \subseteq V(f) \\
&\implies V \subseteq V(f) \qquad \text{by the minimality of } V \\
&\implies f \in I(V).
\end{aligned}$$

Thus, $I(V^*) \subseteq I(V)$ and $I(V) = I(V^*)$. $\quad\square$

For the sake of simplicity, we will focus our attention on a single parametric variable $t$—that is, we will take $m = 1$.

There is an important connection between the parameterizability and the irreducibility of an affine variety. But first we will introduce a useful criterion for the irreducibility of an affine variety $V$ in terms of $I(V)$. We say that an ideal $I$ in a ring $R$ is a *prime ideal* if

$$a, b \in R, \ ab \in I \implies \text{either } a \in I \text{ or } b \in I.$$

For example, in $\mathbb{Z}$ we know that every ideal $I$ is of the form $I = m\mathbb{Z}$ for some $m \in \mathbb{N} \cup \{0\}$. If $m = ab$ with $1 \leq a, b < m$, then $a, b \notin I$ but $ab = m \in I$ and $I$ is not prime. On the other hand, if $m = p$ is prime and $ab \in p\mathbb{Z}$, then $p \mid ab$ so that $p \mid a$ or $p \mid b$ implying that either $a \in I$ or $b \in I$. Thus, $I$ is prime. Consequently, $I = m\mathbb{Z}$ is a prime ideal if and only if $m$ is prime.

**Lemma 5.8.2** *Let $V \subseteq F^n$ be an affine variety. Then $V$ is irreducible if and only if $I(V)$ is a prime ideal.*

**Proof.** See the exercises. $\quad\square$

Now we can demonstrate the link between the parameterizability of an affine variety and its irreducibility.

**Theorem 5.8.3** *Let $F$ be an infinite field and $V \subseteq F^n$ be an affine variety parameterized by*

$$x_i = \frac{p_i(t)}{q_i(t)} \qquad (1 \leq i \leq n)$$

*where $p_i(t), q_i(t) \in F[t]$. Then $V$ is irreducible.*

**Proof.** We will only consider the case of a polynomial parameterization here. The general case, which requires some extra work to clear the denominators, will be left to the exercises. Let

$$V^* = \{(p_1(t), \ldots, p_n(t)) \mid t \in F\}.$$

Then $I(V) = I(V^*)$ so, by Lemma 5.8.2, it suffices to show that $I(V^*)$ is a prime ideal. Let $f, g \in F[x_1, \ldots, x_n]$ and $h = fg$. Then

$$
\begin{aligned}
h = fg \in I(V^*) &\Longrightarrow h(a_1, \ldots, a_n) = 0 \quad \forall\, (a_1, \ldots, a_n) \in V^* \\
&\Longrightarrow h(p_1(t), \ldots, p_n(t)) = 0 \quad \forall\, t \in F \\
&\Longrightarrow h(p_1(t), \ldots, p_n(t)) \quad \text{is the zero polynomial in } t \\
&\qquad \text{since } F \text{ is infinite} \\
&\Longrightarrow f(p_1(t), \ldots, p_n(t))\, g(p_1(t), \ldots, p_n(t)) = 0 \\
&\qquad \text{as polynomials in } t \\
&\Longrightarrow \text{either } f(p_1(t), \ldots, p_n(t)) = 0 \\
&\qquad \text{or} \quad g(p_1(t), \ldots, p_n(t)) = 0 \quad \text{as polynomials} \\
&\Longrightarrow \text{either } f(p_1(t), \ldots, p_n(t)) = 0 \quad \forall\, t \in F \\
&\qquad \text{or} \quad g(p_1(t), \ldots, p_n(t)) = 0 \quad \forall\, t \in F \\
&\Longrightarrow \text{either } f \in I(V^*) \quad \text{or} \quad g \in I(V^*).
\end{aligned}
$$

Thus, $I(V) = I(V^*)$ is a prime ideal and, by Lemma 5.8.2, $V$ is irreducible. $\square$

In one sense, Theorem 5.8.3 just appears to transfer the problem of showing that $V$ is irreducible to showing that $V$ is the smallest affine variety containing $V^*$ (as defined in the proof). This may not be very easy either, but it does provide another tool. Consider the ellipse discussed earlier. There we had a parameterization for which $V^*$ consisted of all points on the ellipse except for $(-3, 0)$. Let $f(x, y) \in \mathbb{R}[x, y]$ be such that $f$ is zero at all points in $V^*$. Then, for any neighborhood $N$ of $(-3, 0)$, there must be a point $(a, b) \in N$ (indeed many points) such that $f(a, b) = 0$. But clearly, $f(x, y)$, viewed as a function in two variables, is continuous. Hence, we must have $f(-3, 0) = 0$.

Consequently, any affine variety that contains $V^*$ must also contain the point $(-3, 0)$ and therefore contains $V(\frac{x^2}{9} + \frac{y^2}{4} - 1)$—that is, $V(\frac{x^2}{9} + \frac{y^2}{4} - 1)$ is the smallest affine variety containing $V^*$ and is therefore irreducible. Clearly, this kind of continuity argument can be applied to many affine varieties in $\mathbb{R}^3$ for which we have a parameterization missing a finite number of points.

We have been concerned in this section with finding a parameterization of an affine variety and with the benefits of having a parameterization. However, we also would like to have an equational basis for affine varieties; they often relate better to our intuition and experience. For example, we instantly recognize $x^2 + y^2 = 0$ as defining a circle, and so on. In some situations we may have a parametric description of a variety and want to generate an equational description. This process is known as *implicitization*. In some situations, it is pretty straightforward just to eliminate the parameters. For example, if we have the parametric equations

$$x = t^2, y = t^3, z = t^4$$

then we clearly have

$$x^2 = z, y = xz$$

so that we could reasonably expect that the equations

$$x^2 - z = 0, xz - y = 0$$

will define our curve. This leads us to one very important application of Groebner bases. For less transparent situations than that just described, Groebner bases provide a way to get equations that define a variety that is initially described by parametric equations. Suppose that we have

$$x_1 = f_1(t_1, \ldots, t_m)$$
$$x_2 = f_2(t_1, \ldots, t_m)$$
$$\ldots = \ldots$$
$$x_n = f_n(t_1, \ldots, t_m).$$

Then we can apply Buchberger's algorithm (Theorem 5.4.7) to the ideal

$$I = \langle f_1(t_1, \ldots, t_m) - x_1, f_2(t_1, \ldots, t_m) - x_2, \ldots f_n(t_1, \ldots, t_m) - x_n \rangle$$

with the lexicographic ordering

$$t_1 > t_2 > \cdots > t_m > x_1 > x_2 \ldots > x_n$$

to obtain a Groebner basis for $I$. Buchberger's algorithm progressively elimi-nates the variables $t_1, t_2, \ldots, t_m$ until we are eventually left with polynomials in just the variables $x_1, \ldots, x_n$. The polynomials in the basis that involve only the variables $x_1, \ldots, x_n$ will generally provide equations for the smallest variety containing the parameterization. Slightly modified, the same technique will work for rational parameterizations.

**Example 5.8.4** Consider the parameterization

$$x = 2 - t^2, y = t - t^3.$$

Let

$$I = \langle t^3 - t + y, t^2 + x - 2 \rangle.$$

With the lexicographic order $t > x > y$, Buchberger's algorithm will yield the following Groebner basis:

$$\{t^3 - t + y, t^2 + x - 2, t(1 - x) + y, ty$$

$$+ (1 - x)(x - 2), y^2 + (x - 1)^2(x - 2)\}.$$

Thus, the smallest variety containing the parameterization is defined by the equation

$$y^2 + (x - 1)^2(x - 2) = 0.$$

**Exercises 5.8**

1. Find parametric equations for the lines in $\mathbb{R}^2$ defined by the equations

   (i) $x = -2$
   (ii) $y = 3$
   (iii) $2x + y - 5 = 0$.

2. Find the equations of the following parametrically defined lines in the form $ax + by + c = 0$.

   (i) $x = 1 + 2t, \quad y = 2 - t.$
   (ii) $x = -2, \qquad y = 1 + t.$

3. Find parametric equations for the circle in $\mathbb{R}^2$ defined by the equation $x^2 + y^2 - 4 = 0$ using

   (i) the point $(-2, 0)$ and the line $x = 0$
   (ii) the point $(-2, 0)$ and the line $x = -3$.

4.  (i)  Find parametric equations for the ellipse $E$ in $\mathbb{R}^2$ defined by the equation $x^2/4 + y^2/9 - 1 = 0$ using the point $(-2, 0)$ and the line $x = 0$.

   (ii)  Show that the parametric equations that you obtain define a bijection from the line $x = 0$ to $E$ with the point $(-2, 0)$ removed.

5.  Find parametric equations for the parabola in $\mathbb{R}^2$ defined by the equation $y^2 - 4(x - 1) = 0$.

6.  Find parametric equations for the hyperbola in $\mathbb{R}^2$ defined by the equation $x^2 - y^2 - 1 = 0$.

7.  Find parametric equations for the curve $y^2 - x(x - 1)^2$ in $\mathbb{R}^2$ using the line $x = 0$ and the point $(1, 0)$.

8.  Let $f(x, y, z) = x^2 + y^2 + (z - 1)^2 - 1 \in \mathbb{R}^3[x, y, z]$. Then $V(f)$ is the surface of a sphere. Use the point $(0, 0, 2) \in V(f)$ and points $(u, v, 0)$ in the $xy$ plane to obtain the following parameterization of $V(f)$:

$$x = \frac{4u}{u^2 + v^2 + 4}, \qquad y = \frac{4v}{u^2 + v^2 + 4}, \qquad z = \frac{2u^2 + 2v^2}{u^2 + v^2 + 4}.$$

9.  Show that the process of parameterizing the ellipse $E$ in Exercise 4 by a line can be reversed. In other words, it is possible to reverse the roles and parameterize the line $x = 0$ by (all but one of) the points on $E$.

10.  Describe all the prime ideals in $F[x]$.

11.  Show that $I = \langle x, y, z \rangle \subseteq F[x, y, z]$ is a prime ideal that is not principal.

12.  Let $V, V_1,$ and $V_2$ be affine varieties in $F^n$, where $V_1$ and $V_2$ are proper subsets of $V$ and $V = V_1 \cup V_2$. Let $f_i, g_j \in F[x_1, \ldots, x_n]$ for $1 \le i \le \ell, \ 1 \le j \le m, \ V_1 = V(f_1, \ldots, f_\ell),$ and $V_2 = V(g_1, \ldots, g_m)$. Establish the following:

   (i)  $\exists \ i_0, j_0$ such that $V \not\subseteq V(f_{i_0})$ and $V \not\subseteq V(g_{j_0})$.

   (ii)  $f_{i_0}, g_{j_0} \notin I(V)$.

   (iii)  $f_{i_0} g_{j_0} \in I(V)$.

   (iv)  $I(V)$ is not prime.

13.  Let $V$ be an irreducible affine variety in $F^n$. Let $f, g \in F[x_1, \ldots, x_n]$. Establish the following:

   (i)  $V \cap V(fg) = (V \cap V(f)) \cup (V \cap V(g))$.

   (ii)  $f, g \notin I(V) \implies fg \notin I(V)$.

   (iii)  $I(V)$ is a prime ideal.

*14. Let $F$ be an infinite field and $V \subseteq F^n$ be an affine variety parameterized by

$$x_i = \frac{p_i(t)}{q_i(t)} \qquad (1 \le i \le n)$$

where $p_i(t), q_i(t) \in F[t]$. Show that $V$ is irreducible.


## 5.9 Intersection Multiplicities

In this section we will consider the manner in which a line can intersect a curve. We begin by considering an example.

Let $f(x, y) = x^2 + y^2 - 2 \in F[x, y]$ and $C = V(f)$. Clearly, $P = (1, 1) \in C$. Consider an arbitrary line $L$ through $P$ that is given by the parametric equations

$$\left. \begin{array}{l} x = 1 + ct, \\ y = 1 + dt, \end{array} \right\} \qquad (5.26)$$

where not both $c$ and $d$ are zero. In fact, since we will want to divide by $c^2 + d^2$ at one point later, we need the slightly stronger assumption that $c^2 + d^2 \neq 0$. This would follow automatically from the assumption that not both $c$ and $d$ are zero in fields such as $\mathbb{Q}$ and $\mathbb{R}$, but this does not necessarily follow in fields such as $\mathbb{Z}_p$ and $\mathbb{C}$. To examine the points of intersection of this line with $C$, we substitute the parametric expressions into $f(x, y)$ to obtain

$$\begin{aligned} f(1 + ct, 1 + dt) &= (1 + ct)^2 + (1 + dt)^2 - 2 \\ &= 1 + 2ct + c^2 t^2 + 1 + 2dt + d^2 t^2 - 2 \\ &= t(2(c + d) + (c^2 + d^2)\, t) \end{aligned}$$

and equate this to zero to obtain

$$t(2(c + d) + (c^2 + d^2)t) = 0.$$

Since we constructed the line $L$ so that it would pass through the point $P$ and since $P$ is given by the value $t = 0$, it is no surprise that one solution is given by $t = 0$. This just reflects the fact that $P$ lies on the curve $C$.

A "second" point of intersection is obtained by setting

$$2(c + d) + (c^2 + d^2)\, t = 0$$

to obtain

$$t = \frac{-2(c+d)}{c^2 + d^2}.$$

Thus, $L$ meets $C$ in two distinct points, except when $c = -d$, in which case

$$f(1 + ct, 1 + dt) = t^2(c^2 + d^2).$$

Equating this to zero, we have

$$t^2(c^2 + d^2) = 0$$

which has a "double" solution at $t = 0$. We describe this by saying that $L$ meets $C$ "twice" at the point $P$. We naturally consider this line to be the tangent to $C$ at $P$ (although later we will give a precise definition of what we mean by a tangent). This leads us to the concept of intersection multiplicity.

Let $f(x, y) \in F[x, y]$ and $P = (a, b) \in C = V(f(x, y))$. Let $L$ be the line given by parametric equations

$$x = a + ct$$

$$y = b + dt.$$

Let

$$f(a + ct, b + dt) = t^m g(t) \tag{5.27}$$

where $m \geq 1$ and $g(0) \neq 0$. Then the *intersection multiplicity* of $L$ with $C$ at $P$ is defined to be $m$.

It is important to appreciate the following:

(i) $f(a + ct, b + dt)$ is a polynomial in $t$.
(ii) When $t = 0$, we have

$$f(a + c \cdot 0, b + d \cdot 0) = f(a, b) = 0$$

since $P \in C$.
(iii) In light of (i) and (ii), the constant term in $f(a + ct, b + dt)$ must be zero. It is for this reason that we know that we must be able to write $f(a + ct, b + dt)$ in the form (5.27) with $m \geq 1$.

As a second example, consider the curve $C = V(f(x, y))$ where

$$f(x, y) = y - x^3 + 6x^2 - 14x + 7 \in F[x, y].$$

Then $P = (2, 5) \in C$. Let $L$ be the line

$$x = 2 + ct$$
$$y = 5 + dt.$$

Thus $L$ and $C$ intersect at the point $P$. To determine the multiplicity of their intersection at $P$, we substitute the parametric description of $L$ into $f(x, y)$ to obtain

$$f(2 + ct, 5 + dt) = dt - c^3 t^3 - 2ct = t(d - 2c - c^3 t^2).$$

Thus, provided $d \neq 2c$, the intersection multiplicity is 1 whereas, if $d = 2c$, then

$$f(2 + ct, 5 + dt) = -c^3 t^3$$

and the multiplicity of intersection is 3.

The two examples considered so far conform to our initial expectations that a line will normally meet a curve at a point of intersection multiplicity one unless we happen to be considering a tangent line where the multiplicity would be higher. However, life is not always quite that simple.

Consider the curve $C = V(f(x, y))$ where $f(x, y) = (y - 2)^2 - (x - 1)^3$. Then $P = (1, 2) \in C$. Let $L$ be the line defined by

$$x = 1 + ct$$
$$y = 2 + dt.$$

Then $L$ passes through $P$. To determine the multiplicity of intersection at the point $P$, consider

$$f(1 + ct, 2 + dt) = d^2 t^2 - c^3 t^3$$
$$= t^2(d^2 - c^3 t).$$

Then, for $d \neq 0$, the multiplicity of intersection is 2 whereas, for $d = 0$, the multiplicity is 3. In particular, every line through $P$ has intersection multiplicity of at least 2.

So far we have seen that there are points on curves such that there is a *unique* line through the point with intersection multiplicity greater than one and that there are points for which *every* line through the point has intersection multiplicity greater than one. It would be natural to expect that with a little bit of work, we could find a point for which there were lines of intersection multiplicity one and other lines of intersection multiplicity greater than one. Somewhat surprisingly, this cannot happen.

To see why, let $f(x, y) \in F[x, y]$ have degree $m$, $(a, b) \in V(f)$, and once again consider $f(a+ct, b+dt)$ as a polynomial in $t$. We know that the constant term is zero, so that we have the factor $t$. We will have a factor of $t^2$ if and only if the coefficient of $t$ is zero. Let

$$f(a + ct, b + dt) = t^2 g(t) + h(c, d)\, t.$$

Since we are wondering if there could be particular choices of $c$ and $d$ so that the coefficient of $t$ is zero, we consider the coefficient of $t$ as a polynomial in $c$ and $d$. To see how we obtain terms in $t$, consider any monomial $c_{ij} x^i y^j$ of $f(x, y)$. Under parametric substitution this becomes

$$c_{ij}(a + ct)^i (b + dt)^j.$$

To get a term from this product that is linear in $t$, we must select the scalar (field element) from all the brackets except one and select $ct$ or $dt$, as appropriate, from the remaining bracket. This yields a term of the form

$$c_{ij} a^{i-1} b^j\, ct \quad \text{or} \quad c_{ij} a^i b^{j-1}\, dt$$

for some $c_{ij}, \in F$. Since $\deg(f) = m$, we have for some $c_i \in F$,

$$f(a + ct, b + dt) = c_m t^m + c_{m-1} t^{m-1} + \cdots + (\gamma c + \delta d)\, t$$

so that the coefficient of $t$ in $f(a + ct, b + dt)$ is of the form

$$\gamma c + \delta d$$

for some $\gamma, \delta \in F$. Thus we find a point with intersection multiplicity greater than one if and only if $\gamma c + \delta d = 0$. This happens if and only if one of the following two cases holds:

*Case* (i) $\gamma = \delta = 0$. Then for all choices of $c$ and $d$ (that is, for all lines $L$ through $P$), $L$ meets $C$ at $P$ with intersection multiplicity at least 2.

*Case* (ii) $\gamma \neq 0$ or $\delta \neq 0$. Without loss of generality, we can assume that $\gamma \neq 0$. Then,

$$c = -\gamma^{-1} \delta d$$

and all solutions are of the form

$$(c, d) = (-\gamma^{-1} \delta d, d)$$
$$= d(-\gamma^{-1} \delta, 1).$$

It follows that the only line through $(a, b)$ yielding multiple solutions is the line determined by the pair $(-\gamma^{-1}\delta, 1)$. Therefore, in this case, there exists a unique line statisfying $\gamma c + \delta d = 0$ and therefore a unique line with intersection multiplicity at least 2.

For different values of $d$, these solutions all determine the *same* line in parametric form (5.26). (Note that $d \neq 0$, since $d = 0$ would imply that $c = 0$, but we cannot have both $c$ and $d$ equal to zero since the equations $x = a + ct$, $y = b + dt$ would no longer define a line.)

Thus either all lines through the point $P$ have intersection multiplicity at least two or else there exists a unique line through $P$ with intersection multiplicity at least two.

We have therefore established the following remarkable result.

**Theorem 5.9.1** *Let $P$ be a point on the curve $C : f(x, y) = 0$ where $f(x, y) \in F[x, y]$. Then exactly one of the following situations must prevail:*

(i) *There exists a unique line through $P$ that intersects $C$ with intersection multiplicity at least 2.*
(ii) *Every line through $P$ meets $C$ with intersection multiplicity at least 2.*

We can therefore make the following definition:

The *tangent* (line) to $C$ at the point $P$ is the unique line through $P$ with intersection multiplicity at least two if such a unique line exists.

In the light of the discussion in this section, it is also reasonable to identify how bad the bad points are on a curve. So we say that a point $P$ on a curve $C$ is a *double point* if every line through $P$ has intersection multiplicity at least two and at least one line intersects $C$ at $P$ with intersection multiplicity exactly 2. We define *triple* and *quadruple* points, and so on, in a similar manner. We refer to this as the *multiplicity of intersection of $C$ at $P$*.

## Exercises 5.9

1. Determine the intersection multiplicity for lines at each point where the curve in Exercise 13 (ii) in section 5.5 crosses the $x$-axis.

2. Determine the intersection multiplicity for lines at each point where the curve in Exercise 14 (ii) in section 5.5 crosses the $x$-axis.

3. Determine the intersection multiplicity for lines at each point where the curve in Exercise 15 (ii) in section 5.5 crosses the $x$-axis. Also determine the intersection multiplicity at the point $(5, 8)$.

4. Find the equation of the tangent (in implicit and parametric form) at the point $(4, 6)$ on the curve $y^2 - x(x - 1)^2 = 0$.

## 5.10  Singular and Nonsingular Points

In light of our discussion on intersection multiplicities in the previous section, the following definition now makes sense.

We define a point $P$ on the curve $f(x, y) = 0$ to be *singular* if every line through $P$ has intersection multiplicity at least two and to be *nonsingular* if there exists a unique line through $P$ with intersection multiplicity at least two.

We say that the curve $C : f(x, y) = 0$ is *singular* if it contains a singular point, and it is *nonsingular* if it has no singular points.

Our goal in this section is to develop a simple test for singularity/nonsingularity.

The test that we are about to develop depends on the idea of formal (partial) differentiation. The idea can be applied to polynomials in an arbitrary number of variables but, to keep the notation simple, we will just work with two variables.

For any monomial

$$m = m(x, y) = a_{\alpha\beta} x^\alpha y^\beta \in F[x, y]$$

we define the *partial derivative* of $m$ *with respect to* $x$ to be

$$\frac{\partial m}{\partial x} = \alpha\, a_{\alpha\beta} x^{\alpha-1} y^\beta.$$

Of course, $\frac{\partial m}{\partial y}$ is defined similarly.

We then extend the definition to an arbitrary polynomial

$$f = f(x, y) = \sum a_{\alpha\beta} x^\alpha y^\beta$$

by

$$\frac{\partial f}{\partial x} = \sum \alpha\, a_{\alpha\beta} x^{\alpha-1} y^\beta.$$

This "formal" partial differentiation satisfies some familiar rules:

The *product rule* :
$$\frac{\partial}{\partial x}(fg) = f\frac{\partial g}{\partial x} + \frac{\partial f}{\partial x} g$$

The *modified chain rule* :
$$\frac{\partial}{\partial t} f(x(t), y(t)) = \frac{\partial f}{\partial x} \cdot \frac{\partial x}{\partial t} + \frac{\partial f}{\partial y} \cdot \frac{\partial y}{\partial t}.$$

The verification of the first of these rules reduces to the rule for the product of monomials. Let

$$f = ax^\alpha y^\beta, \qquad g = bx^\gamma y^\delta.$$

Then

$$f\left(\frac{\partial g}{\partial x}\right) + \left(\frac{\partial f}{\partial x}\right)g = ax^\alpha y^\beta \cdot b\gamma x^{\gamma-1} y^\delta + a\alpha x^{\alpha-1} y^\beta \cdot bx^\gamma y^\delta$$

$$= (\alpha + \gamma)\, abx^{\alpha+\gamma-1} y^{\beta+\delta}$$

whereas

$$\frac{\partial(fg)}{\partial x} = \frac{\partial}{\partial x}(abx^{\alpha+\gamma} y^{\beta+\delta})$$

$$= (\alpha + \gamma)\, abx^{\alpha+\gamma-1} y^{\beta+\delta}.$$

Thus,

$$\frac{\partial(fg)}{\partial x} = f\frac{\partial g}{\partial x} + \frac{\partial f}{\partial x}g.$$

We leave the remainder of the proofs of the product rule and modified chain rule as exercises for you.

**Theorem 5.10.1** *Let $f(x, y) \in F[x, y]$ and $P = (a, b)$ be a point on the curve $C : f(x, y) = 0$. Then $P$ is a singularity if and only if*

$$\frac{\partial f}{\partial x} = \frac{\partial f}{\partial y} = 0$$

*at the point P.*

**Proof.** Consider the intersection of the line $L$ given by the parametric equations

$$x = a + ct$$
$$y = b + dt$$

with $C$. The points of intersection are given by

$$f(a + ct, b + dt) = 0$$

where, of course, one solution is $t = 0$. Thus, $f(a + ct, b + dt)$ is a polynomial in $t$ with zero as the constant term:

$$f(a + ct, b + dt) = c_m t^m + c_{m-1} t^{m-1} + \cdots + c_1 t.$$

Note that $a$ and $b$ are fixed by the point $P$ so that we consider the coefficients $c_i$ as polynomials in $c$ and $d$—that is, as elements of $F[c, d]$.

Thus, the intersection multiplicity at $P$ is greater than one—that is, $P$ is a singular point if and only if $c_1$, as a linear polynomial in $c$ and $d$, is identically zero. The key here is to notice that we can isolate $c_1$ by differentiating $f(a + ct, b + ct)$ with respect to $t$, which yields

$$\frac{\partial}{\partial t} f(a + ct, b + dt) = m c_m t^{m-1} + \cdots + 2 c_2 t + c_1$$

and then evaluating at $t = 0$:

$$\frac{\partial}{\partial t} f(a + ct, b + dt) \bigg|_{t=0} = c_1.$$

From the modified chain rule, we deduce

$$c_1 = \frac{\partial f}{\partial x} c + \frac{\partial f}{\partial y} d \tag{5.28}$$

where the partial derivatives are evaluated at $P$. Since we are considering $c$ and $d$ as variables, equation (5.28) describes $c_1$ as an element of $F[c, d]$. Thus we have (evaluating partial derivatives at $P$)

$P$ is singular $\iff c_1$ is the zero polynomial in $c$ and $d$

$$\iff \frac{\partial f}{\partial x}\bigg|_{t=0} c + \frac{\partial f}{\partial y}\bigg|_{t=0} d \quad \text{is the zero polynomial in } c \text{ and } d$$

$$\iff \frac{\partial f}{\partial x}\bigg|_P = 0 = \frac{\partial f}{\partial y}\bigg|_P. \quad \square$$

**Corollary 5.10.2** *Let $f(x, y) \in F[x, y]$ and $C$ be the curve defined by $f(x, y) = 0$. Then $C$ is nonsingular if and only if there is no point $P \in C$ with*

$$\frac{\partial f}{\partial x}\bigg|_P = \frac{\partial f}{\partial y}\bigg|_P = 0.$$

Consider $f(x, y) = y^2 + x^3 - x^5$ and the curve $C : f(x, y) = 0$. Then any singular point on $C$ must satisfy the equations

$$f(x, y) = 0, \quad \frac{\partial f}{\partial x} = 0 \quad \text{and} \quad \frac{\partial f}{\partial y} = 0,$$

That is,

$$y^2 + x^3 - x^5 = 0$$
$$3x^2 - 5x^4 = 0$$
$$2y = 0.$$

These equations imply

$$y = 0, \quad x^3(1 - x^2) = 0, \quad x^2(3 - 5x^2) = 0.$$

The only common solution is the point $O = (0,0)$. Thus, $C$ has a unique singularity at the origin.

**Exercises 5.10**

1. Show that the following curves in $\mathbb{R}^2$ have no singular points:

   (i) Circle.
   (ii) Ellipse.
   (iii) Parabola.
   (iv) Hyperbola.

2. Find the singular points of the curve $C : x^3 + xy + y^3 = 0$ in $\mathbb{R}^2$.

3. Find the singular points of the curve $C : x^2(x - 1)^2 + y^2(y - 2)^2 = 0$ in $\mathbb{R}^2$.

4. Find the singular points of the curve $C : x^2(x - 2)^2 + y(y - a)^2 = 0$ in $\mathbb{R}^2$.

5. Find the singular points of the curve in $\mathbb{R}^2$ defined by the polynomial $y^2 - x(x^2 - 3x + 2)$.

6. Find the singular points of the curve in $\mathbb{R}^2$ defined by the polynomial $y^2 - x(x^2 - 2x + 1)$.

7. Find the singular points of the curve in $\mathbb{R}^2$ defined by the polynomial $y^2 - (x - 1)^3$.

8. Show that the curve $C : (1 + x^2)^2 + y^2 = 0$ has no singularity in $\mathbb{R}^2$ but does have singularities in $\mathbb{C}^2$.

9. Show that the curve $C : (1 + x^2)^2 + y^2(y + 1) = 0$ has no singularity in $\mathbb{R}^2$ but does have singularities in $\mathbb{C}^2$.

10. Show that the curve $C : x^2(1 + x^2)^2 + y^2 = 0$ has a singularity in $\mathbb{R}^2$ and additional singularities in $\mathbb{C}^2$.

# 6

## Elliptic Curves

Although elliptic curves have been studied for many decades and a wealth of information was gained about them, it is only relatively recently that they have gained a certain level of "notoriety". This has been the result of the application of the theory of elliptic curves to such things as Fermat's Last Theorem, cryptography, and integer factorization. In this chapter we develop some basic properties of elliptic curves, particularly the group structure on an elliptic curve. There are other situations, and also simpler situations, when it is possible to define an algebraic structure on a set of solutions to an equation. One simpler example where one is dealing exclusively with integer solutions occurs in the study of Pell's Equation. This is treated in section 7.7, but you are welcome to digress to that section at this stage, if you wish to do so. To understand the critical intersection properties of elliptic curves, it is necessary to consider them in projective space. For this reason, we devote some time to introducing and working with projective spaces and, in particular, the projective plane.

### 6.1 Elliptic Curves

We define an *elliptic curve* to be a nonsingular curve $E : f(x, y) = 0$ where $f(x, y)$ has the form

$$f(x, y) = y^2 - (x^3 + ax + b) \in F[x, y]$$

and char $F \neq 2, 3$. See section 7.5 for some further comments regarding this form of the equation. Over fields of characteristic 2 or 3, the concept of an elliptic curve can be captured by more general equations.

In the discussion to follow we will depart from our usual practice of denoting the inverse of a nonzero element $b$ in an arbitrary field by $b^{-1}$ and will write $\frac{a}{b}$ sometimes for $ab^{-1}$. The fractional notation can be more convenient and intuitive. For a curve written in this form, we define the *discriminant* to be the expression

$$\Delta = 4a^3 + 27b^2.$$

**Theorem 6.1.1** *The curve* $E : f(x, y) = 0$ *where* $f(x, y) = y^2 - (x^3 + ax + b) \in F[x, y]$, *and* char $F \neq 2, 3$, *is singular if and only if*

$$\Delta = 4a^3 + 27b^2 = 0.$$

**Proof.** By Corollary 5.10.2, $E$ is singular if and only if there exists a solution to the equations

$$f(x, y) = 0, \qquad \frac{\partial f}{\partial x} = 0, \qquad \frac{\partial f}{\partial y} = 0.$$

That is,

$$y^2 - (x^3 + ax + b) = 0 \tag{6.1}$$

$$3x^2 + a = 0 \tag{6.2}$$

$$2y = 0. \tag{6.3}$$

First assume that $(x, y)$ is a solution. Then, since char $F \neq 2, 3$,

$$y = 0, \qquad x^3 + ax + b = 0, \qquad x^2 = -\frac{a}{3}. \tag{6.4}$$

From the third equality in (6.4), we have $3x^3 + ax = 0$ which, when combined with the second equality in (6.4), yields

$$b = 2x^3. \tag{6.5}$$

Combining this with the third equality in (6.4), we now obtain

$$4a^3 + 27b^2 = 4(-3x^2)^3 + 27(2x^3)^2 = 0.$$

Conversely, now assume that $\Delta = 0$. Clearly, $a = 0$ if and only if $b = 0$. So this presents us with two cases to consider:

*Case* (1) $a = b = 0$. Then

$$f(x, y) = y^2 - x^3$$

and it is easily verified that there is a singularity at the origin $(0, 0)$.

*Case* (2) $a \neq 0 \neq b$. From $\Delta = 0$ we must have

$$-\frac{3b}{2a} = \frac{2a^2}{9b}.$$

So consider the point with coordinates

$$x = -\frac{3b}{2a} = \frac{2a^2}{9b}, \qquad y = 0. \tag{6.6}$$

Equation (6.3) is immediately satisfied. Also,

$$x^2 = x \cdot x = -\frac{3b}{2a} \cdot \frac{2a^2}{9b} = -\frac{a}{3}$$

and equation (6.2) is satisfied. Finally,

$$\begin{aligned}
x^3 + ax + b &= x \cdot x^2 + ax + b \\
&= \frac{2a^2}{9b} \cdot \frac{9b^2}{4a^2} + a\left(-\frac{3b}{2a}\right) + b \\
&= \frac{b}{2} - \frac{3}{2}b + b \\
&= 0 \\
&= y^2
\end{aligned}$$

so that (6.1) is satisfied. Thus, the point with coordinates as given in (6.6) is a singular point and the proof is complete. $\quad\Box$

Note that in our proof there were times when we wanted to divide by 2 or by 3. For this reason, the criterion of Theorem 6.1.1 does not apply over fields of characteristic 2 or 3.

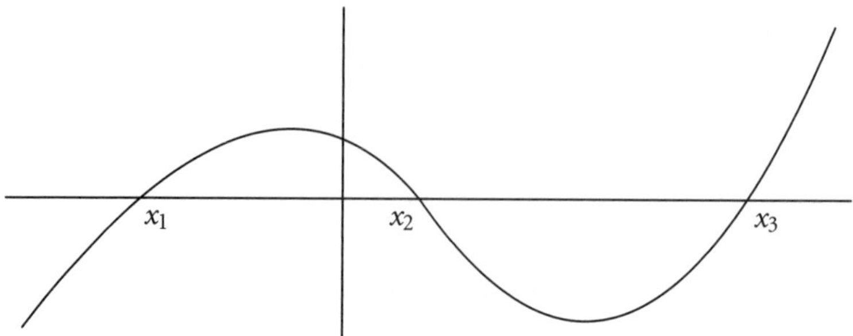


**Figure 6.1**

Let us see how the criterion of Theorem 6.1.1 applies to special cases. We continue to assume that we are working over a field of characteristic $\neq 2, 3$.

*Case* $a = b = 0$. Then $\Delta = 0$ so that the curve $E : y^2 = x^3$ is singular.

*Case* $a = 0,\ b \neq 0$. Then $\Delta = 27b^2 \neq 0$ and $E : y^2 = x^3 + b$ is nonsingular.

*Case* $a \neq 0,\ b = 0$. Then $\Delta = 4a^3 \neq 0$ and $E : y^2 = x^3 + ax$ is nonsingular.

*Case* $F = \mathbb{R}$ and $a > 0$. Then $\Delta = 4a^3 + 27b^2 > 0$. Therefore, $E : y^2 = x^3 + ax + b$ is nonsingular.

When working over an arbitrary field we don't have any convenient way to graph a cubic curve. However, over $\mathbb{R}$ we can obtain a useful representation. So consider the curve

$$E : y^2 = x^3 + ax + b$$

over $\mathbb{R}$. If we require that $E$ be nonsingular, then $4a^3 + 27b^2 \neq 0$. Coincidentally, by Proposition 5.7.1, this is exactly the condition required to ensure that the polynomial $x^3 + ax + b$ has three distinct roots over $\mathbb{C}$. Unfortunately, two of them might be complex. However, if we consider the case when the three distinct roots are real, then we can see what the graph of $C$ must look like. First, the graph of $y = x^3 + ax + b$ must cross the $x$-axis in three distinct points, as in Figure 6.1.

Then $x^3 + ax + b$ is nonnegative in the intervals $x_1 \leq x \leq x_2$ and $x_3 \leq x$. Thus, the points with real coordinates on the curve $C$ will be given by

$$y = \pm\sqrt{x^3 + ax + b}, \qquad x_1 \leq x \leq x_2,\ x_3 \leq x.$$

So that the graph of $C$ will be (roughly) of the form in the following diagram.

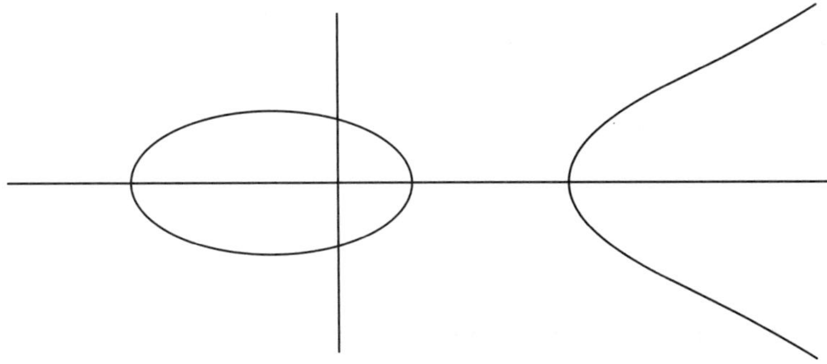

Note that for $y = \pm\sqrt{x^3 + ax + b}$,

$$\frac{dy}{dx} = \pm\frac{3x^2 + a}{2\sqrt{x^3 + ax + b}}$$

where the denominator is zero when $x = x_1, x_2$ or $x_3$. Thus the "slope" of the curve at these points is vertical.

### Exercises 6.1

1. Show that the change of variables described here, when performed in succession, will transform the curve

$$y^2 = px^3 + qx^2 + rx + s \qquad (p \neq 0)$$

over $F$ (where char $(F) \neq 3$) to the form

$$y^2 = x^3 + ax + b.$$

 

(i)  Multiply by $p^2$ and substitute

$$px \to x$$
$$py \to y.$$

(ii)  Substitute

$$x \to x - \frac{1}{3}q$$
$$y \to y.$$

2. For each of the following pairs of values of $a$ and $b$, determine whether the curve $y^2 = x^3 + ax + b$ over $\mathbb{Q}$ is singular or nonsingular.

   (i) $a = 1, \ b = 1$.
   (ii) $a = -1, \ b = 1$.
   (iii) $a = -3, \ b = 2$.
   (iv) $a = -2, \ b = 3$.
   (v) $a = -2^2 3^3, \ b = 2^4 3^3$.
   (vi) $a = -2^2 3^3 m^2, \ b = 2^4 3^3 m^3, \ m \in \mathbb{N}$.

3. For each of the following pairs of values of $a$ and $b$ determine whether the curve $y^2 = x^3 + ax + b$ over (a) $\mathbb{Z}_5$ and (b) $\mathbb{Z}_7$ is singular or nonsingular.

   (i) $a = 0, \ b = 1$.
   (ii) $a = 2, \ b = 3$.
   (iii) $a = 3, \ b = 1$.
   (iv) $a = 1, \ b = 5$.
   (v) $a = 2, \ b = 2$.

*4. Let $F = \mathrm{GF}(2^n)$, $n \in \mathbb{N}$, and $f(x, y) = y^2 - (x^3 + ax^2 + bx + c) \in F[x, y]$. Establish the following:

   (i) Every element of $F$ is a square—that is, for all $\alpha \in F$, there exists $x \in F$ with $x^2 = \alpha$. (Hint: Use induction on $n$.)
   (ii) $f(x, y)$ is singular.

5. Let $F$ be an algebraically closed field of characteristic 3, $f(x, y) = y^2 - (x^3 + ax + b) \in F[x, y]$, and $E : f(x, y) = 0$. Establish the following:

   (i) $a \neq 0 \implies E$ is nonsingular.
   (ii) $a = 0 \implies E$ has 3 singular points.

6. Let $F$ be a field with char $(F) \neq 2$ and let $a \in F^*$. Show that the polynomial $f(x) = x^2 + a$ either has no roots in $F$ or else it has two distinct roots.

7. Let $F$ be a field with char $(F) \neq 3$ and let $a \in F^*$. Show that the polynomial $f(x) = x^3 + a$ cannot have two repeated roots in $F$.

## 6.2 Homogeneous Polynomials

We say that a polynomial $f(x_1, \ldots, x_n) \in F[x_1, \ldots, x_n]$ is a *homogeneous polynomial* (of degree $m$) if

$$f(tx_1, tx_2, \ldots, tx_n) = t^m f(x_1, x_2, \ldots, x_n).$$

For example,

$$x^2 + xy + z^2$$

$$x^3 + x^2 y + y^2 z + z^3$$

are homogeneous polynomials of degrees 2 and 3, respectively, whereas

$$x^2 + xy + z$$

is not homogeneous. Thus, the polynomial $f(x_1, \ldots, x_n)$ is homogeneous if every monomial term has the same degree.

The zero set of a homogeneous polynomial has one very important property. It is closed under scalar multiplication.

**Lemma 6.2.1** *Let $f(x_1, \ldots, x_n) \in F[x_1, \ldots, x_n]$ be a homogeneous polynomial. Then*

$$(a_1, \ldots, a_n) \in V(f), \ t \in F \implies (ta_1, \ldots, ta_n) \in V(f).$$

*In particular, $(0, 0, \ldots, 0) \in V(f)$.*

**Proof.** Let $f$ be homogeneous of degree $m$. Then

$$f(tx_1, \ldots, tx_n) = t^m f(x_1, \ldots, x_n).$$

Hence,

$$
\begin{aligned}
(a_1, \ldots, a_n) \in V(f), \ t \in F &\implies t^m f(a_1, \ldots, a_n) = 0 \\
&\implies f(ta_1, \ldots, ta_n) = 0 \\
&\implies (ta_1, \ldots, ta_n) \in V(f).
\end{aligned}
$$

$\square$

There is an easy way to "homogenize" any polynomial. Let $f(x_1, \ldots, x_n) \in F[x_1, \ldots, x_n]$ have degree $m$. Let $z$ denote another variable. We define

$$f^h(x_1, x_2, \ldots, x_n, z) = z^m f\left(\frac{x_1}{z}, \frac{x_2}{z}, \ldots, \frac{x_n}{z}\right).$$

Then $f^h(x_1, x_2, \ldots, x_n, z)$ will be a homogeneous polynomial in $F[x_1, \ldots, x_n, z]$ of degree $m$. For example, if

$$f(x, y) = x^3 y^2 + xy + 1$$

then

$$f^h(x, y, z) = x^3 y^2 + xyz^3 + z^5$$

whereas if

$$f(x_1, x_2, x_3) = x_1^2 x_2^3 x_3 + x_2^2 + x_2 x_3^2 + x_3^6$$

then

$$f^h(x_1, x_2, x_3, z) = x_1^2 x_2^3 x_3 + x_2^2 z^4 + x_2 x_3^2 z^3 + x_3^6.$$

Note that if $f$ is already homogeneous, then $f^h = f$. For example, if $f(x, y) = x^2 + xy + y^2$, then

$$f^h(x, y, z) = z^2 \left( \left( \frac{x}{z} \right)^2 + \left( \frac{x}{z} \right)\left( \frac{y}{z} \right) + \left( \frac{y}{z} \right)^2 \right)$$

$$= x^2 + xy + y^2$$

$$= f(x, y).$$

The polynomial $f^h(x_1, x_2, \ldots, x_n, z)$ is called the *homogenization* of $f(x_1, \ldots, x_n)$. Note that

$$f^h(x_1, \ldots, x_n, 1) = f(x_1, \ldots, x_n).$$

It comes as no great surprise that there is a close connection between $V(f(x_1, \ldots, x_n))$ and $V(f^h(x_1, \ldots, x_n, z))$, which we now elaborate on in the next lemma.

**Lemma 6.2.2** *Let $f(x_1, \ldots, x_n) \in F[x_1, \ldots, x_n]$ be of degree $m$. Define $\varphi_h :$ $F^n \to F^{n+1}$ by*

$$\varphi_h(a_1, \ldots, a_n) = (a_1, \ldots, a_n, 1).$$

*Let*

$$V(f^h)^* = \{(a_1, \ldots, a_n, b) \in V(f^h) \mid b \neq 0\}.$$

*Define $\varphi_r : V(f^h)^* \to F^n$ by*

$$\varphi_r(a_1, \ldots, a_n, b) = (a_1 b^{-1}, \ldots, a_n b^{-1}).$$

(i) *$\varphi_h$ is injective and $\varphi_h(V(f)) \subseteq V(f^h)$.*
(ii) *$\varphi_r(V(f^h)^*) = V(f)$.*

(iii) $\varphi_r$ *induces the relation* $\sim$ *on* $V(f^h)^*$ *where*

$$(a_1, \ldots, a_n, b) \sim (c_1, \ldots, c_n, d) \iff \exists \lambda \in F^*$$

$$\text{with} \quad (c_1, \ldots, c_n, d) = \lambda(a_1, \ldots, a_n, b).$$

**Proof.** (i) Clearly, $\varphi_h$ is an injective mapping. For any $(a_1, \ldots, a_n) \in V(f)$,

$$f^h(a_1, \ldots, a_n, 1) = f(a_1, \ldots, a_n)$$
$$= 0.$$

Thus, $\varphi_h$ maps $V(f)$ into $V(f^h)^*$.

(ii) Let $(a_1, \ldots, a_n, b) \in V(f^h)^*$. Then $b \in F^*$ and

$$f\varphi_r(a_1, \ldots, a_n, b) = f(a_1 b^{-1}, \ldots, a_n b^{-1}) = b^{-m}(b^m f(a_1 b^{-1}, \ldots, a_n b^{-1}))$$
$$= b^{-m} f^h(a_1, \ldots, a_n, b)$$
$$= b^{-m} \cdot 0$$
$$= 0.$$

Therefore, $\varphi_r$ maps $V(f^h)^*$ to $V(f)$. For any $(a_1, \ldots, a_n) \in V(f)$, we have by part (i)

$$(a_1, \ldots, a_n, 1) = \varphi_h(a_1, \ldots, a_n) \in V(f^h)^*$$

and

$$\varphi_r(a_1, \ldots, a_n, 1) = (a_1, \ldots, a_n).$$

Thus, $\varphi_r$ is surjective.

(iii) Let $(a_1, \ldots, a_n, b), (c_1, \ldots, c_n, d) \in V(f^h)^*$. Then

$$(a_1, \ldots, a_n, b) \sim (c_1, \ldots, c_n, d) \implies \exists \lambda \in F^* \text{ with } (c_1, \ldots, c_n, d)$$
$$= \lambda(a_1, \ldots, a_n, b)$$
$$\implies \varphi_r(c_1, \ldots, c_n, d)$$
$$= (\lambda a_1 (\lambda b)^{-1}, \ldots, \lambda a_n (\lambda b)^{-1})$$
$$= (a_1 b^{-1}, \ldots, a_n b^{-1})$$
$$= \varphi_r(a_1, \ldots, a_n, b).$$

Conversely,

$$\varphi_r(a_1, \ldots, a_n, b) = \varphi_r(c_1, \ldots, c_n, d)$$
$$\Rightarrow (a_1 b^{-1}, \ldots, a_n b^{-1}) = (c_1 d^{-1}, \ldots, c_n d^{-1})$$

$$\Rightarrow a_i b^{-1} = c_i d^{-1} \qquad i = 1, \ldots, n$$

$$\Rightarrow c_i = (db^{-1})a_i \qquad i = 1, \ldots, n$$

$$\Rightarrow (a_1, \ldots, a_n, b) \sim (c_1, \ldots, c_n, d). \quad \square$$

**Example 6.2.3** Let $f(x, y) = x^2 + y^2 - 1 \in \mathbb{R}[x, y]$.

The following diagrams illustrate the relationships between $V(f)$, $V(f^h)$ and $\varphi_h(V(f))$. Note that

$$f^h(x, y, z) = x^2 + y^2 - z^2.$$

In particular, we see that $V(f^h)$ is much "bigger" than either $V(f)$ or $\varphi_h(V(f))$ and that $\varphi_h(V(f)) \subseteq V(f^h)$. Figure 6.2 illustrates the mapping $\varphi_h$. Figure 6.3 illustrates $V(f)$, $V(f^h)$ and $\varphi_h(V(f))$. Figure 6.4 illustrates the mapping $\varphi_r$.

### Exercises 6.2

1. (i) Find an example of a polynomial $f \in \mathbb{R}[x, y]$ such that $V(f)$ is not closed under scalar multiplication (of vectors).
   (ii) Find an example of a homogeneous polynomial $f \in \mathbb{R}[x, y]$ such that $V(f)$ is not closed under addition (of vectors).

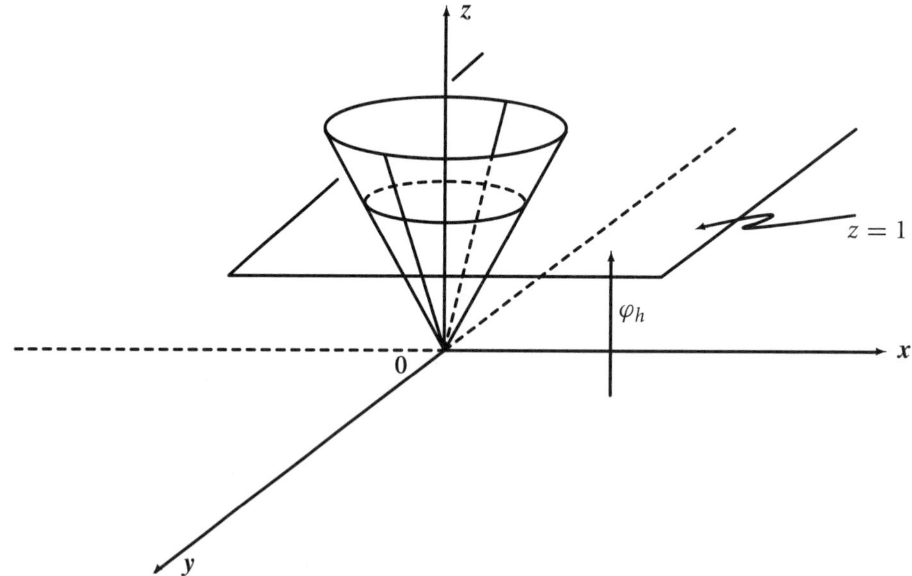


**Figure 6.2**

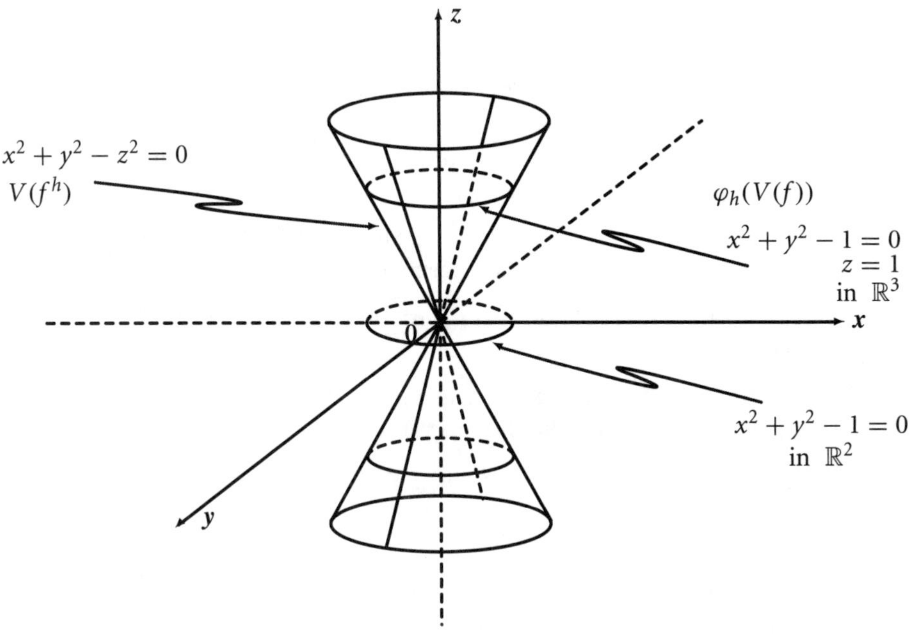


**Figure 6.3**

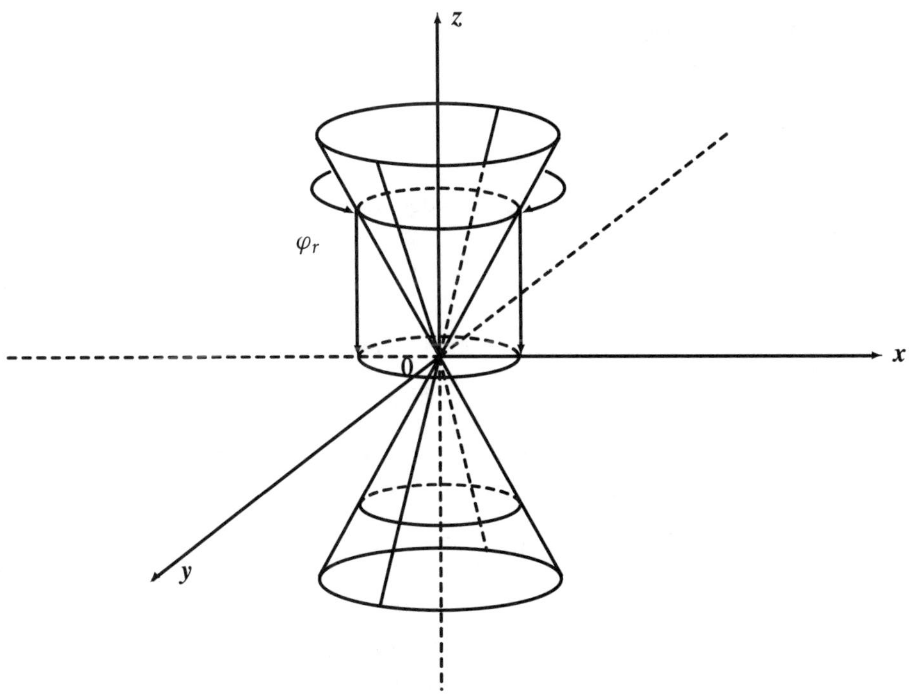


**Figure 6.4**

2. Show that the mapping

$$H : f \to f^h \qquad (f \in F[x_1, \ldots, x_n])$$

respects multiplication of polynomials, but not addition—that is, $(fg)^h = f^h g^h$, but, in general, $(f + g)^h \neq f^h + g^h$.

*3. Let $f, p, q \in F[x_1, \ldots, x_n]$ and $f = pq$. Show that $f$ is homogeneous if and only if $p$ and $q$ are both homogeneous.

*4. Let $f \in F[x_1, \ldots, x_n]$. Show that $f$ is irreducible if and only if $f^h$ is irreducible.

*5. Let $F$ be an algebraically closed field and $f \in F[x_1, \ldots, x_n]$. Show that $V(f)$ is irreducible if and only if $V(f^h)$ is irreducible.

6. Let $f(x, y) = y^2 - x + 1 \in \mathbb{R}[x, y]$.

    (i) Graph $V(f)$.
    (ii) Graph $V(f^h)$.
    (iii) Identify $\varphi_h(V(f))$ in $V(f^h)$.

7. Let $f(x, y) = y^2 - (x - 1)^3 \in \mathbb{R}[x, y]$.

    (i) Graph $V(f)$.
    (ii) Graph $V(f^h)$.
    (iii) Identify $\varphi_h(V(f))$ in $V(f^h)$.

## 6.3 Projective Space

The observations in the previous section, especially Lemma 6.2.1 and Lemma 6.2.2 (iii), lead us to consider the following relation $\sim$ on $F^{n+1} \backslash \{(0, 0, \ldots, 0)\}$:

$$(a_1, \ldots, a_{n+1}) \sim (b_1, \ldots, b_{n+1}) \iff \exists\, t \in F^* \text{ such that } (b_1, \ldots, b_{n+1})$$
$$= t(a_1, \ldots, a_{n+1})$$
$$= (ta_1, \ldots, ta_{n+1}).$$

It is very easy to see that $\sim$ is an equivalence relation. We define the $n$-*dimensional projective space* over $F$ to be

$$\mathbb{P}^n(F) = (F^{n+1} \backslash \{0\}) / \sim$$

that is, the set of $\sim$ classes. We call $\mathbb{P}^2(F)$ the *projective plane (over $F$)*. Let us denote by $[a_1, \ldots, a_{n+1}]$ the $\sim$ class containing $(a_1, \ldots, a_{n+1})$.

For $\mathbb{P}^2(\mathbb{R})$, there is a nice geometric interpretation. For any point $P = (a, b, c) \in \mathbb{R}^3$, $P \neq (0, 0, 0)$, the line through the origin and $P$ consists of the

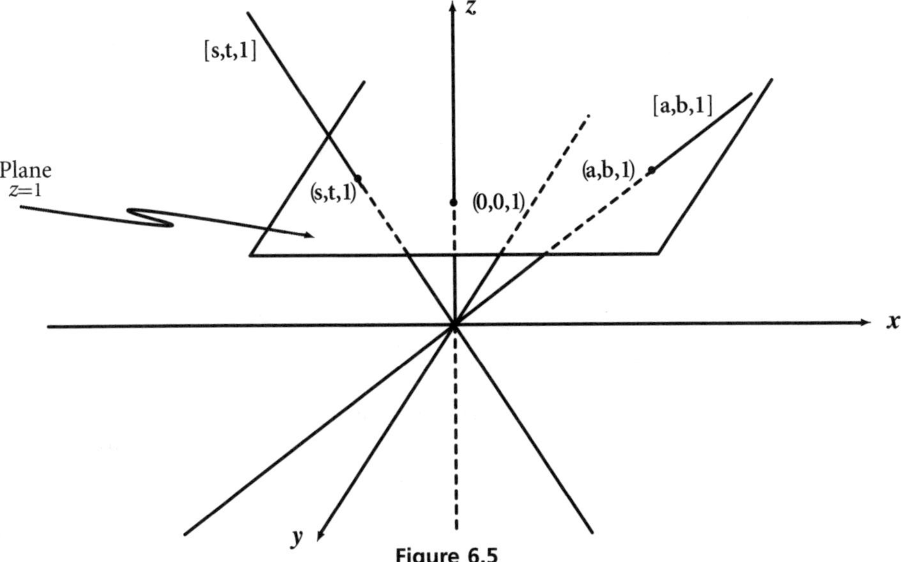


**Figure 6.5**

set of points

$$\{t(a, b, c) \mid (t \in \mathbb{R}\}.$$

However, this is just

$$[a, b, c] \cup \{(0, 0, 0)\}.$$

Thus we can consider any element $[a, b, c] \in \mathbb{P}^2(\mathbb{R})$ as a ray or line through the origin and the point $(a, b, c)$ with the origin deleted. More generally, we define the *line through the origin and the point* $(a_1, \ldots, a_{n+1}) \in F^{n+1}$ to be the set of points

$$\{t(a_1, \ldots, a_{n+1}) \mid t \in F\}.$$

In this way we can view the elements of $\mathbb{P}^n(F)$ as lines in $F^{n+1}$ through the origin (but with the origin deleted).

Now

$$\varphi_h(F^n) = \{(a_1, \ldots, a_n, 1) \mid a_i \in F\}$$

so that each point in $\varphi_h(F^n)$ determines a unique point $[a_1, \ldots, a_n, 1]$ in $\mathbb{P}^2(F)$. Thus, we have an injective mapping $\varphi_\mathbb{P} : F^n \to \mathbb{P}^n(F)$ defined by

$$\varphi_\mathbb{P}(a_1, \ldots, a_n) = [a_1, \ldots, a_n, 1].$$

We can represent these mappings and structures digrammatically as follows:

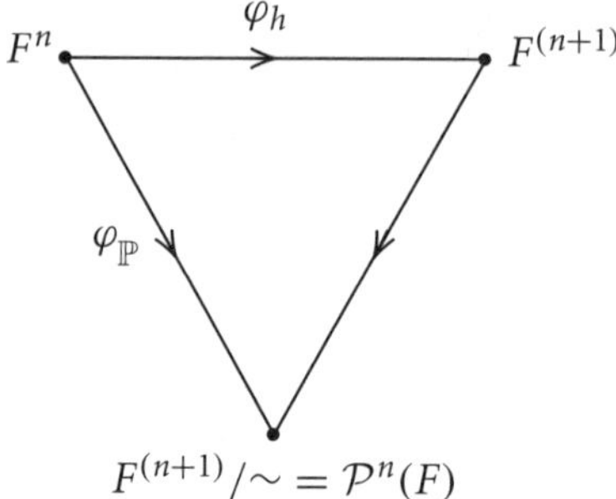



In other words, $\varphi_\mathbb{P}$ is just the composition of the mapping $(a_1, \ldots, a_{n+1}) \longmapsto [a_1, \ldots, a_{n+1}]$ of $F^{n+1}$ to $F^{n+1}/\sim$ with the mapping $\varphi_h$. Now, for any $[a_1, \ldots, a_n, b] \in \mathbb{P}^n(F)$ with $b \neq 0$, we have

$$[a_1, \ldots, a_n, b] = [a_1 b^{-1}, \ldots, a_n b^{-1}, 1]$$

so that

$$\varphi_\mathbb{P}(F^n) = \{[a_1, \ldots, a_n, 1] \mid a_i \in F\}$$
$$= \{[a_1, \ldots, a_n, a_{n+1}] \mid a_i \in F, \ a_{n+1} \neq 0\}.$$

It is sometimes convenient simply to identify $F^n$ with $\varphi_\mathbb{P}(F^n)$. So, if $\varphi_\mathbb{P}(F^n)$ contains all the points of the form $[a_1, \ldots, a_n, a_{n+1}]$, $a_{n+1} \neq 0$, what is new when we pass from $F^n$ to $\mathbb{P}^n(F)$? The answer is: all the points of the form $[a_1, \ldots, a_n, 0]$. These are the points in projective space corresponding to the lines through the origin that lie in the "hyperplane" defined by $z = 0$. We define the elements of the form $[a_1, \ldots, a_n, 0]$ to be *points at infinity*. Note that the elements in the class $[a_1, \ldots, a_n, 0]$ are precisely the elements in the plane $z = 0$ lying on the line through the points $(a_1, \ldots, a_n, 0)$ and the origin. We will get a better intuitive feeling for this and how the points with $z = 0$ relate to the points with $z \neq 0$ when we consider some examples.

Projective spaces provide an interesting perspective on homogeneous polynomials.

**Lemma 6.3.1** *Let $h(x_1, \ldots, x_{n+1}) \in F[x_1, \ldots, x_{n+1}]$ be a homogeneous polynomial in $x_1, \ldots, x_{n+1}$. Then $V(h)$ is a union of $\sim$ classes, together with $(0, \ldots, 0)$.*

**Proof.** Let $(a_1, \ldots, a_{n+1}) \in V(h)$. By Lemma 6.2.1, $t(a_1, \ldots, a_{n+1}) \in V(h)$ for all $t \in F^*$. Thus, $[a_1, \ldots, a_{n+1}] \subseteq V(h)$ and $V(h)$ (excluding the origin) is a union of $\sim$ classes. Also, $(0, \ldots, 0) \in V(h)$ for any homogeneous polynomial $h$.  $\square$

This makes it reasonable to define $[a_1, \ldots, a_{n+1}]$ to be *a zero of the (homogeneous) polynomial* $h(x_1, \ldots, x_{n+1})$ if

$$h(a_1, \ldots, a_{n+1}) = 0$$

and then we write $[a_1, \ldots, a_{n+1}] \in V(h)$. It further suggests that projective space might be a good context in which to study homogeneous polynomials. In particular, Lemma 6.3.1 makes it possible for us to adopt the following conventions. We refer to the elements of $\mathbb{P}^2(F)$ as *points*. For any elements $a, b, c \in F$, not all zero, we call the set

$$\{[x, y, z] \in \mathbb{P}^2(F) : ax + by + cz = 0\}$$

a *projective line* in $\mathbb{P}^2(F)$. With respect to these definitions of points and lines, $\mathbb{P}^2(F)$ then becomes a *projective geometry*. This aspect will be explored in the exercises.

Note that, in $F^3$, $V(ax + by + cz)$ consists of all points on a plane passing through the origin. Thus, the points on the projective line defined by the equation $ax + by + cz = 0$ are precisely the lines passing through the origin (less the origin) that lie in the plane in $F^2$ defined by the same equation $ax + by + cz = 0$. It is then customary to refer to the "line" $ax + by + cz = 0$ in $\mathbb{P}^2(F)$. A particularly important instance of this is when we take $a = b = 0$, $c = 1$, which leads to the "line" $z = 0$. However, the points $[x, y, z] \in \mathbb{P}^2(F)$ with $z = 0$ are precisely the points at infinity. Thus we have the situation where the points at infinity constitute a line in the projective plane $\mathbb{P}^2(F)$.

We saw earlier that, for any $f \in F[x_1, \ldots, x_n]$, we must expect that $V(f^h)$ will be significantly larger than $V(f)$ or $\varphi_h(V(f))$. However, if we work in $\mathbb{P}^n(F)$ instead of $F^{n+1}$, then the situation is quite different.

Consider Example 6.2.3 again. There we see that for each point $[a, b, c] \in V(x^2 + y^2 - z^2)$, since $(a, b, c) \neq (0, 0, 0)$ we must have $c \neq 0$ so that

$$
\begin{aligned}
[a, b, c] &= [ac^{-1}, bc^{-1}, 1] \quad \text{where} \quad (ac^{-1}, bc^{-1}) \in V(x^2 + y^2 - 1) \\
&= \varphi_{\mathbb{P}}(ac^{-1}, bc^{-1}) \\
&\in \varphi_{\mathbb{P}}(V(f)).
\end{aligned}
$$

Thus, in $\mathbb{P}^2(\mathbb{R})$,

$$V(f^h) = V(x^2 + y^2 - z^2) \subseteq \varphi_{\mathbb{P}}(V(f)).$$

On the other hand,

$$
\begin{aligned}
(a, b) \in V(f) &\implies (a, b, 1) \in V(f^h) \\
&\implies [a, b, 1] \in V(f^h).
\end{aligned}
$$

Thus,

$$\varphi_{\mathbb{P}}(V(f)) \subseteq V(f^h)$$

and we have equality

$$\varphi_{\mathbb{P}}(V(f)) = V(f^h).$$

We could express this by saying that in projective space $V(f)$ and $V(f^h)$ come together. Note that there is no solution in this example of the form $[*, *, 0]$, since $(a, b, 0) \in V(f^h)$ forces $a = b = 0$ and $[0, 0, 0]$ is not an element of $\mathbb{P}^2(\mathbb{R})$.

However, life is not always quite that simple. Indeed, if that were always the case, then there would be a good deal less interest in pursuing the added complication of $\mathbb{P}^2(F)$ over $F^n$.

**Example 6.3.2** Let $f(x, y) = x + y - 1 \in \mathbb{R}[x, y]$. Then $f(x, y) = 0$ is just the equation of a line and

$$f^h(x, y, z) = x + y - z.$$

(1) $V(f^h)$ is the plane in $\mathbb{R}^3$ passing through the origin and containing the line $x + y - 1 = 0$, $z = 1$.

(2) $\varphi_h(V(f))$ is exactly the line $x + y - 1 = 0$, $z = 1$ (see Fig. 6.6).

(3) $V(f^h) \supseteq \varphi_{\mathbb{P}}(V(f))$
$\phantom{(3) V(f^h)} = \{[a, b, 1] \mid (a, b) \in V(f)\}.$

This captures all points $(a, b, c) \in V(f^h)$ with $c \neq 0$. However,

$$(x, y, 0) \in V(f^h) \iff x + y = 0.$$

Thus, $V(f^h)$ also contains the line

$$x + y = 0, \quad z = 0$$

or the point $[1, -1, 0]$ in $\mathbb{P}^2(\mathbb{R})$ (see Fig. 6.6). In other words, $V(f^h)$ contains one extra point in $\mathbb{P}^2(F)$ in this case.

One way to visualize the introduction of the point $[1, -1, 0]$ is as the result of a limiting process. We can describe the result of lifting the line $x + y - 1 = 0$ into $\mathbb{R}^3$ either by the equations

$$x + y - 1 = 0, \quad z = 1$$

or in parametric form

$$x = 1 - t, \quad y = t, \quad z = 1.$$

When we transform these points into $\mathbb{P}^2(\mathbb{R})$, they assume the form

$$[1 - t, t, 1] = [1 - \frac{1}{t}, -1, -\frac{1}{t}]$$

$$\rightarrow [1, -1, 0]$$

as $t \rightarrow \infty$ (see Fig. 6.6).

Thus we can view the point $[1, -1, 0]$ as arising from the process of taking the limit of a sequence of points in $\varphi_{\mathbb{P}}(V(f))$. In fact, this is no accident and is not an isolated situation. It is, for that reason, that a deeper study of polynomials and their associated varieties requires a mixture of both algebraic and analytical techniques.

Intuitively, we think of the point $[1, -1, 0]$ as being added to our initial line

$$x + y - 1 = 0, \quad z = 1$$

at "infinity." The only problem, when trying to visualize it, is that it is added at both "ends" of our line. So perhaps we have to view the line as being closed into a big infinite loop.

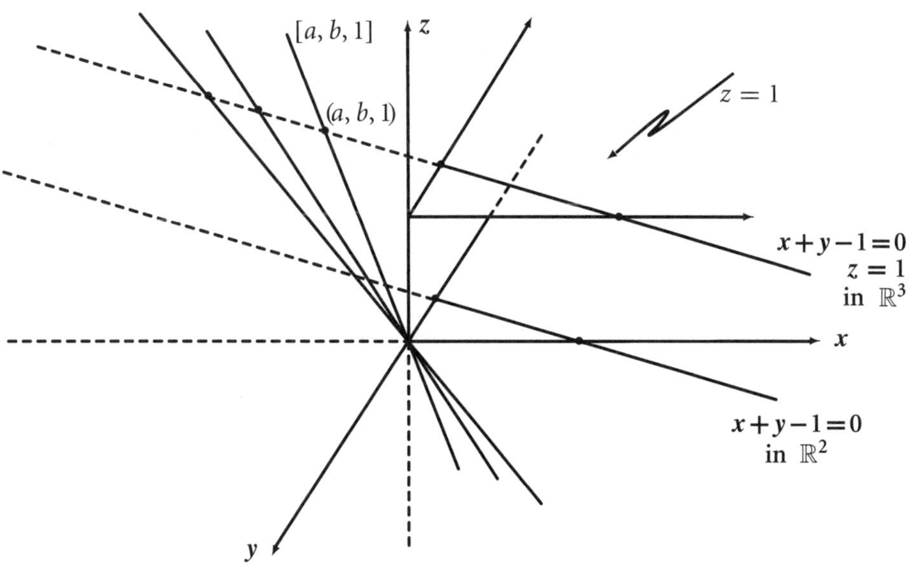


**Figure 6.6**

When we homogenized a circle, we obtained no extra points in $V(f^h)$. When we homogenized a line, we obtained one extra point. It is possible to have more. Let

$$f(x, y) = \frac{1}{4}x^2 - y^2 - 1 \in \mathbb{R}[x, y].$$

Then $f(x, y) = 0$ is the equation of a hyperbola, and

$$f^h(x, y, z) = \frac{1}{4}x^2 - y^2 - z^2.$$

For the solutions with $z \neq 0$, we have

$$\varphi_{\mathbb{P}}(V(f)) = \{[a, b, 1] \mid (a, b) \in V(f)\}.$$

However, when $z = 0$ we obtain additional solutions from

$$\frac{1}{4}x^2 - y^2 = 0$$

namely,

$$\frac{1}{2}x - y = 0, \qquad \frac{1}{2}x + y = 0.$$

These provide two additional points in $\mathbb{P}^2(\mathbb{R})$:

$$[2, 1, 0] \qquad \text{and} \qquad [2, -1, 0].$$

This process is illustrated in Figure 6.7.

It is important to note that in these examples, the extra points added were points at infinity. As we now see, this is the only way that new points can be introduced.

**Proposition 6.3.3** *Let $f \in F[x_1, \ldots, x_n]$. Then*

$$\varphi_{\mathbb{P}}(V(f)) = \{[a_1, \ldots, a_{n+1}] \in V(f^h) \mid a_{n+1} \neq 0\}.$$

**Proof.** Let us denote the set on the right by $S$. Then

$$(a_1, \ldots, a_n) \in V(f) \implies \varphi_{\mathbb{P}}(a_1, \ldots, a_n) = [a_1, \ldots, a_n, 1] \in V(f^h).$$

Hence,

$$\varphi_{\mathbb{P}}(V(f)) \subseteq S.$$

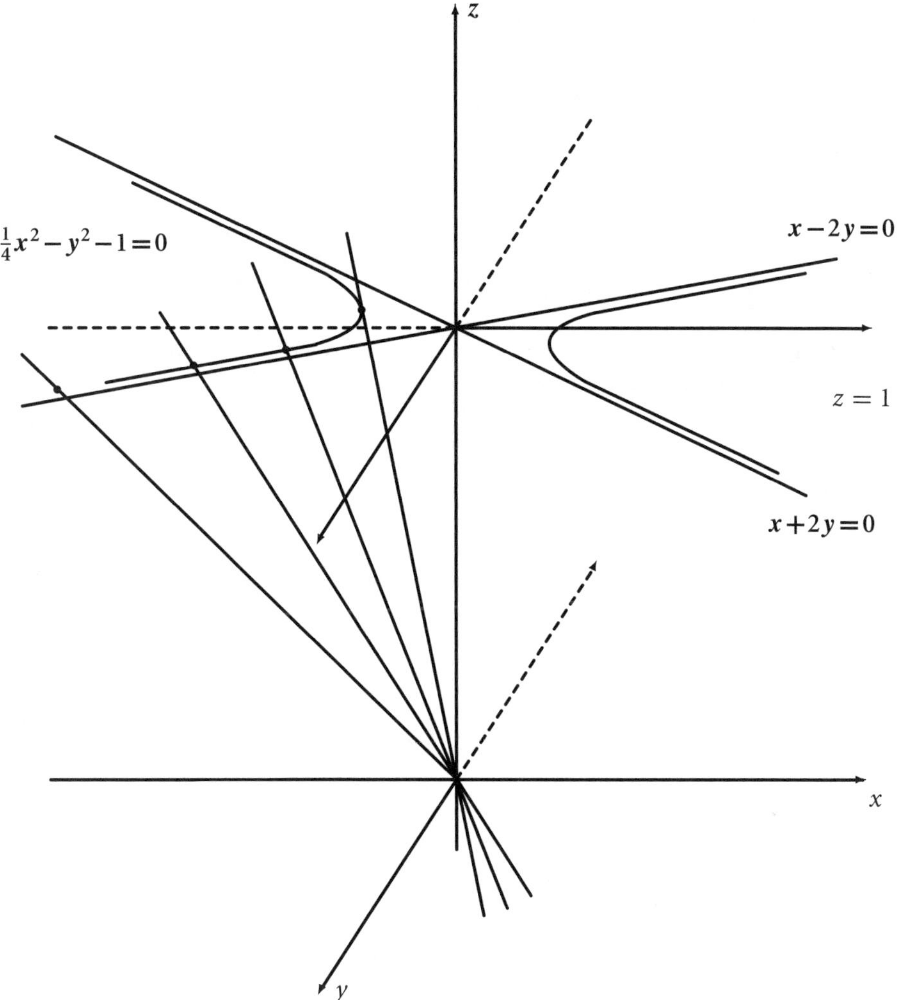


**Figure 6.7**

Conversely,

$$[a_1, \ldots, a_{n+1}] \in S \implies a_{n+1} \neq 0$$

$$\implies [a_1, \ldots, a_{n+1}] = [a_1 a_{n+1}^{-1}, \ldots, a_n a_{n+1}^{-1}, 1]$$

and, if $m = \deg(f)$,

$$f(a_1 a_{n+1}^{-1}, \ldots, a_n a_{n+1}^{-1}) = (a_{n+1}^{-1})^m (a_{n+1})^m f(a_1 a_{n+1}^{-1}, \ldots, a_n a_{n+1}^{-1})$$

$$= (a_{n+1}^{-1})^m f^h(a_1, \ldots, a_n, a_{n+1})$$

$$= (a_{n+1}^{-1})^m \cdot 0$$

$$= 0.$$

Therefore, $(a_1 a_{n+1}^{-1}, \ldots, a_n a_{n+1}^{-1}) \in V(f)$ and

$$[a_1, \ldots, a_n, a_{n+1}] \in \varphi_{\mathbb{P}}(V(f)).$$

Thus,

$$S \subseteq \varphi_{\mathbb{P}}(V(S))$$

and equality prevails. $\quad\square$

**Example 6.3.4** Let us consider what happens with an elliptic curve as we move from affine to projective space. Let

$$f(x, y) = y^2 - (x^3 + ax + b) \in F[x, y].$$

Then

$$f^h(x, y, z) = y^2 z - (x^3 + axz^2 + bz^3) \in F[x, y, z].$$

By Proposition 6.3.3,

$$V(f^h) = \varphi_{\mathbb{P}}(V(f)) \cup \{[x, y, 0] \mid (x, y, 0) \in V(f^h), \ (x, y) \neq (0, 0)\}.$$

Now

$$[x, y, 0] \in V(f^h) \iff -x^3 = 0.$$

Thus we have a triple point where $x = 0$. Since $[0, y, 0] = [0, 1, 0]$ for all $y \in F^*$, we see that $V(f^h)$ has exactly one new point at "infinity"—namely, $[0, 1, 0]$.

It is also important to notice that the line $z = 0$ intersects $V(f^h)$ in $[0, 1, 0]$ with multiplicity 3. This means that $z = 0$ is the tangent to $V(f^h)$ at $[0, 1, 0]$ and also that the multiplicity of intersection is 3.

We can confirm parametrically that the line at infinity ($z = 0$) is the unique line through the point $[0, 1, 0]$ with multiplicity of intersection with $V(f^h)$ greater than one. In other words, the line at infinity is the tangent to $V(f^h)$ at the point $[0, 1, 0]$.

A line in $\mathbb{P}^2(F)$ consists of all the projective points contained in a plane $W$, say, through the origin in $F^3$. A plane through the origin in $F^3$ is just a subspace of dimension 2 and is therefore spanned by any two independent vectors $u = (x_1, y_1, z_1)$, $v = (x_2, y_2, x_2)$ in $W$. Then the points of $W$ are just the linear combinations

$$su + tv = s(x_1, y_1, z_1) + t(x_2, y_2, x_2) = (sx_1 + tx_2, sy_1 + ty_2, sz_1 + tz_2)$$

of $u$ and $v$, where $s, t \in F$. This yields the following parametric description of $W$:

$$x = sx_1 + tx_2$$
$$y = sy_1 + ty_2$$
$$z = sz_1 + tz_2$$

in terms of the parameters $s$ and $t$. Now, to ensure that $W$ passes through the point $[0, 1, 0]$ (equivalently, for $W$ as a subspace of $F^3$ to contain the line $[0, 1, 0]$), we can choose $(x_1, y_1, z_1) = (0, 1, 0)$ so that the parametric equations become

$$x = tx_2$$
$$y = s + ty_2$$
$$z = tz_2.$$

Substituting these expressions into $f^h$, we obtain

$$f^h(tx_2, s + ty_2, tz_2) = (s + ty_2)^2 tz_2 - ((tx_2)^3 + atx_2(tz_2)^2 + b(tz_2)^3)$$
$$= t[(s + ty_2)^2 z_2 - t^2(x_2^3 + ax_2 z_2^2 + bz_2^3)].$$

We have a common factor $t$, as expected, corresponding to the point of intersection given by $t = 0$—namely, $x = 0, y = s, z = 0$ or the point $[0, s, 0] = [0, 1, 0]$. The only way that we can obtain a multiple factor $t^m$ with $m > 1$ from $f^h(tx_2, s + ty_2, tz_2)$, remembering that we are considering $f^h(tx_2, s + ty_2, tz_2)$ as a polynomial in $s$ and $t$, is if $z_2 = 0$, and then

$$f^h(tx_2, s + ty_2, tz_2) = -t^3 x_2^3.$$

Thus we obtain a unique line with intersection multiplicity greater than one when $z_2 = 0$. However, that is precisely the case when all the points $0, (0, 1, 0)$ and $(x_2, y_2, 0)$ lie in the plane $z = 0$ (in $F^3$) or the line $z = 0$ in $\mathbb{P}^2(F)$.

An important feature of the projective plane is that curves that did not intersect might intersect when homogenized. To see this, consider two parallel lines in $\mathbb{R}^2$:

$$x + y - 1 = 0, \qquad x + y - 2 = 0.$$

Clearly, these do not intersect in $\mathbb{R}^2$. Let

$$f(x, y) = x + y - 1, \qquad g(x, y) = x + y - 2.$$

Then

$$f^h(x, y, z) = x + y - z, \qquad g^h(x, y, z) = x + y - 2z.$$

In $\mathbb{P}^2(\mathbb{R})$, $f^h$ has solutions

$$[t, 1 - t, 1] \quad \text{and} \quad [1, -1, 0]$$

whereas $g^h$ has solutions

$$[t, 2 - t, 1] \quad \text{and} \quad [1, -1, 0].$$

Thus, $f^h = 0$ and $g^h = 0$ "intersect" in the point $[1, -1, 0]$.

Now, by Lemma 6.2.2, the point $[t, 1 - t, 1]$ in $\mathbb{P}^2(\mathbb{R})$ corresponds to the points $(t, 1 - t)$ in $\mathbb{R}^2$ whereas the point $[t, 2 - t, 1]$ in $\mathbb{P}^2(\mathbb{R})$ corresponds to the points $(t, 2 - t)$ in $\mathbb{R}^2$. Thus we could think of $f^h(x, y, z) = 0$ and $g^h(x, y, z) = 0$ as being lines in $\mathbb{R}^2$ but with a point $[1, -1, 0]$ being added "at infinity" as their point of intersection.

If we repeat this process with all lines in $\mathbb{R}^2$, we arrive at a geometric interpretation of $\mathbb{P}^2(\mathbb{R})$ as being obtained from $\mathbb{R}^2$ by the adjunction of points at infinity—one point for each family of parallel lines or one point for each line through the origin.

Further remarks:

(1) When we extend a line

$$L: a_1 x + b_1 y + c_1 = 0$$

to the projective plane

$$L^h: a_1 x + b_1 y + c_1 z = 0$$

we are adding to $L$ a single point $[-b_1, a_1, 0]$ "at infinity."

(2) When we extend an elliptic curve

$$E: y^2 = x^3 + ax + b$$

to the projective plane

$$E^h: y^2 z = x^3 + axz^2 + bz^3$$

we are also adding to $E$ a single point $[0, 1, 0]$ at infinity.

When $b_1 = 0$, $L$ and $E$ will intersect at the point $[0, a_1, 0] = [0, 1, 0]$ "at infinity."

(3) An elliptic curve $E$ is nonsingular and so there exists a tangent at every point on $E$ in $F^2$. In $\mathbb{P}^2(F)$, we have one additional point $[0, 1, 0]$. Substituting $z = 0$ into

$$E^h : y^2 z = x^3 + axz^2 + bz^3$$

we obtain

$$x^3 = 0.$$

Thus, $[0, 1, 0]$ is a triple point of intersection of the line $z = 0$ with $E^h$.

(4) Let $P = (x_1, y_1)$. Then the line in $\mathbb{P}^2(F)$ through $P = [x_1, y_1, 1]$ and $[0, 1, 0]$ is

$$x = x_1 z$$

which is the extension to $\mathbb{P}^2(F)$ of the line

$$x = x_1.$$

### Exercises 6.3

1. Show that $\sim$ does not respect $+$.

2. Let $(x_1, y_1, z_1)$, $(x_2, y_2, z_2) \in F^3$ be such that $(x_1, y_1, z_1) \neq (0, 0, 0) \neq (x_2, y_2, z_2)$ and

$$x_1 y_2 = x_2 y_1, \qquad x_1 z_2 = x_2 z_1, \qquad y_1 z_2 = y_2 z_1.$$

   Show that $[x_1, y_1, z_1] = [x_2, y_2, z_2]$.

3. Let $h = ax + by + cz \in F[x, y, z]$ and $[x_1, y_1, z_1] = [x_2, y_2, z_2]$ in $\mathbb{P}^2[F]$. Show that $(x_1, y_1, z_1) \in V(h) \Leftrightarrow (x_2, y_2, z_2) \in V(h)$.

4. Let $f(x, y) = \frac{1}{4}x^2 - y^2 - 1 \in \mathbb{R}[x, y]$.

   (i) Using the point $(2, 0)$ and the line $x = 0$, derive the following parametric representation for the hyperbola $f(x, y) = 0$:

$$x = \frac{2(1 + t^2)}{t^2 - 1}, \qquad y = \frac{2t}{t^2 - 1}.$$

   (ii) The parameterization in (i) provides the following parameterization of $\varphi_{\mathbb{P}}(V(f))$:

$$\left[ \frac{2(1 + t^2)}{t^2 - 1}, \frac{2t}{t^2 - 1}, 1 \right].$$

Use this parameterization to derive the two points in $V(f^h)$ that do not lie in $\varphi_{\mathbb{P}}(V(f))$ as limits of points in $\varphi_{\mathbb{P}}(V(f))$.

*5. Let $f(x, y) = -x^2 + \frac{y^2}{9} - 1 \in \mathbb{R}[x, y]$.

(i) Show that in $\mathbb{P}^2(\mathbb{R})$, the difference between $V(f^h)$ and $\varphi_{\mathbb{P}}(V(f))$ consists of the two points $[1, 3, 0]$ and $[1, -3, 0]$.

(ii) Using the point $(0, 3)$ and the line $y = 0$, derive the following parameterization for $V(f)$:

$$x = \frac{2t}{1 - t^2}, \qquad y = \frac{3(t^2 + 1)}{t^2 - 1}.$$

(iii) Use the resulting parameterization

$$\left[ \frac{2t}{1 - t^2}, \ \frac{3(t^2 + 1)}{t^2 - 1}, \ 1 \right]$$

of $\varphi_{\mathbb{P}}(V(f))$ to derive the points $[1, 3, 0]$ and $[1, -3, 0]$ as limits of points in $\varphi_{\mathbb{P}}(V(f))$.

6. Let $f(x, y) = y^2 - 2(x - 1) \in \mathbb{R}[x, y]$.

(i) Show that in $\mathbb{P}^2(\mathbb{R})$, the difference between $V(f^h)$ and $\varphi_{\mathbb{P}}(V(f))$ is the single extra point $[1, 0, 0]$.

(ii) Using the point $(1, 0)$ and the line $x = 0$, derive the following parameterization for $V(f)$:

$$x = 1 + \frac{2}{t^2}, \qquad y = \frac{2}{t}.$$

(iii) Use the resulting parameterization

$$\left[ 1 + \frac{2}{t^2}, \ -\frac{2}{t}, \ 1 \right]$$

of $\varphi_{\mathbb{P}}(V(f))$ to derive the point $[1, 0, 0]$ as a limit of points in $\varphi_{\mathbb{P}}(V(f))$.

7. Let $f(x, y) = y^2 - (x - 1)(x - 2)^2 \in \mathbb{R}[x, y]$.

(i) Show that in $\mathbb{P}^2(\mathbb{R})$, the difference between $V(f^h)$ and $\varphi_{\mathbb{P}}(V(f))$ is the single extra point $[0, 1, 0]$.

(ii) Using the point $(2, 0)$ and the line $x = 1$, derive the following parameterization of $V(f)$:

$$x = 1 + t^2, \qquad y = -t(t^2 - 1).$$

(iii)  Use the resulting parameterization

$$[1 + t^2, \; -t(t^2 - 1), \; 1]$$

of $\varphi_{\mathbb{P}}(V(f))$ to derive the point $[0, 1, 0]$ as a limit of points in $\varphi_{\mathbb{P}}(V(f))$.

8. Find the equation of the line (in homogeneous form) through each of the following pairs of points in $\mathbb{P}^2(F)$:

   (i)  $[1, 0, 1]$, $[1, 2, 1]$.
   (ii)  $[0, 1, 1]$, $[2, 1, 2]$.
   (iii)  $[2, -1, 1]$, $[1, 1, 0]$.
   (iv)  $[2, 1, 0]$, $[1, 2, 0]$.

9. Show that all points "at infinity" in $\mathbb{P}^2(F)$ lie on a single line.

## 6.4  Intersection of Lines and Curves

One of the big advantages of working in projective space over an algebraically closed field as opposed to affine space over an arbitrary field is that it regularizes the pattern of intersection between curves.

Let $f(x, y) \in F[x, y]$. Then, $C : f(x, y) = 0$ is

| | | | | |
|---|---|---|---|---|
| a | *line* | if | $\deg(f(x, y)) = 1$ | |
| a | *conic* (or quadratic) | if | $\deg(f(x, y)) = 2$ | |
| a | *cubic* | if | $\deg(f(x, y)) = 3,$ | and so on. |

We extend these definitions to $\mathbb{P}^2(F)$. Let $h(x, y, z) \in F[x, y, z]$ be a homogeneous polynomial. Then

$$C = \{[x, y, z] \in \mathbb{P}^2(F) \mid h(x, y, z) = 0\}$$

is

| | | | |
|---|---|---|---|
| a | *line* | if | $\deg(h(x, y, z)) = 1$ |
| a | *conic* | if | $\deg(h(x, y, z)) = 2$ |
| a | *cubic* | if | $\deg(h(x, y, z)) = 3.$ |

In $F^2$, a line $L$ can intersect another line in zero or one point; it can intersect a conic in zero, one, or two points; and it can intersect a cubic in zero, one, two, or three points. However, adding the assumption that $F$ is algebraically closed, working over projective space and counting multiplicities, the picture changes significantly. We consider the two cases that are most important for us later.

**Intersection of Two Lines.** Consider two distinct lines

$$L_1 : a_1 x + b_1 y + c_1 = 0 \ (a_1, b_1 \text{ not both } 0)$$
$$L_2 : a_2 x + b_2 y + c_2 = 0 \ (a_2, b_2 \text{ not both } 0).$$

In $F^2$, $L_1$ and $L_2$ may have one point of intersection or no points of intersection. Now consider $\mathbb{P}^2(F)$. There we have the homogenized forms

$$L_1^h : a_1 x + b_1 y + c_1 z = 0 \tag{6.7}$$

$$L_2^h : a_2 x + b_2 y + c_2 z = 0. \tag{6.8}$$

Without loss of generality, we may assume that $a_1 \neq 0$. Then, from (6.7),

$$x = -a_1^{-1}(b_1 y + c_1 z) \tag{6.9}$$

and, substituting this into (6.8),

$$-a_2 a_1^{-1}(b_1 y + c_1 z) + b_2 y + c_2 z = 0$$

so that

$$(-a_2 a_1^{-1} b_1 + b_2)y + (-a_2 a_1^{-1} c_1 + c_2)z = 0. \tag{6.10}$$

Let $\alpha = (-a_2 a_1^{-1} b_1 + b_2)$ and $\beta = (-a_2 a_1^{-1} c_1 + c_2)$. Then

$$\alpha - \beta = 0 \implies a_2 = a_1^{-1} a_2 a_1, b_2 = a_1^{-1} a_2 b_1, c_2 = a_1^{-1} a_2 c_1$$
$$\implies (a_2, b_2, c_2) = \lambda(a_1, b_1, c_1) \text{ with } \lambda = a_1^{-1} a_2$$
$$\implies L_1^h, L_2^h \text{ are the same line.}$$

Hence, either $\alpha \neq 0$ or $\beta \neq 0$. Let us assume that $\alpha \neq 0$. Then, from (6.10),

$$y = -\alpha^{-1}\beta z$$

and, from (6.9),

$$x = -a_1^{-1}(b_1(-\alpha^{-1}\beta z) + c_1 z) = (a_1^{-1} b_1 \alpha^{-1}\beta - a_1^{-1} c_1)z.$$

Therefore, there is a unique point of intersection at

$$[(a_1^{-1} b_1 \alpha^{-1}\beta - a_1^{-1} c_1)z, -\alpha^{-1}\beta z, z] = [a_1^{-1} b_1 \alpha^{-1}\beta - a_1^{-1} c_1, -\alpha^{-1}\beta, 1].$$

Thus, we have actually established the following result.

**Proposition 6.4.1** *Any two distinct lines in $\mathbb{P}^2(F)$ meet in exactly one point.*

We shall leave the consideration of the intersection of lines with conics to the exercises and pass on to the consideration of the intersection of a line and an elliptic curve.

**Intersection of a Line with an Elliptic Curve.** Consider the following line and elliptic curve

$$L : a_1 x + b_1 y + c_1 = 0 \ (a_1, b_1 \text{ not both zero})$$
$$E : y^2 - (x^3 + ax + b) = 0.$$

(For the current discussion, it is not necessary that $E$ be nonsingular.) In $F^2$, $L$ and $E$ may have zero, one, two, or three points of intersection. Now consider $\mathbb{P}^2(F)$. Here we have the homogenized forms

$$L^h : a_1 x + b_1 y + c_1 z = 0$$
$$E^h : y^2 z - (x^3 + axz^2 + bz^3) = 0.$$

*Case* (i) $b_1 \neq 0$. Suppose that there is a solution with $z = 0$. As a point on $E^h$ we must then have $x^3 = 0$, or $x = 0$, so that from $L^h$ we have $b_1 y = 0$ or $y = 0$. However, $[0, 0, 0]$ is not a point in $\mathbb{P}^2(F)$. Consequently, if $b_1 \neq 0$, then there are no solutions of the form $[*, *, 0]$. But, for $z \neq 0$,

$$[x, y, z] = [xz^{-1}, yz^{-1}, 1]$$

so we need only consider solutions with $z = 1$, which returns us to finding the points of intersection of $L$ and $E$ in $F^2$. Since $b_1 \neq 0$, we have $y = -b_1^{-1}(a_1 x + c_1)$ from $L$, which we can substitute into $E$ to obtain an equation of the form

$$x^3 + \alpha x^2 + \beta x + \gamma = 0.$$

Since $F$ is algebraically closed, we obtain 3 solutions (counting multiplicities) corresponding to 3 points of intersection in $\mathbb{P}^2(F)$.

*Case* (ii) $b_1 = 0, a_1 \neq 0$. Then, from the equation for $L^h$, we obtain

$$x = -a_1^{-1} c_1 z. \tag{6.11}$$

Substituting into $E^h$, we find that

$$\begin{aligned}
0 &= y^2 z - ((-a_1 c_1 z)^3 + a(-a_1 c_1 z)z^2 + bz^3) \\
&= z[y^2 - z^2(-a_1^3 c_1^3 - aa_1 c_1 + b)] \\
&= z[y^2 + z^2(a_1^3 c_1^3 + aa_1 c_1 - b)].
\end{aligned}$$

As expected, one solution is given by $z = 0$. In this case, by (6.11), we also have $x = 0$, so that $[0, y, 0] = [0, 1, 0]$ is a solution.

Now consider the other possibility that

$$y^2 + z^2(a_1^3 c_1^3 + aa_1 c_1 - b) = 0 \qquad (6.12)$$

and let $\alpha = (a_1^3 c_1^3 + aa_1 c_1 - b)$.

*Case* (iia) If $\alpha = 0$, then by (6.12) we get $y^2 = 0$, yielding the double point

$$[x, 0, z] = [-a_1^{-1} c_1 z, 0, z] = [-a_1^{-1} c_1, 0, 1].$$

Thus, we have a single point and a double point of intersection, giving a total of three points of intersection, counting multiplicities.

*Case* (iib) Now suppose that $\alpha \neq 0$. Since the field $F$ is algebraically closed, there exists an element $\beta \in F$ with $\beta^2 = \alpha$. From (6.12) we then have $y = \pm\beta z$, which yields two distinct solutions if $\mathrm{char}(F) \neq 2$ and a repeated solution if $\mathrm{char}(F) = 2$—namely,

$$[-a_1^{-1} c_1 z, \pm\beta z, z] = [-a_1^{-1} c_1, \pm\beta, 1].$$

Thus, in all cases, counting multiplicities, we have three points of intersection.

**Theorem 6.4.2** *Let $F$ be an algebraically closed field. Then any line and curve of the form $y^2 = x^3 + ax + b$ will have three points of intersection in $\mathbb{P}^2(F)$, counting multiplicities.*

One immediate consequence of Theorem 6.4.2 is the following corollary, which will be of importance to us when discussing groups on elliptic curves.

**Corollary 6.4.3** *Let $F$ be an algebraically closed field. Let $L_i : L_i(x, y, z) = 0$, $1 \leq i \leq 3$, be three (not necessarily distinct) lines and let*

$$C : L_1(x, y, z)L_2(x, y, z)L_3(x, y, z) = 0.$$

*Let $E$ be an elliptic curve. Then, counting multiplicities, $E^h$ and $C^h$ have 9 points in common in $\mathbb{P}^2(F)$.*

In fact, what we have just established in Theorem 6.4.2 is a very special case of a much general and deeper theorem that is beyond the scope of this book:

**Theorem 6.4.4 (Bézout's Theorem)** *Let $F$ be an algebraically closed field and $C_i : f_i(x, y, z) = 0$ $(i = 1, 2)$ be curves in $\mathbb{P}^2(F)$ of degrees $m$ and $n$, respectively,* *and with no irreducible components in common. Then, with appropriate counting of multiplicities, $C_1$ and $C_2$ have exactly $mn$ points in common in $\mathbb{P}^2(F)$.*

Bézout's Theorem can be particularly helpful when the conclusion is violated. In other words, if the curves $C_i : f_i(x, y, z) = 0$ $(i = 1, 2)$ have more than $mn$ points in common, then they must have a common component. This could lead to a factorization of $f_1$ or $f_2$. Moreover, this can be applied over fields that are not algebraically closed, since by passing to the projective space over a (possibly larger) algebraically closed field, the number of points of intersection can only increase (see section 6.7.)

Now, since we are working with points in the projective plane and can identify triples that are nonzero multiples of each other, we know that the only values of $z$ that we have to consider are $z = 0$, $z = 1$. So it is tempting just to substitute in these values and see what we get. If we get a complete set of distinct points, then there is no problem. However, whenever there are repetitions, we have to be more careful.

When searching for additional points of intersection in $\mathbb{P}^2(F)$, it will sometimes happen that we will want to count a point with $z = 0$ as a repeated point. If, for example, we arrived at the point where we wished to solve the equation

$$z^m g(x, y, z) = 0$$

where $z$ is not a factor of $g(x, y, z)$, and if the choice of $z = 0$ led to solutions $(x_0, y_0, 0)$ different from $(0, 0, 0)$, then we would consider each of these solutions as $m$-times repeated.

### Exercises 6.4

1. Let $\mathcal{F}$ be a family of (at least two) parallel lines in $\mathbb{R}^2$. Show that they intersect at a common point in $\mathbb{P}^2(\mathbb{R})$.

2. Let $P \in \mathbb{P}^2(\mathbb{R})$ but $P \notin \varphi_h(\mathbb{R}^2)$, where $\varphi_h$ is defined as in Lemma 6.2.2. Show that there exist two parallel lines in $\mathbb{R}^2$ that intersect in $P$.

3. Let $C : x^2 + y^2 - 1 = 0$. Determine the number of points of intersection of $C$ with each of the following lines in the contexts indicated:

    (i) $x + y - 1 = 0$ in $\mathbb{R}^2$.
    (ii) $x + y - \sqrt{2} = 0$ in $\mathbb{R}^2$.
    (iii) $ix - y + 2 = 0$ in $\mathbb{C}^2$ and in $\mathbb{P}^2(\mathbb{C})$.
    (iv) $ix + y = 0$ in $\mathbb{C}^2$ and in $\mathbb{P}^2(\mathbb{C})$.

4. Let $C : x^2/16 + y^2 - 1 = 0$. Determine the number of points of intersection of $C$ with each of the following lines in the contexts indicated:

    (i) $x - y = 0$ in $\mathbb{R}^2$.
    (ii) $y - 1 = 0$ in $\mathbb{R}^2$.

(iii) $y - 3 = 0$ in $\mathbb{R}^2$ and in $\mathbb{C}^2$.
(iv) $x - 4iy = 0$ in $\mathbb{C}^2$ and in $\mathbb{P}^2(\mathbb{C})$.

5. Let $C : y^2 - 4x = 0$. Determine the number of points of intersection of $C$ with each of the following lines in the contexts indicated:

   (i) $x - y - 2 = 0$ in $\mathbb{R}^2$.
   (ii) $x - 2y + 4 = 0$ in $\mathbb{R}^2$.
   (iii) $x - y + 2 = 0$ in $\mathbb{R}^2$ and in $\mathbb{C}^2$.
   (iv) $y - 3 = 0$ in $\mathbb{R}^2$, $\mathbb{C}^2$ and in $\mathbb{P}^2(\mathbb{C})$.

6. Let $C : x^2/4 - y^2/9 - 1 = 0$. Determine the number of points of intersection of $C$ with each of the following lines in the contexts indicated:

   (i) $x - 3 = 0$ in $\mathbb{R}^2$.
   (ii) $x - 1 = 0$ in $\mathbb{R}^2$ and in $\mathbb{C}^2$.
   (iii) $3x - 2y = 0$ in $\mathbb{R}^2$, $\mathbb{C}^2$ and in $\mathbb{P}^2(\mathbb{C})$.
   (iv) $3x - 2y + 2 = 0$ in $\mathbb{R}^2$, $\mathbb{C}^2$ and in $\mathbb{P}^2(\mathbb{C})$.

7. Let $C : (x - y + 1)(2x - y + 1) = 0$. Determine the number of points of intersection of $C$ with each of the following lines in the contexts indicated:

   (i) $x - 2y + 1 = 0$ in $\mathbb{R}^2$.
   (ii) $2x - 2y + 1 = 0$ in $\mathbb{R}^2$, $\mathbb{C}^2$ and in $\mathbb{P}^2(\mathbb{C})$.

8. Let $C : (x - y)(x - y + 1) = 0$. Determine the number of points of intersection of $C$ with each of the following lines in the contexts indicated:

   (i) $2x - y + 1 = 0$ in $\mathbb{R}^2$.
   (ii) $2x - 2y - 3 = 0$ in $\mathbb{R}^2$, $\mathbb{C}^2$ and in $\mathbb{P}^2(\mathbb{C})$.

9. Let $C : (x - 1)^2 + (y - 1)^2 = 0$. Determine the number of points of intersection of $C$ with each of the following lines in the contexts indicated:

   (i) $x - 2y + 1 = 0$ in $\mathbb{R}^2$.
   (ii) $x - y + 1 = 0$ in $\mathbb{R}^2$ and in $\mathbb{C}^2$.
   (iii) $x - iy - 1 = 0$ in $\mathbb{R}^2$, $\mathbb{C}^2$ and in $\mathbb{P}^2(\mathbb{C})$.

10. Let $C : y^2 - (x - 1)(x - 4)(x - 5) = 0$. Determine the number of points of intersection of $C$ with each of the following lines in the contexts indicated:

    (i) $x - y - 1 = 0$ in $\mathbb{R}^2$.
    (ii) $x - y = 0$ in $\mathbb{C}^2$.
    (iii) $x - 2 = 0$ in $\mathbb{R}^2$, in $\mathbb{C}^2$ and in $\mathbb{P}^2(\mathbb{C})$.
    (iv) $x = 0$ in $\mathbb{R}^2$, in $\mathbb{C}^2$ and in $\mathbb{P}^2(\mathbb{C})$.
    (v) $x - 1 = 0$ in $\mathbb{R}^2$, $\mathbb{C}^2$ and in $\mathbb{P}^2(\mathbb{C})$.

11. Let $f(x, y) = x^2 - y^2 + 1$, $g(x, y) = x^2 - y^2 + 2$. Discuss the number of points of intersection of $C_1 : f(x, y) = 0$, $C_2 : g(x, y) = 0$ in the following contexts:

   (i) $\mathbb{R}^2$.
   (ii) $\mathbb{C}^2$.
   (iii) $\mathbb{P}^2(\mathbb{C})$.

12. Let $f(x, y) = x^3 - 2x^2 y - 3xy^2 + x + 2$, $g(x, y) = x^3 - 2x^2 y - 3xy^2 + y + 1 \in \mathbb{C}[x, y]$. Determine the points of intersection of $C_1 : f(x, y) = 0$ and $C_2 : g(x, y) = 0$ in $\mathbb{P}^2(\mathbb{C})$.

## 6.5 Defining Curves by Points

In this section we consider the question of how many points it takes to define a curve. We are interested in curves in the projective plane defined by homogeneous polynomials. Since it is the simplest, extends what we already know, and illustrates the general case nicely, we begin with lines.

**Lemma 6.5.1** *Let $P_1 = [x_1, y_1, z_1]$, $P_2 = [x_2, y_2, z_2] \in \mathbb{P}^2(F)$ be distinct points. Then there is a unique line in $\mathbb{P}^2(F)$ passing through $P_1$ and $P_2$—namely,*

$$(y_2 z_1 - y_1 z_2)x + (x_1 z_2 - x_2 z_1)y + (-x_1 y_2 + x_2 y_1)z = 0. \qquad (6.13)$$

**Proof.** By straightforward subsitution we can verify that $P_1$ and $P_2$ lie on the line given in (6.13). Now consider the general equation of a line in $\mathbb{P}^2(F)$:

$$L : ax + by + cz = 0.$$

Then $L$ passes through $P_1$ and $P_2$ if and only if

$$ax_1 + by_1 + cz_1 = 0$$
$$ax_2 + by_2 + cz_2 = 0.$$

We wish to solve these equations for $a, b, c$ (with $a, b, c$ not all equal to zero). Now, not all of $x_1, y_1, z_1$ can be zero since $[0, 0, 0]$ is not a point in $\mathbb{P}^2(F)$. So, without loss of generality, we may assume that $x_1 \neq 0$. Then we have

$$x_1 a + y_1 b + z_1 c = 0 \qquad (6.14)$$

$$(y_2 - x_2 x_1^{-1} y_1)b + (z_2 - x_2 x_1^{-1} z_1)c = 0. \qquad (6.15)$$

If $y_2 = x_2 x_1^{-1} y_1$ and $z_2 = x_2 x_1^{-1} z_1$, then $x_2 x_1^{-1} \neq 0$ (since $P_2 \neq [0, 0, 0]$) and

$$(x_2, y_2, z_2) = x_2 x_1^{-1}(x_1, y_1, z_1)$$

which would imply that

$$P_2 = [x_2, y_2, z_2] = [x_1, y_1, z_1] = P_1.$$

However, by hypothesis, $P_1 \neq P_2$. Therefore, one of

$$p = y_2 - x_2 x_1^{-1} y_1, \qquad q = z_2 - x_2 x_1^{-1} z_1$$

must be nonzero. Let us assume that $q \neq 0$. Then, from (6.14) and (6.15), we have

$$c = -q^{-1} p b$$

$$a = -x_1^{-1}(y_1 b - q^{-1} p z_1 b)$$

$$= -x_1^{-1}(y_1 - q^{-1} p z_1) b$$

and the equation for $L$ becomes

$$-x_1^{-1}(y_1 - q^{-1} p z_1) b x + b y - q^{-1} p b z = 0.$$

We may assume that $b \neq 0$ because, if $b = 0$, then $a = c = 0$ and we no longer have the equation of a line. Removing the common factor $b$ and multiplying by $x_1 q\ (\neq 0)$, we obtain

$$-(y_1 q - p z_1) x + x_1 q y - x_1 p z = 0$$

or

$$(y_2 z_1 - x_2 x_1^{-1} y_1 z_1 - y_1 z_2 + x_2 x_1^{-1} y_1 z_1) x + x_1 q y - x_1 p z = 0$$

which simplifies to

$$(y_2 z_1 - y_1 z_2) x + (x_1 z_2 - x_2 z_1) y + (-x_1 y_2 + x_2 y_1) z = 0.$$

Since the line given by this equation does indeed pass through the points $P_1$ and $P_2$, it follows that this is the unique line passing through these two points. $\square$

Next we consider curves defined by homogeneous polynomials of degree $n$. We need to know the number of monomial terms in such a polynomial. We restrict our attention to polynomials in three variables.

**Lemma 6.5.2** *Let $f(x, y, z) \in F[x, y, z]$ be a homogeneous polynomial of degree $n$. Then the largest possible number of monomial terms in $f(x, y, z)$ is $\frac{(n+1)(n+2)}{2}$.*

**Proof.** For each $i$ with $0 \leq i \leq n$, consider the number of monomials of degree $n$ (the coefficients are not important here) of the form $x^i y^j z^k$. We must have

$$j + k = n - i$$

which means that there are precisely $n - i + 1$ possible choices for the pair $(j, k)$. Tabulating the numbers of possibilities, we have

| $i$ | $\# \, (j, k)$ |
| --- | --- |
| $n$ | 1 |
| $n - 1$ | 2 |
| $n - 2$ | 3 |
| $\vdots$ | |
| 0 | $n + 1$. |

Therefore, the largest possible number of monomial terms is

$$1 + 2 + 3 + \cdots + (n + 1) = \frac{(n + 1)(n + 2)}{2}. \quad \square$$

For each $n \in \mathbb{N}$, put the monomials of degree $n$ in some order,

$$m_1(x, y, z), \ m_2(x, y, z), \ldots, m_k(x, y, z)$$

where $k = \frac{(n+1)(n+2)}{2}$. Now define a mapping $\mu_n : \mathbb{P}^2(F) \to F^k$ by

$$\mu_n([a, b, c]) = (m_1(a, b, c), \ m_2(a, b, c), \ldots, m_k(a, b, c)).$$

**Theorem 6.5.3** *Let $n \in \mathbb{N}$, $k = \frac{(n+1)(n+2)}{2}$, and $P_1, \ldots, P_{k-1}$ be points in $\mathbb{P}^2(F)$. Then there exists a curve defined by a nonzero homogeneous polynomial of degree $n$ that passes through the points $P_1, \ldots, P_{k-1}$. The curve is unique if and only if the vectors $\mu_n(P_i)$, $1 \leq i \leq k - 1$, are independent in $F^k$.*

**Proof.** Let $P_i = [x_i, y_i, z_i]$. Let $f(x, y, z) \in F[x, y, z]$ be a homogeneous polynomial of degree $n$ with unspecified coefficients. With the notation introduced prior to the theorem, $f(x, y, z)$ has $k$ monomial terms $c_j m_j(x, y, z)$, $1 \leq j \leq k$. In order that the curve $C : f(x, y, z) = 0$ passes through the points $P_i$, we must have

$$f(x_i, y_i, z_i) = 0 \qquad (1 \leq i \leq k - 1).$$

Thus, we have $k - 1$ homogeneous linear equations in the $k$ unknowns $c_j$. Hence, there must exist a nonzero solution. To be more precise, let $A$ be the $(k-1) \times k$ matrix with $i$th row equal to $\mu_n([a_i, b_i, c_i])$. Let $X = [c_1, \ldots, c_k]^T$. Then any nonzero solution to the homogeneous system of equations

$$AX = 0$$

provides the coefficients for a nonzero homogeneous polynomial $f$ with $P_i \in V(f)$, $i = 1, \ldots, k - 1$.

If the vectors $\mu_k([x_i, y_i, z_i])$ are independent, then rank $(A) = k - 1$ and the dimension of the solution space is 1. Thus, any two nonzero solutions are scalar multiples of each other so that the resulting polynomials are scalar multiples of each other and they define the same curve.

On the other hand, if the vectors $\mu_k([x_i, y_i, z_i])$ are not independent, then rank $(A) < k - 1$ and so the dimension of the solution space for $AX = 0$ is at least 2. Thus, there will be two nonzero polynomials $f_1, f_2$ that are not scalar multiples of each other with $P_i \in V(f_1) \cap V(f_2)$, $i = 1, \ldots, k - 1$.  $\square$

Consider the case of conics. Here, $n = 2$ so that the number of distinct monic monomials is

$$k = \frac{(2+1)(2+2)}{2} = 6.$$

Let us order these six monomials as follows:

$$x^2, \quad xy, \quad xz, \quad y^2, \quad yz, \quad z^2$$

and define $\mu_2$ accordingly. Theorem 6.5.3 tells us that there will be a conic passing through any five points in $\mathbb{P}^2(F)$.

**Example 6.5.4** Consider the following points in $\mathbb{P}^2(\mathbb{Q})$:

$$P_1 = [1, 0, 0], \quad P_2 = [0, 1, 0], \quad P_3 = [0, 0, 1], \quad P_4 = [1, 0, 1], \quad P_5 = [1, 1, 1].$$

Then, in matrix form,

$$A = \begin{bmatrix} \mu_2(P_1) \\ \vdots \\ \mu_2(P_5) \end{bmatrix} = \begin{bmatrix} 1 & 0 & 0 & 0 & 0 & 0 \\ 0 & 0 & 0 & 1 & 0 & 0 \\ 0 & 0 & 0 & 0 & 0 & 1 \\ 1 & 0 & 1 & 0 & 0 & 1 \\ 1 & 1 & 1 & 1 & 1 & 1 \end{bmatrix}.$$

Using the first row, we reduce $A$ to

$$\begin{bmatrix} 1 & 0 & 0 & 0 & 0 & 0 \\ 0 & 0 & 0 & 1 & 0 & 0 \\ 0 & 0 & 0 & 0 & 0 & 1 \\ 0 & 0 & 1 & 0 & 0 & 1 \\ 0 & 1 & 1 & 1 & 1 & 1 \end{bmatrix}$$

and further reduce it, using other rows, to

$$B = \begin{bmatrix} 1 & 0 & 0 & 0 & 0 & 0 \\ 0 & 0 & 0 & 1 & 0 & 0 \\ 0 & 0 & 0 & 0 & 0 & 1 \\ 0 & 0 & 1 & 0 & 0 & 0 \\ 0 & 1 & 0 & 0 & 1 & 0 \end{bmatrix}$$

which clearly has rank 5, so that the vectors $\mu_2(P_1), \ldots, \mu_2(P_5)$ are independent and the solution is unique. Reducing $A$ as before, we solve

$$B \begin{bmatrix} c_1 \\ c_2 \\ \vdots \\ c_6 \end{bmatrix} = 0$$

to obtain

$$c_1 = c_3 = c_4 = c_5 = 0, \qquad c_2 = -c_5.$$

Thus, the unique conic passing through $P_1, \ldots, P_5$ is

$$-c_5 xy + c_5 yz = 0$$

or

$$xy - yz = 0.$$

There is one situation when it is obvious from the geometry that five distinct points in $\mathbb{P}^2(F)$ do not determine a unique conic. Suppose that $P_1, P_2, P_3,$ and $P_4$ are distinct collinear points—say, $P_1, P_2, P_3, P_4 \in V(l)$—where $l(x, y, z) \in F[x, y, z]$ is homogeneous of degree 1. Let $P_5$ be a fifth distinct point and $l_i \in F[x, y, z]$, $i = 1, 2, 3, 4$, be such that $V(l_i)$ is the line determined by $P_i$ and $P_5$ $(1 \leq i \leq 4)$. Then,

$$P_1, P_2, \ldots, P_5 \in V(ll_1) \cap V(ll_2) \cap \ldots \cap V(ll_4)$$

where the $V(ll_i)$ are distinct conics.

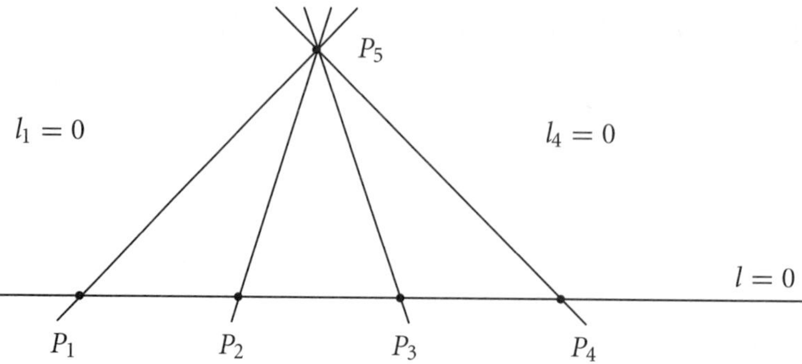


However, we will see in section 6.7 that, provided no four of the points are collinear, then five distinct points will determine a unique conic. With the help of Bézout's Theorem, we can determine one set of circumstances under which we obtain a unique conic without resorting to the calculation of the vectors $\mu_n(P_i)$ and Theorem 6.5.3.

**Corollary 6.5.5** *Let F be an algebraically closed field. Let $P_1, \ldots, P_5 \in \mathbb{P}^2(F)$ be distinct points and $f(x, y, z) \in F[x, y, z]$ be an irreducible homogeneous polynomial of degree 2 such that the curve $C : f(x, y, z) = 0$ passes through $P_1, \ldots, P_5$. Then C is the unique conic through $P_1, \ldots, P_5$.*

**Proof.** Let $D : g(x, y, z) = 0$ be another curve of degree 2, with $g(x, y, z) \in F[x, y, z]$ homogeneous, and passing through $P_1, \ldots, P_5$. Since $f$ is irreducible and $F$ is algebraically closed, $V(f)$ is irreducible by Corollary 5.6.4. We now consider two cases:

*Case* (i) $V(f) \subseteq V(g)$. By Corollary 5.6.4 (i), $f$ divides $g$. Since $\deg(f) = \deg(g) = 2$, there must exist $\lambda \in F^*$ with $g = \lambda f$.

*Case* (ii) $V(f) \nsubseteq V(g)$. Then, since $V(f)$ is irreducible, $V(f)$ and $V(g)$ have no irreducible component in common and, by Bézout's Theorem, $|V(f) \cap V(g)| = 4$, which is a contradiction.

Thus, we must have $g = \lambda f$, for some $\lambda \in F^*$, and $V(g) = V(f)$ so that $C$ is unique. $\square$

**Exercises 6.5**

1. Show that the criterion in Theorem 6.5.3 is independent of the representatives chosen for the elements of $\mathbb{P}^2(F)$.

2. Show that, for all $n \in \mathbb{N}$, there exists an irreducible homogeneous polynomial $f(x, y, z) \in F[x, y, z]$ of degree $n$.

3. Use the preceding exercise and Bézout's Theorem to show that, in general, $n^2$ points are insufficient to determine a curve in $\mathbb{P}^2(F)$ of degree $n \in \mathbb{N}$ uniquely.

4. (i) Show that $n^2 > \frac{(n+1)(n+2)}{2}$ for all $n \in \mathbb{N}$, $n \geq 4$.
   (ii) Reconcile (i), Theorem 6.5.3, and the observation in Exercise 3.

5. Let $P = [x_1, y_1, z_1]$, $Q = [x_2, y_2, z_2]$ be distinct points in $\mathbb{P}^2(F)$. Show that $\mu_2(P)$ and $\mu_2(Q)$ are independent vectors in $F^6$.

6. Let $P = [x_1, y_1, z_1]$, $Q = [x_2, y_2, z_2]$ be distinct points in $\mathbb{P}^2(F)$. Use Theorem 6.5.3 to show that there exists a unique line $ax + by + cz = 0$ through the points $P, Q$.

7. Find a conic passing through the points

$$[1, 0, 1], \quad [1, 0, -1], \quad [1, 1, 0], \quad [1, -1, 0], \quad [0, 1, 0]$$

   in $\mathbb{P}^2(\mathbb{Q})$.

8. Show that the points
   $[0, 0, 1], [1, 1, 1], [2, 2, 1], [3, 3, 1], [1, -1, 1] \in \mathbb{P}^2(\mathbb{Q})$ fail the test in Theorem 6.5.3. Find two distinct conics passing through these points.

9. Show that the conics that you found in Exercise 8 are reducible.

10. Let $F$ be an algebraically closed field, $f, g \in F[x, y, z]$ be homogeneous polynomials of degree $n$, and $P_1, \ldots, P_{n^2+1}$ be distinct points in $V(f) \cap V(g)$. Show that if $f$ is irreducible, then there exists $\lambda \in F^*$ with $g = \lambda f$.

## 6.6 Classification of Conics

The material in this section is not required in later sections, but the fact that we are able to classify conics so precisely does focus the spotlight on the natural next step—cubics.

From elementary calculus, we are all familiar with the classification of conics (conic sections) in $\mathbb{R}^2$ into circles, ellipses, hyperbolas, parabolas, pairs of lines, and points. Somewhat unexpectedly, the classification is much simpler in the projective plane $\mathbb{P}^2(\mathbb{R})$. Throughout this section, we will work over the field $\mathbb{R}$.

Recall that by a *conic* we mean a curve in $\mathbb{P}^2(F)$ defined by an equation of the form

$$f(x, y, z) = 0$$

where $f(x, y, z) \in F[x, y, z]$ is a homogeneous polynomial of degree 2. The most general form for the polynomial $f(x, y, z)$ is

$$f(x, y, z) = ax^2 + by^2 + cz^2 + dxy + exz + fyz.$$

Since we are assuming that $F = \mathbb{R}$ in this section, we may rewrite this as

$$f(x, y, z) = ax^2 + by^2 + cz^2 + 2dxy + 2exz + 2fyz$$

(where we have simply written $d, e$, and $f$ as $2(\frac{d}{2})$, $2(\frac{e}{2})$, and $2(\frac{f}{2})$). Using matrix notation, we can then write

$$f(x, y, z) = [x, y, z] \begin{bmatrix} a & d & e \\ d & b & f \\ e & fc \end{bmatrix} \begin{bmatrix} x \\ y \\ z \end{bmatrix}$$

$$= [x, y, z] \, A \, [x, y, z]^T$$

where

$$A = \begin{bmatrix} a & d & e \\ d & b & f \\ e & f & c \end{bmatrix}$$

is a symmetric matrix. Now by standard matrix theory, any symmetric matrix is "diagonalizable" (this is known variously as the *Spectral Theorem* or the *Principal Axis Theorem*). In other words, there exists an orthogonal matrix $M$ $(M^{-1} = M^T$ where $M^T$ denotes the transpose of the matrix $M$), such that

$$M^{-1} A M = \text{diag} \, \{d_1, d_2, d_3\}.$$

Hence, if we perform the change of variables

$$\begin{bmatrix} x \\ y \\ z \end{bmatrix} = M \begin{bmatrix} x' \\ y' \\ z' \end{bmatrix}$$

then we obtain

$$f(x, y, z) = [x, y, z] \, A \begin{bmatrix} x \\ y \\ z \end{bmatrix}$$

$$= [x', y', z'] \, M^T A M \begin{bmatrix} x' \\ y' \\ z' \end{bmatrix}$$

$$= [x', y', z'] \, M^{-1} A M \begin{bmatrix} x' \\ y' \\ z' \end{bmatrix}$$

$$= [x', y', z'] \, \mathrm{diag}\,\{d_1, d_2, d_3\} \begin{bmatrix} x' \\ y' \\ z' \end{bmatrix}$$

$$= d_1 (x')^2 + d_2 (y')^2 + d_3 (z')^2.$$

Since $V(f_1) = V(-f_1)$, we may assume that $d_i \geq 0$ for at least two values of $i$. By means of a simple exchange of variables, we can also assume that $d_1 > 0$. Then we can perform a further change of variable $x'' = e_1 x'$, where $e_1^2 = d_1$, to eliminate the coefficient $d_1$. We can similarly eliminate $d_2$ and $d_3$ if either of them is nonzero. By a further change of variables we can also interchange the variables (for example, $x = y''$, $y = z''$, $z = x''$). Thus, by a suitable change of variables, we can transform $f(x, y, z)$ into one of the following forms:

(1) $X^2 = 0.$

(2) $X^2 + Y^2 = 0.$

(3) $X^2 - Y^2 = 0.$

(4) $X^2 + Y^2 + Z^2 = 0.$

(5) $X^2 + Y^2 - Z^2 = 0.$

Now the combined effect of the various changes of variables has still been linear. In other words, there is a $3 \times 3$ matrix $C$ such that

$$\begin{bmatrix} x \\ y \\ z \end{bmatrix} = C \begin{bmatrix} X \\ Y \\ Z \end{bmatrix}.$$

In addition, each of the transformations is reversible. In other words, the matrix $C$ is invertible and

$$\begin{bmatrix} X \\ Y \\ Z \end{bmatrix} = C^{-1} \begin{bmatrix} x \\ y \\ z \end{bmatrix}.$$

Let

$$C^{-1} = D = \begin{bmatrix} d_{11} & d_{12} & d_{13} \\ d_{21} & d_{22} & d_{23} \\ d_{31} & d_{32} & d_{33} \end{bmatrix}.$$

Now let us have a look at this resulting categorization of conics into five classes:

(1) $X^2 = 0$. This corresponds to

$$(d_{11}x + d_{12}y + d_{13}z)^2 = 0$$

the product of a line with itself.

(2) $X^2 + Y^2 = 0$.
This defines the single point $[0, 0, 1]$ in $\mathbb{P}^2(\mathbb{R})$ and corresponds to the point of intersection of the two lines

$$d_{11}x + d_{12}y + d_{13}z = 0, \qquad d_{21}x + d_{22}y + d_{23}z = 0.$$

(3) $X^2 - Y^2 = 0$.
This factorizes as

$$(X - Y)(X + Y) = 0$$

and consists of the two lines $X - Y = 0$, $X + Y = 0$.

(4) $X^2 + Y^2 + Z^2 = 0$.
The only solution in real numbers is $X = Y = Z = 0$. However, $[0, 0, 0]$ is not a point in $\mathbb{P}^2(\mathbb{R})$, and so there is no solution in this case.

(5) $X^2 + Y^2 - Z^2 = 0$.
Putting $Z = 0$ would force $X = Y = 0$. Again, since $[0, 0, 0]$ is not a point in $\mathbb{P}^2(\mathbb{R})$, there is no solution with $Z = 0$. Thus, we may assume that $Z = 1$ and then the equation becomes the equation of a circle:

$$X^2 + Y^2 = 1.$$

## 6.7 Reducible Conics and Cubics

Here we give some consideration to reducible conics and cubics. If a homogeneous cubic polynomial is reducible, then one of the factors must be linear. Since elliptic curves are defined by irreducible polynomials, we were able to get by without considering this situation. However, since so much of our discussion concerns the number of points of intersection of curves, it is interesting to see how this has a bearing on the factorization of polynomials. Bézout's Theorem tells us that a line and a cubic with no common component will have three points of intersection (over an algebraically closed field). So, what if a line and a cubic have four points in common? By Bézout's Theorem, they must have a common component. However, a line is clearly irreducible and so can have only one component—namely, itself. Thus the line must be a part of the cubic curve.

**Theorem 6.7.1** *Let $F$ be an algebraically closed field and $f(x, y, z)$, $g(x, y, z)$, $h(x, y, z) \in F[x, y, z]$ be homogeneous polynomials of degree $1, 2$, and $3$, respectively. Let the corresponding line, conic and cubic in $\mathbb{P}^2(F)$ be*

$$
\begin{aligned}
L \ &: f(x, y, z) = 0 \\
C_g &: g(x, y, z) = 0 \\
C_h &: h(x, y, z) = 0.
\end{aligned}
$$

(i) *If $L$ and $C_g$ have three points in common (counting multiplicities), then $f(x, y, z)$ divides $g(x, y, z)$ and $L$ is a component of $C_g$.*

(ii) *If $L$ and $C_h$ have four points in common (counting multiplicities), then $f(x, y, z)$ divides $h(x, y, z)$ and $L$ is a component of $C_h$.*

**Proof.** (i) By Bézout's Theorem, Theorem 6.4.4, we see that $L$ and $C_h$ must have a component in common. However, $L$ is a line, $f(x, y, z)$ is irreducible, and so $V(f)$ is also irreducible. Hence, $V(f) \subseteq V(C_h)$ and, by Corollary 5.6.4, $f$ divides $h$.

Part (ii) follows similarly. $\square$

Our proof of Theorem 6.7.1 depended critically on Bézout's Theorem, which only works in $\mathbb{P}^2(F)$ over an algebraically closed field. However, with slightly stronger hypotheses we can consider a slightly weaker result, but with a more elementary proof.

**Theorem 6.7.2** *Let $f(x, y, z)$, $g(x, y, z)$, $h(x, y, z) \in F[x, y, z]$ be homogeneous polynomials of degrees $1, 2$, and $3$, respectively. Let the corresponding line, conic, and cubic in $\mathbb{P}^2(F)$ be*

$$
\begin{aligned}
L \ &: f(x, y, z) = 0 \\
C_g &: g(x, y, z) = 0 \\
C_h &: h(x, y, z) = 0.
\end{aligned}
$$

(i) *If $L$ and $C_g$ have three distinct points in common, then $f(x, y, z)$ divides $g(x, y, z)$ and $L$ is a component of $C_g$.*
(ii) *If $L$ and $C_h$ have four distinct points in common, then $f(x, y, z)$ divides $h(x, y, z)$ and $L$ is a component of $C_h$.*

**Proof.** We will prove (ii) and allow you to construct the parallel argument for (i). Let

$$f(x, y, z) = lx + my + nz$$

$$h(x, y, z) = ax^3 +$$

$$bx^2 y + cx^2 z +$$

$$dxy^2 + exyz + hxz^2 +$$

$$ky^3 + py^2 z + syz^2 + tz^3.$$

We will assume that $l \neq 0$. Similar arguments apply if $m \neq 0$ or $l = m = 0$ and $n \neq 0$. Then $l^{-1} f(x, y, z)$ determines the same line, and so we may assume that $l = 1$. By Theorem 5.3.2, we can write

$$h(x, y, z) = (x + my + nz)q(x, y, z) + r(y, z) \qquad (6.16)$$

for some $q(x, y, z) \in F[x, y, z], r(x, y) \in F[y, z]$. Moreover, the division process in Theorem 5.3.2 ensures that $q$ is homogeneous of degree 2 and that $r$ is either equal to zero or is homogeneous of degree 3. Hence, $r(y, z) = h(x, y, z) - (x + my + nz)q(x, y, z)$ can be written as

$$r(y, z) = \alpha y^3 + \beta y^2 z + \gamma yz^2 + \delta z^3 \qquad (6.17)$$

for some $\alpha, \beta, \gamma, \delta \in F$. Now let $P_i = [x_i, y_i, z_i]$, $1 \leq i \leq 4$, be four distinct points lying on both $L$ and $C$. Then, $f(x_i, y_i, z_i) = 0 = h(x_i, y_i, z_i)$ and, from (6.16) and (6.17), we obtain four equations in $\alpha, \beta, \gamma, \delta$:

$$y_i^3 \alpha + y_i^2 z_i \beta + y_i z_i^2 \gamma + z_i^3 \delta = 0 \qquad (6.18)$$

for $1 \leq i \leq 4$. So suppose that $z_i \neq 0$ for all $i$. Divide the $i$th equation in (6.18) by $z_i^3$ and write $v_i = y_i/z_i$. Then we have the system of equations

$$v_i^3 \alpha + v_i^2 \beta + v_i \gamma + \delta = 0 \qquad (1 \leq i \leq 4).$$

Let the matrix of coefficients be $A$. Then, by Lemma 5.2.7,

$$\det(A) = \prod_{1 \leq i < j \leq 4} (v_i - v_j)$$

which will be nonzero provided $v_i \neq v_j$ for all $i, j$. However,

$$v_i = v_j \implies \frac{y_i}{z_i} = \frac{y_j}{z_j}$$

$$\implies y_j = \frac{z_j}{z_i} y_i = \lambda y_i$$

$$\text{where } \lambda = \frac{z_j}{z_i}$$

$$\implies x_j = l x_j = -m y_j - n z_j \text{ (recalling that } l = 1)$$

$$= \lambda(-m y_i - n z_i)$$

$$= \lambda x_i$$

$$\implies x_j = \lambda x_i, \ y_j = \lambda y_i, \ z_j = \lambda z_i.$$

Thus, $v_i = v_j$ would imply that $[x_i, y_i, z_i] = [x_j, y_j, z_j]$ in $\mathbb{P}^2(\mathbb{R})$, which is a contradiction. Hence, $v_i \neq v_j$ and $\det(A) \neq 0$. It follows that the system (6.18) has a unique solution

$$\alpha = \beta = \gamma = \delta = 0$$

and, from (6.16), we have

$$h(x, y, z) = f(x, y, z) q(x, y, z)$$

as required.

Now consider the case where $z_i = 0$ for some $i$. A line (other than $z = 0$) contains only one point at infinity and therefore the remaining $z_j$ must be nonzero. Also, $y_i \neq 0$, since no point of the form $(x_i, 0, 0)$ can lie on $L$ (recall that $[0, 0, 0] \notin \mathbb{P}^2(\mathbb{R})$). Hence, the corresponding equation in (6.18) will show that $\alpha = 0$. There then remain three other equations in $\beta, \gamma, \delta$ together with three points with third components nonzero. We may now proceed as in the first case. $\square$

In section 6.5, we saw that there is a conic through any five points in $\mathbb{P}^2(F)$, but that the conic would not be unique if four of the points are collinear. We will now see that if no four of the points are collinear then the conic is indeed unique.

**Theorem 6.7.3** *Let $F$ be an algebraically closed field and $P_1, \ldots, P_5 \in \mathbb{P}^2(F)$. Let $f, g \in F[x, y, z]$ be homogeneous polynomials of degree 2 with $P_1, \ldots, P_5 \in V(f) \cap V(g)$.*

    (i) *$f$ is irreducible if and only if no three of the $P_i$ are collinear.*
    (ii) *If $f$ is irreducible, then $V(f) = V(g)$.*
    (iii) *If $f$ is reducible and no four of the $P_i$ are collinear, then $V(f) = V(g)$.*

**Note.** We do not assume that the points $P_1, \ldots, P_5$ are distinct. If, for instance, $P_1 = P_2$, then whenever we have two curves intersecting at $P_1 = P_2$, we count it as a double point of intersection and so on. However, we will not emphasize this point in the proof.

**Proof.** (i) Let $f$ be irreducible. Suppose that three of the points lie on a line

$$L \; : \; l(x, y, z) = 0.$$

By Theorem 6.7.1, $L$ must divide $f$, which contradicts the assumption that $f$ is irreducible. Hence, no three of the points are collinear.

Conversely, if $f$ is reducible, then there must exist linear polynomials $l_1, l_2$ with $f = l_1 l_2$ and $V(f) = V(l_1) \cup V(l_2)$. Hence, either three of the points lie in $V(l_1)$ or three lie in $V(l_2)$. Thus, if $f$ is reducible, then three of the points must be collinear.

(ii) By Corollary 5.6.4, we know that $V(f)$ is irreducible. If $V(f) \not\subseteq V(g)$, then $V(f)$ and $V(g)$ have no irreducible component in common and, by Bézout's Theorem, $|V(f) \cap V(g)| = 2 \times 2 = 4$, which is a contradiction. Hence, we can assume that $V(f) \subseteq V(g)$. Then, by Corollary 5.6.4, $f$ divides $g$. But, $\deg(f) = \deg(g) = 2$. Hence there exists $\lambda \in F^*$ with $g = \lambda f$ so that $V(f) = V(g)$.

(iii) Now suppose that $f$ is reducible. If $g$ is irreducible, then by part (ii), with the roles of $f$ and $g$ reversed, we have $f = \lambda g$, for some $\lambda \in F^*$, which would imply that $f$ is also irreducible, which is a contradiction. Hence we can assume that $g$ is also reducible.

Since $\deg(f) = \deg(g) = 2$, there must be homogeneous linear polynomials $l_1, l_2, m_1, m_2 \in F[x, y, z]$ of degree 1 with

$$f = l_1 l_2, \quad g = m_1 m_2.$$

Since $P_1, \ldots, P_5 \in V(f) = V(l_1) \cup V(l_2)$, either $l_1$ or $l_2$ must contain three of the points $P_1, \ldots, P_5$. Without loss of generality, we can assume that $P_1, P_2, P_3 \in V(l_1)$. Since no four of the points are collinear, we must have $P_4, P_5 \in V(l_2)$. Now, since $P_1, P_2, P_3 \in V(g)$, either $V(m_1)$ contains two of them or $V(m_2)$ contains two of them. Again, without loss of generality, we can assume that two of them lie in $V(m_1)$. However, two distinct lines intersect at, at most, one point. Hence, $V(l_1) = V(m_1)$. Since $P_4, P_5 \notin V(l_1) = V(m_1)$, we must have $P_4, P_5 \in V(m_2)$. Hence, since $V(l_2)$ and $V(m_2)$ are also lines, $V(l_2) = V(m_2)$. Now $l_1, l_2, m_1$, and $m_2$ are all linear and therefore irreducible. Hence, by Corollary 5.6.4 (i), there are $\lambda, \mu \in F^*$ such that

$$m_1 = \lambda l_1, \quad m_2 = \mu l_2$$

from which we conclude that $g = m_1 m_2 = (\lambda l_1)(\mu l_2) = \lambda \mu l_1 l_2 = \lambda \mu f.$ $\quad \square$

**Exercises 6.7**

1. Let $F$ be an algebraically closed field and $f(x, y, z)$, $l(x, y, z) \in F[x, y, z]$ be homogeneous polynomials of degree $n$ and 1, respectively. Let $|V(f) \cap V(l)| \geq n + 1$. Show that $l$ divides $f$.

2. Let $F$ be an algebraically closed field and $f(x, y, z)$, $k(x, y, z) \in F[x, y, z]$ be homogeneous polynomials of degree $n$ and 2, respectively, where $n \geq 3$. Let $P_1, \ldots, P_{2n+1} \in V(f) \cap V(k)$ be such that no four of the points are collinear. Show that $k$ divides $f$.

*3. Let $F$ be an algebraically closed field and $f(x, y, z) \in F[x, y, z]$ be homogeneous polynomials of degree 3 and suppose that no four of the points $P_1, \ldots, P_{10} \in V(f)$ are collinear and no eight lie on a conic. Show that $V(f)$ is the unique cubic through the points $P_1, \ldots, P_{10}$.

4. In each of the following cases, show that any conic passing through the points in $\mathbb{P}^2(F)$ listed must be reducible, without calculating the conics:

   (i)   $[1, 0, 0]$,  $[1, 1, 1]$,  $[1, 0, 1]$,  $[0, 1, -1]$,  $[0, 0, 1]$.

  (ii)  $[0, 1, 0]$,  $[1, 1, 1]$,  $[1, 0, 0]$,  $[1, 0, 1]$,  $[1, 1, 0]$.

 (iii)  $[1, 1, 0]$,  $[0, 1, 0]$,  $[1, 1, 1]$,  $[0, 1, 1]$,  $[2, 2, 3]$.

## 6.8 The Nine-Point Theorem

The main result in this section shows that there are interesting consequences from the restriction on the number of points of intersection with elliptic curves. This will be critical when considering the associative law for groups on elliptic curves. Those readers willing to take the associative law on faith may skip this section though, by doing so, they will miss a lovely piece of geometry. A general homogeneous cubic in $F[x, y, z]$ has up to 10 monomial terms so that we need at least nine points to specify it uniquely. If we have only eight points then there will be independent solutions in Theorem 6.5.3. In some circumstances, two solutions will be sufficient to describe all the solutions. All that we are doing is taking advantage of the fact that the set of solutions to the system of equations $AX = 0$ in Theorem 6.5.3 is a subspace of $F^k$ and, if that subspace is of dimension two, then it must have a basis of two vectors.

**Theorem 6.8.1 (The Nine-Point Theorem)** *Let $F$ be an algebraically closed field and $f_i(x, y, z) \in F[x, y, z]$, $i = 1, 2, 3$ be homogeneous cubics such that the curves $V(f_1)$ and $V(f_2)$ have no common irreducible components. Let $P_1, \ldots, P_8 \in V(f_1) \cap V(f_2) \cap V(f_3)$. Then,*

    (i) *no four of the points $P_1, \ldots, P_8$ are collinear*

   (ii) *there exist $\lambda, \mu \in F$ with*

$$f_3 = \lambda f_1 + \mu f_2$$

  (iii) *$P_9 \in V(f_1) \cap V(f_2) \implies P_9 \in V(f_3)$.*

**Proof.** For the sake of simplicity, we will assume that the points $P_1, \ldots, P_8$ are distinct although, with careful attention to the multiplicity of intersections, we may dispense with this assumption.

(i) If four of the points—say $P_1, \ldots, P_4$—lie on a line $L$, then by Theorem 6.7.1 (ii), $L$ is a common component of $V(f_1)$ and $V(f_2)$, which is a contradiction.

(ii) Let

$$L : l(x, y, z) = 0 \qquad \text{be a line through } P_1 \text{ and } P_2$$
$$C : k(x, y, z) = 0 \qquad \text{be a conic through } P_3, \ldots, P_7.$$

Such a line and curve exist by Theorem 6.5.3. Since no four of the points $P_3, \ldots, P_7$ are collinear, it follows from Theorem 6.7.3, that $C$ is uniquely determined. There are now three cases depending on whether $P_8$ lies on $L$, or on $C$, or on neither.

*Case* (1). $P_8$ *lies on* $L$. Let $Q_1 = [x_1, y_1, z_1]$ be a point on $L$ that is different from $P_1, P_2$, and $P_8$. Let $Q_2 = [x_2, y_2, z_2]$ be a point that is on neither $L$ nor $C$ (see the following diagram.) Consider the equations

$$cf_3(x_1, y_1, z_1) + c_1 f_1(x_1, y_1, z_1) + c_2 f_2(x_1, y_1, z_1) = 0 \qquad (6.19)$$
$$cf_3(x_2, y_2, z_2) + c_1 f_1(x_2, y_2, z_2) + c_2 f_2(x_2, y_2, z_2) = 0. \qquad (6.20)$$

As a system of two homogeneous linear equations in the three unknowns $c, c_1, c_2$, there is a nonzero solution, which we will just denote by $c, c_1$, and $c_2$. Let

$$d(x, y, z) = cf_3(x, y, z) + c_1 f_1(x, y, z) + c_2 f_2(x, y, z) \qquad (6.21)$$

and $D = V(d)$. By the construction of $D$, we have $P_1, \ldots, P_8, Q_1, Q_2 \in V(d)$. If $d(x, y, z) = 0$ and $c \neq 0$, then

$$f_3(x, y, z) = -c^{-1} c_1 f_1(x, y, z) - c^{-1} c_2 f_2(x, y, z)$$

and we have the desired result. All other cases lead to a contradiction. If $d(x, y, z) = 0$ and $c = 0$, then at least one of $c_1$ and $c_2$ is nonzero (say, $c_1$), so that

$$f_1(x, y, z) = c_1^{-1} c_2 f_2(x, y, z)$$

which contradicts the assumption that $f_1$ and $f_2$ have no common component.

We will now show that the assumption $d(x, y, z) \neq 0$ also leads to a contradiction. Since $d(x, y, z) \neq 0$, $V(d)$ is a line, conic, or cube. From (6.21) and the hypothesis, we have that $P_1, \ldots, P_8 \in V(d)$, and from (6.19) and (6.20) we have that $Q_1, Q_2 \in V(d)$.

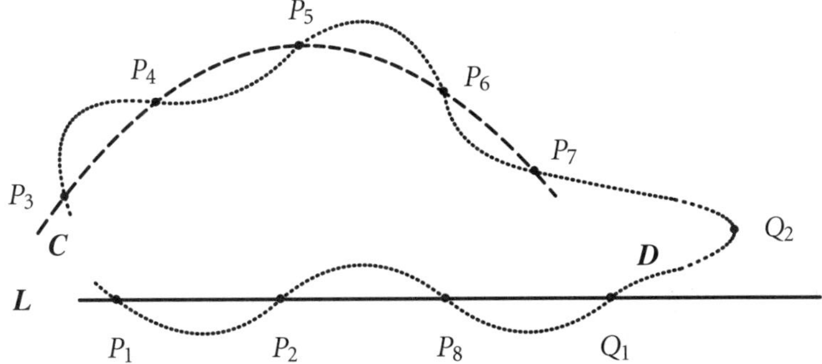


Now $L$ and $V(d)$ have four points ($P_1$, $P_2$, $P_8$, and $Q_1$) in common. Hence, by Theorem 6.7.1 (ii), $l(x, y, z)$ divides $d(x, y, z)$—say,

$$d(x, y, z) = l(x, y, z)\, q(x, y, z)$$

where, since $V(d)$ is a line, conic, or cubic, either $q$ is a constant or $V(q)$ must be a line or conic. We now have that $P_1, \ldots, P_8 \in D = V(d)$, $P_1, P_2, P_8 \in L = V(l)$, but no four of $P_1, \ldots, P_8$ are collinear by part (i). It follows that $P_3, P_4, P_5, P_6, P_7 \in V(q)$. However, $C$ is the unique conic through these points, so that $V(q) = C$. But this implies that

$$D = V(l) \cup V(q)$$
$$= L \cup C$$

which does not contain $Q_2$ by the way that $Q_2$ was chosen. Thus, we have a contradiction, since $D$ was specifically constructed to pass through $Q_2$.

*Case* (2). $P_8$ *lies on* $C$. Let $Q_1 = [x_1, y_1, z_1]$ be a point on $C$ different from $P_3, \ldots, P_8$ and again choose $Q_2 = [x_2, y_2, z_2]$ to be a point that is neither on $L$ or $C$. Let $d(x, y, z)$ and $D = V(d)$ be defined as before. As in Case (1), if $d(x, y, z) = 0$, we either have the desired result or a contradiction. So again it suffices to show that $d(x, y, z) \neq 0$ leads to a contradiction.
We break the argument into two cases.

*Case* (2a). $C$ *irreducible.* We have $|C \cap D| \geq 7 > 2 \times 3 = 6$ so that, by Bézout's Theorem, $C$ and $D$ must have a component in common. However, $C$ is irreducible. Hence, $C \subseteq D$ and, by Corollary 5.6.4, $k(x, y, z)$ divides $d(x, y, z)$—say,

$$d(x, y, z) = k(x, y, z)\, m(x, y, z)$$

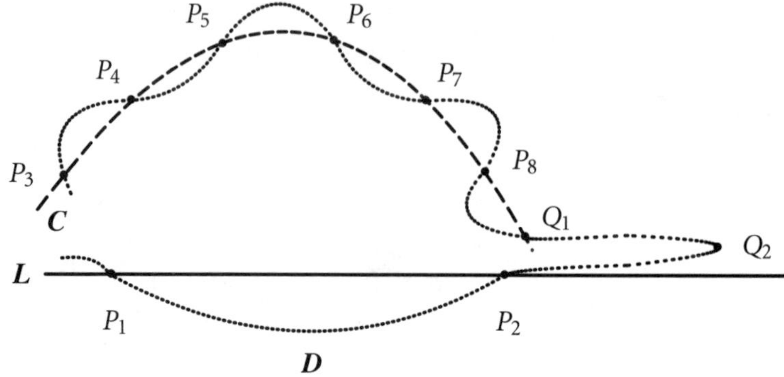


where, matching degrees, we see that $m$ is homogeneous and linear. Thus, $M = V(m)$ is a line and

$$D = C \cup M.$$

Now, by Bézout's Theorem, if $P_1 \in C$, then we can argue

$$P_1 \in C \Longrightarrow |C \cap V(f_1)| = |\{P_1, P_3, \ldots, P_8\}| \ \geq 7 > 6$$
$$\Longrightarrow C \ \text{and} \ V(f_1) \ \text{have a common component}$$
$$\Longrightarrow C \subseteq V(f_1) \ \text{since } C \text{ is irreducible}$$

and, similarly, $C \subseteq V(f_2)$. This contradicts the hypothesis that $V(f_1)$ and $V(f_2)$ have no common components. Hence, $P_1 \notin C$ and, likewise, $P_2 \notin C$. But, $P_1, P_2 \in D$ so that we must have $P_1, P_2 \in M$. Since two points determine a line, hence $M = L$ and

$$D = C \cup L.$$

However, $D$ contains $Q_2$ whereas $C \cup L$ does not. Thus we have a contradiction again.

*Case* (2b). *C reducible.* Since $C$ is a conic, the only possibility is that $k(x, y, z) = l_1(x, y, z)l_2(x, y, z)$ and $C = L_1 \cup L_2$ where $L_1 = V(l_1)$ and $L_2 = V(l_2)$ are lines. No four of $P_3, \ldots, P_8$ are collinear, so three (say, $P_3, P_4, P_5$) lie on $L_1$ and three (say, $P_6, P_7, P_8$) lie on $L_2$. Since $Q_1 \in C$ by choice, we may assume that $Q_1 \in L_1$. Then, $|L_1 \cap D| \geq 4$ and by Theorem 6.7.1 (ii), $l_1$ divides $d$—say,

$$d = l_1 q_1$$

where $V(q_1)$ must be a conic. Then, $P_6, P_7, P_8 \in L_2 \cap V(q_1)$ and, by Theorem 6.7.1 (i), $l_2$ divides $q_1$—say,

$$q_1 = l_2 l_3$$

where $l_3$ must be linear. Thus,

$$d = l_1 l_2 l_3.$$

Since no four of $P_1, \ldots, P_8$ are collinear, $P_1, P_2 \notin L_1 \cup L_2$. Hence, $P_1, P_2 \in V(l_3)$ and, since a line is defined by two points,

$$V(l_3) = L.$$

Thus,

$$D = L_1 \cup L_2 \cup L = C \cup L$$

where $Q_2 \in D$ but $Q_2 \notin C \cup L$. Thus we again have a contradiction.

*Case* (3). $P_8$ *lies on neither C nor L.* We divide this case into two subcases.

*Case* (3a). *One of $P_3, \ldots, P_7$ lies on L.* By part (i) of the theorem, no more than one can lie on $L$. Without loss of generality, we may assume that $P_3 \in L$. Choose $Q_1 \in L$ and distinct from $P_1, P_2$, and $P_3$. Choose $Q_2 \in C$ such that $Q_2 \notin L$ and $Q_2$ is not collinear with any three of the four points $P_4, P_5, P_6$, and $P_7$. Then no four of the points $P_4, P_5, P_6, P_7$, and $Q_2$, are collinear so that, by Theorem 6.7.3, $C$ is uniquely determined by the points $P_4, P_5, P_6, P_7, Q_2$. Let $d(x, y, z)$ and $D = V(d)$ be defined as in Case (1). If $d(x, y, z) = 0$, then, as before, we either obtain the desired conclusion or a contradiction, so that it remains to show that we also obtain a contradiction if $d(x, y, z) \neq 0$.

First we see that $L$ and $D$ have four points $(P_1, P_2, P_3, Q_1)$ in common so that $L$ must be a component of $D$ (Theorem 6.7.1 (ii)). Let

$$d(x, y, z) = l(x, y, z)\, q(x, y, z)$$

where $V(q)$ must be a conic.

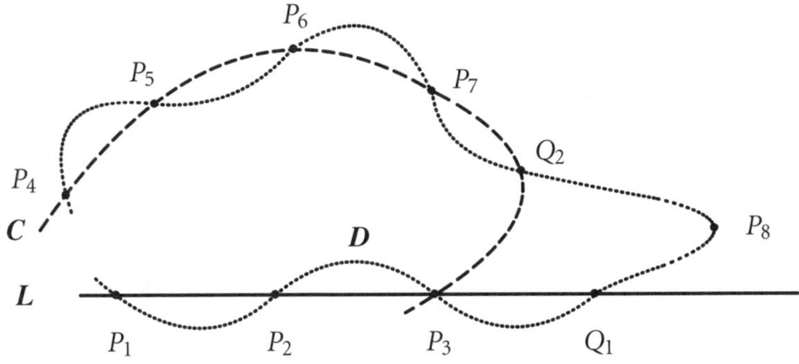

Since no four of $P_1, \ldots, P_8$ are collinear, none of $P_4, \ldots, P_7 \in L$ and, by choice, $Q_2 \notin L$. Hence, $P_4, \ldots, P_7, Q_2 \in V(q)$. Thus, $V(q)$ and $C$, both conics, have five points in common, which implies that they have an irreducible component in common. If either $V(q)$ or $C$ is irreducible, then it follows that both must be irreducible and that $V(q) = C$. If they have a linear component in common, then both are products of two lines. Since no four of $P_4, \ldots, P_7, Q_2$ are collinear, $V(q)$ and $C$ must be the product of the same two lines and therefore $C = V(q)$. Hence,

$$D = V(l) \cup V(q) = L \cup C$$

where $P_8 \in D$ but $P_8 \notin L \cup C$. Thus we have a contradiction.

*Case* (3b). *None of $P_3, \ldots, P_7$ lies on L.* Let $Q_1$ and $Q_2$ be two distinct points on $L$ distinct from $P_1$ and $P_2$. Let $d(x, y, z)$ and $D = V(d)$ be defined as in Case (1). Again, if $d(x, y, z) = 0$, then we either obtain the desired result or a contradiction, so that it remains to show that $d(x, y, z) \neq 0$ leads to a contradiction. As before, $C$ is uniquely determined by the points $P_3, \ldots, P_7$.

In this case, $L$ and $D$ have four points in common so that $L$ must be a component of $D$ (Theorem 6.7.1 (ii)). Let

$$d(x, y, z) = l(x, y, z)\, q(x, y, z)$$

where $V(q)$ must be a conic.

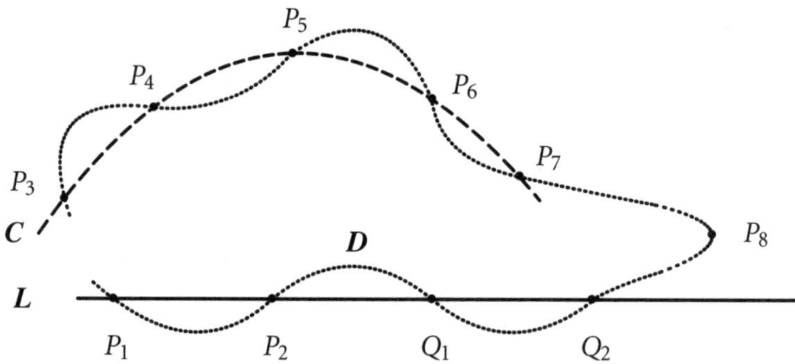


Then, $P_3, \ldots, P_7 \in V(q) \cap C$. By the uniqueness of $C$, we must have $V(q) = C$. Thus,

$$D = V(l) \cup V(q) = L \cup C$$

where $P_8 \in D$ but $P_8 \notin L \cup C$, a contradiction.

(iii) This part follows immediately from part (ii). $\quad\square$

## 6.9 Groups on Elliptic Curves

Let us consider an elliptic curve:

$$E : y^2 = x^3 + ax + b, \qquad 4a^3 + 27b^2 \neq 0$$

or (in homogeneous terms) of the form

$$E : y^2 z = x^3 + axz^2 + bz^3, \qquad 4a^3 + 27b^2 \neq 0.$$

There are several important points to observe here.

(1) The condition $4a^3 + 27b^2 \neq 0$ ensures that $E$ is nonsingular and that there exists a tangent at each point.

(2) From Exercise 11 in section 5.2, $f(x, y) = y^2 - (x^3 + ax + b)$ is irreducible and, therefore, so also is $f^h$.

(3) Substituting $z = 0$ into the homogeneous form of the equation for $E$, the equation reduces to $x^3 = 0$. Thus we have a triple point of intersection at $[0, 1, 0]$. In other words, the tangent to $E$ at $[0, 1, 0]$ has a third point of intersection with $E$ at $[0, 1, 0]$.

For the time being, we will assume that we are working within an algebraically closed field $F$ and that $E \subseteq \mathbb{P}(F^2)$, although the difference between $E$ as a projective curve and $E$ as an affine curve is just the one point at infinity.

Now fix a point $\odot$, which we will call the *base point*, on $E$. We define a binary operation $+$ on $E$ as follows.

Let $P$ and $Q$ be arbitrary points on $E$. Let $L$ denote the line through $P$ and $Q$. If $P = Q$, then we take $L$ to be the tangent at $P$. By Theorem 6.4.2, $L$ and $E$ intersect in three points. Let $P \circ Q$ denote the third point. Now let the line through $P \circ Q$ and $\odot$ meet $E$ in $R$. Then we define

$$P + Q = R$$

or

$$p + Q = (P \circ Q) \circ \odot.$$

Several possible configurations are presented in the next few diagrams:

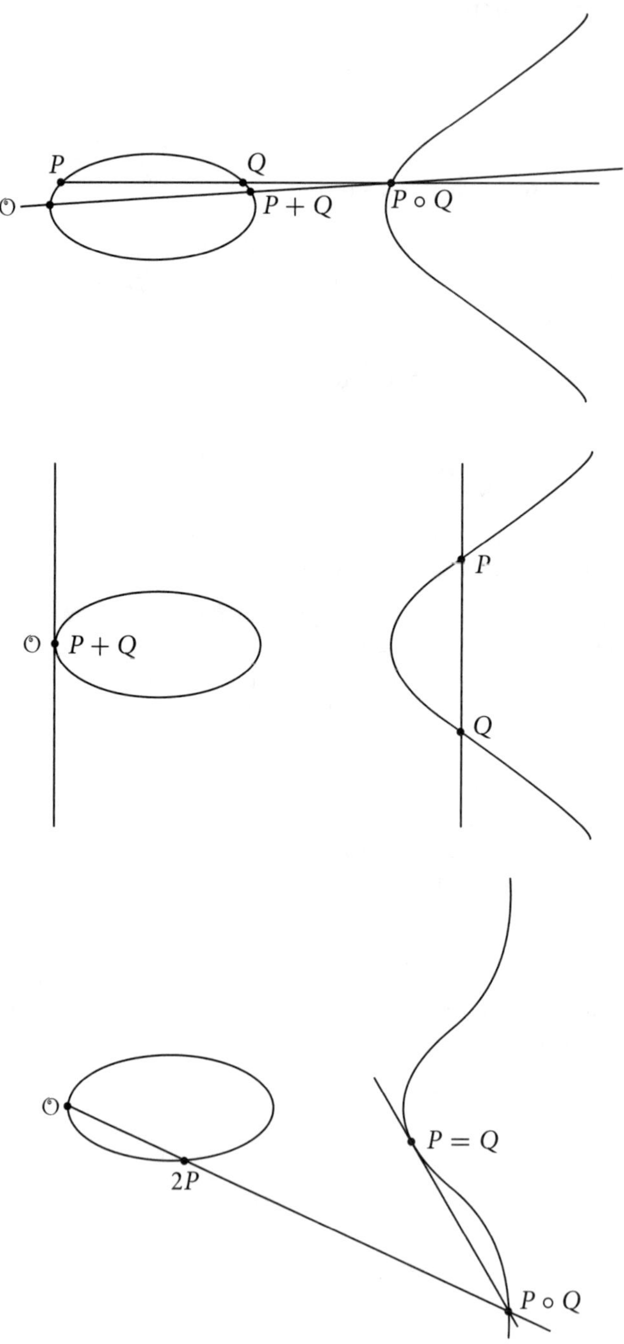

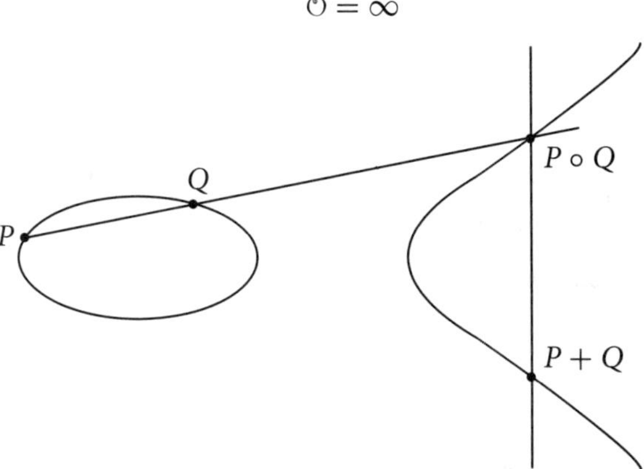



It is the following fact that makes elliptic curves so interesting to geometers and cryptographers alike.

**Theorem 6.9.1** $(E, +)$ *is an abelian group.*

**Proof.** Clearly, $+$ is commutative.

**Identity.** Let $P \in E$. The line through $\mathcal{O}$ and $P$ then intersects $C$ at a third point which we denote by $Q$. Turning this around, the third point on $E$ and the line through $Q$ and $\mathcal{O}$ is just $P$. Thus,

$$P + \mathcal{O} = (P \circ \mathcal{O}) \circ \mathcal{O} = Q \circ \mathcal{O} = P$$

and $\mathcal{O}$ is the identity.

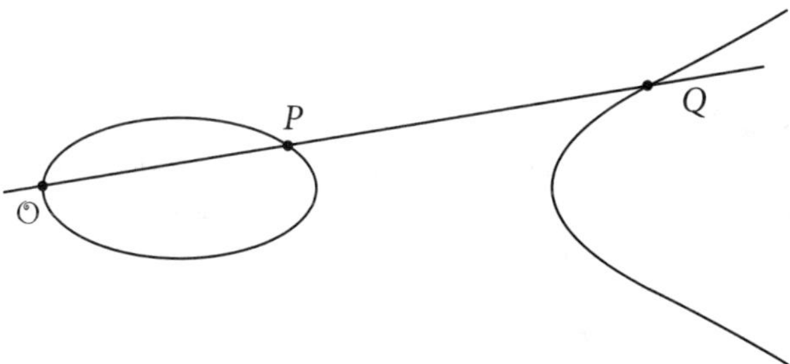



**Inverse.** Let $L$ be the tangent at $\mathcal{O}$. Then $L$ must meet $E$ at a third point $S$, say. Now let the line through $P$ and $S$ intersect $E$ in a third point $P'$. Then the third point on $E$ and the line through $P$ and $P'$ is $S$ and the third point on $E$ and the line through $S$ and $\mathcal{O}$ is $\mathcal{O}$. Thus,

$$P + P' = \mathcal{O}$$

and $P'$ is the inverse of $P$.

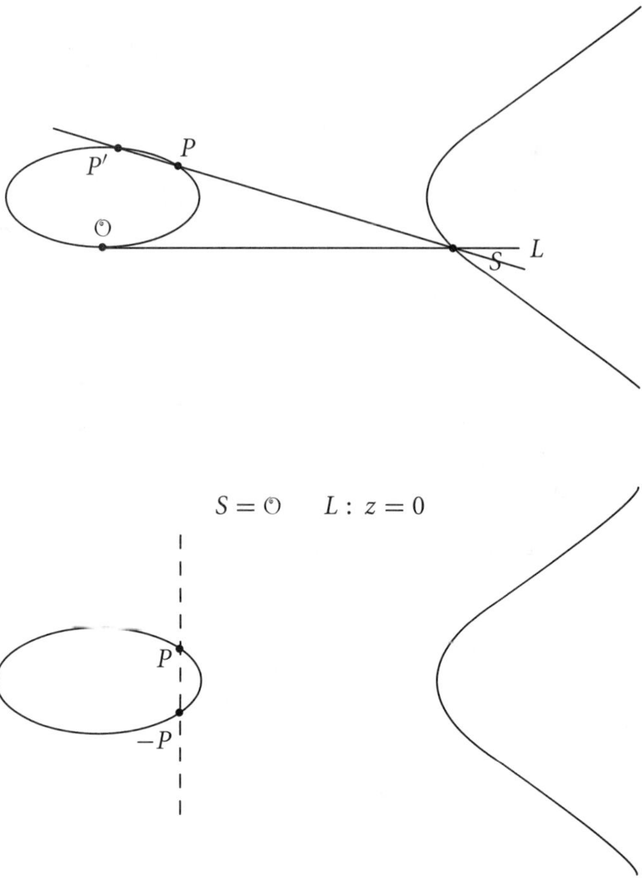


**Associativity.** It would be so easy to overlook the issue of associativity. In so many circumstances, it follows easily because we are dealing with numbers, matrices, or composition of functions. Here it is not at all clear that the associative law will hold, and its verification depends on the (far from obvious) Nine-Point Theorem. Let $P, Q, R$ be any points on the curve. Let the line $L_1$ through $P$ and $Q$ meet $E$ in a third point $S$, the line $L_4$ through $S$ and $\odot$ meet $E$ in a third point $T$, and the line $L_2$ through $T$ and $R$ meet $E$ also in $U$. Then $P + Q = T$ and $(P + Q) + R$ is the third point on $E$ and the line through $\odot$ and $U$.

Similarly, let $V$ denote the third point on $E$ and the line $L_5$ through $Q$ and $R$, $W$ be the third point on $E$ and the line $L_3$ through $\odot$ and $V$, and $U'$ be the third point on $E$ and the line $L_6$ through $P$ and $W$. Then, $Q + R = W$ and $P + (Q + R)$ is the third point on $E$ and the line through $\odot$ and $U'$.

Thus we will have $P + (Q + R) = (P + Q) + R$ if we can show that $U = U'$.

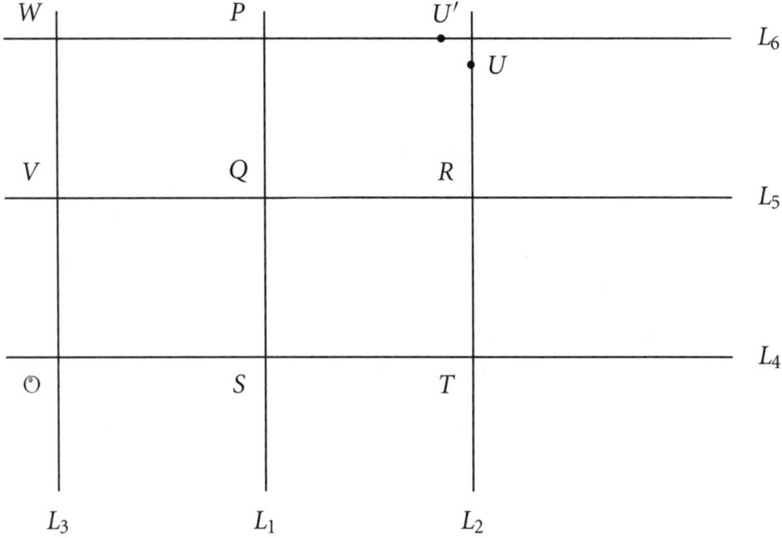


For each $i$, let $l_i(x, y, z) = 0$ define the line $L_i$. To keep the notation simple, we will write $l_i = l_i(x, y, z)$. Let

$$f_1(x, y, z) = l_1 l_2 l_3$$
$$f_2(x, y, z) = l_4 l_5 l_6.$$

Since $E$ is irreducible, $C_1 : f_1(x, y, z) = 0$ and $E$ are cubic curves with no common component and they intersect in the nine points $\odot, P, Q, R, S, T, U, V, W$. On the other hand, the curve $C_2 : f_2(x, y, z) = 0$ also passes through the eight points $\odot, P, Q, R, S, T, V$, and $W$. Hence, by Theorem 6.8.1 (the Nine-Point Theorem), $C_2$ must also pass through the ninth point, $U$, in the intersection of $E$ and $C_1$. However, $E$ and $C_2$ also intersect in $U'$. Since, in total, $E$ and $C_2$ intersect in exactly nine points, it follows that $U'$ must equal $U$. (Note that other possibilities, such as $U = R$ and $L_6$ also passing through $R$ won't work because we count multiplicities of intersection. $L_6$ would then have four points of intersection with $E$ and therefore divide $E$, which would be a contradiction.) Therefore, $P + (Q + R) = (P + Q) + R$ and the operation $+$ is associative.

Note that the points $\odot, P, \ldots, W$ may not all be distinct, nor need the lines all be distinct. However, this presents no problems, when the multiplicities are counted appropriately. $\square$

### Exercises 6.9

1. Show that the operation $\circ$ is not associative. (A proof "by diagram" is sufficient).

2. For any points $P, Q, R, S$ on $E$, establish the following:

   (i) $P \circ Q = Q \circ P$.

   (ii) $(P \circ Q) \circ P = Q$.

   (iii) $((P \circ Q) \circ R) \circ S = P \circ ((Q \circ S) \circ R)$. (Hint: Set up a diagram of intersecting lines as in Theorem 6.9.1 and apply Theorem 6.8.1.)

3. Let $(E, +)$ be a group on $E$ relative to the point $\mathcal{O}$ and $(E, +')$ be the group on $E$ relative to the point $\mathcal{O}'$. The goal of this exercise is to show that $(E, +)$ and $(E, +')$ are isomorphic. Let $\mathbb{O} = \mathcal{O} \circ \mathcal{O}'$. Define a mapping $\varphi : E \to E$ by: $\varphi(P) = \mathbb{O} \circ P$.

   (i) Show that $\varphi$ is bijective.

   (ii) Show that, for all points $P, Q$, $(\mathbb{O} \circ Q) \circ (P \circ \mathcal{O}') = P + Q$. (Hint: $\mathbb{O} = \mathcal{O} \circ \mathcal{O}'$ and use Exercise 2 (iv).)

   (iii) Show that $\varphi(P) +' \varphi(Q) = \mathbb{O} \circ ((P \circ \mathcal{O}') \circ (\mathbb{O} \circ Q))$. (Hint: Use Exercise 2 (iv).)

   (iv) Show that $\varphi(P) +' \varphi(Q) = \varphi(P + Q)$.

## 6.10 The Arithmetic on an Elliptic Curve

We now want to make some basic observations regarding the arithmetic in the group of a nonsingular elliptic curve:

$$E : y^2 = x^3 + ax + b, \quad 4a^3 + 27b^2 \neq 0$$

over an algebraically closed field $F$, with char $F \neq 2, 3$. We have seen that such a curve has one point $[0, 1, 0]$ "at infinity" in the projective plane. It turns out, rather surprisingly perhaps, that it simplifies some of the calculations if we choose that point at infinity as the point $\mathcal{O}$. So let us set $\mathcal{O} = [0, 1, 0]$. Since all the other points of $E$ lie in $F^2$, we will perform most of our calculations in $F^2$ and only resort to $\mathbb{P}^2(F)$ when necessary.

One useful consequence of the choice of $\mathcal{O} = [0, 1, 0]$ is the following.

**Lemma 6.10.1** *Let the line $x = c$ intersect $E$ in $F^2$ at the points $P$ and $Q$, where $P$ and $Q$ are not necessarily distinct. Then $P + Q = \mathcal{O}$ and $Q = -P$. Consequently, if $P = (x_1, y_1)$, then $-P = (x_1, -y_1)$.*

**Proof.** Substituting $x = c$ into the equation for $E$, we obtain the quadratic equation (in $y$)

$$y^2 = c^3 + ac + b$$

which has just two roots that must correspond to $P$ and $Q$. Thus, the third point of intersection of the line $x = c$ with $E$ within $\mathbb{P}^2(F)$ is $\mathcal{O}$. Now the line $L$ joining $\mathcal{O}$ to $\mathcal{O}$ is the tangent line $z = 0$ so that the third point of intersection of $L$ with $E$ is also $\odot$ (since in $\mathbb{P}^2(F)$, the intersection of $z = 0$ and $y^2 z - (x^3 + axz^2 + bz^3)$ is given by $x^3 = 0$, yielding a triple point of intersection at $[0, 1, 0]$). Thus,

$$P + Q = \odot$$

as required.

To establish the final claim, let $P = (x_1, y_1) \in E$. Then we also have $Q = (x_1, -y_1) \in E$ so that the line $x = x_1$ meets $E$ at the points $P$ and $Q$ (or a repeated point $P = Q$ if $y_1 = 0$). By the preceding argument, $Q = -P$. $\quad\square$

In the calculations below, it will again be convenient to use the familiar fractional notation $\frac{a}{b}$ ($a \in F$, $b \in F^*$) to denote $ab^{-1}$, even though our field $F$ may not be a subfield of $\mathbb{C}$.

**Theorem 6.10.2** *Let $E : y^2 = x^3 + ax + b$ be an elliptic curve over an algebraically closed field $F$ and let $(E, +)$ be the group of $E$ with $\odot = [0, 1, 0]$, the point at infinity. Let $P, Q \in E$. Then $P + Q$ can be described as follows:*

(i) *If $P = \odot$ or $Q = \odot$, then $P + Q = Q$ und $P + Q = P$, respectively. For the remaining cases, we can assume that $P, Q \in F^2$—say*
$$P = (x_1, y_1), \quad Q = (x_2, y_2).$$
(ii) *If $x_1 = x_2$ and either $y_1 \neq y_2$ or $y_1 = y_2 = 0$, then*

$$P + Q = \odot.$$

*In the remaining cases, $P + Q = R$ will be a point in $F^2$. So let*
$$R = (x_3, y_3).$$
(iii) *If*

$$x_1 = x_2, \quad y_1 = y_2 \neq 0$$

*then*

$$x_3 = -2x_1 + \left(\frac{3x_1^2 + a}{2y_1}\right)^2,$$

$$y_3 = -y_1 + \left(\frac{3x_1^2 + a}{2y_1}\right)(x_1 - x_3).$$

(iv) *If $x_1 \neq x_2$, then*

$$x_3 = -x_1 - x_2 + \left(\frac{y_2 - y_1}{x_2 - x_1}\right)^2$$

$$y_3 = -y_1 + \left(\frac{y_2 - y_1}{x_2 - x_1}\right)(x_1 - x_3).$$

**Proof.** (i) This follows immediately from the fact that $\mathcal{O}$ is the identity element for $(E, +)$.

(iia) $x_1 = x_2$ and $y_1 \neq y_2$. Then $P$ and $Q$ are distinct points on the line $L : x = x_1 (= x_2)$. By Lemma 6.10.1,

$$P + Q = \mathcal{O}.$$

(iib) $x_1 = x_2$ and $y_1 = y_2 = 0$. In this case, the line $x = x_1 (= x_2)$ meets $E$ at $P = Q$ with multiplicity 2, and the result again follows from Lemma 6.10.1.

(iii) $x_1 = x_2$, $y_1 = y_2 \neq 0$. Here, $P = Q$ and so we must first describe the tangent to $E$ at $P$. Consider the intersection of the line

$$L : x = x_1 + ct, \quad y = y_1 + dt$$

with $E$. Substituting $L$ into the equation for $E$, we obtain

$$(y_1 + dt)^2 = (x_1 + ct)^3 + a(x_1 + ct) + b$$

which reduces to

$$c^3 t^3 + t^2 (3c^2 x_1 \quad d^2) + t(3cx_1^2 + ac - 2dy_1) = 0.$$

Now we have the tangent when $t$ is a double root—that is, when the coefficient of $t$ is zero, exactly when

$$(3x_1^2 + a)c - 2y_1 d = 0$$

or, since $y_1 \neq 0$,

$$d = \frac{3x_1^2 + a}{2y_1} c.$$

Thus, $L$ is given by

$$x = x_1 + ct, \qquad y = y_1 + \frac{3x_1^2 + a}{2y_1} ct$$

or, in implicit form (by eliminating $t$)

$$y = y_1 + \frac{3x_1^2 + a}{2y_1}(x - x_1)$$

$$= \frac{3x_1^2 + a}{2y_1} x + \left( y_1 - \frac{3x_1^2 + a}{2y_1} x_1 \right)$$

$$= mx + c \tag{6.22}$$

where

$$m = \frac{3x_1^2 + a}{2y_1} \quad \text{and} \quad c = y_1 - \frac{3x_1^2 + a}{2y_1}\, x_1.$$

To find the third point of intersection, $R = (x_3, y_3)$ of $L$ with $E$, we substitute (6.22) into the equation for $E$:

$$(mx + c)^2 = x^3 + ax + b$$

or

$$x^3 - m^2 x^2 + (a - 2mc)x + b - c^2 = 0$$

which will have three roots in $F$—namely, $x_1$, $x_1$, and $x_3$. Hence,

$$x^3 - m^2 x^2 + (a - 2mc)x + b - c^2 = (x - x_1)(x - x_1)(x - x_3).$$

Equating the coefficients of $x^2$, we obtain

$$-m^2 = -2x_1 - x_3$$

so that

$$x_3 = m^2 - 2x_1 = -2x_1 + \left(\frac{3x_1^2 + a}{2y_1}\right)^2$$

and, from (6.22),

$$y_3 = y_1 + \frac{3x_1^2 + a}{2y_1}\,(x_3 - x_1).$$

The line $L'$ joining $\odot$ to $R = (x_3, y_3)$ is then $x = x_3$. To obtain the third point of intersection of $L'$ with $E$, we substitute $x = x_3$ into the equation for $E$:

$$y^2 = (x_3)^3 + ax_3 + b.$$

However, $(x_3, y_3)$ is a point on $E$. Hence, $(x_3)^3 + ax_3 + b = (y_3)^2$ so that we have

$$y^2 = (y_3)^2$$

and $y = \pm y_3$. Thus, the third point is given by $(x_3, -y_3)$ and (iii) holds.

    (iv) $x_1 \neq x_2$. The line

$$L: (y_2 - y_1)(x - x_1) = (x_2 - x_1)(y - y_1)$$

passes through $P$ and $Q$, and so is the line determined by $P$ and $Q$. Let the third point of intersection of $L$ with $E$ be $(x_3, y_3)$. From the equation for $L$,

$$y = \frac{y_2 - y_1}{x_2 - x_1} x + \frac{x_2 y_1 - x_1 y_2}{x_2 - x_1}$$

$$= mx + c$$

where

$$m = \frac{y_2 - y_1}{x_2 - x_1}, \qquad c = \frac{x_2 y_1 - x_1 y_2}{x_2 - x_1}.$$

Substituting in the equation for $E$ to find all the points of intersection, we obtain, as before,

$$(mx + c)^2 = x^3 + ax + b$$

or

$$x^3 - m^2 x + (a - 2mc)x + b - c^2 = 0.$$

This equation will have three solutions in $F$, two of which are $x_1$ and $x_2$, and the third will be $x_3$. Thus we must have

$$x^3 - m^2 x + (a - 2mc)x + b - c^2 = (x - x_1)(x - x_2)(x - x_3).$$

Equating the coefficients of $x^2$, we obtain

$$-m^2 = -x_1 - x_2 - x_3$$

or

$$x_3 = m^2 - x_1 - x_2$$

and, from the equation for $L$, we obtain

$$y_3 = mx_3 + c.$$

Now, as in part (iii), the line joining $\mathcal{O}$ to $(x_3, y_3)$ will meet $E$ in a third point $(x_3, -y_3)$. Thus,

$$x_3 = \quad m^2 - x_1 - x_2$$

$$y_3 = -mx_3 - c$$

$$= -y_1 + \frac{y_2 - y_1}{x_2 - x_1}(x_1 - x_3)$$

as claimed.   $\square$

The usefulness of Theorem 6.10.2 in relation to performing calculations in the group of an elliptic curve is pretty obvious. However, it has another wonderful consequence.

**Corollary 6.10.3** *Let $F_1$ be a subfield of an algebraically closed field $F$ and let $E : y^2 = x^3 + ax + b$ be an elliptic curve over $E$ with $a, b \in F_1$. Let $\mathcal{O} = [0, 1, 0]$ and*

$$E(F_1) = \{(x, y) \in E : x, y \in F_1\} \cup \{\mathcal{O}\}.$$

*Then $(E(F_1), +)$ is a subgroup of $(E, +)$.*

**Proof.** Let $P, Q \in E(F_1)$ where $P = (x_1, y_1)$ and $Q = (x_2, y_2)$. Then $x_1$, $x_2, y_1, y_2 \in F_1$ and it follows immediately from Theorem 6.10.2 that $P + Q \in E(F_1)$. Also it follows from Lemma 6.10.1, that $-P \in E(F_1)$. Thus, $(E(F_1), +)$ is a subgroup. $\square$

This is very nice, but for any given choice of $F$ and $F_1$ (for example, $F = \mathbb{C}$, $F_1 = \mathbb{Q}$), it may be a challenge to determine whether $E(F_1)$ contains any element other than $\mathcal{O}$.

The next lemma contains some useful facts concerning elements of low order.

**Lemma 6.10.4** *Let $P = (x, y)$ be a point other than the identity on the elliptic curve $E : y^2 = x^3 + ax + b$ over a field $F$ with $\mathrm{char}(F) \neq 2, 3$ and let $\mathcal{O} = [0, 1, 0]$.*

   (i)  *$P$ has order 2 if and only if $y = 0$.*
   (ii)  *$P$ has order 3 if and only if $y \neq 0$ and $P + P = (x, -y)$.*
   (iii)  *$P$ has order 4 if and only if $y \neq 0$ and $P + P = (*, 0)$.*

**Proof.** (i) We have

$$P \text{ is of order } 2 \Leftrightarrow P + P = \mathcal{O}$$
$$\Leftrightarrow P = -P$$
$$\Leftrightarrow (x, y) = (x, -y)$$
$$\Leftrightarrow y = -y$$
$$\Leftrightarrow 2y = 0$$
$$\Leftrightarrow y = 0.$$

(ii) We have

$$P \text{ is of order } 3 \Leftrightarrow 3P = \mathcal{O}$$

$$\Leftrightarrow P + P = -P = (x, -y).$$

(iii) We have

$$P \text{ is of order } 4 \Leftrightarrow 4P = 0, \quad 2P \neq \mathcal{O}$$

$$\Leftrightarrow P + P = -(P + P) \neq \mathcal{O}$$

$$\Leftrightarrow P + P = (*, 0). \quad \square$$

After all the work that we have done, it turns out to be relatively straightforward to calculate groups on elliptic curves over small finite fields.

**Example 6.10.5** Let $E : y^2 = x^3 + x + 4$ over $\mathbb{Z}_{11}$. Then

$$\Delta = 4 \cdot 1 + 27 \cdot 4^2 = 436 = 7 \qquad (\text{in } \mathbb{Z}_{11}).$$

Therefore, $E$ is nonsingular and $(E(\mathbb{Z}_{11}), +)$ is a group. To find all the elements of $E(\mathbb{Z}_{11})$, all we have to do is calculate all the possible values of $x^3 + x + 4$ as $x$ ranges over $\mathbb{Z}_{11}$ and then all the possible values for $y^2$ as $y$ ranges over $\mathbb{Z}_{11}$, and take all pairs $(x, y)$ where these values agree and then include $\mathcal{O}$, the point at infinity.

| $x, y$ | $x^2, y^2$ | $x^3$ | $x^3 + x + 4$ |
|--------|------------|-------|---------------|
| 0 | 0 | 0 | 4 |
| 1 | 1 | 1 | 6 |
| 2 | 4 | 8 | 3 |
| 3 | 9 | 5 | 1 |
| 4 | 5 | 9 | 6 |
| 5 | 3 | 4 | 2 |
| 6 | 3 | 7 | 6 |
| 7 | 5 | 2 | 2 |
| 8 | 9 | 6 | 7 |
| 9 | 4 | 3 | 5 |
| 10 | 1 | 10 | 2 |

We see that the values that $x^3 + x + 4$ and $y^2$ have in common are

$$1, 3, 4, 5.$$

Therefore, matching pairs of values of $x$ and $y$ we find that

$$E = \{(0, 2), (0, 9), (2, 5), (2, 6), (3, 1), (3, 10), (9, 4), (9, 7), \mathcal{O}\}.$$

Thus, $(E, +)$ is an abelian group of order 9 and is therefore isomorphic to either $\mathbb{Z}_9$ or $\mathbb{Z}_3 \times \mathbb{Z}_3$.

We can determine which of these groups we actually have by considering the number of elements of order 3 ($\mathbb{Z}_3 \times \mathbb{Z}_3$ has eight such elements whereas $\mathbb{Z}_9$ has only two).

Let $P$ be a point on $E$. By Lemma 6.10.4, $P$ is a point of order 3 if and only if $P = (x, y)$ is such that $y \neq 0$ and $P + P = (x, -y)$. Calculating the first coordinate of $P + P$ according to Theorem 6.10.2 (iii), this yields

$$-2x + \left(\frac{3x^2 + 1}{2y}\right)^2 = x$$

or

$$(3x^2 + 1)^2 = 12xy^2$$

$$= 12x(x^3 + x + 4).$$

This simplifies (over $\mathbb{Z}_{11}$) to

$$8x^4 + 5x^2 + 7x + 1 = 0.$$

The only roots of this equation in $\mathbb{Z}_{11}$ are 3 and 7, which, when substituted into $x^3 + x + 4$, yield the values 1 and 2. As $y^2 = 2$ has no solution in $\mathbb{Z}_{11}$, and $1^2 = 10^2 = 1$, it follows that the only candidates for points of order 3 are

$$(3, 1), \quad (3, 10).$$

Consequently, $(E, +) \cong \mathbb{Z}_9$ and any point other than $\mathcal{O}$, $(3, 1)$, and $(3, 10)$ is a generator.

**Exercises 6.10**

1. Let $E : y^2 = x^3 + ax + b$ be an elliptic curve.

   (i)  Show that there are, at most, three elements in $(E, +)$ of order 2.
   (ii)  Show that there are, at most, eight elements in $(E, +)$ of order 3.
   (iii)  Show that there are, at most, 12 elements in $(E, +)$ of order 4.

2. Use Exercises 1 (i) and 1(ii) to identify certain groups that never occur as the groups of elliptic curves.

3. Determine the group of each of the following elliptic curves over the field $\mathbb{Z}_5$. Take the "base point" $\mathcal{O}$ to be the point at infinity.

   (i)  $y^2 = x^3 - 1$.
   (ii)  $y^2 = x^3 + 1$.
   (iii)  $y^2 = x^3 + x$.

(iv) $y^2 = x^3 + 4x + 1$.
(v) $y^2 = x^3 - x$.
(vi) $y^2 = x^3 + x + 1$.

4. Determine the group of the following elliptic curve over the field $\mathbb{Z}_7$. Take the "base point" $\circlearrowleft$ to be the point at infinity.

(i) $y^2 = x^3 + 4x + 1$.
(ii) $y^2 = x^3 + x + 4$.

## 6.11 Results Concerning the Structure of Groups on Elliptic Curves

There is a vast amount of literature concerning groups on elliptic curves, and in this section we will list just a few of the more important general results. So, throughout, we will assume that we are dealing with an elliptic curve

$$E : y^2 = x^3 + ax + b, \qquad 4a^3 + 27b^2 \neq 0$$

over a field $F$, which will be specified in each case, and we take the base point to be $\circlearrowleft = [0, 1, 0]$.

**Theorem 6.11.1** (L. J. Mordell, 1922) *Let the elliptic curve $E$ be defined over $\mathbb{Q}$. Then the group $(E(\mathbb{Q}), +)$ is finitely generated and is therefore isomorphic to a group of the form*

$$\mathbb{Z} \times \mathbb{Z} \times \cdots \times \mathbb{Z} \times F$$

*where the number of copies of $\mathbb{Z}$ is finite and $F$ is a finite (abelian) group.* $\square$

Mordell's Theorem was subsequently (1928) extended by A. Weil to a much larger class of groups defined on curves. The number of times that $\mathbb{Z}$ appears as a factor is called the (Mordell-Weil) *rank* of $(E, +)$.

**Theorem 6.11.2** (B. Mazur, 1977, 1978) *Let the elliptic curve $E$ be defined over $\mathbb{Q}$. Then the subgroup of $(E(\mathbb{Q}), +)$ consisting of the elements of finite order is isomorphic to one of the following 15 groups:*

$$\mathbb{Z}_n, \qquad n = 1, 2, 3, \ldots, 10, 12$$
$$\mathbb{Z}_2 \times \mathbb{Z}_{2n}, \qquad n = 1, 2, 3, 4$$

*and each of these groups does occur.*

The next result, from E. Lutz and T. Nagell, brings the calculation of the groups of some simple elliptic curves over $\mathbb{Q}$ into the realm of what is reasonably manageable even by hand calculations.

**Theorem 6.11.3** (Lutz-Nagell, 1937)  *Let the elliptic curve $E$ be defined over $\mathbb{Q}$ and $a, b \in \mathbb{Z}$. If $P = (x, y)$ is a point on $E$ of finite order, then $x, y \in \mathbb{Z}$ and either $y = 0$ or $y^2$ divides $\Delta$.*

The condition presented in Theorem 6.11.3 is a necessary condition for $P$ to be of finite order, but it is not a sufficient condition.

**Theorem 6.11.4** (i)  *If the elliptic curve $E$ is defined over $\mathbb{R}$, then $(E, +)$ is isomorphic to either*

$$\text{where } \mathbb{R}/\mathbb{Z} \quad \text{or} \quad (\mathbb{R}/\mathbb{Z}) \times \mathbb{Z}_2.$$

(ii)  *If the elliptic curve $E$ is defined over $\mathbb{C}$, then there exist $\alpha, \beta \in \mathbb{C}$ such that $(E, +)$ is isomorphic to*

$$\mathbb{C}/\langle \alpha, \beta \rangle.$$

*where $\langle \alpha, \beta \rangle = \mathbb{Z}\alpha + \mathbb{Z}\beta$.*

We present just two results concerning the groups of elliptic curves over finite fields. The first result tells us that the number of points on an elliptic curve over $\mathrm{GF}(p^n)$ is approximately $p^n$.

**Theorem 6.11.5** (H. Hasse, 1930s)  *Let $E$ be an elliptic curve over $\mathrm{GF}(p^n)$. Then*

$$p^n + 1 - 2\sqrt{p^n} \le |E| \le p^n + 1 + 2\sqrt{p^n}.$$

The second result identifies the group structure of an elliptic curve over $\mathrm{GF}(p^n)$.

**Theorem 6.11.6**  *Let $E$ be an elliptic curve over $\mathrm{GF}(p^n)$. Then there exist integers $n_1, n_2$ such that*

$$(E, +) \cong \mathbb{Z}_{n_1} \times \mathbb{Z}_{n_2}$$

*where $n_2$ divides $n_1$ and $p^n - 1$.*

For any abelian group $G$, recall (Exercise 32 in section 4.4) that we denote by Tor $(G)$ the subgroup consisting of the elements of finite order and call it the *torsion subgroup* of $G$. It is clear that Theorem 6.11.3 is a very important tool when it comes to identifying the torsion subgroup of an elliptic curve over $\mathbb{Q}$. For any elliptic curve $E$ over $\mathbb{Q}$, with $a, b \in \mathbb{Z}$ and discriminant $\Delta$, let

$$LN(E) = \{(x, y) \in E \mid y = 0 \ \text{ or } \ y^2 \text{ divides } \Delta\}$$

and call the elements of $LN(E)$, *LN elements*. By Theorem 6.11.3, we know that

$$\text{Tor}\,(E, +) \subseteq LN(E) \cup \{0\}.$$

In general, we do not get equality here. Nevertheless, for small values of $a$ and $b$, it is a straightforward matter to determine the values of $y$ for which $y^2$ divides $\Delta$. For each such value $y = y_0$, we then wish to find the roots of

$$x^3 + ax + (b - y_0^2) = 0.$$

Since $a$, $b - y_0^2 \in \mathbb{Z}$, it follows from Proposition 2.6.7 that the integer roots must be divisors of $b - y_0^2$. This greatly simplifies the search for $LN$ elements.

**Example 6.11.7** Let $E : y^2 = x^3 - 5x + 2$ over $\mathbb{Q}$. Then $\Delta = -4 \cdot 5^3 + 27 \cdot 4 = 4(27 - 125) = -4 \cdot 98 = -2^3 \cdot 7^2$. Let $(x, y) \in LN(E)$. Then,

$$y = 0,\ \pm 1,\ \pm 2,\ \pm 7,\ \pm 14.$$

Now we must see if there are any points on $E$ with these values of the $y$-coordinate.

(1) $y = 0$. We have

$$x^3 - 5x + 2 = (x - 2)(x^2 + 2x - 1)$$

where $x^2 + 2x - 1$ has no rational roots. Thus, the only integer value of $x$ for which $x^3 - 5x + 2 = 0$ is $x = 2$. Thus, $(2, 0) \in LN(E)$.

(2) $y = \pm 1$. Then,

$$x^3 - 5x + 1 = 0$$

has no integer solutions.

(3) $y = \pm 2$. Here,

$$4 = x^3 - 5x + 2 \Rightarrow x^3 - 5x - 2 = 0$$
$$\Rightarrow (x + 2)(x^2 - 2x - 1) = 0$$

where $x^2 - 2x - 1$ has no integer roots. Thus, the only root of $x^3 - 5x - 2$ is $x = -2$. Hence, $(-2, \pm 2) \in LN(E)$.

(4) $y = \pm 7$. Then,

$$49 = x^3 - 5x + 2$$

or

$$x^3 - 5x - 47 = 0$$

has no integer solutions.

(5) $y = \pm 14$. Then,

$$196 = x^3 - 5x + 2$$

or

$$x^3 - 5x - 194 = 0$$

has no integer solutions.

Thus, $LN(E) = \{(2, 0), (-2, \pm 2)\}$.

Now $(2, 0)$ is an element of order 2, so $(2, 0) \in \text{Tor}\,(E, +)$. On the other hand, $(-2, +2) + (-2, +2)$ has first coordinate equal to

$$4 + \left(\frac{12 - 5}{4}\right)^2 = 4 + \frac{49}{16}$$

which is not an integer. Therefore, $(-2, 2) \notin \text{Tor}\,(E)$. Similarly, $(-2, -2) \notin \text{Tor}\,(E)$. Thus,

$$\text{Tor}\,(E) = \{\mathbb{O},\ (2, 0)\}$$

and $\text{Tor}\,(E)$ is isomorphic to $\mathbb{Z}_2$.

In some examples it is necessary *to calculate larger multiples of elements in LN(E) than two to see* that they are not, in fact, elements of $\text{Tor}\,(E)$.

**Example 6.11.8** Let $E : y^2 = x^3 - 3x + 3$. Then $\Delta = 27.5$ and Theorem 6.11.3 indicates that the $y$-coordinates of torsion elements of $E$ will be in the set $y = 0, \pm 1, \pm 3$. On checking, we find that

$$LN(E) = \{(1, \pm 1), (-2, \pm 1)\}.$$

Let $P = (1, 1)$ and $P + P = (x_1, y_1)$. Then,

$$x_1 = -2 + \left(\frac{3 \cdot 1 - 3}{2 \cdot 1}\right)^2 = -2$$

$$y_1 = -1 + \frac{3 \cdot 1 - 3}{2 \cdot 1}(1 + 2) = -1.$$

Thus, $P + P = (-2, -1) \in LN(E)$. Let $4P = (x_2, y_2)$. Then

$$x_2 = 4 + \left(\frac{12 - 3}{-2}\right)^2 = 4 + \left(\frac{9}{2}\right)^2$$

so that $x_2$ is not an integer. Hence, by Theorem 6.11.3, $4P \notin \mathrm{Tor}\,(G)$. There-fore, $P$ and $P + P$ are not elements of $\mathrm{Tor}\,(G)$ either. Similarly, $(1, -1)$ and $(-2, \pm 1) \notin \mathrm{Tor}\,(G)$. Thus we are reduced to $\mathrm{Tor}\,(G) = \{\mathcal{O}\}$.

### Exercises 6.11

1. Determine the torsion subgroup of the groups on each of the following elliptic curves over the rational numbers. Take the "base point" $O$ to be the point at "infinity".

$$\begin{aligned}
&\text{(i)} && y^2 = x^3 + 1. \\
&\text{(ii)} && y^2 = x^3 - 1. \\
&\text{(iii)} && y^2 = x^3 + x. \\
&\text{(iv)} && y^2 = x^3 - x. \\
&\text{(v)} && y^2 = x^3 + x + 1. \\
&\text{(vi)} && y^2 = x^3 + x - 1. \\
&\text{(vii)} && y^2 = x^3 - x + 1. \\
&\text{(viii)} && y^2 = x^3 - x - 1. \\
&\text{(ix)} && y^2 = x^3 + x + 2. \\
&\text{(x)} && y^2 = x^3 + 5x + 2. \\
&\text{(xi)} && y^2 = x^3 + 5x - 2.
\end{aligned}$$

# 7

# Further Topics Related to Elliptic Curves

In this chapter we consider several topics relating to elliptic curves. We begin with two important areas of applications: cryptography and prime factorization. The next three sections consider ideas about elliptic curves and about curves related to elliptic curves. We consider curves of the form $y^2 = x^3 + ax + b$ that are not elliptic curves, then we look at the concept of birational equivalence in the light of which we see that elliptic curves are not as special as they might at first appear. Next we consider the genus of a curve that places elliptic curves in a hierarchy of curves. In the very last section, we consider the solutions to a well-known family of Diophantine equations.

## 7.1 Elliptic Curve Cryptosystems

An *elliptic curve cryptosystem* is a variant of the *ElGamal* public key cryptosystem. The components of an ElGamal cryptosystem are

(1) A group $G$

(2) An element $g \in G$

(3) An integer $a \in \mathbb{N}$

(4) $h = g^a$.

Thus, the key space is

$$\mathcal{K} = \{(G, g, a, h) \mid G \text{ is a finite group}, g \in G, \ a \in \mathbb{N}, \ h = g^a\}$$

or, since $h$ is determined by $g$ and $a$, simply

$$\mathcal{K} = \{(G, g, a) \mid G \text{ is a finite group}, g \in G, \ a \in \mathbb{N}\}.$$

The public components of the system are

$$G, g, h = g^a$$

while $a$, the key to the system, is kept secret. The message space for the ElGamal system is $G$ so that plain text messages must first be converted into elements in $G$ or strings of elements in $G$. The encrypted message space is $G \times G$.

A message $x \in G$ is encrypted as follows:

e1: Select a random integer $b < |g|$.
e2: Compute $y_1 = g^b$, $y_2 = xh^b$.
e3: Send the cipher text $(y_1, y_2)$. Thus,

$$e(x) = (y_1, y_2).$$

The recipient and owner of the key then decodes the message as follows:

$$d\,(e(x)) = d\,(y_1, y_2)$$
$$= y_2 (y_1^a)^{-1}.$$

This does indeed recover the original message, since

$$d\,(e(x)) = y_2 (y_1^a)^{-1}$$
$$= xh^b ((g^b)^a)^{-1}$$
$$= xg^{ab} (g^{ab})^{-1}$$
$$= x.$$

The security of the ElGamal cryptosystem relies entirely on the fact that the knowledge of

$$g \quad \text{and} \quad g^a$$

does not immediately reveal the value of the exponent $a$. The following problem:

Given an element $g$ in a group $G$ and an element $h \in \langle g \rangle$, find $n \in \mathbb{N}$ such that

$$h = g^n$$

is known as the *discrete log problem*.

The discrete log problem is considered to be computationally hard in general. However, in particular instances, for example, if $G = \mathbb{Z}_n$ (see the exercises), it is known to be easy. So it is important to choose the group $G$, the element $g$, and the integer $a$ with great care.

In theory, the ElGamal public key encryption scheme can be implemented in any finite group, provided that the discrete log problem is considered to be intractable. This is thought to be the case with groups on elliptic curves. So first let us consider a literal adaptation of the ElGamal scheme to elliptic curves. The components are

(1) An elliptic curve $E$ over $\mathbb{Z}_p$, where $p$ is a large prime

(2) A point $P \in E$, of large order

(3) An very large integer $a \in \mathbb{N}$

(4) $Q = aP$.

The key space in this case is

$$\mathcal{K} = \{(E, P, a, Q) \mid P \in E, \ Q = aP\}.$$

The public components of the system are

$$E, \ P, \ Q = aP.$$

The message space consists of pairs of points on $E$—in other words, it is a subset of $(\mathbb{Z}_p \times \mathbb{Z}_p)^2$.

A message $x \in E$ is encrypted as follows:

e1: Select a random integer $b$ less than the order of $P$.
e2: Compute $y_1 = bP$, $y_2 = x + bQ$.
e3: Send the cipher text $(y_1, y_2)$. Thus,

$$e(x) = (y_1, y_2).$$

The recipient and owner of the key then decode the message as follows:

$$d\,(e(x)) = d\,(y_1, y_2)$$
$$= y_2 - ay_1.$$

This recovers the original message, since

$$d\,(e(x)) = y_2 - ay_1$$
$$= x + bQ - a(bP)$$
$$= x + abP - abP$$
$$= x.$$

One modification of this scheme, divised by Menezes-Vanstone, using elliptic curves runs as follows.

The components are

(1) A large prime $p$

(2) An elliptic curve $E$ over $\mathbb{Z}_p$

(3) A point $P \in E$ for which the discrete log problem in $\langle P \rangle$ is considered intractable

(4) An integer $a \in \mathbb{N}$

(5) $Q = aP$.

The key space is

$$\mathcal{K}\{(E, P, a, Q) \mid P \in E, \ Q = aP\}.$$

The public components of the system are

$$E, \ P, \ p, \ Q = aP.$$

The difference here is that, the message space is $\mathbb{Z}_p^* \times \mathbb{Z}_p^*$.

A message $x = (x_1, x_2) \in \mathbb{Z}_p^* \times \mathbb{Z}_p^*$ is encrypted as follows:

e1: Select a random integer $b$ less than the order of $P$.
e2: Compute

$$R = bP,$$
$$bQ = (q_1, q_2)$$
$$y_1 = q_1 x_1 \bmod p$$
$$y_2 = q_2 x_2 \bmod p.$$

($b$ should be chosen to avoid the (unlikely) possibility that either $q_1 = 0$ or $q_2 = 0$.)
e3: Send the cipher text

$$e(x) = (R, y_1, y_2).$$

The recipient and owner of the key then decode the original message as follows. The first step is to determine the values of $q_1$ and $q_2$ with the information available. This can be done by computing $aR$ since

$$
\begin{aligned}
aR &= abP \\
&= baP \\
&= bQ \\
&= (q_1, q_2).
\end{aligned}
$$

Now the message can be decrypted by

$$
\begin{aligned}
d\,(e(x)) &= d\,(R, y_1, y_2) \\
&= (y_1 q_1^{-1} \bmod p, \quad y_2 q_2^{-1} \bmod p) \\
&= (x_1, x_2) \\
&= x.
\end{aligned}
$$

A slight disadvantage of the ElGamal and elliptic curve cryptosystems relative to other cryptosystems is the way that they expand the original message. However, elliptic curve cryptosystems have short key lengths and, consequently, more modest memory requirements, which is an important feature for situations in which memory and processing power are limited. An important area of applications is in smart cards.

In terms of security, the discrete log problem, on which the security of elliptic curve cryptography depends, is believed to be as difficult as the factorization problem for large integers (which is the basis of the security of the RSA system). Combined with the high precision arithmetic employed on elliptic curves, this makes elliptic curve cryptography a serious competitor to the RSA system. We refer the reader to [Men] for further information on this topic.

### Exercises 7.1

1. Let $g$ be a generator for $(\mathbb{Z}_n, +)$, $m \in \mathbb{N}$ and $h = mg$. Show how to solve for $m$ using the Euclidean algorithm.

2. Construct the $\log_2$ table for $\mathbb{Z}_{11}^*$, that is, for each element $y$ of $\mathbb{Z}_{11}^*$, solve $y = 2^x$ for $x$.

3. Construct the $\log_\alpha$ table for $\mathrm{GF}(9) = \mathbb{Z}_3[y]/(1 + y^2)$, where $\alpha = 1 + y$; that is, for each element $y$ of $\mathrm{GF}(9)$, solve $y = a^x$ for $x$.

## 7.2 Fermat's Last Theorem

Everyone who has studied a little Euclidean geometry is familiar with the equation

$$
x^2 + y^2 = z^2 \tag{7.1}
$$

as it expresses the fact that, in a right-angled triangle with sides of lengths $x, y,$ and $z$ where $z$ is the length of the hypotenuse, the sum of the areas of the squares on the two shorter sides (of lengths $x$ and $y$) is equal to the area of the square on the hypotenuse (Pythagoras' Theorem). Triples $[x, y, z]$ that are solutions to the equation (7.1) are known as *Pythagorean triples*. It is not difficult to find Pythagorean triples consisting of positive integers— for example, $[3, 4, 5]$ and $[5, 12, 13]$. It is also easy to see that if $[x, y, z]$ is a Pythagorean triple and if $\lambda \in \mathbb{N}$, then $[\lambda x, \lambda y, \lambda z]$ is also a Pythagorean triple. Thus, equation (7.1) has infinitely many integer solutions. Indeed, equation (7.1) has infinitely many independent solutions, that is, solutions that are not multiples of each other. An algorithm for generating Pythagorean triples will be developed in the exercises.

The outcome is dramatically different if we increase the exponent in equation (7.1) to consider the equations $x^3 + y^3 = z^3$, $x^4 + y^4 = z^4$, and so on. In the 17th century, the French mathematician P. de Fermat studied these equations and came to the conclusion that there were no integer solutions to the equations of the form

$$x^n + y^n = z^n \qquad (n \geq 3)$$

with $x, y, z$ all nonzero. Fermat did not record a solution; at least, no solution has been found in his papers. However, in a copy of the translation into latin by C. G. Bachet of Diophantus' "Arithmetica", Fermat wrote the following words in the margin opposite Problem II.8: "I have a truly marvelous demonstration of this proposition, which this margin is too narrow to contain" (see ([Boy], page 354) and ([Sing], page 69)). As this was the last major claim remaining unresolved among the many assertions that Fermat made without providing detailed proofs, it became known as *Fermat's Last Theorem.* Not until the 1990s were mathematicians able to verify Fermat's famous claim. Elliptic curves played a critical role.

The link between Fermat's Last Theorem and elliptic curves was observed by G. Frey in 1984. Suppose that $a, b, c \in \mathbb{N}$ are such that

$$a^n + b^n = c^n \qquad (n \geq 3). \qquad (7.2)$$

Let $A = a^n$, $B = b^n$, and consider the curve

$$E : y^2 = x(x - A)(x + B)$$

or

$$y^2 = x^3 + (B - A)x^2 - ABx.$$

We will refer to this as the *Frey elliptic curve.* This is not in the standard form for an elliptic curve that we have been using. However, let us perform the change of variables $y \to y,\ x \to x - \frac{1}{3}D$ where $D = B - A$. Then,

$$x^3 + Dx^2 - ABx \ \longrightarrow\ \left(x - \frac{1}{3}D\right)^3 + D\left(x - \frac{1}{3}D\right)^2 - AB\left(x - \frac{1}{3}D\right)$$

$$= x^3 - Dx^2 + \frac{D^2}{3}x - \frac{D^3}{27}$$

$$+ Dx^2 - \frac{2}{3}D^2x + \frac{D^3}{9} - ABx + \frac{ABD}{3}$$

$$= x^3 - \left(\frac{1}{3}D^2 + AB\right)x + \left(\frac{2}{27}D^3 + \frac{ABD}{3}\right).$$

Thus, with respect to the new coordinates, our curve is

$$E : y^2 = x^3 + px + q$$

where

$$p = -\left(\frac{1}{3}D^2 + AB\right), \qquad q = \frac{2}{27}D^3 + \frac{1}{3}ABD.$$

In this form, $E$ has discriminant

$$\Delta = 4p^3 + 27q^2$$

$$= -4\left[\frac{1}{27}D^6 + \frac{1}{3}ABD^4 + A^2B^2D^2 + A^3B^3\right]$$

$$+ 27\left[\frac{4}{27^2}D^6 + 2 \cdot \frac{2D^3}{27}\frac{ABD}{3} + \frac{1}{9}A^2B^2D^2\right]$$

$$= -A^2B^2D^2 - 4A^3B^3$$

$$= -A^2B^2[D^2 + 4AB]$$

$$= -A^2B^2[B^2 - 2BA + A^2 + 4AB]$$

$$= -A^2B^2[A^2 + 2AB + B^2]$$

$$= -A^2B^2(A + B)^2$$

$$= -a^{2n}b^{2n}(a^n + b^n)^2$$

$$= -(a^2b^2c^2)^n$$

where the last step follows from (7.2). Thus we have the following Proposition.

**Proposition 7.2.1** *If there exists a nontrivial solution to*

$$x^n + y^n = z^n$$

*then there exists a Frey elliptic curve E such that $\Delta = -t^n$ for some integer $t > 1$.*

To indicate, very briefly, how this ties in with the ultimate proof of Fermat's Last Theorem, we need to introduce a few concepts. Our definitions are not completely precise, but they will suffice to give you a general idea of the major stepping stones along the way. Let $\mathbb{C}^+$ denote the upper half of the complex plane

$$\mathbb{C}^+ = \{x + iy \mid y > 0\}.$$

Now consider the following set of quotients of one linear polynomial by another in $\mathbb{C}(z)$:

$$\Gamma = \left\{ \frac{az + b}{cz + d} \mid a, b, c, d \in \mathbb{Z}, ab - cd = 1 \right\}.$$

Each element of $\Gamma$ determines a mapping $z \longrightarrow \frac{az+b}{cz+d}$ of $\mathbb{C}^+$ to $\mathbb{C}^+$. Furthermore, each of these mappings is a permutation of $\mathbb{C}^+$ and together they form an infinite group of permutations (under the usual composition of permutations), which is known as the *modular group*. The group $\Gamma$ has many subgroups. Particularly noteworthy are the following. For any positive integer $n$, let

$$\Gamma(n) = \left\{ \frac{az + b}{cz + d} \mid a \equiv d \equiv \pm 1 \bmod n, b \equiv c \equiv 0 \bmod n \right\}$$

Then $\Gamma(n)$ is a normal subgroup of $\Gamma$ such that the quotient $\Gamma/\Gamma(n)$ is finite. Any subgroup $M$ of $\Gamma$ such that $\Gamma(n) \subseteq M$, for some positive integer $n$, is called a *congruence subgroup*. The detailed verification of these observations concerning $\Gamma$ is left to the exercises. Next we say that a function $f : \mathbb{C}^+ \longrightarrow \mathbb{C}^+$ is a *modular function* if

(i) there exist functions $g, h : \mathbb{C}^+ \longrightarrow \mathbb{C}^+$ that are differentiable everywhere such that $f = g/h$ "almost everywhere"
(ii) $f(\gamma z) = f(z)$ for all $\gamma \in \Gamma$.

In addition, if $M$ is a congruence subgroup and the function $f : \mathbb{C}^+ \longrightarrow \mathbb{C}^+$ satisfies the condition (i) above and also

(ii) $_M f(\gamma z) = f(z)$ for all $\gamma \in M$

then we say that $f$ is $M$-*modular*. Clearly if $f$ is modular then it is $M$-modular for all congruence subgroups $M$. Finally, one last "*modular*" definition. We say that

$$E : y^2 = x^3 + ax + b$$

is a *modular elliptic curve* if, for some congruence subgroup $M$ of $\Gamma$, there exist nonconstant $M$-modular functions $f$ and $g$ such that

$$x = f(t), y = g(t) \quad (t \in \mathbb{C}^+)$$

provides a parameterization of $E$.

The proof of Fermat's Last Theorem was then completed in the following steps. The possible link to modular curves was raised in the 1950s and 1960s, when Shimura/Taniyama conjectured that every elliptic curve with rational coefficients is modular. Frey surmised that the Frey elliptic curve, if it existed, would not be modular, and this was confirmed in 1986 by K. Ribet. Finally, in 1994, A. Wiles established that a class of elliptic curves (which would include the Frey elliptic curve, if it existed) does indeed consist of modular elliptic curves. Consequently, the Frey elliptic curve cannot exist and Fermat's Theorem was validated. The final stages were accompanied by a high drama suitable to such a long-standing and famous assertion. The principal character, A. Wiles, secluded himself for seven years to work exclusively on this challenge and then, emerging from his seclusion, presented a series of four lectures on his solution to experts only to find that there was a gap in one of his arguments. However, with further work together with his student R. Taylor, he was able to bridge the gap and the proof was finally complete. The Shimura/Taniyama conjecture was established a few years later in its full generality.

### Exercises 7.2

1. Let $[x, y, z]$ be a Pythagorean triple. Establish the following:

    (i) $(x, y) = (x, z) = (y, z) = (x, y, z)$.

    (ii) There exists a Pythagorean triple $[x_1, y_1, z_1]$ such that

    $$\text{(a)} \quad (x_1, y_1, z_1) = 1, \qquad \text{(b)} \quad [x, y, z] = (x, y, z)[x_1, y_1, z_1].$$

    (When (a) holds, we say that $[x_1, y_1, z_1]$ is a *primitive Pythagorean triple*.)

2. Let $[x, y, z]$ be a primitive Pythagorean triple. Establish the following:

    (i) $z$ is odd. (Equivalently, $x$ and $y$ have opposite parity.)

    (ii) $\left( \frac{z+x}{2}, \frac{z-x}{2} \right) = 1$.

3. Let $[x, y, z]$ be a primitive Pythagorean triple with $x$ odd and $y$ even. Establish the following:

    (i) $\frac{z+x}{2} \cdot \frac{z-x}{2} = \left( \frac{y}{2} \right)^2$.

    (ii) There exist $a, b \in \mathbb{N}$ with

    $$\frac{z+x}{2} = a^2, \qquad \frac{z-x}{2} = b^2, \qquad a > b, \qquad (a, b) = 1.$$

(iii) $x = a^2 - b^2, \quad y = 2ab, \quad z = a^2 + b^2.$
(iv) $a$ and $b$ have opposite parity.

4. Let $a, b \in \mathbb{N}$ be such that $a > b, \quad (a, b) = 1$, and $a$ and $b$ have opposite parity. Let

$$x = a^2 - b^2, \quad y = 2ab, \quad z = a^2 + b^2.$$

Show that $[x, y, z]$ is a primitive Pythagorean triple.

5. Show that $\Gamma$ is a group of permutations of $\mathbb{C}^+$.

6. Show that $\Gamma(n)$ is a normal subgroup of $\Gamma$ for each positive integer $n$.

## 7.3 Elliptic Curve Factoring Algorithm

There are two related problems that one can consider regarding a large odd integer $n$:

(1) Is $n$ a prime number?

(2) Can we find a proper prime factor?

It would appear that the first question is more tractable than the second, at least in the sense that there is a primality testing algorithm with a maximum running time that can be expressed as a polynomial in the number of digits in the target number. This algorithm was discovered by M. Agrawal, N. Kayal and N. Saxena [AKS] and improvements have been made to the running speed since then [LP].

The idea for the AKS algorithm begins with the simple observation that, for any integer $n > 1$, $n$ is a prime number if and only if

$$(x - a)^n = x^n - a \text{ in } \mathbb{Z}_n[x], \text{ for all elements } a \text{ in } \mathbb{Z}_n.$$

However, as an algorithm, this would require testing with a large number of values of $a$. Now let $f(x)$ be a polynomial with integer coefficients and let $I$ be the ideal in $\mathbb{Z}[x]$ generated by $n$ and $f(x)$. If $n$ is a prime number, then it is easy to see that

$$(x - a)^n - (x^n - a) \text{ is contained in } I \text{ for all integers } a. \tag{7.3}$$

However, the striking achievement of Agrawal, Kayal and Saxena was to show that, in the converse direction, with a suitably chosen (cyclotomic) polynomial $f(x)$, there exists an integer $K$, that is relatively small compared with $n$, such that if (7.3) holds for every integer a between 1 and $K$, then

Either (1) $n$ is divisible by a prime less than $K$, or
(2) $n$ is a power of a prime.

From this it is then possible to determine whether or not $n$ is a prime.

Turning to the second question at the beginning of this section, its importance to modern commercial transactions cannot be overestimated. Indeed, the very security of the RSA encryption scheme depends on the difficulty of the second question. However, one of the more successful algorithms for tackling the second question, developed by Lenstra, uses elliptic curves, and we will describe it here.

We begin with a purely number–theoretical approach called the $(p-1)$ *method* that was developed by J. Pollard. It is based on the following result.

**Lemma 7.3.1** *Let $n$ be an odd integer and $p$ be a prime number that divides $n$. Let $p - 1 = q_1^{\alpha_1} \cdots q_k^{\alpha_k}$ where the $q_i$ are distinct prime numbers, $\alpha_i \in \mathbb{N}$, and let $a \in \mathbb{N}$ be such that $q_i^{\alpha_i} \leq a$, for all $i$. Let $b \in [0, n-1]$ be such that $b \equiv 2^{a!} \mod n$.*

> (i) $b \geq 1$.
> (ii) $p$ *divides* $\gcd(b-1, n)$.
> (iii) *If $b > 1$, then* $\gcd(b-1, n)$ *is a nontrivial proper divisor of $n$.*

**Proof.** (i) Since $n$ is odd, it is clear that $2^{a!}$ is not a multiple of $n$. Hence, $b \not\equiv 0 \mod n$ and $1 \leq b \leq n - 1$.

(ii) Since $p$ divides $n$, it must also divide $b - 2^{a!}$ so that

$$b \equiv 2^{a!} \mod p. \tag{7.4}$$

From the definition of $a$, it follows that $p - 1$ divides $a!$. So let $c \in \mathbb{N}$ be such that

$$a! = (p-1)\,c.$$

Then, from (7.4), and Fermat's Theorem (Theorem 1.11.3),

$$b \equiv \left(2^{(p-1)}\right)^c \mod p$$

$$\equiv 1^c \mod p$$

$$= 1 \mod p.$$

Thus, $p$ divides $b - 1$ and (ii) follows.

(iii) By the definition of $b$, we have that $b < n$. Hence, by (ii), we have

$$p \leq \gcd(b-1, n) < n$$

and the claim follows. $\quad\square$

Note that the only reason for choosing $a!$ in the previous lemma is to ensure that $p - 1$ divides $a!$.

The factoring algorithm based on Lemma 7.3.1 then proceeds as follows:

(1) Guess a value of $a$.

(2) Compute $b$ using modular arithmetic at each step, calculating

$$2^2 \bmod n, \quad (2^2)^3 \bmod n, \quad (2^{2.3})^4 \bmod n, \ \ldots \ .$$

(3) Use the Euclidean algorithm to compute $d = \gcd(b - 1, n)$.

(4) If $d \neq 1, n$, then $d$ is a nontrivial proper divisor. Otherwise, repeat the algorithm with a different choice of $a$.

At first sight it might appear that the algorithm will guarantee success. However, there are two points at which it can break down.

(1) It may happen that $b = 1$, so that $b - 1 = 0$ and $\gcd(b - 1, n) = n$. This would happen, for example, if $n = pm$, $(p, m) = 1$ and $\varphi(m)$ divides $a!$ (see the exercises).

(2) Since the prime factors of $n$ are unknown at the beginning of the algorithm, the choice of $a$ is a guess. If there is no prime factor $p$ of $n$ for which $p - 1$ divides $a!$, then $b$ might not be congruent to $1 \bmod p$ and so we cannot conclude that $\gcd(b - 1, n)$ is a nontrivial factor of $n$. In this regard, the algorithm works best if $n$ has a prime factor $p$ for which $p - 1$ has relatively small prime factors.

One way to view the $(p - 1)$ algorithm is as a clever way to search for a nonzero noninvertible element in $\mathbb{Z}_n$—namely, the element $b - 1$. Once found, the $\gcd(b - 1, n)$ is then a nontrivial proper factor of $n$. Lenstra observed that it is possible to use elliptic curves to perform a similar kind of search.

The starting point is an elliptic curve

$$E : y^2 = x^3 + ax + b$$

over $\mathbb{Q}$ with $a, b \in \mathbb{Z}$, together with a point $P = (x_0, y_0) \in E$ where $x_0, y_0 \in \mathbb{Z}$, $y_0 \neq 0$. Such pairs are fairly easy to generate: Choose $a, x_0, y_0 \in \mathbb{Z}$ arbitrarily, and then set $b = y_0^3 - x_0^3 - ax_0$.

Next choose an integer $k$ with many divisors in the manner that $a!$ was chosen in the $(p - 1)$ method.

Now compute the point $kP$. One way to do this in an efficient manner is to write $k$ to the base 2 as

$$k = a_0 + a_1 2 + \cdots + a_t 2^t \qquad (a_i = 0, 1)$$

and then successively compute

$$a_t P, \quad 2a_t P, \quad a_{t-1}P + 2a_t P, \quad 2(a_{t-1}P + 2a_t P), \ \ldots \ .$$

This is known as the *double and add method.* (See section 1.12, where the same approach to calculating large exponents was described, except that there it was called *square and multiply.*)

The crucial feature here is that we do not perform the arithmetic in $\mathbb{Q}$, but in $\mathbb{Z}_n$. What is particularly interesting about this method is that we are hoping that the computation breaks down in the "right" way.

Consider what happens at each step. Suppose that we have computed the $m$th point $Q$ without encountering any problems and that we examine the computation of the $(m + 1)$st point. We must perform one of the following calculations:

   (a)  Compute $2Q$.
   (b)  Compute $P + Q$.

Let $Q = (u, v)$.

*Case (a)*   This case has, itself, two subcases depending on whether $v = 0$ or $v \neq 0$.

*Case (a1)*   $v = 0$. In this case the algorithm has failed and we try again with a different pair $(E, P)$ or a different choice of $k$.

*Case (a2)*   $v \neq 0$. Then we try to compute the coordinates $x_{2Q}, y_{2Q}$ of $2Q$ by the formulas (from Theorem 6.10.2)

$$x_{2Q} = -2u + \left(\frac{3u^2 + a}{2v}\right)^2$$

$$y_{2Q} = \quad -v + \left(\frac{3u^2 + a}{2v}\right)(u - x_{2Q}).$$

To perform this calculation in $\mathbb{Z}_n$ we need to be able to calculate an inverse for $2v$. Since we are assuming that $n$ is odd, this can be done if and only if $(v, n) = 1$. In other words, the calculation breaks down if and only if $(v, n) > 1$.

*Case (b)*   If $P = Q$, then we are back in Case (a). So suppose that $P \neq Q$. If $x_0 = u$ then the algorithm has failed and we start over. If $x_0 \neq u$, then we try to compute $P + Q = (x_{P+Q}, y_{P+Q})$ as

$$x_{P+Q} = -x_0 - u + \left(\frac{v - y_0}{u - x_0}\right)^2$$

$$y_{P+Q} = -v + \left(\frac{v - y_0}{u - x_0}\right)^2 (x_0 - x_{P+Q}).$$

Again, to be able to perform this calculation, it is necessary that $u - x_0$ be invertible in $\mathbb{Z}_n$—that is, that $(u - x_0, n) = 1$. Thus, the calculation breaks down if and only if $(u - x_0, n) > 1$.

The algorithm is successful if the previous calculation breaks down, since then we obtain a nontrivial proper divisor of $n$, and either $1 < \gcd(v, n) < n$ (in Case (a2)) or $1 < \gcd(u - x_0, n) < n$ (in Case (b)).

To see what is happening "behind the scenes", so to speak, with this algorithm, suppose that $p$ is a proper prime factor of $n$ and that we have been lucky enough to choose a value of $k$ for which the order, $m$ say, of $E$ as an elliptic curve over $\mathbb{Z}_p$ divides $k$. Since $mP = 0$, over $\mathbb{Z}_p$, we must also have $kP = 0$. Imagine that, in parallel with the previous calculation for $kP$ over $\mathbb{Z}_n$, we perform the "same" calculations over $\mathbb{Z}_p$. Since $kP = 0$ over $\mathbb{Z}_p$, there must be a first step where

$$\text{either} \quad 2Q = 0 \quad \text{or} \quad P + Q = 0.$$

In the first instance, $y_Q = 0 \pmod p$ and in the second, $y_Q = -y_P \pmod p$ and $x_Q = x_P \pmod p$. Thus,

$$p \mid y_Q \quad \text{or} \quad p \mid (x_Q - x_P)$$

so that $\gcd(y_Q, n) > 1$ or $\gcd(x_Q - x_P, n) > 1$.

This corresponds exactly to the situations in which the calculations broke down in the algorithm.

Thus, our objective in the choice of $k$ is to capture the order of $E$ as a group over $\mathbb{Z}_p$ for some prime divisor $p$ of $n$. Hasse's Theorem can be of some help here in making suitable choices.

One last point. To know that the points on $E$ considered as an elliptic curve over $\mathbb{Z}_p$ form a group, we need to know that $E$ is nonsingular. So, at the very beginning, we would like to know that $(4a^3 + 27b^2, n) = 1$. If we are lucky enough to choose values of $a$ and $b$ so that

$$1 < \gcd(4a^3 + 27b^2, n) < n$$

then we are done and there is no need to proceed with the algorithm. If $\gcd(4a^3 + 27b^2, n) = n$, then we choose another pair $(E, P)$.

Elliptic curves can also be used to test the primality of an integer $n$. It is a test that is applied to integers that are thought to be prime. If successful, the test then confirms that the integer is indeed prime. For the details, see Koblitz [Kob].

## Exercises 7.3

1. Let $n = pm$ where $\gcd(p, m) = 1$. Show that the $(p - 1)$ method fails if $p - 1$ and $\varphi(m)$ both divide $a!$.

2. Let $n$ be a positive integer, $n > 1$. Show that $n$ is a prime number if and only if

$$(x - a)^n = x^n - a \text{ in } \mathbb{Z}_n[x], \text{ for all elements } a \text{ in } \mathbb{Z}_n.$$

## 7.4 Singular Curves of Form $y^2 = x^3 + ax + b$

After all the attention that we have given to nonsingular curves of the form $y^2 = x^3 + ax + b$, it is only natural to wonder what can be said about singular curves of the same form. As it turns out, the singularity on such a curve is its weakness and can be used to parameterize the curve by a line.

So, throughout this section, let

$$E : y^2 = x^3 + ax + b$$

be a singular curve over an algebraically closed field $F$ of characteristic different from 2 and 3. Then

$$\Delta = 4a^3 + 27b^2 = 0.$$

There are two cases to consider:

*Case* (i) $a = b = 0$.
*Case* (ii) $a \neq 0, \ b \neq 0$.

In Case (i), the curve reduces to

$$y^2 = x^3.$$

This is easily seen to be parameterized by the equations

$$x = t^2, \quad y = t^3 \qquad (t \in F).$$

We now consider Case (ii). From the proof of Theorem 6.1.1, we know that $E$ has a singularity at the point $S = (\alpha, 0)$ where

$$\alpha = -\frac{3b}{2a} = \frac{2a^2}{9b}.$$

We also saw that in this situation

$$3\alpha^2 + a = 0. \tag{7.5}$$

Equipped with this information, consider an arbitrary line through $S$ in parametric form

$$x = \alpha + ct$$
$$y = dt.$$

Subsituting into the equation for $E$ we obtain

$$d^2 t^2 = (\alpha + ct)^3 + a(\alpha + ct) + b$$
$$= \alpha^3 + a\alpha + b + (3\alpha^2 + a)ct + 3\alpha c^2 t^2 + c^3 t^3$$
$$= t^2(3\alpha c^2 + c^3 t) \qquad \text{(by (7.5) and since } S \in E).$$

Rearranging, we obtain

$$t^2(d^2 - 3\alpha c^2 - c^3 t) = 0$$

so that $S$ is a double point of intersection for all lines with one exception. We have a triple point of intersection at $S$ for any line satisfying

$$d^2 - 3\alpha c^2 = 0$$

or

$$d^2 = 3\alpha c^2 \tag{7.6}$$

which, in general, has two solutions. Thus, every line through $S$, with the exception of the two lines satisfying (7.6), will meet $E$ in a unique third point. So consider the line $L_t$ through $S$ and the point $(0, t)$:

$$\frac{0 - t}{\alpha - 0} = \frac{y - t}{x - 0}$$

or

$$y = \alpha^{-1}(\alpha - x)\, t. \tag{7.7}$$

Now we know that $x = \alpha$ must be a repeated root of $x^3 + ax + b$ (since $\frac{\partial}{\partial x}(x^3 + ax + b) = 0$ when $x = \alpha$), so to find the third root $\beta$, we can write

$$x^3 + ax + b = (x - \alpha)^2(x - \beta).$$

Equating the coefficients of $x^2$, we find

$$-2\alpha - \beta = 0 \quad \text{or}$$
$$\beta = -2\alpha.$$

Thus,

$$x^3 + ax + b = (x - \alpha)^2(x + 2\alpha).$$

Substituting (7.7) into $E$, we obtain

$$\alpha^{-2}(\alpha - x)^2 t^2 = (x - \alpha)^2(x + 2\alpha)$$

or

$$(x - \alpha)^2(x + 2\alpha - \alpha^{-2}t^2) = 0.$$

Thus, the third point of intersection of $L_t$ with $E$ is given by

$$x = -2\alpha + \alpha^{-2}t^2 \tag{7.8}$$

which, substituted into (7.7), yields

$$y = \alpha^{-1}(3\alpha - \alpha^{-2}t^2)\, t$$
$$= 3t - \alpha^{-3}t^3. \tag{7.9}$$

From (7.6) we know that the lines with slope $m = \frac{d}{c}$ where $m^2 = 3\alpha$ have a triple point of intersection with $E$ at $S$. Now the slope of the line in (7.7) is $-\alpha^{-1}t$, so that this corresponds to

$$-\alpha^{-1}t = m$$

or

$$t = -m\alpha.$$

Both the possible values of $t$, when subsituted into (7.8) and (7.9), yield the point $S$.

Thus, if we retain one value—say, $t = m\alpha$—but reject the other, then the equations (7.8) and (7.9) provide a parameterization of $E$ in terms of the line $x = 0$ with the point $(0, -m\alpha)$ deleted. This provides a complete description of $E$.

If $P$ is a singular point on $E$, then every line through $P$ has intersection multiplicity at $P$ of at least two. Hence it is not possible to define $P + Q$, for any point $Q$ on $E$, in the manner of section 6.9 for elliptic curves. So we cannot construct a group on $E$ as we did in the elliptic case. However, if we just consider the nonsingular points on $E$, then it is possible to introduce a group structure based on the geometry of the curve. Over $\mathbb{Q}$, depending on the nature of the singularity, one obtains either the additive group of rational numbers $(\mathbb{Q}, +)$ or the multiplicative group of nonzero rational numbers $(\mathbb{Q}^*, \cdot)$. Either way, the group is infinite and not finitely generated. For details of these constructions, see Silverman and Tate ([ST], III.7).

**Exercises 7.4**

1. Let the curve $f(x, y) = 0$ be given by the parametric equations (7.8) and (7.9). Show that this is a singular curve of the form $y^2 = x^3 + ax + b$.

2. Show that a singular curve of the form $y^2 = x^3 + ax + b$ has just one singular point.

## 7.5 Birational Equivalence

Following our consideration of conics and cubics in sections 6.6, 6.7, and 6.8, it might appear that we focused our attention on a very special class of cubics—namely, the elliptic curves. However, as we will see in this section, elliptic curves really represent a much larger class of curves.

Let $f(x_1, \ldots, x_n)$, $g = (x_1, \ldots, x_n) \in F[x_1, \ldots, x_n]$. Then the curves

$$C_1 : f(x_1, \ldots, x_n) = 0 \quad \text{and} \quad C_2 : g(x_1, \ldots, x_n) = 0$$

are said to be *birationally* equivalent if there exist

$$\alpha_i(x_1, \ldots, x_n), \ \beta_i(x_1, \ldots, x_n), \ \gamma_i(x_1, \ldots, x_n), \ \delta_i(x_1, \ldots, x_n) \in F[x_1, \ldots, x_n]$$

for $1 \le i \le n$, such that

(1) for all $P = (a_1, \ldots, a_n) \in C_1$ (with the possible exception of a finite number of points) the point with coordinates

$$x_i = \frac{\alpha_i(a_1, \ldots, a_n)}{\beta_i(a_1, \ldots, a_n)} \qquad (1 \le i \le n)$$

is a point on $C_2$, and all but, at most, a finite number of points on $C_2$ are obtained in this way;

(2) for all points $Q = (b_1, \ldots, b_n) \in C_2$ (with the possible exception of a finite number of points) the point with coordinates

$$x_i = \frac{\gamma_i(b_1, \ldots, b_n)}{\delta_i(b_1, \ldots, b_n)} \qquad (1 \le i \le n)$$

is a point on $C_1$ and all but, at most, a finite number of points on $C_1$ are obtained in this way.

Roughly speaking, birationally equivalent curves can be thought of as parameterizing each other. As we have seen in earlier sections (and will see in the following exercises) the degrees of such curves can be quite different. Intuitively, birationally equivalent curves should have similar properties. We are now able to indicate the central role of elliptic curves in the study of cubics. By a *rational point* we will mean a point with coordinates in $\mathbb{Q}$.

Let $f(x, y) \in \mathbb{Q}[x, y]$ be a cubic polynomial such that the curve $C$ : $f(x, y) = 0$ is nonsingular and has a rational point. We will show, modulo a few assumptions to simplify the calculations, that the curve $f(x, y) = 0$ is birationally equivalent to a curve of the form

$$D : y^2 = x^3 + ax + b.$$

Basically, we will assume, without comment, that any divisor that appears in the calculation is, indeed, nonzero.

Let $P$ be a rational point on $C$. Since $C$ is nonsingular, there exists a tangent line $L$ at $P = (x_0, y_0)$. Consider $L$ as given in parameterized form

$$x = x_0 + ct$$
$$y = y_0 + dt.$$

If we substitute these values into $f(x, y)$, we will obtain a cubic polynomial $f(x_0 + ct,\ y_0 + dt) = f_1(t)$ in $t$. Then $L$ meets $C$ exactly where $f_1(t) = 0$. Since $L$ is a tangent to $C$ at $t = 0$, we know that $t$ must be a double root so that

$$f_1(t) = t^2 f_2(t)$$

where $f_2(t)$ is linear. If we solve $f_2(t) = 0$, we find a second (third, if counting multiplicities) point of intersection of $L$ with $C$. Call this point $Q$.

Since the coordinates of $P$ are rational and the coefficients of $f(x, y)$ are rational, it follows that the coefficients of $L$ are rational. Hence, the coefficients of $f_1(t)$ are rational and therefore those of $f_2(t)$ are rational. Consequently, the second value of $t$ is rational and finally we see that the coordinates of $Q$ are also rational.

We can now perform a simple change of variables to make $Q$ the origin with respect to the new variables. If $Q$ is the point $(q_1, q_2)$, then this is equivalent to working with the polynomial $g(x, y) = f(x - q_1,\ y - q_2)$ instead of $f(x, y)$. However, $f(x, y)$ and $g(x, y)$ are clearly birationally equivalent. So there is no loss in simply assuming that $Q$ is the origin to start with and that $f(0, 0) = 0$. This means that the constant term in $f(x, y)$ is zero.

Now consider any line $y = tx$ through the origin ($Q$). This meets $C$ where

$$f(x, tx) = 0.$$

Since there is no constant term in $f(x, y)$, it follows that every term in $f(x, y)$ is divisible by either $x$ or $y$ and, therefore, that every term in $f(x, tx)$ is divisible by $x$. Hence,

$$f(x, tx) = xf^*(x, t)$$

where the highest power of $x$ in $f^*(x, t)$ is at most 2 and the highest power of $t$ is at most 3. However, $t^3$ can only occur if there is a term $ky^3$ in $f(x, y)$. This would contribute a term $kt^3x^3$ to $f(x, tx)$ and a term $kt^3x^2$ to $f^*(x, t)$. Thus we can write

$$f^*(x, t) = A(t)x^2 + B(t)x + C(t)$$

where $A(t)$ is a polynomial in $t$ of degree, at most, 3. Similar arguments show that $B(t)$ is a polynomial in $t$ of degree, at most, 2 and $C(t)$ is a polynomial in $t$ of degree, at most, 1.

Now every point on $C$, other than $Q$, must lie on the intersection of $C$ with some line $y = tx$ and so is a solution to

$$f^*(x, t) = A(t)x^2 + B(t)x + C(t) = 0 \tag{7.10}$$

for some value of $t$. Conversely, any solution to (7.10) will determine a point on $C$. Therefore, (7.10) gives an alternative polynomial description of $C$, except for the point $Q$.

Let $y = t_0 x$ correspond to the line $PQ$. Since $PQ$ is the tangent at the point $P$, it follows that $x_0$ is a double zero of $f(x, t_0 x)$. Now suppose that, in addition, $P = Q = (0, 0)$, then we have a triple point at $P$ so that $x_0 = 0$ is a triple root of

$$f(x, t_0 x) = xf^*(x, t_0) = x(A(t_0)x^2 + B(t_0)x + C(t_0))$$

which implies that $f(x, t_0 x) = x^3 A(t_0)$. However, this can only happen if $f(x, y)$ is homogeneous of degree 3, which would imply that $f(x, y)$ has a singularity at $(0, 0)$, contradicting our assumption that $f(x, y)$ is nonsingular. Hence, we may assume that $P \neq Q$ and that $x_0 \neq 0$. However,

$$0 = f(x_0, t_0 x_0) = x_0 f^*(x_0, t_0)$$

$$= x_0(A(t_0)x_0^2 + B(t_0)x_0 + C(t_0))$$

where $x_0 \neq 0$, yet $x_0$ is a double root (since $PQ$ is the tangent at $P$). So we can deduce that $x_0$ is a double root of the quadratic equation

$$A(t_0)x^2 + B(t_0)x + C(t_0) = 0.$$

For that to happen, we must have

$$B(t_0)^2 - 4A(t_0)C(t_0) = 0. \tag{7.11}$$

For any point on $C$, as given by (7.10),

$$\left(A(t)x + \frac{1}{2}B(t)\right)^2 = A(t)^2 x^2 + A(t)B(t)x + \frac{1}{4}B(t)^2$$

$$= A(t)[A(t)x^2 + B(t)x] + \frac{1}{4}B(t)^2$$

$$= A(t)[-C(t)] + \frac{1}{4}B(t)^2 \qquad \text{by (7.10)}$$

$$= \frac{1}{4}(B(t)^2 - 4A(t)C(t)). \qquad (7.12)$$

Now make the change of variable

$$t = t_0 + \frac{1}{z}. \qquad (7.13)$$

Then the constant term (relative to $1/z$) in $(A(t)x + \frac{1}{2}B(t))^2$ will, from (7.12), be

$$\frac{1}{4}(B(t_0)^2 - 4A(t_0)C(t_0))$$

which is just zero, by (7.11). Hence, since $B(t)$ is a quadratic in $t$ and $C(t)$ is linear in $t$,

$$\frac{1}{4}\left(B\left(t_0 + \frac{1}{z}\right)^2 - 4A\left(t_0 + \frac{1}{z}\right)C\left(t_0 + \frac{1}{z}\right)\right)$$

$$= c_4\frac{1}{z^4} + c_3\frac{1}{z^3} + c_2\frac{1}{z^2} + c_1\frac{1}{z} \qquad \text{(for some } c_i \in \mathbb{Q}\text{)}$$

$$= \frac{1}{z^4}(c_4 + c_3 z + c_2 z^2 + c_1 z^3)$$

$$= \frac{1}{z^4}D(z)$$

where $D(z)$ is a polynomial in $z$ of degree, at most, 3 with coefficients in $\mathbb{Q}$. By Section 5.8, there exists a substitution

$$z = \alpha X + \beta, \qquad X = \frac{1}{\alpha}(z - \beta) \qquad (7.14)$$

for which

$$D(z) = D(\alpha X + \beta) = X^3 + aX + b.$$

Now set

$$Y = \left(A(t)x + \frac{1}{2}B(t)\right)(\alpha X + \beta)^2. \tag{7.15}$$

Then, from (7.12) and (7.13), we obtain

$$
\begin{aligned}
\frac{Y^2}{(\alpha X + \beta)^4} &= \frac{\left(A(t)x + \frac{1}{2}B(t)\right)^2 (\alpha X + \beta)^4}{(\alpha X + \beta)^4} \\
&= \left(A(t)x + \frac{1}{2}B(t)\right)^2 \\
&= \frac{1}{4}(B(t)^2 - 4A(t)C(t)) \\
&= \frac{1}{4}\left(B\left(t_0 + \frac{1}{z}\right)^2 - 4A\left(t_0 + \frac{1}{z}\right)C\left(t_0 + \frac{1}{z}\right)\right) \\
&= \frac{1}{z^4}D(z) \\
&= \frac{1}{(\alpha X + \beta)^4} \cdot (X^3 + aX + b)
\end{aligned}
$$

from which it follows that

$$D : Y^2 = X^3 + aX + b$$

a cubic curve of the desired form.

The final task is to make it a bit more transparent that the curves $C$ and $D$ really are birationally equivalent.

First, from (7.13), (7.14), and $y = tx$, we have

$$
\begin{aligned}
X &= \frac{1}{\alpha}(z - \beta) = \frac{1}{\alpha}\left(\frac{1}{t - t_0} - \beta\right) \\
&= \frac{1}{\alpha}\left(\frac{1}{y/x - t_0} - \beta\right)
\end{aligned}
$$

a rational expression in $x$ and $y$; and from (7.15), it follows that $Y$ is also a rational expression in $x$ and $y$. Moreover, the whole argument leading to the definitions of $X$ and $Y$ shows that whenever $(x, y)$ is a point on $C$, then $(X, Y)$ will be a point on $D$.

On the other hand, from (7.13) and (7.14),

$$t = t_0 + \frac{1}{z} = t_0 + \frac{1}{\alpha X + \beta}$$

$$= \frac{t_0 \alpha X + (t_0 \beta + 1)}{\alpha X + \beta} \tag{7.16}$$

whereas from (7.15)

$$x = \frac{1}{A(t)} \left( \frac{Y}{(\alpha X + \beta)^2} - \frac{1}{2} B(t) \right). \tag{7.17}$$

Combining (7.16) and (7.17), we obtain $x$ as a rational expression in $X$ and $Y$. Combining (7.16) and (7.17), with $y = tx$, we also obtain $y$ as a rational combination of $X$ and $Y$.

We leave it as an exercise for you to show that if $(X, Y)$ is a point on the curve $D$, then $(x, y)$, where $x$ and $y$ are defined as noted earlier, is a point on $C$.

It can also be shown that $D$ is, in general, nonsingular, but this lies beyond the scope of this discussion. $\quad\square$

**Exercises 7.5**

1. Show that the mappings

$$\theta : (x, 0) \longrightarrow (x^4, x^3)$$

$$\varphi : (x, y) \longrightarrow \left( \frac{y^3}{x^2}, 0 \right)$$

establish the birational equivalence of the curves

$$C_1 : y = 0, \qquad C_2 : y^4 - x^3 = 0$$

over $\mathbb{R}$.

2. Show that the mappings

$$\theta : (x, 0) \longrightarrow (x^5, x^2)$$

$$\varphi : (x, y) \longrightarrow \left( \frac{x}{y^2}, 0 \right)$$

establish the birational equivalence of the curves

$$C_1 : y = 0, \qquad C_2 : y^5 - x^2 = 0.$$

## 7.6 The Genus of a Curve

Associated with every curve is a nonnegative integer called its *genus*. This is a very important invariant in the analysis of curves. There are various possible definitions of the genus, some of which are topological in nature. We begin with a result that puts a bound on the number of singularities that certain curves can have.

**Proposition 7.6.1** *Let $C$ be an irreducible curve of degree $n$ over an algebraically closed field $F$ and let $C$ have $m$ singularities. Then*

$$m \leq \frac{(n-1)(n-2)}{2}.$$

**Proof.** We argue by contradiction. Suppose that

$$m > \frac{(n-1)(n-2)}{2}.$$

Clearly, lines are nonsingular and it is left as an exercise to show that irreducible conics are nonsingular. So we may assume that $n \geq 3$. By hypothesis, $C$ has (at least) $\frac{(n-1)(n-2)}{2} + 1$ singularities—say, $P_1, P_2, \ldots, P_{\frac{(n-1)(n-2)}{2}+1}$. Choose an additional $n - 3$ points on $C$ to obtain $P_1, \ldots, P_{k-1}$ points all told where we define $k$ by the equation

$$k - 1 = \frac{(n-1)(n-2)}{2} + 1 + n - 3$$

so that

$$\begin{aligned}
k &= \frac{(n-1)(n-2)}{2} + 1 + n - 2 \\
&= \frac{n^2 - 3n + 2 + 2 + 2n - 4}{2} \\
&= \frac{n^2 - n}{2} \\
&= \frac{(n-1)n}{2} \\
&= \frac{((n-2)+1)((n-2)+2)}{2}.
\end{aligned}$$

Hence, by Theorem 6.5.3, there exists a curve, $D$ say, of degree $n - 2$ that passes through the points $P_1, \ldots, P_{k-1}$. Since $C$ is irreducible and $\deg(D) = n - 2 < n = \deg(C)$, it follows that $C$ and $D$ have no components in common. Consequently, by Bézout's Theorem, $C$ and $D$ have $n(n-2)$ points of intersection, counting multiplicities. However, the multiplicity of intersection between two curves at a singular point on one of the curves is at least two. Hence, counting multiplicities, the number of points of intersection of $C$ and $D$ is at least

$$2 \times \left( \frac{(n-1)(n-2)}{2} + 1 \right) + n - 3$$

$$= n^2 - 3n + 2 + 2 + n - 3$$

$$= n^2 - 2n + 1$$

$$> n^2 - 2n$$

which contradicts Bézout's Theorem. Therefore, the result follows.  $\square$

For any irreducible curve $C$ of degree $n$ with $m$ singularities, all of which are double points, we define

$$g(C) = \frac{(n-1)(n-2)}{2} - m.$$

By Proposition 7.6.1, $g(C)$ is a nonnegative integer. The next result spells out the key invariance property of $g(C)$.

**Theorem 7.6.2** Let $C$ be an irreducible curve that is birationally equivalent to two irreducible curves $C_1$ and $C_2$, each of which is such that all of its singularities are double points. Then $g(C_1) = g(C_2)$.

This result makes the following definition possible. Let $C$ be an irreducible curve that is birationally equivalent to an irreducible curve $D$, all of whose singularities are double points. Then the *genus* of $C$ is $g(D)$. Of course, if all the singularities of $C$ are double points, then we can take $D = C$.

**Example 7.6.3** For any line $L$ or irreducible ($=$ nondegenerate) conic $C$,

$$g(L) = 0 = g(C).$$

**Example 7.6.4** For any elliptic curve $E$,

$$g(E) = 1.$$

We close with some interesting facts concerning the classification of curves by their genus that highlight the special position of elliptic curves.

(1) Every curve of genus 0 is birationally equivalent to either a line or a conic.

(2) Any curve of genus 1 over $\mathbb{Q}$ with a rational point is birationally equivalent to a nonsingular cubic (and therefore an elliptic curve).

(3) If a curve over $\mathbb{Q}$ with genus 0 has one rational point then it has infinitely many rational points.

(4) There exist curves over $\mathbb{Q}$ with genus 1 that have finitely many rational points and there exist curves over $\mathbb{Q}$ with genus 1 that have infinitely many rational points.

(5) Any curve over $\mathbb{Q}$ with genus greater than 1 has only finitely many rational points.

### Exercises 7.6

1. Let $E : y^2 = x^3 + ax + b$ be such that $a$ and $b$ are nonzero and $\Delta = 0$. Show that $g(E) = 0$.

2. Let $C_n : x^n + y^n = a^n$ over $\mathbb{C}$  $(n \geq 3, a \neq 0)$. Show that $g(C_n) = \frac{(n-1)(n-2)}{2}$.

## 7.7 Pell's Equation

In the previous chapter we saw how the points on an elliptic curve can be endowed with a natural group structure. The study of the algebraic structure of curves sits squarely in the area of algebraic geometry and there is a vast amount of information available in the many texts on the subject. In conclusion, we present one more well-known example of a different character and of ancient origins, where the set of points on a curve permits a very natural group structure.

When we consider equations of the form $f(x_1, \ldots, x_n) = 0$, where $f \in \mathbb{Z}[x_1, \ldots, x_n]$ and are interested in finding *integer* solutions, it is customary to refer to the equations as Diophantine equations. As the name might suggest, such equations have been studied ever since the days of ancient Greece. The organized study of the integer solutions of such equations belongs in the domain of number theory. It will suit our purposes here to consider just one example—namely, equations of the form $x^2 - dy^2 = 1$ where $d$ is a positive integer and not a square. A simple example would be $x^2 - 5y^2 = 1$. This has a solution $x = 9, y = 4$, for instance, but has many, indeed infinitely many, solutions. Part of the interest in equations of this form derived from the fact that, as the values of the solutions $x, y$ increase, the ratios $x/y$ provide better and better approximations to the square root of $d$. For example, the solutions $(x, y) = (9, 4), (161, 72), (682, 305)$ provide the following approximations to $\sqrt{5}$:

$$(9, 4) \longrightarrow 2.25$$

$$(161, 72) \longrightarrow 2.236111$$

$$(682, 305) \longrightarrow 2.236065574$$

$$\sqrt{5} = 2.236067978\ldots.$$

Equations of this sort were of interest to Archimedes, who posed a famous problem known as *the cattle problem*, which was not solved until the 18th century. These equations were also of interest to Indian mathematicians, and solutions to the equations for values of $d = 2, 3$ were known to ancient Greek and Indian mathematicians. Brahmagupta found a general method to solve such equations in the seventh century. A similar solution was found by Brouncker in the 17th century. So it was something of an accident that Euler happened to call such an equation *Pell's equation* and the name has stuck ever since. To prove that Pell's equation does have nontrivial integer solutions (that is, different from $x = \pm 1, y = 0$) for all values of $d$ would be too much of a digression for us at this point. So we refer you to [L], [McCo] or [NZ] for a proof of that fact and more details of the history of the equation. We will assume that nontrivial integer solutions do exist. Since each solution consists of a pair of values, one for $x$ and one for $y$, we will refer to the solutions as ordered pairs $(x, y)$ such that $x^2 - dy^2 = 1$. Note that there is no solution with $x = 0$ and that if $(x, y)$ is a solution to Pell's equation, then so also are $(\pm x, \pm y)$. Turning this around, we have that the set of integer solutions is

$$S = \{(\pm x, \pm y) \mid x^2 - dy^2 = 1, \ x > 0, \ y \geq 0\}.$$

We call any solution $(x, y)$ with $x, y > 0$ a *positive solution.*

Before proceeding to a description of the positive solutions, we need to consider the following subset of $\mathbb{R}$: For a positive integer $d$, not a perfect square, let

$$\mathbb{Q}(\sqrt{d}) = \{a + b\sqrt{d} \mid a, b \in \mathbb{Q}\}.$$

It is a simple exercise to show that $\mathbb{Q}(\sqrt{d})$ is a field. In fact, $\mathbb{Q}(\sqrt{d})$ is just a more intuitive way of describing the field

$$\mathbb{Q}[z]/(z^2 - d).$$

Since $\pm\sqrt{d}$ are the two roots of $x^2 - d$, it follows that the mapping

$$\varphi : a + b\sqrt{d} \longrightarrow a - b\sqrt{d} \qquad (a, b \in \mathbb{Q})$$

is an automorphism of $\mathbb{Q}(\sqrt{d})$.

**Lemma 7.7.1** *Let $(a, b)$ be a positive solution for which a is the smallest possible value (in $\mathbb{N}$). Then*

   (i) *$b$ is also the smallest possible value in $\mathbb{N}$ for all positive solutions*

   (ii) *if $(x, y)$ is a solution with $1 < x + y\sqrt{d}$, then $(x, y)$ is a positive solution and*

$$x + y\sqrt{d} \ \geq\ a + b\sqrt{d}.$$

**Proof.** (i) Let $(x, y)$ be any positive solution with $x, y \in \mathbb{N}$. By the choice of $(a, b)$ we have $x \geq a$. Also,

$$x^2 - dy^2 = 1 = a^2 - db^2$$

so that

$$d\,(y^2 - b^2) = x^2 - a^2 \geq 0.$$

Hence, $y^2 \geq b^2$ and $y \geq b$.

   (ii) We have

$$(x + y\sqrt{d})(x - y\sqrt{d}) = x^2 - dy^2 = 1$$

so that we also have $x - y\sqrt{d} > 0$ and

$$x - y\sqrt{d} = (x + y\sqrt{d})^{-1} < 1.$$

Hence,

$$x \ = \frac{1}{2}(x + y\sqrt{d}) + \frac{1}{2}(x - y\sqrt{d}) \geq \frac{1}{2} + 0 \ > 0$$

$$y\sqrt{d} = \frac{1}{2}(x + y\sqrt{d}) - \frac{1}{2}(x - y\sqrt{d}) > \frac{1}{2} - \frac{1}{2} = 0.$$

Therefore, $(x, y)$ is a positive solution. It now follows from part (i) that $x \geq a$ and $y \geq b$. Therefore,

$$x + y\sqrt{d} \ \geq\ a + b\sqrt{d}. \quad \square$$

   In light of Lemma 7.7.1 (i), it seems reasonable to refer to the solution $(a, b)$ as the smallest positive solution.

**Theorem 7.7.2** *Let $d$ be a positive integer and not a perfect square. Let $(a, b)$ be the smallest positive solution to $x^2 - dy^2 = 1$. Then*

$$P = \{(a_n, b_n) \mid (a + b\sqrt{d})^n = a_n + b_n\sqrt{d}\}$$

*is the set of all positive solutions. (Note that $(a_1, b_1) = (a, b)$.)*

**Proof.** Applying the automorphism $\varphi$ of $\mathbb{Q}(\sqrt{d})$, introduced earlier in this section, to

$$a_n + b_n\sqrt{d} = (a + b\sqrt{d})^n$$

we obtain

$$a_n - b_n\sqrt{d} = (a - b\sqrt{d})^n.$$

Hence,

$$\begin{aligned}
a_n^2 - db_n^2 &= (a_n + b_n\sqrt{d})(a_n - b_n\sqrt{d}) \\
&= (a + b\sqrt{d})^n(a - b\sqrt{d})^n \\
&= (a^2 - b^2 d)^n \\
&= 1.
\end{aligned}$$

Thus, every element of $P$ is a solution.

Now let $(x, y)$ be any positive solution and suppose that $(x, y) \notin P$. Since $a, b, d, x, y \in \mathbb{N}$, it follows that there must exist an integer $n$ with

$$(a + b\sqrt{d})^n \le x + y\sqrt{d} < (a + b\sqrt{d})^{n+1}.$$

However, $(a + b\sqrt{d})^n = a_n + b_n\sqrt{d}$ so that $(a + b\sqrt{d})^n = x + y\sqrt{d}$ would imply that $(x, y) = (a_n, b_n) \in P$, which would be a contradiction. Hence we must have

$$(a + b\sqrt{d})^n < x + y\sqrt{d} < (a + b\sqrt{d})^{n+1}.$$

Multiplying by $(a - b\sqrt{d})^n$ (which we know must be positive) we obtain

$$\begin{aligned}
1 = (a^2 - d\,b^2)^n &= (a + b\sqrt{d})^n(a - b\sqrt{d})^n \\
&< (x + y\sqrt{d})(a - b\sqrt{d})^n \\
&< (a + b\sqrt{d})^{n+1}(a - b\sqrt{d})^n = (a + b\sqrt{d})(a^2 - d\,b^2)^n \\
&= a + b\sqrt{d}.
\end{aligned}$$

Let $(x+y\sqrt{d})(a-b\sqrt{d})^n = p+q\sqrt{d}$. Applying the automorphism $\varphi$, we get

$$p - q\sqrt{d} = (x - y\sqrt{d})(a + b\sqrt{d})^n$$

so that

$$
\begin{aligned}
p^2 - dq^2 &= (p + q\sqrt{d})(p - q\sqrt{d}) \\
&= (p + q\sqrt{d})(x - y\sqrt{d})(a + b\sqrt{d})^n \\
&= (x + y\sqrt{d})(x - y\sqrt{d})(a - b\sqrt{d})^n(a + b\sqrt{d})^n \\
&= (x^2 - dy^2)(a^2 - db^2)^n \\
&= 1.
\end{aligned}
$$

Thus, $(p, q)$ is a solution with

$$1 < p + q\sqrt{d} < a + b\sqrt{d}$$

which contradicts Lemma 7.7.1 (ii). Therefore, $(x, y) \in P$ and $P$ is the set of all positive solutions. $\square$

It is evident from the previous arguments that there is a strong connection between the solutions to Pell's equation and the numbers of the form $x + y\sqrt{d}$. Let

$$G = \{x + y\sqrt{d} \mid x^2 - dy^2 = 1\}.$$

Then the mapping

$$\theta : (x, y) \rightarrow x + y\sqrt{d}$$

is a bijection of $S$ (the set of solutions to $x^2 - dy^2$) to $G$. This makes the following observation especially interesting.

**Lemma 7.7.3** $(G, \cdot)$ *is a subgroup of* $(\mathbb{Q}(\sqrt{d}), \cdot)$.

**Proof.** Exercise. $\square$

With the help of the bijection $\theta$, we can now pull the group structure of $G$ back to $S$ by defining

$$
\begin{aligned}
(x_1, x_2) * (y_1, y_2) &= \theta^{-1}(\theta(x_1, x_2) \cdot \theta(y_1, y_2)) \\
&= \theta^{-1}((x_1 + x_2\sqrt{d})(y_1 + y_2\sqrt{d}))
\end{aligned}
$$

$$= \theta^{-1}(x_1 y_1 + x_2 y_2 d + (x_1 y_2 + x_2 y_1)\sqrt{d})$$
$$= (x_1 y_1 + x_2 y_2 d, \; x_1 y_2 + x_2 y_1).$$

In this way, we endow $S$ with a natural, although not very obvious, group structure. As we will now see, the structure of $G$ is fairly simple.

**Theorem 7.7.4** $G \cong \mathbb{Z}_2 \times \mathbb{Z}$.

**Proof.** First recall that $1, -1 \in G$ trivially since

$$1 = 1^2 - d \cdot 0^2 = (-1)^2 - d \cdot 0^2$$

so that $1 = 1 + 0 \cdot \sqrt{d}$, $-1 = -1 + 0 \cdot \sqrt{d} \in G$. Let $H = \{1, -1\}$. Clearly, $H$ is a subgroup of $G$. Let $(a, b)$ be the smallest positive solution for $x^2 - dy^2 = 1$, let $\alpha = a + b\sqrt{d}$, and let

$$K = \langle \alpha \rangle = \{\alpha^n \mid n \in \mathbb{Z}\}.$$

Then $K$ is also a subgroup of $G$ and, since $G$ is abelian, $H$ and $K$ are both normal subgroups of $G$. Clearly, $H \cap K = \{1\}$. To show that $G$ is the (internal) direct product of $H$ and $K$, it only remains to show that $G = HK$.

Let $x + y\sqrt{d} \in G$. We only consider the case where $x < 0$, with the case where $x > 0$ being similar. Then,

$$x + y\sqrt{d} = (-1)(-x - y\sqrt{d})$$

where $-x > 0$.

*Case* (i) $-y > 0$. Then $(-x, -y)$ is a positive solution and so, by Theorem 7.7.2, there exists $n \in \mathbb{N}$ with $-x - y\sqrt{d} = \alpha^n$. Therefore,

$$x + y\sqrt{d} = (-1)(-x - y\sqrt{d})$$
$$= (-1)\alpha^n \in HK.$$

*Case* (ii) $-y < 0$. Then,

$$(-x - y\sqrt{d})(-x + y\sqrt{d}) = x^2 - dy^2 = 1$$

so that

$$-x - y\sqrt{d} = (-x + y\sqrt{d})^{-1}$$

where $(-x, y)$ is a positive solution. By Theorem 7.7.2, there exists $n \in \mathbb{N}$ with

$$-x + y\sqrt{d} = \alpha^n.$$

Therefore,

$$x + y\sqrt{d} = (-1)(-x - y\sqrt{d})$$
$$= (-1)(-x + y\sqrt{d})^{-1}$$
$$= (-1)\alpha^{-n} \in HK.$$

Thus, in all cases, $x + y\sqrt{d} \in HK$. Hence, $G$ is the (internal) direct product of $H$ and $K$. Consequently,

$$G \cong H \times K.$$

However, it is clear that $H$ is cyclic of order 2 whereas $K$ is an infinite cyclic group. Hence, $H \cong \mathbb{Z}_2$ and $K \cong \mathbb{Z}$. Therefore, $G \cong \mathbb{Z}_2 \times \mathbb{Z}$, as claimed.  $\square$

**Corollary 7.7.5**  $S \cong \mathbb{Z}_2 \times \mathbb{Z}$.

**Exercises 7.7**

1. Let $(a, b) \in \mathbb{N} \times \mathbb{N}$ be a positive solution to the equation $x^2 - dy^2 = 1$ $(d \in \mathbb{N})$ for which $b$ is the smallest possible value (in $\mathbb{N}$). Show that $a$ is also the smallest possible positive value for any solution.

2. Show that $(9, 4)$ is the smallest positive solution to the equation $x^2 - 5y^2 = 1$.

# References


[AKS]  Agrawal, M., N. Kayal and N. Saxena. "Primes is in $P$," Annals of Mathematics, 160,(2004), 7817–93.

[Bon]  Boneh, D. "Twenty Years of Attacks on the RSA Cryptosystem," Notices of the American Mathametical Society, 46 (1999), 203–12.

[Boy]  Boyer, C. B. *A History of Mathematics* (2nd Edition), John Wiley & Sons, New York, 1991.

[BC]  Brown, J. W. and R. V. Churchill, *Complex Variables and Applications*, McGraw Hill, New York, 2004.

[CLO]  Cox, D., J. Little, and D. O'Shea. *Ideals, Varieties and Algorithms*, Springer-Verlag, New York, 1992.

[CLRS]  Cormen, T. H., C. E. Leiserson, R. L. Rivest, and C. Stein. *Introduction to Algorithms*, Second Edition, MIT Press and McGraw-Hill, 2001.

[Dau]  Dauben, J. "Review of 'The Universal History of Numbers and The Universal History of Computing, Volumes I and II' by Georges Ifrah," Notices of the American Mathematical Society, 49 (2002), 32–38, 211–216.

[Di]  Dickson, L. E. *History of the Theory of Numbers*, Volume II, Chelsea Publishing Co., New York, 1966, 575–78.

[Ful]  Fulton, W. *Algebraic Curves—An Introduction to Algebraic Geometry*, Addison-Wesley, New York, 1989.

[HW]  Herstein, I. N. and D. J. Winter. *Matrix Theory and Linear Algebra*, Macmillan Publishing Co, New York, 1988.

[Hod]  Hodges, A. *Alan Turing: The Enigma*, Vintage, London, 1992.

[Hul]  Hulek, K. C. B. *Elementary Algebraic Geometry*, American Mathematical Society, Providence, 2003.

[Kob]  Koblitz, N. *A Course in Number Theory and Cryptography*, Springer, New York, 1994.

[L]     Lenstra, H. W., Jr. "Solving the Pell Equation," Notices of the American Mathematical Society, 49 (2002), 182–92.

[LP]     Lenstra, H. W. Jr. and C. Pomerance. "Primality Testing with Gaussian Periods," http://www.math.dartmouth.edu/~carlp/PDF/complexity12.pdf, preliminary version, 2005.

[McCa]     McCarthy, P. J. *Algebraic Extensions of Fields*, Blaisdell, Waltham, Massachusetts, 1966.

[McCo]     McCoy, N. H. *The Theory of Numbers*, Collier-McMillan, London, 1965.

[Mil]     Miller, R. A. "The Cryptographic Mathematics of Enigma," http://www.nsa.gov/about/cryptologic_heritage/center_crypt_history/publications/wwii.shtml, 2001.

[Men]     Menezes, A. J. *Elliptic Curve Public Key Cryptosystems*, Kluwer Academic Publishers, Boston, 1993.

[Nav]     Navarro, G. "On the Fundamental Theorem of Finite Abelian Groups," American Mathematical Monthly, February (2003), 153–154.

[Nel]     Nelson, B. *Punched Cards to Bar Codes*, Helmers Publishing. Peterborough, NH, 1997.

[Neu]     Neuman, P. M. "A Lemma that is not Burnside's," Math. Scientist, New York, 1979 (4), 1331–41.

[Nic]     Nicholson, W. K. *Elementary Linear Algebra*, McGraw-Hill Ryerson, New York, 2001.

[NZ]     Niven, I., and H. S. Zuckerman. *An Introduction to the Theory of Numbers*, John Wiley, New York, 1960.

[Pal]     Palmer, R. C. *The Bar Code Book*, Trafford Publishing, Victoria, BC, 2007.

[Rej]     Rejewski, M. "How Polish Mathematicians Deciphered the Enigma," Annals of the History of Computing 3, 1981, 2132–33.

[Pat]     Patterson, W. *Mathematical Cryptology for Computer Scientists and Mathematicians*, Rowman & Littlefield, New York 1987.

[RSA]     Rivest, R. L., A. Shamir, and L. Adleman. "A Method for Obtaining Digital Signatures and Public-Key Cryptosystems," Communications of the ACM, 21 (1978), 120–26.

[Sing]     Singh, S. and K. Ribet. *Fermat's Last Stand*, Scientific American, Munn & Co., New York, November 1997, 68–73.

[ST]     Silverman, J., and J. Tate. *Rational Points on Elliptic Curves*, Springer-Verlag, New York, 1992.

[Ste]     Stepanov, S. A. *Arithmetic of Algebraic Curves*, Monographs in Contemporary Mathematics, Consultants Bureau, New York, 1994.

[Sti]     Stinson, D. R. *Cryptography Theory and Practice*, CRC Press, New York, 1995.

[Van]     Vanstone, S. A., and R. C. van Oorschot. *An Introduction to Error Correcting Codes with Applications*, Kluwer Academic Publishers, Boston, 1992.

[Var]     Varsanyi, G. *Assignments for Vibrational Spectra of Seven Hundred Benzene Derivatives*, J.Wiley & Sons, New York, 1974.

[W]     Wells, D. *The Penguin Book of Curious and Interesting Puzzles*, Penguin Books, London U.K. 1992.

[Zel]     Zelinsky, D. *A First Course in Linear Algebra*, Academic Press, New York, 1968.

# Index